Essentials of Nuclear Medicine

Springer

London
Berlin
Heidelberg
New York
Barcelona
Budapest
Hong Kong
Milan
Paris
Santa Clara
Singapore
Tokyo

Malcolm V. Merrick

Essentials of Nuclear Medicine

Second Edition

With 198 illustrations, 13 in colour

Springer

Dr Malcolm V. Merrick MA, BM, BCh(Oxon), MSc(London), FRCP(Edinburgh), FRCR, DMRD
Consultant in Nuclear Medicine and Radiology, Head of Nuclear Medicine, Western General Hospital, Edinburgh and Royal Infirmary of Edinburgh, UK
Consultant in Nuclear Medicine, Royal Hospital for Sick Children, Edinburgh, UK
Senior Lecturer in Medicine, University of Edinburgh, UK

ISBN-13:978-3-540-76205-8

British Library Cataloguing in Publication Data
Merrick, M.V.
Essentials of nuclear medicine
1. Nuclear medicine
I. Title
616'. 07575
ISBN-13:978-3-540-76205-8 e-ISBN-13:978-1-4471-0907-5
DOI: 10.1007/978-1-4471-0907-5

Library of Congress Cataloging-in-Publication Data
A catalog record for this book is available from the Library of Congress.

First edition published by Churchill Livingston 1984

Typeset by MPG Design, Blandford Forum

28/3830-543210 Printed on acid-free paper

Contents

Introduction

Nuclear Medicine has been defined as 'the medical speciality that utilises the nuclear properties of radioactive and stable nuclides to make diagnostic evaluations of anatomical or physiological conditions of the body and to provide therapy with unsealed radioactive sources' (Board of Trustees of the Society of Nuclear Medicine, 1983). These impinge on almost every branch of medical practice. It is this combination of breadth of clinical application with so many branches of the basic sciences including physics, inorganic chemistry, biochemistry, physiology and computer sciences which makes this such a challenging field.

If reliable clinical information is to be obtained, it is essential to have a clear and accurate understanding of the scientific basis of the methods employed, their strengths and their limitations. This book describes clinically useful *in vivo* diagnostic and therapeutic applications of radioactive nuclides. There are many excellent textbooks of physics applied to medicine, physiology and biochemistry and it is assumed that the reader is familiar with relevant aspects of the basic sciences but, where appropriate, attention is drawn to particular points pertinent to the topic under discussion. The order of chapters approximates to the frequency with which the commoner investigations are requested. Thus, bone scintigraphy is the most frequently requested single isotope investigation in the majority of departments and is discussed in Chapter 1. Anyone new to this field is strongly recommended to read Chapter 12 (Selected Topics in Basic Sciences) before proceeding to the clinical chapters, both to familiarize themselves with scientific terms not commonly encountered in other branches of medicine and to obtain an insight into both strengths and limitations of techniques described.

Within each chapter the account relevant to those indications which occur more often is at greater length than and precedes that of the less common. Some older tests which are now rarely performed are briefly included, as the principles they illustrate may be applicable in other fields such as CT or MRI. There is, moreover, a tendency for wheels to be reinvented. Underlying physiological and biochemical mechanisms and technical factors relevant to proper performance and interpretation precede the clinical accounts. The description of radiopharmaceuticals and technical factors precedes the description of findings in specific diseases.

The strength of nuclear medicine is its ability to measure function relatively non-invasively. Its weakness is usually, if naively, cited as the relatively poor anatomical detail it provides compared with radiology, MRI or ultrasound. When anatomical information alone is required, these techniques undoubtedly give better information. On the other hand, inferences about function made from shape are often misleading. The relationship between function and appearance is usually indirect. A dilated renal pelvis is not always obstructed, nor is a small heart necessarily a healthy one. Moreover the functional resolution which can be obtained by imaging radio-labelled tracers far exceeds any other *in vivo* imaging technique, for example differentiating pre-synaptic from post-synaptic dysfunction of receptors in the basal ganglia.

Nuclear medicine has developed to provide techniques to measure, rather than guess, function.

In many situations, just imaging the distribution of an administered radioactive tracer provides as much information as is necessary for patient care. Correct interpretation nevertheless requires an understanding of the physical, physiological and biochemical processes underlying the image. Frequently, the value or reliability of this information can be enhanced if a measurement is made, for example of the amount of radioactivity which has been taken up, the rate at which uptake occurs or the rate of discharge. It is essential to understand how such measurements can be made and especially the limitations of any technique. In some cases, measurements are made without imaging. Imaging and measurement are integral interdependent components of the whole, and nuclear medicine cannot be applied properly if they are regarded as mutually exclusive, or one is regarded as ancillary to the other.

Whether practised by nuclear medicine specialists, diagnostic radiologists or physicians, the objectives of this volume are to make accessible the information about those tests which are available, how to perform them and how to interpret the results. Techniques are described concisely, as there is usually more than one way of performing any particular investigation, depending on available facilities. The objective has been to provide the reader with an understanding of what can or cannot be achieved, what meaning can be attributed to the findings and to indicate, as a starting point, methods which the author has found satisfactory. Provided that the principles are understood, these may then be changed or adapted to particular local or clinical circumstances.

There is properly concern about the hazards of radiation. No investigation is without some attendant cost and it is the duty of the physician or other medical attendant to weigh any possibility of harm against the benefit or hazard of proceeding without the information any test may provide. Used properly, nuclear medicine can provide benefits vastly greater than the minute dangers associated with the very small doses of radiation involved in the majority of tests.

Acknowledgements

I should like to express my thanks to my wife for her help in the preparation of the manuscript and to the radiographers at the Western General Hospital and Royal Hospital for Sick Children, Edinburgh for their meticulous care and assistance over many years. It is a pleasure to acknowledge both preceding and succeeding generations. There is a saying that if we see further than our teachers it is because we are pygmies standing on the shoulders of a giant. I would like particularly to express my debt to that great iconoclast, the late Dr Norman Veall. Thanks are also due to Dr K.W. Chan (Hong Kong) and Dr M. Pagou (Pyreus) for their thoughtful and constructive criticisms of earlier drafts. I would also like to thank Mary Fox for editing and managing the production of this book and Martin Phipps of MPG Design for the design, layout and typesetting of the entire work.

1 Skeleton

Bone scintigraphy is the most frequently performed imaging investigation in the majority of nuclear medicine facilities. Radioisotopes are also employed to study a number of other aspects of bone physio-pathology and metabolism including blood flow, mineral density, calcium content and turnover rate.

Background

Bone is an heterogeneous organized tissue of collagen fibres and hydroxyapatite crystals dispersed in an amorphous matrix. Hydroxyapatite is a complex, relatively insoluble crystal comprised of calcium, phosphate and hydroxide ions. Once crystals have reached an appropriate size they are stabilized by surface incorporation of pyro-phosphate. The arrangement of the various components of bone is not random but is determined and adjusted by specialized cells, osteoblasts, in response to current mechanical requirements. Strength is determined by interactions between all components. Bone, like all living tissues, is constantly renewed and modified to meet changing requirements. The rate of bone turnover tends to increase with age. Osteoclasts are the cells responsible for removing damaged or time-expired bone. One of the characteristic features of osteoclasts is secretion of an alkaline phosphatase enzyme which hydrolyses pyrophosphate, facilitating dissolution of crystals. Radiopharmaceuticals may substitute for calcium (strontium, barium), hydroxide (fluoride), pyrophosphate (bisphosphonates) or may attach to the organic phase (possibly technetium).

Radiopharmaceuticals

Technetium-99m

Simple salts of technetium do not localize in the skeleton, but complexes of technetium with a number of phosphorus-containing compounds are concentrated in bone. These form the basis of all skeletal imaging agents in current use. There are four classes of compound, two of which are employed clinically.

Bisphosphonates (diphosphonates)

Bisphosphonate is the approved generic term for the compounds previously known as diphosphonates. Their general structure has two phosphoric acid residues linked by carbon (Fig. 1.1).

They differ in the other groups linked to the carbon atom. The simplest is methylene bisphosphonate (MDP), in which R and R′ are both

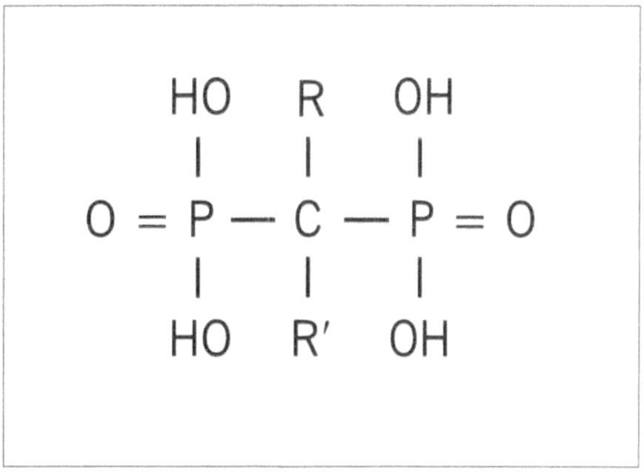

Fig. 1.1 General structural formula of bisphosphonates. They differ in substitution at R and R′.

hydrogen. More than 50% of administered activity concentrates in bone, the rest being rapidly cleared by glomerular filtration. There are considerable variations between individuals in rate of clearance from soft tissues. These are comparable in magnitude with the differences between compounds available as radio-pharmaceuticals; in the absence of specific modifying factors uptake tends to be higher in younger subjects. A higher percentage of administered methane-hydroxy-bisphosphonate (MHDP) activity is taken up into skeleton than of MDP or ethane-hydroxy-bisphosphonate (editronate, EHDP) but the contrast between normal and abnormal regions of the skeleton is not significantly different. There is marginally lower concentration in the skeleton of editronate than of the other compounds and the rate of clearance from soft tissue is slightly slower but no difference has been demonstrated in lesion detection rate.

A number of other compounds have been evaluated including 2,3-dicarboxypropane 1,1-diphosphonic acid (DPD), which is also available as a radiopharmaceutical kit. The only compound in the group shown to have a clear, albeit small, advantage is dimethyl-amino-bisphosphonate (DMAD). Like PYP (see below) this gives a lower contrast between normal bone and soft tissue than MDP and related compounds, but higher contrast between normal and abnormal areas of the skeleton. It improves lesion detection rate by about 5% compared with other agents in clinical use but is not available in a pharmaceutical formulation.

Editronate and number of other bisphosphonates (pamidronate, clodronate) are used therapeutically for reducing bone turnover rate when this is pathologically increased, thus increasing bone mineral density. Requirements for therapeutic use differ in a number of respects from those of a diagnostic agent, in particular good absorption following oral administration.

Further reading:

Journal of Nuclear Medicine 1996; ***37****: 815–8*

Inorganic phosphates

Pyrophosphate (PYP), formed by condensation of two molecules of orthophosphate, is the only useful member of this class. Early papers also refer to longer chain polymers, polyphosphates. These are unstable in solution, hydrolysing at room temperature with release of orthophosphate and PYP, the latter being the active ingredient. Residual polyphosphate tends to form a technetium-labelled colloid taken up by the reticulo-endothelial system, including bone marrow. The principal distinction between Tc-PYP and the bisphosphonates is that bone uptake and contrast between normal bone and soft tissue are both lower with PYP, although contrast between normal and abnormal areas of bone is higher. No difference in sensitivity has been demonstrated between PYP and any of the bisphosphonates currently in use but, because it delineates normal skeletal anatomy less clearly, PYP is now used mainly for diagnosis of ischaemic soft tissue, especially following myocardial infarction (a property shared with the bisphosphonates, page 132) and for red-cell labelling (page 204).

Tetraphosphonates

Ethylene-diamine-tetramethylene phosphate (EDTMP) is one of a number of tetraphosphonates which has been evaluated in animals but not in man. They are cleared from soft tissues more rapidly than the bisphosphonates, but have not been used clinically.

Iminobisphosphonates

These have a P–N–P core which, like the P–C–P core of the bisphosphonates, is resistant to hydrolysis by alkaline phosphatase. The only compound in this group to have been used clinically is the simplest, in which a hydrogen atom is attached to the nitrogen. No advantage has been demonstrated over available bisphosphonates.

Alkaline earth metals

Many ions, principally those of metals, localize in the skeleton. Some are analogues of calcium, in others deposition may be regarded as a means of detoxification – removal from the circulation of potentially harmful unphysiological substances which are not readily excreted.

Calcium

^{45}Ca has a half-life of 165 days and emits only β-particles. ^{47}Ca has a half-life of 4.5 days and emits both β-particles and 1.3 MeV γ rays. Both isotopes have been used for short-term measurements of calcium turnover in plasma, urine and stool. ^{47}Ca has been employed in conjunction with whole-body counting. ^{49}Ca is produced by neutron activation of natural calcium when measuring total body calcium. It has a half-life of 8.8 min, emitting γ-rays of >3 MeV and β-particles. There are no calcium radio-isotopes which are suitable for imaging.

Strontium

Following intravenous injection strontium is cleared from soft tissues more slowly than bisphosphonates. Optimal contrast for imaging is obtained at 24 h. Excretion is predominantly renal, but up to 20% of administered activity is excreted via the liver and may remain in colon for some days. Strontium has a large number of radioactive isotopes, none of which has physical characteristics well suited for imaging with a gamma camera and many are β-emitters with high absorbed radiation dose. It is no longer used for imaging, although much early bone scintigraphy was performed with ^{85}Sr or ^{87m}Sr. ^{85}Sr, which has a half-life of 65 days, has also been used to study long-term bone mineral turnover with whole-body counting, as an analogue of calcium. It is not ideal, having poorer absorption from the gut and preferential excretion. ^{89}Sr is used therapeutically for treatment of metastatic bone cancer whereas ^{90}Sr is one of the biologically important contaminants if radioactivity leaks from a nuclear reactor.

Barium and radium

Barium has many radioactive isotopes, several of which have been used for imaging, but none has any advantage over the bisphosphonates, which are cheaper and more readily available. It is a poorer analogue of calcium for metabolic studies than is strontium.

The remaining member of this group, radium, is an extremely radio-toxic bone-seeking α-particle emitter.

Other bone-seeking radiotracers

Metal ions

Gallium and indium have similar but somewhat unusual properties. When administered carrier-free to normal subjects they are bound by transferrin, a β_1-globulin present in plasma whose physiological function is transport of iron. They have their highest concentrations in those tissues which concentrate iron, especially bone marrow and liver. Tracer concentrations of gallium and indium isotopes concentrate in bone only if transferrin has previously been saturated with large quantities of a trivalent metal ion such as Fe3+, stable gallium, indium, gadolinium or scandium. This has been evaluated in small numbers of patients but, because the quantities of trivalent carrier required to saturate transferrin are dangerously close to toxic levels, this application is obsolete.

Further reading:

*British Journal of Radiology 1975; **48**: 327–51*

Rhenium is a close chemical analogue of technetium and a rhenium-labelled bisphosphonate (^{186}Re-HEDP) has been used therapeutically as an alternative to ^{89}Sr. Radioisotopes of a number of rare earth metals have been evaluated for diagnosis and one (^{153}Sm-EDTMP) for therapy.

Fluorine

Because of its fast blood clearance, high bone uptake and rapid renal clearance, ^{18}F would be the ideal radiopharmaceutical for skeletal imaging but for the high energy of its γ-ray (511 keV, resulting from positron annihilation) and limited availability. Because of its short half-life (100 min) distribution from the source of production is restricted. It is rapidly cleared from soft tissue giving higher contrast in the skeleton between abnormal and normal areas and lower soft-tissue background than the bisphosphonates. Excellent images can be obtained with positron emission tomography but cost renders it impractical outside centres with facilities for PET.

Haemopoietic system

White cells (see also page 208)

White cell scintigraphy is of limited value in the

investigation of osteomyelitis, especially in the axial skeleton where infective tissue may replace bone marrow. Affected areas can thus appear either photon-rich or photon-deficient. The difference between normal and abnormal areas may be small compared with normal inter-subject variability and difficult to identify with confidence. Uptake is also seen where there is inflammation without infection, for example in traumatized neuropathic joints.

Further reading:
Journal of Nuclear Medicine 1991; ***32****: 349–56, 1861–5*

Bone marrow

See page 210.

Factors affecting uptake of skeletal imaging agents

The mechanisms underlying deposition of phosphate derivatives in bone are incompletely understood. *In vitro* both PYP and bisphosphonates attach to the surface of hydroxyapatite crystals (similar to those naturally present in bone) by chemisorption. Crystals which have been so treated are less reactive, do not grow when placed in a saturated solution of appropriate composition and dissolve in water less readily than untreated crystals. The rationale for use of bisphosphonates diagnostically and for treatment of conditions in which rate of bone turnover is abnormal is that PYP, an important naturally occurring inhibitor of crystal growth, is hydrolysed to orthophosphate by alkaline phosphatases. Bisphosphonates, which stabilize crystal surfaces similarly, are resistant to enzymic hydrolysis and thus have a longer duration of action, slowing or preventing bone resorption. There is some evidence that, *in vivo*, while bisphosphonate is associated with the mineral phase, technetium may become detached and bind to the organic matrix.

Further reading:
Journal of Nuclear Medicine 1993; ***34****: 104–8*

Regional skeletal blood flow is an important influence on distribution of uptake; calcium content is not. Oral bisphosphonates at normal therapeutic dose levels have no effect on ^{99m}Tc uptake or distribution, but large intravenous doses for emergency treatment of hypercalcaemia may temporarily reduce uptake of technetium-labelled tracer doses by normal bone. In one case described, bone metastases were nevertheless still visualized. Increased blood flow is usually associated with a locally increased rate of bone turnover and new bone formation, reduced calcium content and increased uptake of scintigraphic agents. Conditions in which blood supply to bone is decreased, for example avascular necrosis, are associated with increased bone mineral density but decreased isotope uptake. In most situations ischaemia provokes a reaction in adjacent healthy tissue, leading to hyperaemia and ultimately removal of devitalized tissue. This is always accompanied by an inflammatory response and increased uptake of bone-seeking agents around the ischaemic area, which may be masked by this surrounding hyperactivity. In extreme cases, for example late osteopetrosis where bone mineral density is enormously increased, uptake of bone-seeking isotopes is reduced or absent, indicating impaired osteoclastic activity and reduced or absent perfusion. It is thus not straightforward to differentiate the effects of local changes in blood flow from those in mineral turnover rate. A further complication is the low (<40%) single pass extraction efficiency of the technetium agents. They are therefore unsuitable for measurement of bone blood flow. In practice the scintigraphic appearances are probably a consequence of interactions between several, or indeed possibly all, of these factors.

Technique

Activity

The usual administered activity in the adult is 400–800 MBq (5.5–11 MBq/kg). Higher activities are required for pinhole and SPECT than for planar imaging and in patients unable to remain still for long enough to obtain adequate images when the lower activity is administered. Patients should be advised to drink at least half a litre of clear fluid in the interval between injection and imaging and to void several times, to minimize both the volume of residual urine in the bladder and the concentration of excreted radioactivity it contains. Under experimental conditions hydration slightly decreases skeletal uptake, but in practice impaired visualization of the sacro-iliac regions and pelvis caused by the higher concentration of residual activity in the urinary tract of dehydrated patients more than outweighs this.

Timing

There are considerable individual variations in rate of clearance from soft tissue of injected radiotracer. Although images can be obtained within 1 h of injection, residual soft-tissue activity reduces contrast and increases the risk of missing clinically important low contrast abnormalities. The interval should not be less than 2 h; optimal contrast is usually achieved at 3–6 h. Longer intervals tend to be advantageous in older and obese subjects. Good quality images can be obtained up to 24 h, although at this time 15 min may be required to collect 100 000 counts using a general purpose collimator. The low count rate, due principally to radioactive decay, necessitates prolonged imaging times. Nevertheless these images may be clinically useful because the contrast between metastases or osteomyelitis and normal bone is often higher in late images and the residue in the urinary tract is low, except in patients with chronic retention and overflow. Late images are particularly useful in patients with urinary diversions, in whom they may be the only way of visualizing the pelvis and lumbar spine clear of excreted activity in an ileal loop bladder.

Projections and collimation

Imaging should be performed with a gamma camera having a large field of view and equipped with a high resolution collimator (page 305). The number of projections required depends on the field of view of the camera, which should be not less than 40 cm in its minimum dimension to include both shoulders of most adult males. When the field of view is large enough to include both shoulders, thorax and upper lumbar spine, administered activity should be adequate to obtain one million counts from the posterior thoracic projection in less than 5 min. All other projections should be acquired for the same time, not equal numbers of counts, particularly the pelvis where many counts come from excreted urinary activity in the bladder. Imaging for equal total number of counts will result in low count densities from skeleton, especially in patients with urinary retention. Many gamma cameras are capable of scanning to obtain whole-body images. Two techniques are employed, continuous scanning and 'step and shoot'. In principle the latter is less likely to be associated with degradation in image quality. The commonest technical problem associated with the use of a scanning whole-body camera is maintaining an anterior detector consistently close to the patient. Image quality can be seriously degraded by careless technique.

Choice of projections should be adapted to the clinical problem under investigation. The two principal categories are local conditions, such as trauma or prostheses which are loose or infected, and systemic conditions including metabolic and endocrine disease, Paget's disease, metastatic malignant neoplasm and haematogenous osteomyelitis. When investigating a local condition, for example a painful prosthesis, views should be obtained of the whole of the area in question, including both ends of the prosthesis and any joint with which it interacts. Oblique or lateral projections are sometimes helpful. If no abnormality is found it is advisable to image the spine, in case pain is referred. Supplementary pinhole images may provide additional information.

The entire axial skeleton must be imaged when investigating systemic conditions such as metastatic disease, haematogenous osteomyelitis or metabolic bone disease. As a minimum, anterior and posterior views of the thorax, abdomen and pelvis are required. In most cases of suspected metastatic disease (the commonest single indication for performing bone scintigraphy) there is little point in routinely imaging distal to the mid-humerus and mid-femur unless the patient has symptoms referable to these areas.

A solitary or occult metastasis more distally is extremely rare. On those occasions when it is necessary to image the feet (for example Paget's disease) two projections at right angles should be obtained to localize any abnormality. For display, if multiple spot images have been obtained they should all be normalized to the same peak counts per pixel. It is often necessary to over-range pixels in the urinary tract, but no part of the skeleton should be over-range.

Occult involvement of skull and facial bones in the absence of metastases elsewhere in the skeleton is rare except in patients with primary tumours of the head and neck. Imaging the skull when staging primary cancers arising in the trunk detects more 'false positives' due to benign conditions such as sinusitis, which have then to be investigated before they can be eliminated, than malignant deposits and is not recommended. When imaging of the head is required, SPECT (page 318) is the procedure of choice, acquiring 128 projections with a pixel size of 5 mm or less (page 315). If SPECT is not available the skull is usually best seen on lateral projections but it is often necessary to obtain additional anterior, posterior, oblique or pinhole views. Very high resolution (pinhole or SPECT) images are sometimes useful. Pinhole images of the skeleton are usually best obtained with a 3–4 mm pinhole and the patient prone or supine, depending on the area to be imaged, for 10–15 min at 5–10 cm pinhole to skin distance.

Further reading:

Journal of Nuclear Medicine 1991; ***32****: 2332–41*

The multi-phase study

Three- or four-phase examinations have been recommended if the differential diagnosis includes osteomyelitis. These comprise:

First pass

A dynamic study of the area of interest during the first pass of an injected bolus of radioactivity. A high sensitivity collimator is essential to obtain adequate counting statistics. As circulation time from injection site to area of interest is difficult to predict accurately, 2 s frames should be acquired for up to 2 min, to ensure that all of the first pass is included. The first pass usually occupies less than 15 s from arrival in the field of view. This is most useful when two sides can be compared and is not helpful in regions obscured by overlying blood vessels, for example most of the spine.

Blood pool

A further frame should be taken 2–3 min after injection, at which time most of the activity is still in the circulation but fairly uniformly mixed. This demonstrates regional differences in blood volume and possibly in vascular permeability.

Late phase

This should be imaged not less than 2 h after administration, to demonstrate regional distribution within the skeleton.

Very late phase

This should be obtained 18–24 h after injection, when extra-skeletal activity and especially residual activity in the urinary tract is at its lowest.

The dynamic phase rarely adds useful information and can on occasion be misleading, for example in patients with limb ischaemia associated with peripheral vascular disease, when flow is often restricted by asymmetrical arterial stenosis. The blood-pool phase often provides useful additional information about local inflammation and should be obtained whenever the differential diagnosis includes osteomyelitis but does not differentiate between bone and soft-tissue pathology. Conventional bone images are always required to differentiate bone from soft-tissue involvement. Contrast is higher at 18–24 h than at 4 h both in osteomyelitis and most metastases but not osteoarthritis; the distinction is however not sufficiently reliable or useful to justify routine additional imaging on the following day.

Equipment

A high resolution collimator increases precision of localization of photon-rich areas and assists differentiation between anatomical variants and pathological accumulations, for example by distinguishing the inferior angle of the scapula from a deposit or fracture in an adjacent rib, but will not reliably improve differential diagnosis of an abnormality (page 305). There is no evidence that improved resolution of planar imaging (from 8 mm FWHM to <4 mm) increases the number of high count-rate abnormalities detected. The improved contrast and localization achieved with SPECT or a pinhole collimator results in detection of some additional lesions and may also assist assessment of their clinical significance. The relative clinical roles of these latter have not been adequately defined. Better resolution and contrast should in principle increase the chances of detecting photon-deficient abnormalities, although this has not been confirmed by a systematic clinical trial.

Many cameras have a whole-body scanning option and may offer 'body contouring' to maintain a constant distance between anterior collimator and patient. Some centres perform a fast initial examination when using a scanning gamma camera followed by local views of suspicious areas. This technique is not recommended. Although many lesions are detected using a low count density whole-body survey many others, especially those of low contrast or which are photon-deficient, are missed. If the low count survey is negative, it follows that a high count density whole-body survey must follow in every case. Once an abnormality has been detected, additional high count density views rarely give information of practical or clinical value, except to eliminate differences due to statistical noise. Especially if the population includes asymptomatic patients with early disease, in whom a high incidence of true negatives is anticipated, a rapid survey is unreliable. Although there will be many true negatives, there will also be more false negatives unless the standard high count density images are obtained in every case, in which case the value of the scout survey is lost.

Normal appearance

In the normal adult, there is symmetrical uptake about the spine; uptake in the left and right halves of the body are virtually mirror images (Fig.1.2).

In the posterior projection the scapula, ribs, spine and pelvis are clearly identified. Lumbar and lower thoracic vertebrae are individually resolved, upper thoracic less clearly and cervical not at all unless a pinhole collimator is employed. Pedicles, laminae and spinous processes should be clearly visible. In thin subjects the anterior portions of the ribs may be seen faintly in the posterior projection, giving rise to increased count rate where they cross the posterior parts. In muscular subjects the insertions of the posterior scalene muscles are visible and are usually symmetrical. There is slightly higher uptake around the ankles, knees, shoulders, elbows and wrists than the mid-shafts of the adjacent long bones. In children and adolescents there is high uptake at the epiphyseal plates, the intensity and pattern varying with both age and sex. An often rather diffuse increase in count rate in the anterior projection of the cervical spine if present is due to uptake in the laryngeal cartilages. Uptake in the thyroid, identified by its characteristic shape, occurs only if the radiopharmaceutical preparation is faulty and contains appreciable free pertechnetate. This can be confirmed by visualization also of gastric mucosa (page 11).

Uptake in the skull should be symmetrical about the mid-line antero-posterior axis. It is usually highest in the region of the ethmoids and, in children, at the basi-sphenoid suture. Sutures and venous lakes may continue to be visualized, often faintly, into adulthood (Fig. 1.3). There is commonly a diffuse and sometimes slightly asymmetrical increase in uptake over the vault of the skull in older subjects, in some cases due to Paget's disease although this may not be confirmed radiographically.

The facial bones and base of skull are difficult to visualize adequately on planar images, but are seen clearly using SPECT (Fig. 1.4).

Planar imaging of the mandible is rarely helpful and dental disease may be difficult to distinguish from infiltrating neoplasm. The best criterion is comparison in SPECT slices with uptake in cervical vertebral bodies, but these must not be confused with lateral masses where

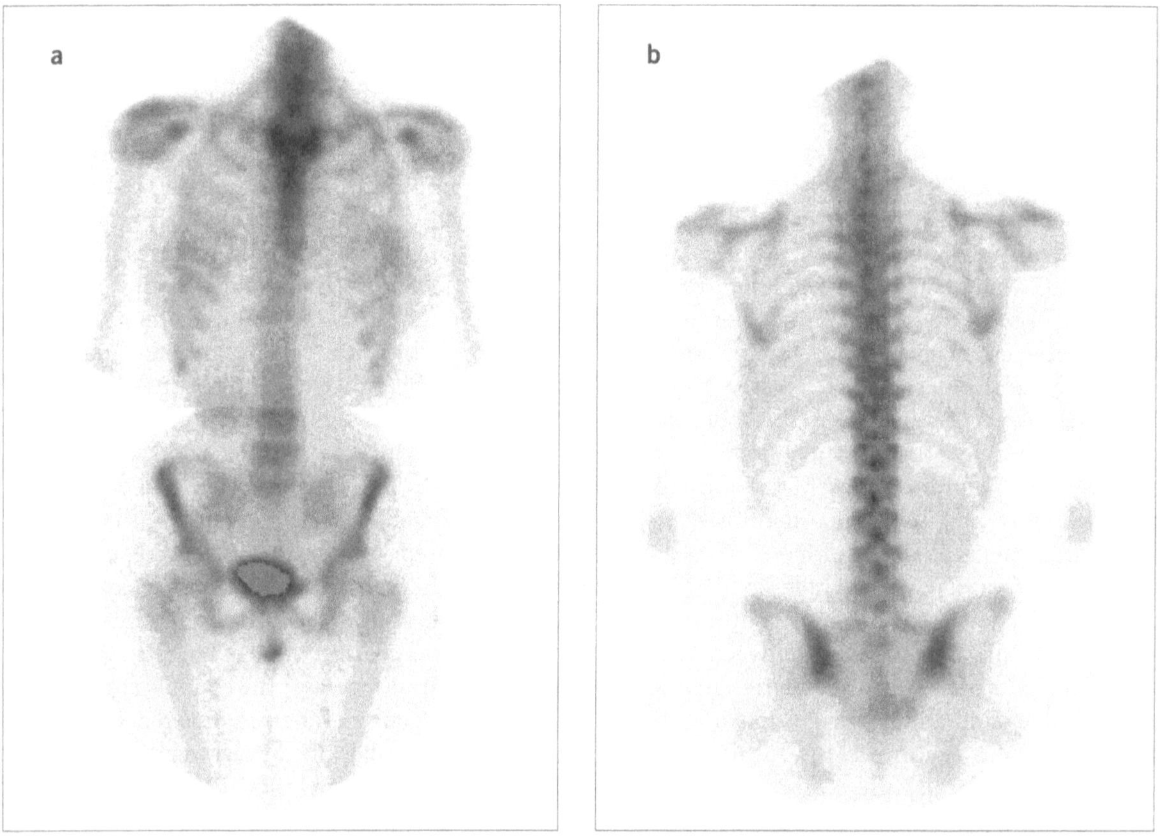

Fig. 1.2 Normal whole body distribution of any bone-imaging agent, in this case methylene bisphosphonate (MDP). Note symmetry about the mid-line and even gradation along spine. (a) Anterior, (b) posterior.

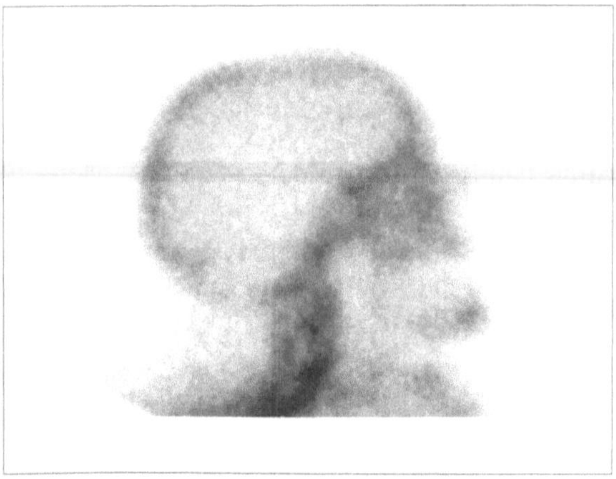

Fig. 1.3 Normal lateral skull.

degenerative or arthritic changes are commonly observed. Tumour infiltration of mandible is associated with uptake of higher intensity than that in the bodies of upper cervical vertebrae.

Using a pinhole collimator (page 305) it is possible to reproduce some angled radiographic projections such as Townes' (of the occiput), Walter's (of the facial bones) or oblique projections such as Stenver's (of the mastoid).

Soft tissue uptake

Skeletal imaging agents are excreted via the urinary tract, which is always visualized. There is always some residual uptake in other soft tissues. Abnormally high accumulations, especially in lung and stomach, may be observed in patients with hyperparathyroidism and vitamin D intoxication. This usually reverts to normal when the precipitating cause is removed. Uptake is commonly seen in recently damaged or infarcted soft tissues including brain, heart, other damaged or ischaemic muscle, burns and in the chest wall following defibrillation (possibly due to muscle damage). PYP is commonly employed for diagnosis of myocardial infarction (page 132) but all bisphosphonates share this property. Myocardial uptake of MHDP, but not of MDP, is occasionally observed in elderly males with carcinoma of prostate without evidence of current or previous ischaemic heart disease. This finding appears benign and should not be overinterpreted. Myositis ossificans, whether or not associated with muscle trauma, is usually readily visible even when radiological changes are minimal. Oblique projections or SPECT are useful to distinguish soft tissue from bone uptake. Ossification of tendon insertions is a recognized complication of toxicity due to vitamin A and its analogues such as isotretinoin. This may be associated with increased uptake of bone-seeking tracers. Accumulation of bone-imaging agents occurs occasionally in a wide variety of soft-tissue metastases (Fig. 1.5) and some primaries, especially from breast (page 271). Uptake into normal breast is also common.

Locally increased concentrations are found in some pleural effusions and ascites, both benign and malignant, in primary and secondary amyloidosis, soft tissues included within a radiotherapy treatment volume and renal parenchyma following radiation or chemotherapy. Cutaneous, subcutaneous and muscular uptake, separately or in combination, have been described in dermatomyositis and periarticular uptake in progressive systemic sclerosis. Diffuse lung uptake is associated with pulmonary microlithiasis (Fig. 1.6). The diagnosis may be confirmed by x-ray computed tomography (CT). Uptake in thyroid and gastric mucosa usually indicates a faulty preparation containing free pertechnetate.

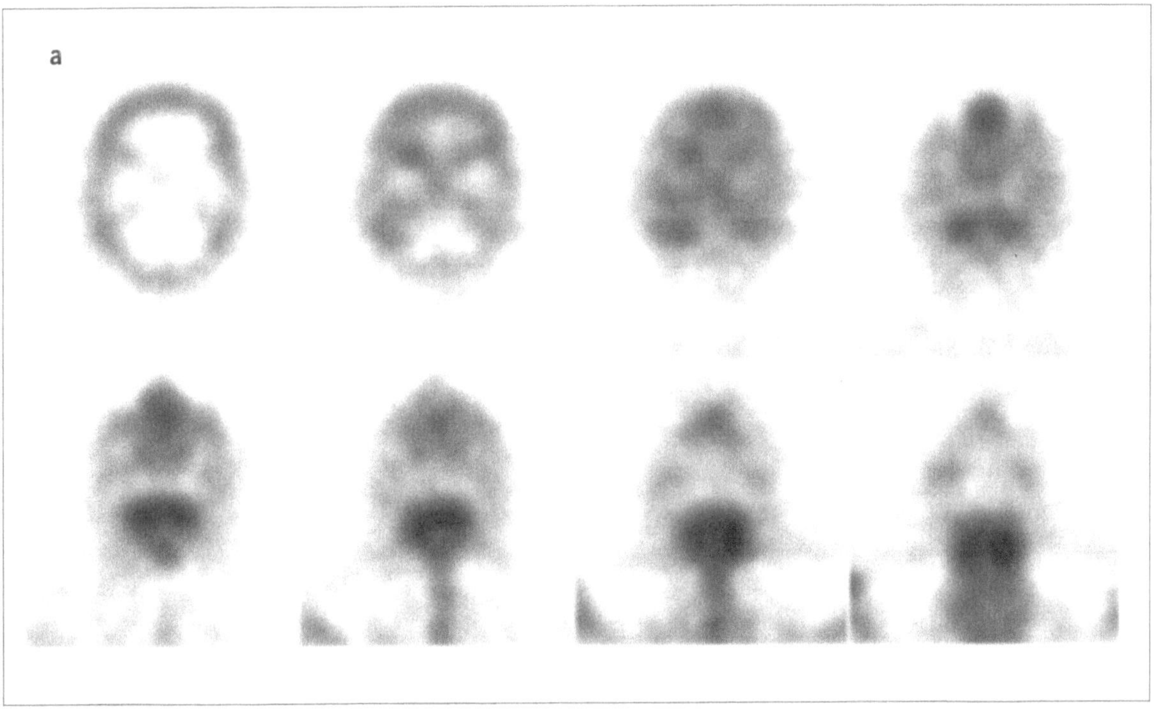

Fig. 1.4 Transverse (a) and coronal (b) SPECT slices through base of skull and facial bones. Note floor of middle fossa (sphenoid and temporal bone), atlanto-occipital joints, orbits and maxilla. There is osteoarthritis of the left neurocentral joint at C3/4. (c) Same subject. Sections in plane of body of mandible, at an angle of approximately 30° to (a). Note body of mandible (right-hand frame) compared with hard palate and maxilla (left-hand frame). (**b,c** *overleaf*)

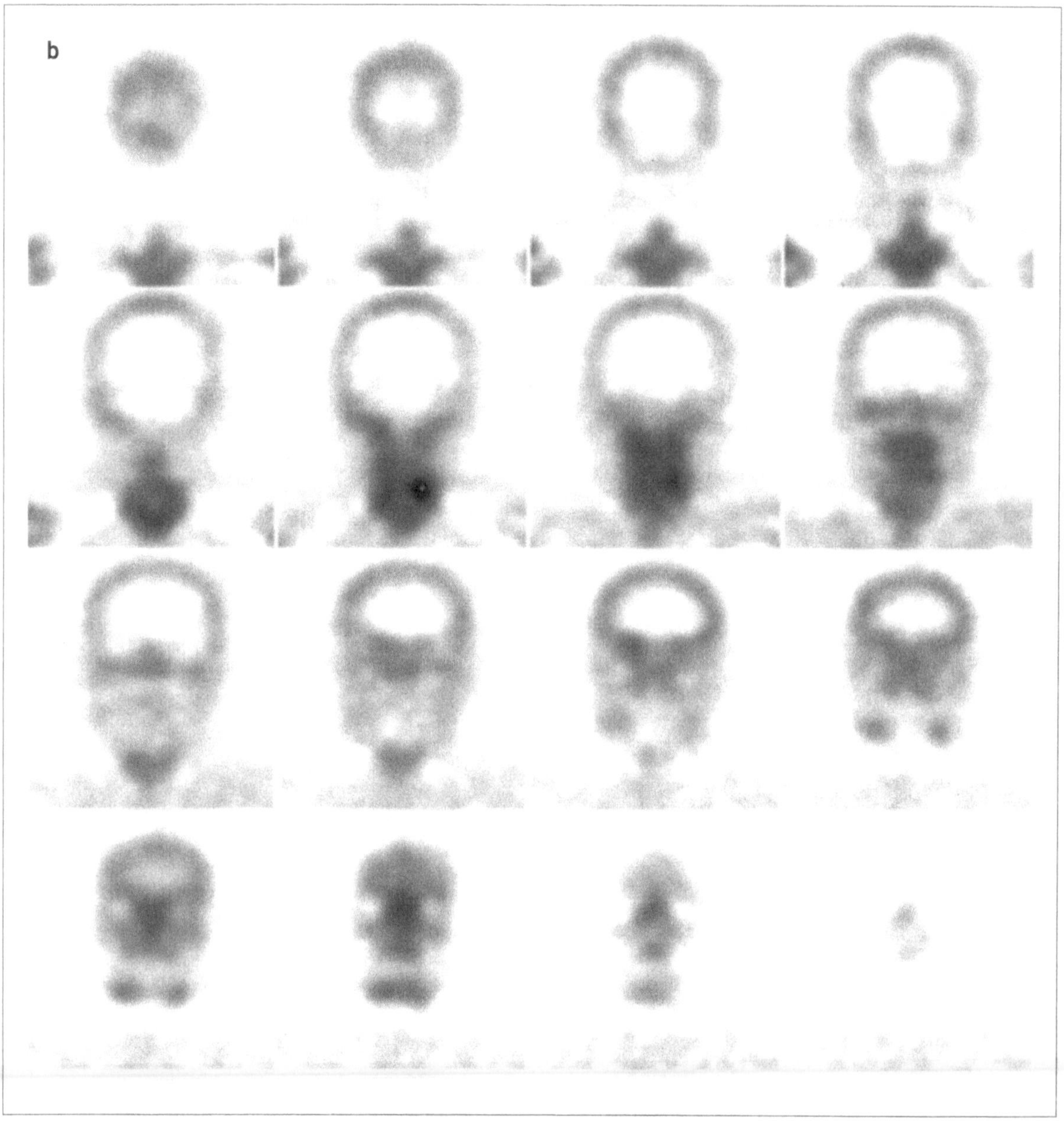

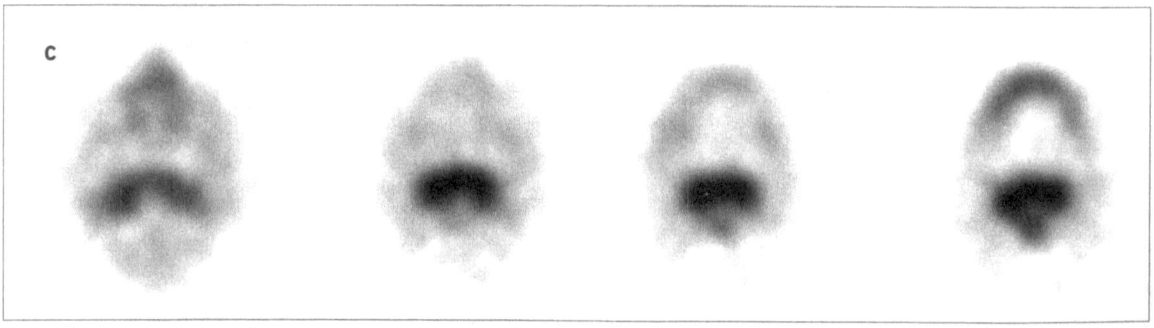

Fig. 1.4 *continued*

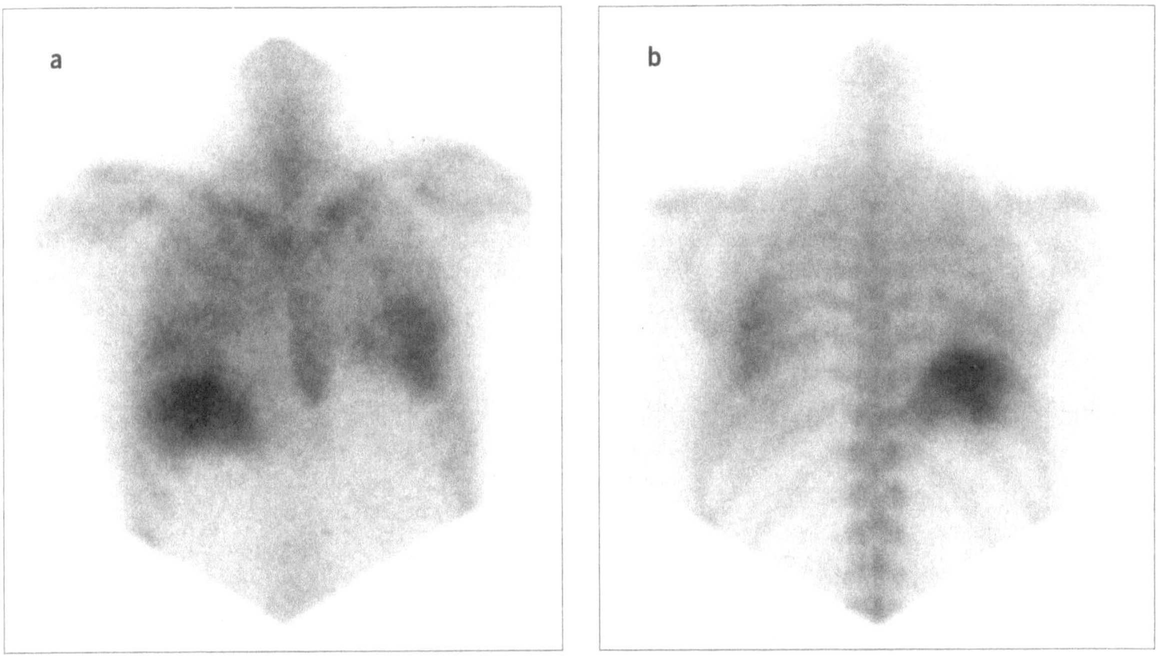

Fig. 1.5 Soft-tissue uptake in liver and a pleural deposit. (a) Anterior, (b) posterior.

Bisphosphonates do not visibly concentrate in placenta or uterus up to 18 weeks but uptake has been identified in both placenta and foetal skeleton at 30 to 32 weeks.

Further reading:
Journal of Nuclear Medicine 1996; ***37****: 469–71*

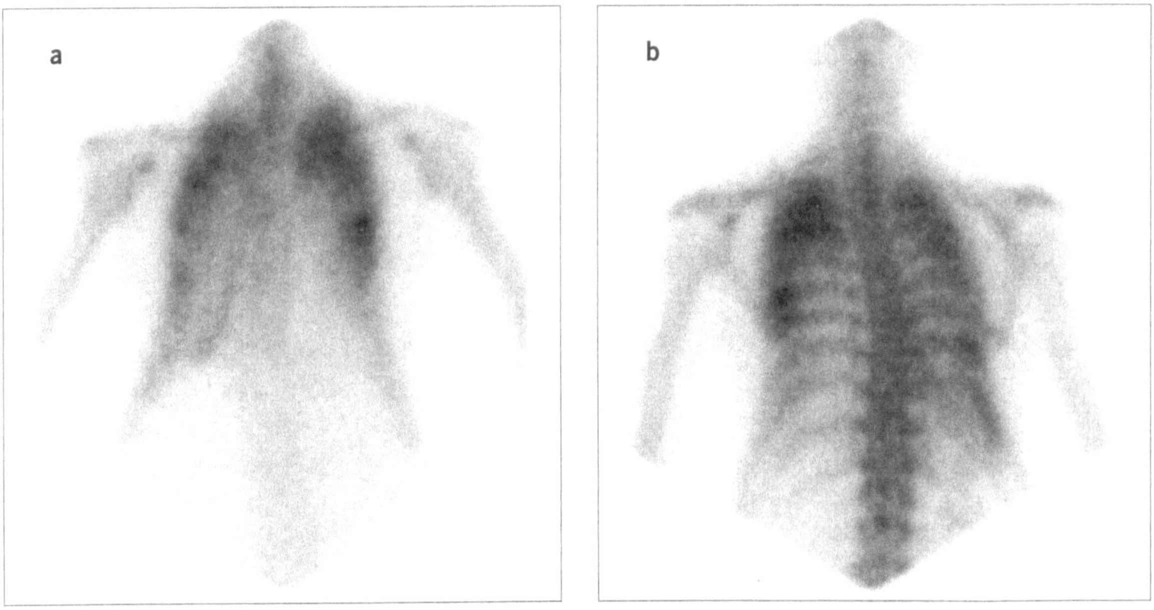

Fig. 1.6 Pulmonary microlithiasis. (a) Anterior, (b) posterior.

The abnormal skeletal scintigram

The majority of focal abnormalities (metastases, fractures, infection, Paget's disease, primary bone tumours) appear as areas of increased uptake (Fig. 1.7). A few are photon deficient (Fig. 1.8).

It is principally because of these latter that high count density and high resolution images are required. A photon-deficient lesion is less likely to be observed in a low count density or low resolution image because of the inherently lower contrast. Osteoblastic activity is present in photon-rich lesions, reduced or absent in photon-deficient ones. With some large destructive deposits there may be a rim of increased osteoblastic activity with a central photon-deficient area where bone and osteoblasts have been destroyed. Areas of infarction are initially photon-deficient, for example in Perthe's disease, but by the time avascular necrosis is suspected in the adult there is commonly considerable reactive change associated with osteoclasis and repair occurring from the periphery.

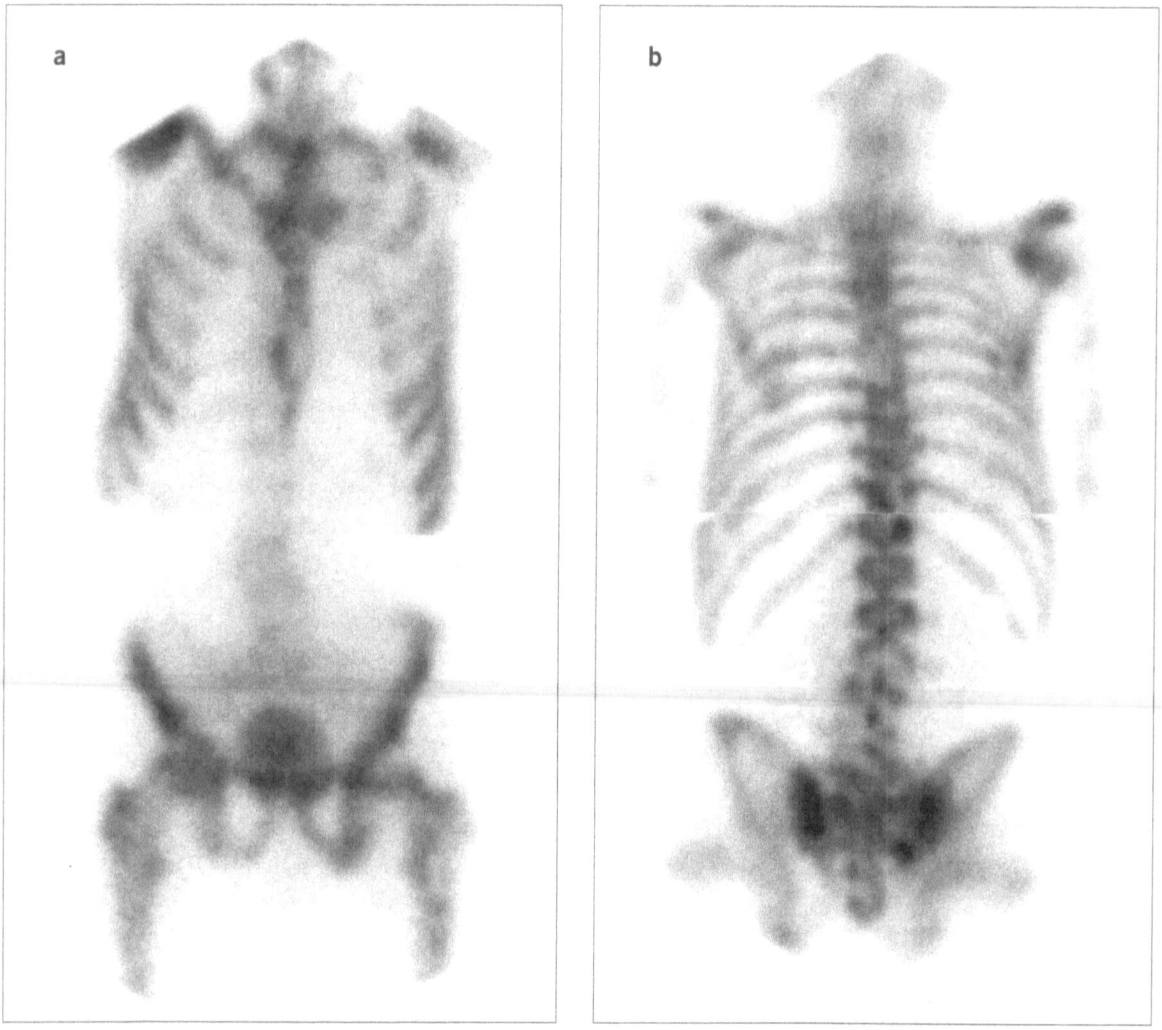

Fig. 1.7 Abnormal bone scintigraphy. (a) Composite anterior projection, (b) composite posterior projection. Uneven uptake in ribs and multiple focal areas of increased uptake in spine, pelvis and limbs due to metastases, in this case from a carcinoma of breast. Few were identifiable on plain radiographs.

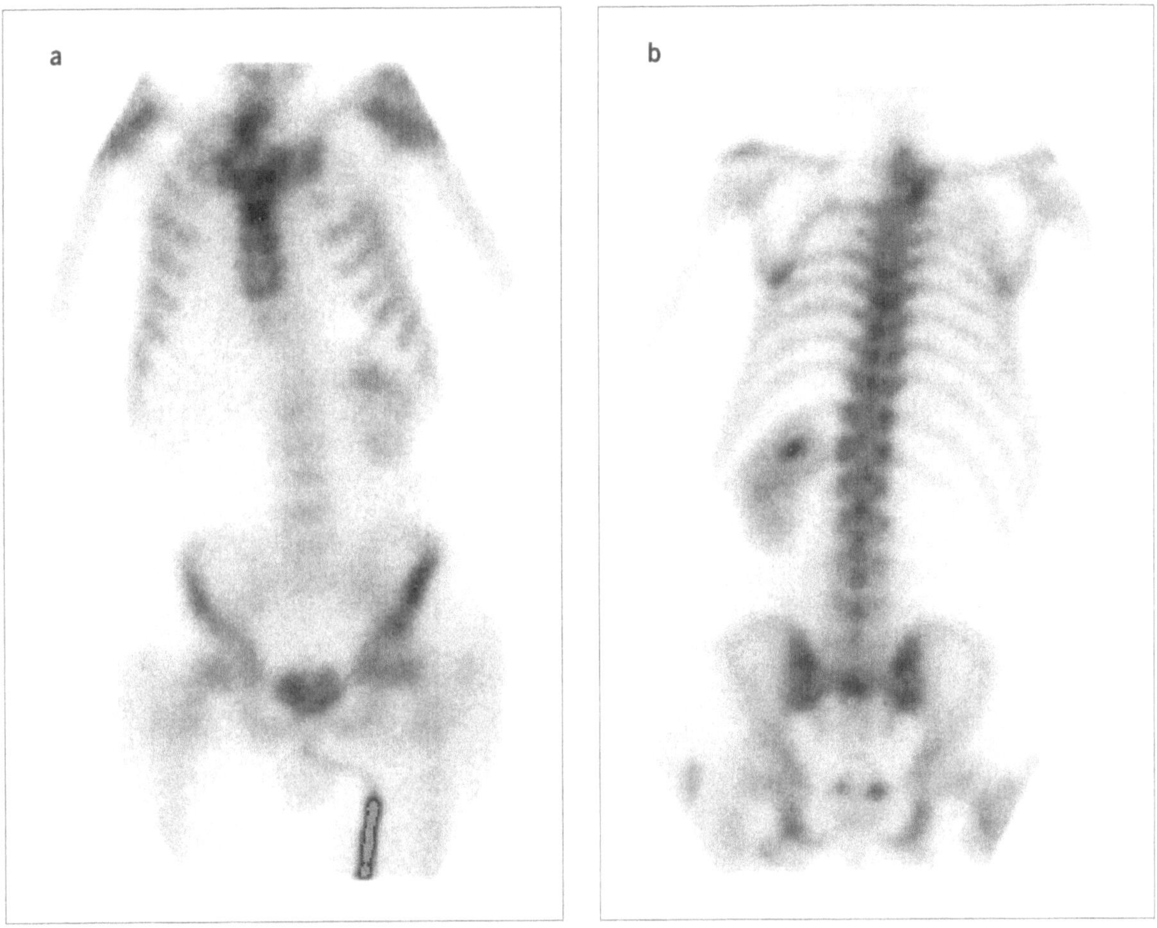

Fig. 1.8 Photon-deficiency in right 1st, 2nd and 4th ribs and in T3 due to lytic deposit from adenocarcinoma of kidney. Previous right nephrectomy. Increased uptake in right 3rd and 5th ribs.

There is no simple correlation between radiographic and scintigraphic appearance, hardly surprising as the two give different (albeit complementary) information. Radiographs show the amount of mineral, principally calcium, interposed between the x-ray source and the film. Skeletal scintigraphy indicates local blood flow or osteoblastic reaction to neoplasm, infection or injury and is not a direct indicator of any underlying pathological process. Following haematogenous implantation of metastasis or infection into bone marrow, resorption is accelerated, principally from the surface of trabeculae. Some tumours may directly provoke osteoblastic activity, thus producing sclerosis; in others osteoclastic activity stimulates osteoblastic repair. Both destruction and bone repair proceed simultaneously at the majority of sites. An abnormality is visible on plain radiographs only when there is a gross disturbance of the equilibrium between removal and replacement, predominant osteoclasis producing an osteolytic lesion and dominance of repair resulting in sclerosis.

An osteolytic deposit in the body of a vertebra cannot be identified on plain radiographs until half of the trabecular bone has been destroyed. A defect in the cortex of a vertebral body must be 1 cm in diameter to be reliably visualized on plain radiographs. Smaller lesions can be detected by tomography and in plain radiographs of peripheral bones. It is thus evident that only relatively large lesions can be detected in the axial skeleton by conventional radiography. Whatever the radiographic appearance, both processes occur in the majority of skeletal deposits. They are thus detected scintigraphically irrespective of which predominates. Scintigraphy, although more sensitive than plain-film radiography, has relatively poor specificity.

The scintigraphic appearance by itself is rarely sufficiently characteristic to be acceptable as proof of metastatic disease, especially if there are only one or two abnormalities. A pattern similar to that in Fig. 1.7 is found occasionally in ankylosing spondylitis. In many patients with scintigraphic abnormalities, review of the radiographs will confirm the diagnosis. In others benign processes such as osteoarthritis or Paget's disease account for scintigraphic abnormalities. However, in a significant minority no plain radiographic abnormality is found to account for a scintigraphic lesion. In the absence of multiple randomly distributed foci associated with collateral evidence, for example a histologically confirmed primary with a predilection for skeletal metastasis, it is essential to obtain confirmation (CT, magnetic resonance imaging (MRI) or if necessary biopsy) before commencing treatment. Even in patients with a high *a priori* risk there should be reluctance to accept absence of a benign explanation as sufficient evidence for metastasis.

Better morphological detail can be obtained with pinhole collimation. This may reduce the need for radiographic confirmation but rarely eliminates it. Metastatic cancer appears on pinhole images of vertebrae as a homogeneous diffuse increase in count rate (Fig. 1.9) or as short-segment accumulations along the end-plates. In contrast compression fracture produces increased uptake along the entire length of the end-plate. In practice the distinction is commonly imprecise.

Disc spaces are unaffected in patients with cancer but reduced in patients with tuberculous or septic spondylitis. Tuberculosis is associated with increased uptake throughout the vertebral

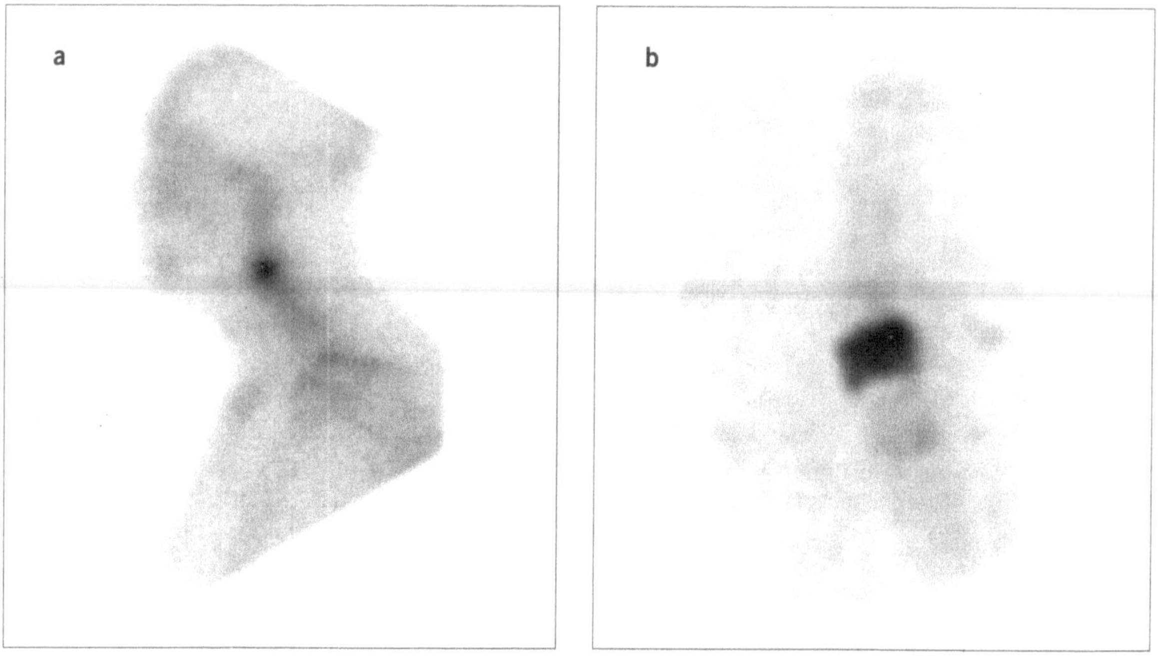

Fig. 1.9 (a) Conventional and (b) pinhole lateral views showing increased uptake throughout C7 and D1 due to metastatic disease. Note that because magnification in pinhole images depends on pinhole-to-object distance, the two projections have different magnifications. Plain radiographs of this area were inconclusive.

body, pyogenic spondylitis with increased uptake in adjacent end-plates separated by a narrowed disc. These findings accord well with classical radiographic descriptions but their reliability have yet to be independently confirmed for scintigraphy. A report of 'hot spots' without further radiographic or other correlation and which does not contain a clinically relevant conclusion (such as metastases, arthropathy, trauma, infection) is of no value.

Response to treatment

Chemotherapy and hormone manipulation

Chemotherapy or hormone interventions may slow or arrest tumour growth, allowing repair to proceed. When there is a favourable tumour response, osteolytic or radiologically occult lesions commonly become sclerotic and therefore more readily visible radiographically. Skeletal scintigraphy during the repair phase shows continuing or increased osteoblastic activity (flare) for 3–6 months after initiation of therapy of sensitive tumours, and in some cases with an increased number of lesions visible. Subsequently, once repair is complete but not necessarily with the restoration of normal trabecular architecture, there is no increase in bone turnover rate and therefore no scintigraphic abnormality for as long as the lesion remains inactive. Thus the bone scan may revert to normal while the radiographic appearances become more obviously abnormal.

Skeletal scintigraphy is an indicator of a response to tumour and does not display the tumour itself. The relationship between tumour and response is complex and indirect. Treatment may affect bone metabolism in other ways, for example the temporary malaise caused both by radiotherapy and much chemotherapy may induce the patient to be less active. Immobility by itself leads to a decreased rate of bone formation and increased resorption, especially of trabecular bone, with rapid loss of bone mass and reduced uptake of bone-seeking tracers. Tumour control is rarely permanent and ultimately there is usually recrudescence of tumour activity, reflected by increased scintigraphic activity but not always associated with change in the radiographic appearance. There is thus no simple or constant correlation between the radiographic and scintigraphic appearances of a bone metastasis. This must be borne in mind when considering skeletal scintigraphy for assessment of response to treatment. It is unreasonable to expect skeletal scintigraphy to give a quantitative measure of response.

Further reading:

European Journal of Nuclear Medicine 1994; ***21:*** *377–80*

Radiotherapy

A similar sequence of events is seen following radiotherapy, although the mechanism may be different. Radiotherapy induces hyperaemia acutely in most tissues, bone being no exception. Scintigraphy performed during or immediately after a course of radiotherapy therefore reveals increased uptake in the volume treated. Subsequently, radiation-induced vasculitis produces a permanent impairment of the blood supply, possibly also associated with reduced osteoblastic function. Scintigraphy performed after the acute reaction has settled shows reduced uptake in the irradiated area which persists for life (Fig. 1.10).

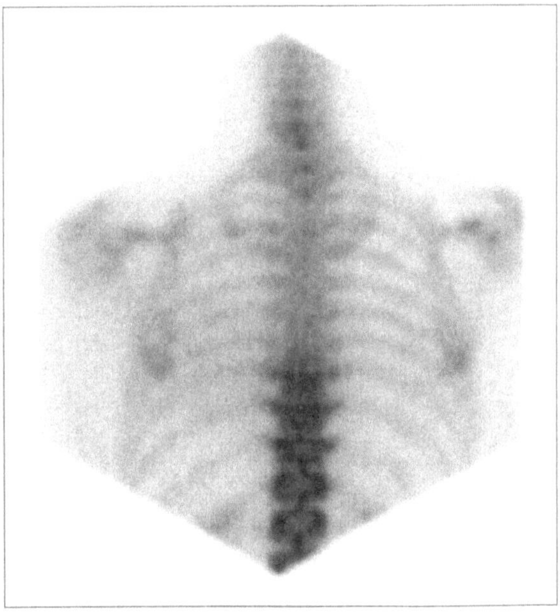

Fig. 1.10 Reduced uptake in upper thoracic spine due to radiotherapy for Hodgkin's disease some years before.

A characteristic feature of radiation-induced changes is their non-anatomical linear boundaries. A common error is to interpret irradiated areas as normal, particularly if these are extensive, for example after treatment of seminoma or lymphoma, and misinterpret normal uptake in adjacent untreated areas as increased due to deposits. Soft-tissue uptake with a linear boundary corresponding to the beam edge is also occasionally seen in irradiated tissues.

Further reading:

Journal of Nuclear Medicine 1992; ***33****: 1780–2*

Sources of error

Count-rate is higher where inferior angle of scapula overlies the ninth rib. This is usually symmetrical except in the presence of a scoliosis or other anomaly. It is easy to overlook an abnormality at this site or to over-report a non-significant asymmetry. Some of the pelvis, most commonly the pubis, is often partially or completely concealed in standard projections by residual excreted activity in the urinary tract. Even when the bladder is almost empty it can be difficult to determine whether a small area of high uptake is residual activity in urethra, vagina or bladder or whether it is in bone. Three approaches may be used to solve this problem. If the residual volume in the bladder is large and much of the pelvis is concealed, it is sometimes helpful to re-image the following morning, by which time the bladder is likely to be empty of radioactive urine (Fig. 1.11). If there is still residual activity (as in some patients with outflow obstruction) there may be no alternative to bladder catheterization. When the residual urinary volume is fairly small, a pelvic outlet view, projecting the pubis clear of the bladder may be helpful (Fig. 1.12). Because of distance from the detector, this is better for detecting abnormalities in the ischium and the inferior pubic rami than in the superior pubic ramus.

A pronounced lumbar lordos may give unusual prominence to lower lumbar vertebrae in the anterior projection and to lower thoracic vertebrae in the posterior, especially in thin subjects. There is usually higher uptake in the concavity of a scoliosis due to asymmetrical transmission of body weight. This may be associated with asymmetry of the pelvis, hip joints and lower limbs. The manubrio-sternal

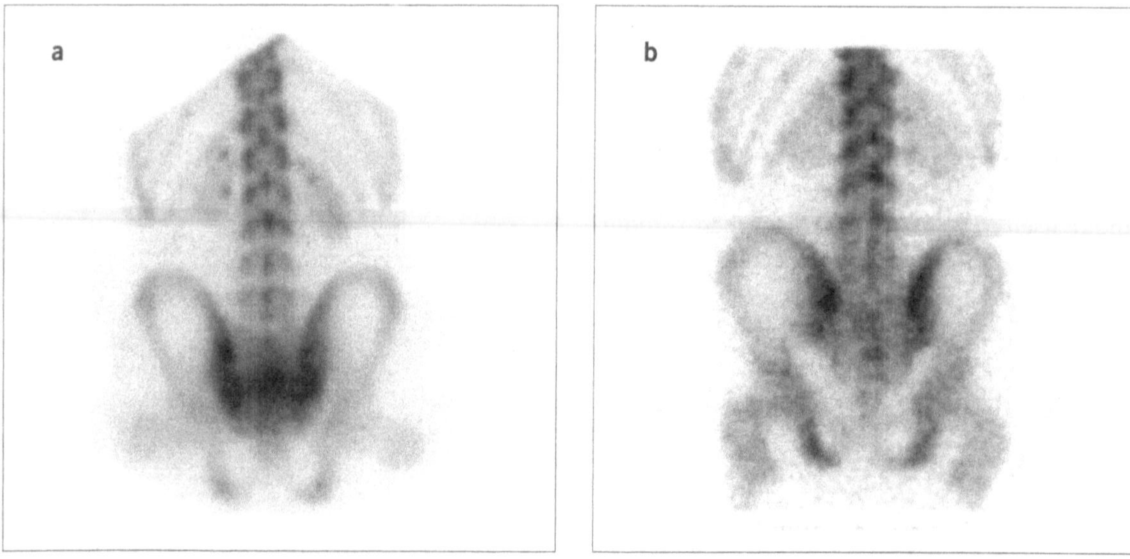

Fig. 1.11 (a) Apparently high uptake in upper sacrum in patient with previous anterior resection of carcinoma of rectum and low back pain. (b) Late view, 18 h after injection when bladder is completely empty confirms high count rate was due to shine-through from full bladder.

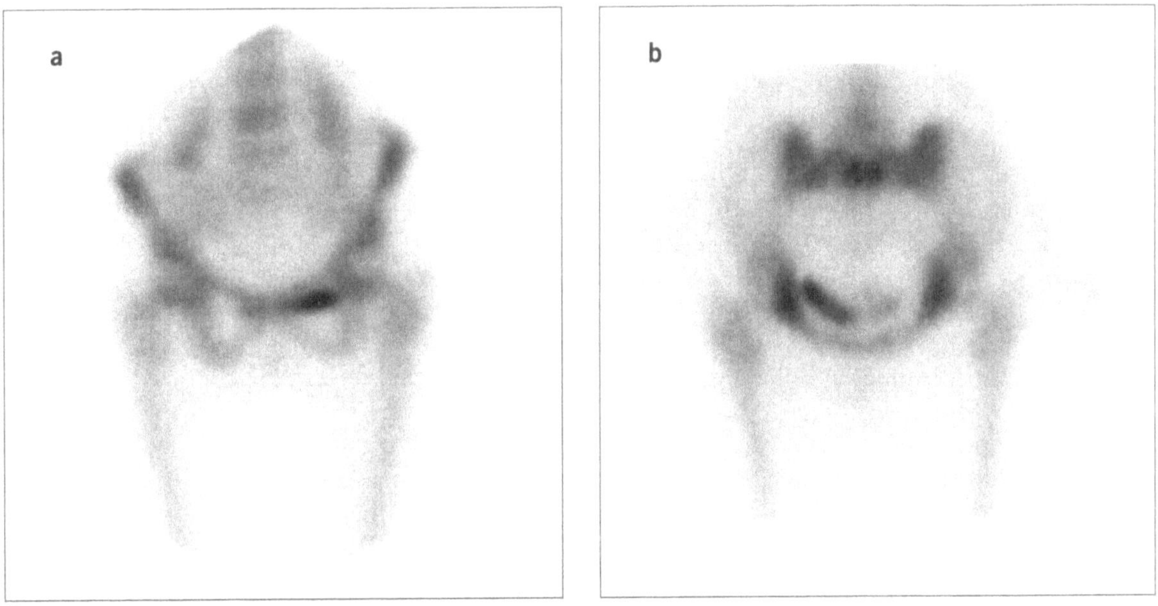

Fig. 1.12 (a) It is uncertain in this anterior projection if there is an abnormal focus in left superior pubic ramus. (b) Pelvic outlet (orientation as if posterior projection, so sides reversed). Activity is in bladder, not bone.

joint can often be separately resolved (Fig. 1.13) and must not be confused with abnormal uptake in the manubrium or body of the sternum, which is usually due to tumour.

Apart from these anatomical pitfalls, the most frequent cause of a false negative is an examination of inadequate technical quality. Common faults include:

- patient not close enough to collimator
- movement
- inadequate statistical or photographic quality
- part of the skeleton obscured by artefacts or activity in the urinary tract
- incorrect positioning
- part of the skeleton not imaged.

False positives can be due to misinterpretation of radiographs or attempting to report scintigraphy without reference to radiographs. Multiple randomly distributed foci, although most often due to disseminated metastatic disease, can be due to osteomalacia if mainly in ribs or ankylosing spondylitis if confined to spine. Artefacts due to activity, usually urinary,

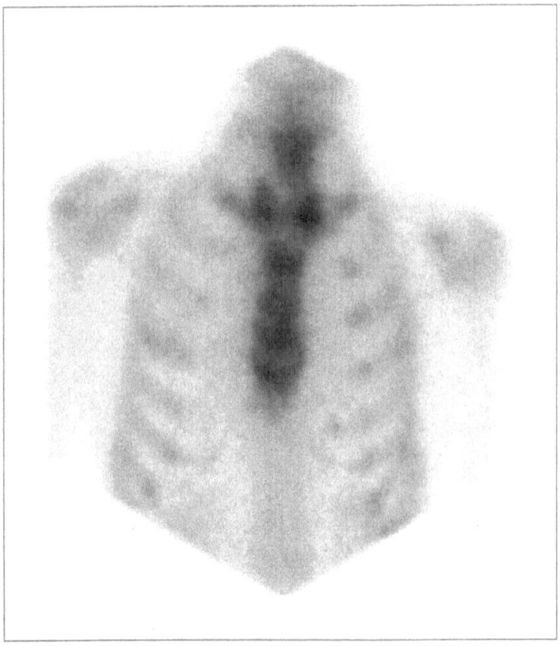

Fig. 1.13 High uptake in sterno-clavicular and sterno-manubrial joints in patient with chronic obstructive airways disease. This is a common finding and must not be confused with metastases in the sternum. It is sometimes possible to see erosions of these joints on radiographs.

contaminating skin or clothing can both simulate and conceal deposits. When in doubt the patient should be requested to remove all clothing, including underclothing and change into an hospital gown. An important cause of a false negative, especially in patients with prostatic carcinoma, is failure to recognize a 'super scan', in which there is symmetrically increased uptake throughout the skeleton (Fig. 1.14). This appearance is seen, less commonly, in patients with other primary tumours including breast or lung and in metabolic bone disease. The scintigraphic appearance is of very high skeletal uptake, low soft-tissue background and faint or absent renal visualization. Exceptionally good quality images, especially if soft-tissue background is unusually low and the kidneys cannot be seen, should always arouse this suspicion.

More commonly, uptake, although high throughout the skeleton, is uneven; particularly in ribs and long bones (Fig. 1.15). A patchy distribution of radioactivity even without clear focal abnormalities is always suggestive of disseminated bony metastases. In both of these situations the diagnosis is usually readily confirmed by re-evaluation of radiographs.

Metastatic disease

Carcinoma of prostate

Skeletal metastases occur in most patients with carcinoma of the prostate. In many they are present at presentation. The classical radiological appearance is of multiple sclerotic deposits, but many untreated patients with no radiographic abnormalities have deposits confirmed on skeletal

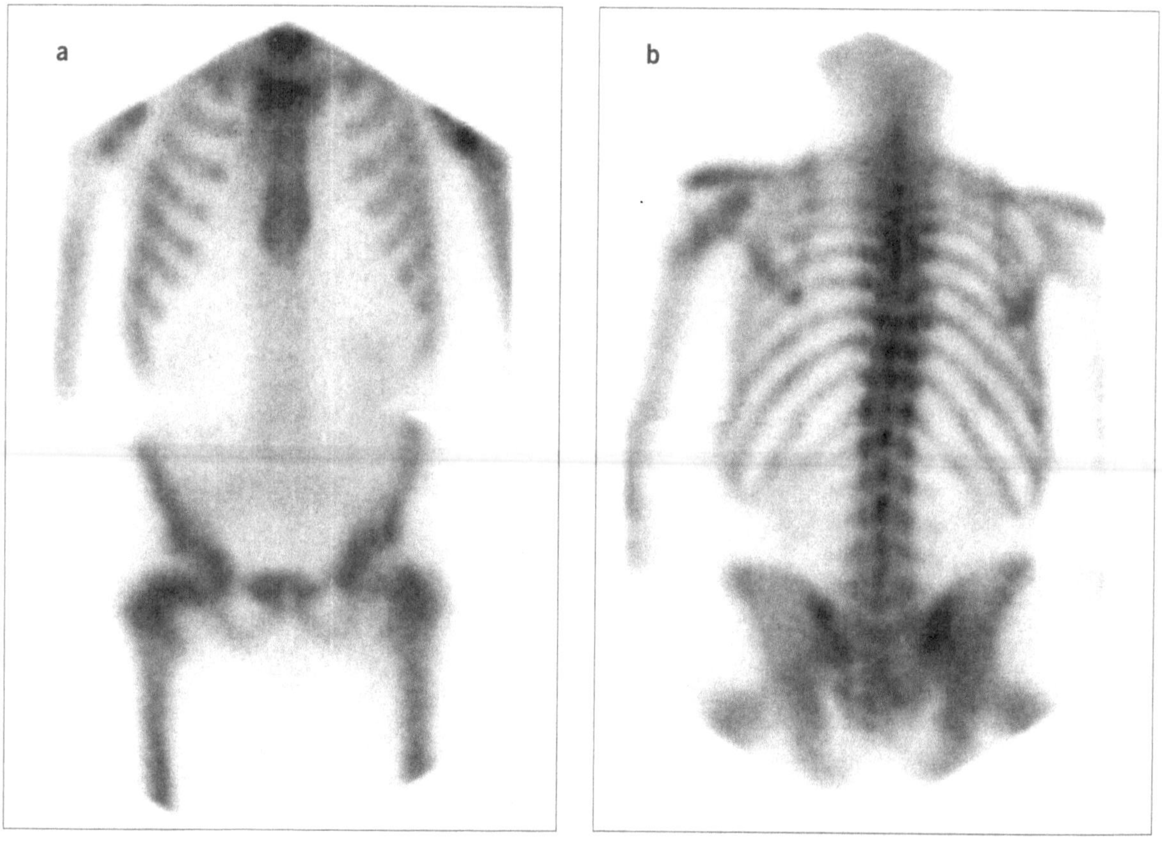

Fig. 1.14 Superscan in untreated patient with carcinoma of prostate at presentation.Composite images. (a) Anterior, (b) posterior.

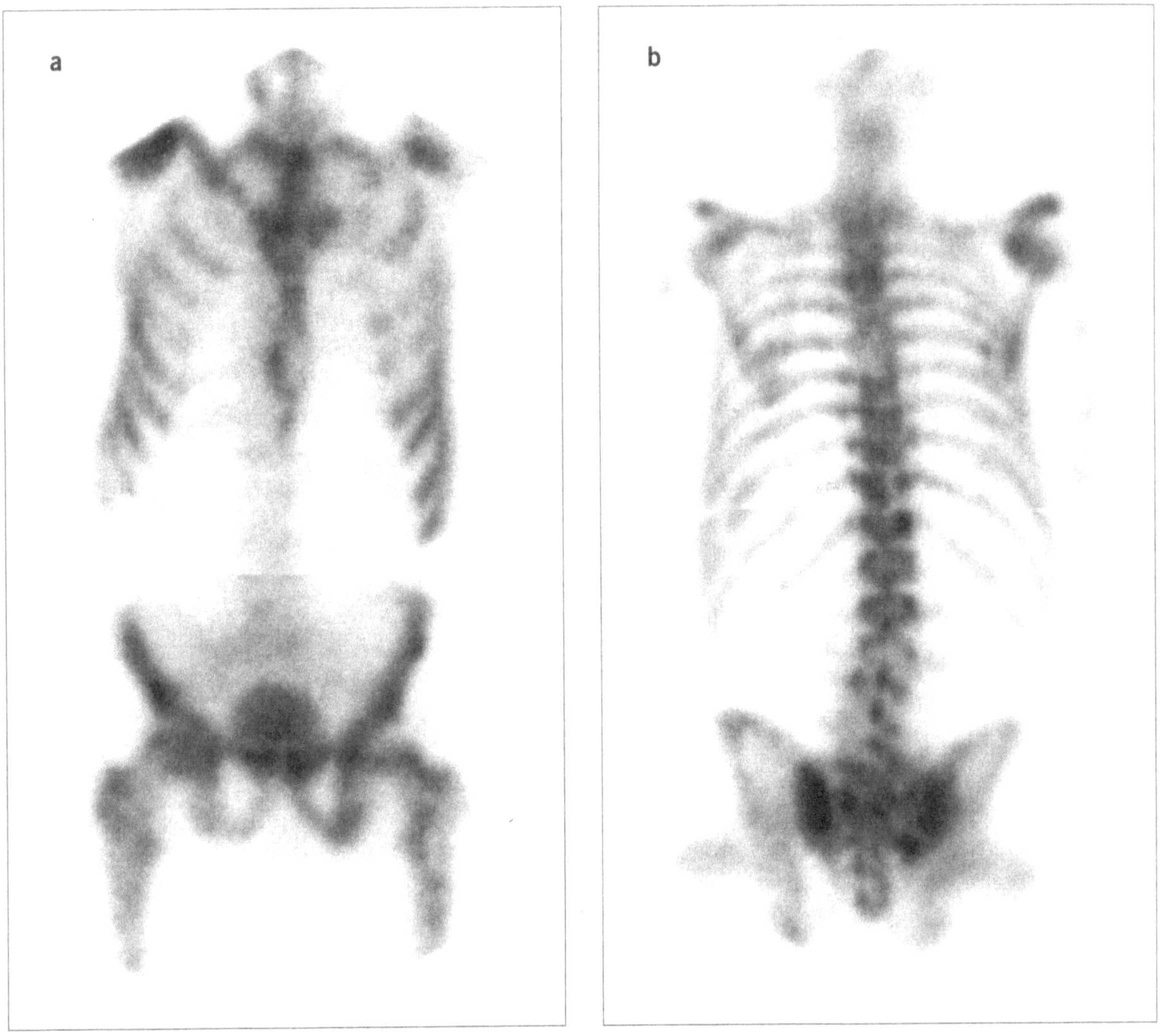

Fig. 1.15 Another patient with carcinoma of prostate: uptake is less uniform than in Fig. 1.14 but skeletal uptake is so high that urinary tract is not visible at 3 h. (a) Anterior, (b) posterior.

scintigraphy. Metastases may occur anywhere in the skeleton, the highest prevalence being in the pelvis and lumbar spine. Scintigraphic abnormalities commonly precede radiological confirmation of disease, which may be difficult to obtain without resort to CT, MRI or, possibly, pinhole imaging because of the high prevalence of osteoarthritis and (in some populations) Paget's disease in this age group. The prevalence of skeletal deposits increases with advancing clinical stage, rising from approximately 10% in T0 tumours (those diagnosed on microscopy of tissue removed with a clinical diagnosis of benign prostatic hypertrophy) to 80% in patients with T4 disease or at post mortem. Bone metastases are uncommon (< 2%) when the PSA (prostate-specific antigen) is < 8 ng/ml, the incidence rising with PSA level. PSA is however not specific for prostatic cancer and is raised in prostatitis and other benign disease. The prognosis of patients with bone metastases is poorer than of those without. There is a correlation between extent of skeletal involvement and survival, more extensive skeletal involvement being evidence of a significantly increased tumour burden.

False negative scintigraphy in untreated patients with prostatic carcinoma is rare except when due to technical deficiencies, especially residual radioactivity in the bladder or underclothing obscuring parts of the pelvis. The pubic rami in particular are difficult to visualize adequately. For practical purposes a normal and technically adequate examination excludes skeletal metastases, irrespective of the PSA or acid phosphatase concentrations. If the examination is technically inadequate no reliance can be placed on negative findings.

Purely osteolytic deposits are extremely rare in carcinoma of the prostate and photon-deficient deposits have not been recorded in the absence of photon-rich deposits elsewhere in the skeleton. Following endocrine therapy or chemotherapy the tumour and its metastases may pass into remission, with increasing radiographic sclerosis but return of the scintigraphic appearance to normal. The commonest pattern is for some lesions to become quiescent whereas others become more active; new ones may appear. When assessing response this is defined as progression. Changes in scintigraphic appearance are often more dramatic than those of the radiographs. It is thus essential to know the endocrine and treatment status of the patient when interpreting the examination. Because of the large number of possible influences it is difficult to use scintigraphy as an index of response to treatment.

Indications

The principal indications for skeletal scintigraphy in patients with carcinoma of prostate are:

- screening those with a raised PSA but without histological confirmation; they may have benign prostatic disease
- staging histologically confirmed tumours
- obtaining an estimate of total tumour burden
- follow-up, if there is a rise in serum PSA or if new symptoms referable to the skeleton develop
- before treatment if radical prostatectomy or radical radiotherapy is contemplated, even if PSA is undetectable.

Further reading:
*Journal of Urology 1991; **145**: 313–8*
*European Journal of Nuclear Medicine 1995; **22**: 207–11*

Carcinoma of breast

The reported incidence of occult skeletal metastases in recent series of patients with T0 mammary carcinomas is negligible: it is less than 1% in T1, 3% in T2, 8% in T3 and 13% in T4. Considering clinical stage rather than the local extent gives a similar picture, with a negligible detection rate of bone metastases in patients with clinical stage 1, 3% in stage 2, 7% stage 3 but almost 50% in stage 4. This is lower than much of the older literature suggests, possibly because of more stringent definition of staging criteria. The prevalence of occult skeletal disease in patients with local or soft tissue recurrence is high (over 30%). Following treatment, scintigraphic abnormalities disappear less commonly than in carcinoma of the prostate but a total scintigraphic remission is occasionally observed after endocrine ablation or chemotherapy. The high generalized uptake of the 'super-scan' is rarely seen in carcinoma of the breast. Soft-tissue uptake is less often observed in normal than in affected breasts but is of limited diagnostic value.

Skeletal deposits may become overt within months of an entirely normal study, implying a false negative rate of up to 7%, irrespective of clinical stage. The distribution of lesions is for the most part similar to other tumours, the commonest sites being, in descending order, ribs, vertebrae, pelvis, limbs and skull. Sternal involvement is more common than in most other tumours, possibly due to extension from internal mammary nodes. Approximately 90% of solitary or adjacent rib lesions are due to minor trauma, especially if these have a linear or curvilinear distribution and should not be diagnosed as metastatic unless there is unequivocal radiological confirmation. Many are due to minor trauma with no associated radiographic abnormality. On the other hand over 90% of solitary sternal lesions are due to metastases or direct local spread. These can sometimes be visualized on plain lateral radiographs of the sternum but often require computed tomography for confirmation. Solitary asymptomatic deposits in the peripheral skeleton are rare and it is reasonable not to image skull or distal to mid-femur or mid-humerus of asymptomatic patients with no central metastases. A more extensive survey is indicated in the presence of local symptoms or if there is an unexplained elevation of serum alkaline phosphatase. Photopoenic deposits account for less than 1% of all skeletal metastases in this disease.

Indications

Because of the therapeutic implications, bone scintigraphy is indicated for staging patients with T1, T2 or T3 tumours, those categorized as clinical stage 2 or 3, patients with loco-regional recurrence, raised alkaline phosphatase and if there are local symptoms for which no definite cause has been found. Prophylactic pinning and radiotherapy of occult deposits may prevent pathological fracture.

Further reading:
Journal of Cancer Research and Clinical Oncology 1990; ***11****: 486–91*

Carcinoma of bronchus

Depending on criteria of selection for investigation, scintigraphy provides evidence of occult bone metastases in 50% or more of patients at the time of presentation, irrespective of histology. Mean survival is less than one year in patients without evidence of skeletal involvement and little more than half this in untreated patients with skeletal secondaries. Solitary peripheral deposits are less rare than in carcinoma of the breast or prostate and it is therefore advisable to examine the whole skeleton, especially if there are localizing symptoms. Osteoblastic activity provoked by metastases from bronchogenic carcinoma tends to be less marked than in breast or prostate secondaries (Fig. 1.16). Images of high resolution and high count density are thus required to identify these often relatively low contrast lesions.

When deciding whether or not an abnormality requires radiographic confirmation, the threshold of suspicion must be set at a lower level than with prostatic or breast tumours. Osteolytic (photon-deficient) deposits are more common and may be present in up to 5% of patients, another

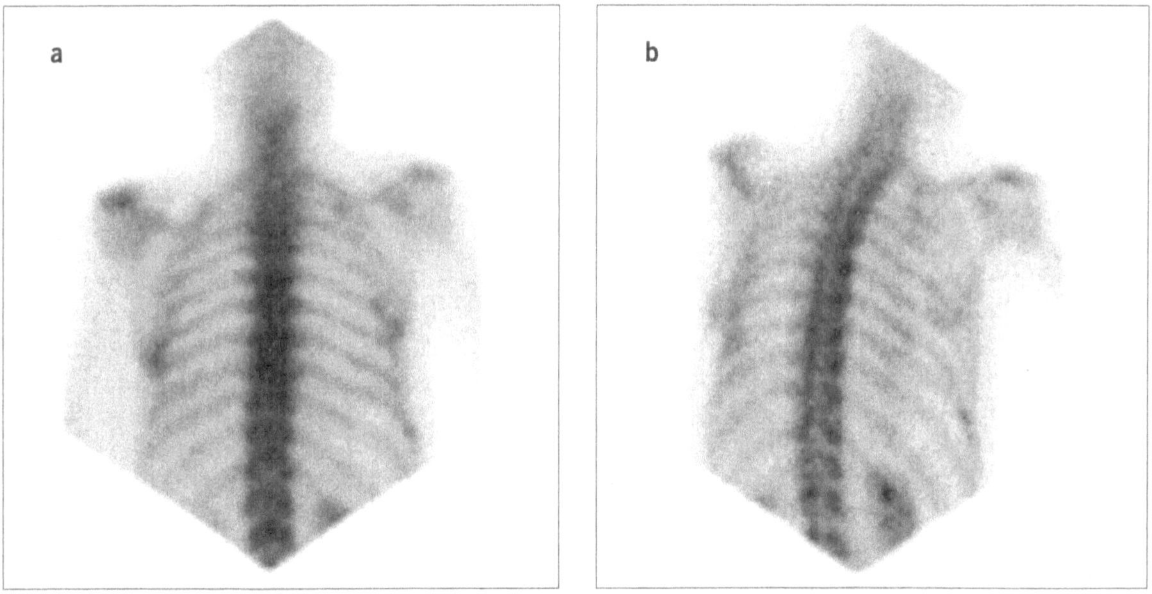

Fig. 1.16 In contrast to Figs 1.14 and 1.15 this solitary deposit in the right 9th rib from a small cell carcinoma of bronchus is provoking little reaction and is only faintly visible in the posterior projection (a). It is better seen on the oblique (b). There is also osteoarthritis of the thoracic spine.

indication for high count density images. Post mortem and follow-up studies indicate a false negative rate of less than 6%, due to small tumours which do not provoke an osteoblastic response. MRI has a higher lesion detection rate but in view of the cost differential and high pick-up rate of bone scintigraphy the latter remains the preferred initial procedure. Because of the low contrast of many lesions and relatively high prevalence of photon-deficient deposits, adherence to stringent technical standards is essential if the false-negative rate is to be minimized. Soft-tissue uptake following radio-therapy is a common finding in the irradiated volume.

Indications

Skeletal scintigraphy is an essential precursor to radical surgery or radical radiotherapy. It should be performed if bone or root pain is suspected but a cause cannot be found on plain radiographs. As with breast cancer, radiotherapy and pinning of an occult metastasis may prevent fracture.

Further reading:

British Journal of Radiology 1986; ***59****: 1185–94*

Myeloma

At least 50% of deposits are wholly osteolytic and provoke no osteoblastic response. Scintigraphy thus appears normal in at least half of patients with skeletal involvement. Most painful deposits are visualized, but this may be because pain is due to fracture. However uptake is detectable in some deposits in the absence of fracture, although there is some divergence of opinion how commonly this occurs. Because of the low sensitivity, skeletal scintigraphy is not recommended as a routine investigation in this condition but is useful in differential diagnosis of unexplained bone pain, by indicating the most promising site for tomography or biopsy. Although myeloma has a higher prevalence than any other tumour of deposits that do not provoke an osteoblastic response, because it is less common than bronchogenic carcinoma the latter is probably the commoner cause for a photopoenic deposit. Photopoenic deposits can however occur due to metastases from any primary site.

Indications

The principal indications for bone imaging of patients with myeloma is to assess whether painful deposits have sufficient osteoblastic activity to justify therapy with one of the bone-seeking agents and if prophylactic pinning is contemplated.

Further reading:

Radiologica Medica 1988; ***76****: 311–5*

Carcinoma of bladder

Skeletal metastases are rare in differentiated (G1/2) or locally early (T1/2) disease, in which the false positive rate of scintigraphy exceeds the true positive rate. Metastases occur with sufficient frequency in more advanced disease to justify skeletal scintigraphy before radical surgery or if radical radiotherapy is contemplated for T3 and T4 and G3 or higher grade tumours. Wholly osteolytic deposits are rare, but solitary peripheral deposits, although uncommon, occur with sufficient frequency to justify total body scintigraphy if there are symptoms referable to the periphery. As with all other tumours it is essential to obtain radiographic confirmation, especially of solitary lesions. This commonly requires CT or MRI. Excreted tracer clears slowly from an ileal bladder and may conceal part of the lumbar spine and pelvis. This has usually cleared by the following morning, when an unobstructed view of the entire skeleton can be obtained.

Further reading:

Clinical Radiology 1985; ***36****: 77–9*

Carcinoma of kidney

Asymptomatic deposits occur infrequently in clinically early disease. Up to 10% of skeletal metastases are osteolytic and photon-deficient. More commonly large deposits have a photon-deficient centre but an (often faint) photon-rich rim where both resorption and repair are occurring (Fig. 1.17). High count density images are therefore essential and normal scintigraphy does not preclude the need for radiography if there are localized symptoms. Scintigraphy is not indicated routinely, but many surgeons feel it is helpful when radical surgery is contemplated.

Further reading:
British Journal of Urology 1983; ***55****: 171–3*

Carcinoma of thyroid (see also page 163)

Differentiated skeletal metastases from papillary thyroid carcinoma are better detected using ^{131}I, which also shows soft-tissue deposits, than by bone imaging. However not all deposits, even of well differentiated tumours, take up iodine (Fig. 5.13). Bone imaging is indicated if skeletal metastases are suspected but iodine imaging is not indicated, for example because of the high radiation dose associated with ^{131}I, or if iodine is not available or if iodine imaging proves negative.

Further reading:
European Journal of Nuclear Medicine 1993; ***20****:1168–74*

Lymphoma

Skeletal deposits occur in a high percentage of patients with advanced disease. Solitary peripheral deposits, although rare, are somewhat less uncommon than with most other tumours; wholly osteolytic deposits are less rare in non-Hodgkin's than in Hodgkin's lymphomas. A whole-body, high count density study is therefore essential. Staging is rarely altered by skeletal scintigraphy, which does not form part of staging protocols.

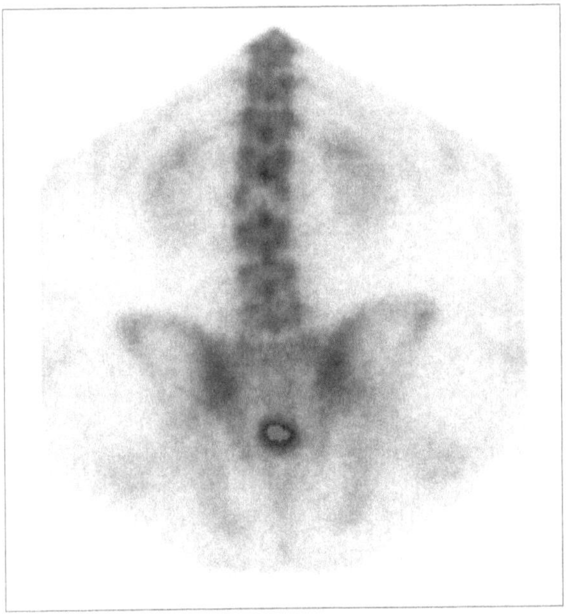

Fig. 1.17 Osteolytic deposit from adenocarcinoma of kidney has destroyed much of the body of L2, which appears photon-deficient. Posterior elements are preserved.

Mastocytosis

This is a rare disorder of mast cell proliferation involving many organs including skin, bone and liver. 70% of patients have skeletal involvement, most commonly affecting the skull and axial skeleton. In most patients the course is indolent, but some follow a more malignant course. The radiological appearances may be lytic, sclerotic or mixed and diffuse or circumscribed. Scintigraphically the appearances are equally variable. Abnormalities may be unifocal, multifocal or diffuse, giving a superscan. The clinical role of bone scintigraphy may be to assess the extent of disease activity in the skeleton.

Further reading:
Journal of Nuclear Medicine 1994; ***35****:1471–5*

Neurofibromatosis

Neurofibromatosis type 1 (von Recklinghausen's disease) is one of the commoner autosomal dominant inherited conditions. The prevalence is said to be 1 in 3000 live births. Many tissues can be affected. Skeletal involvement may be due to pressure from adjacent soft tissue tumours, growth anomalies such as hemi-hypertrophy or bone dysplasias. Abnormal uptake of bone seeking agents is commonly observed both in affected bones and in soft-tissue tumours, including neurofibromas, plexiform neuromas and neurofibrosarcomas.

Further reading:
Journal of Nuclear Medicine 1996; ***36****: 1178–93*

Other metastatic malignancies

Other forms of malignant disease do not frequently give rise to occult skeletal metastases and routine skeletal scintigraphy is therefore not indicated. However when there are local symptoms, irrespective of the primary tumour and appropriate radiographic projections do not confirm a definite cause for the pain, scintigraphy should be performed. The relative role of scintigraphy, CT and MRI depends on local availability, but in general scintigraphy is a less expensive procedure than the others and may enable a limited CT or MRI examination to be targeted, thus obtaining definitive diagnosis at lower cost. Some photon-deficient deposits are better delineated by SPECT (Fig. 1.18)

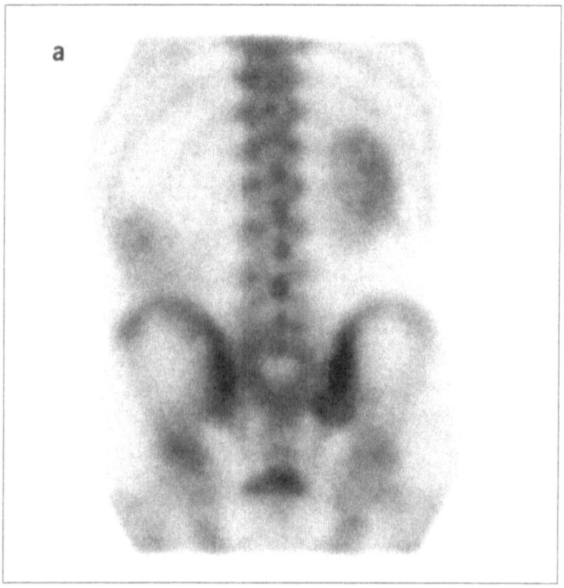

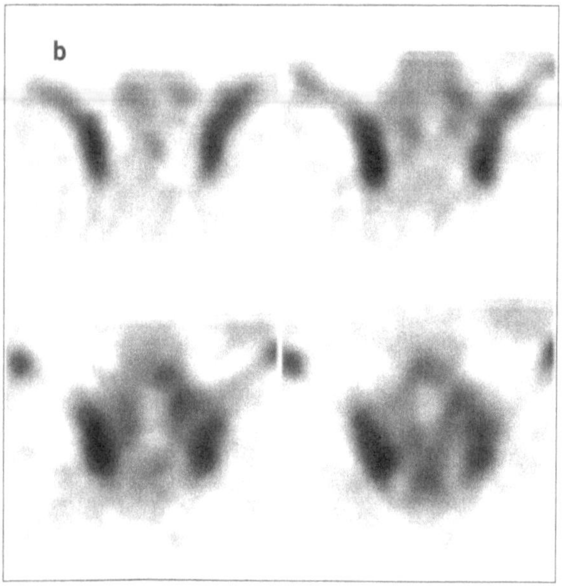

Therapy

Although it is not possible to eradicate malignant disease in bone with radioactive tracers, it is often possible to provide significant symptomatic relief to patients with widespread skeletal metastases provided these are provoking an osteoblastic response. Because of this latter requirement treatment is most commonly effective in carcinoma of prostate, but is sometime useful in carcinoma of breast and other tumours. A high local radiation dose to bone is achieved with β-emitting isotopes such as ^{89}Sr, ^{153}Sm, ^{167}Dy or ^{186}Re. Because of competition between bone clearance, renal clearance and soft-tissue uptake adequate differential concentration can only be achieved when the tumour burden is large. The radiation dose to marrow is high and marrow depression is often the dose-limiting toxic effect. Pre-existing marrow suppression is an important contra-indication.

^{89}Sr has a longer half-life than the other isotopes and emits only β-particles. The other

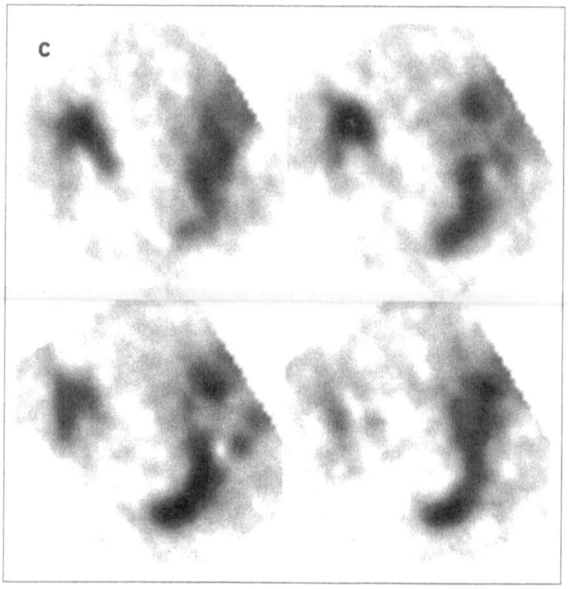

Fig. 1.18 (a) Osteolytic secondary in upper sacrum from melanoma. Slight reaction is visible on (b) oblique coronal SPECT slices. (c) Sagittal sections show destruction of upper sacrum.

isotopes emit both β and γ. The longer half-life of ^{89}Sr results in a lower dose rate to achieve the same administered absorbed dose. Onset of pain relief is slower with it than with the other preparations but the duration of pain relief may be longer. Usual administered activity of ^{89}Sr is 1.5–2.2 MBq/kg by intravenous injection into a freely running infusion, to minimize the risk of inadvertent extravasation.

Further reading:
Journal of Nuclear Medicine 1996; ***37****: 881–4*

Malignant primary bone tumours

Osteogenic sarcoma

This tumour is commoner in males and is unusual in having two peaks of incidence, one in adolescence, at or just before closure of the epiphyses and a second in the seventh and eighth decades in association with Paget's disease. There is always greatly increased uptake in and around the tumour and slightly increased uptake in the rest of an affected limb distal to tumour. Uptake is sometimes seen in soft tissue metastases to lung and liver but this is uncommon.

Computed tomography or MRI are undoubtedly preferred techniques for delineation of the primary tumour and detection of distant metastases. Plain radiographic appearances are usually diagnostic whereas CT or MRI demonstrate soft tissue as well as bony extent of tumour. Scintigraphy rarely adds useful information concerning the primary disease.

It is however sometimes of value to detect metastases, especially in other bones. Thallium is taken up by many osteogenic sarcoma, but also by many benign bone tumours and in myositis ossificans. It is thus of no value for differential diagnosis.

Further reading:
Skeletal Radiology 1990; ***19****: 165–72*

Ewing's sarcoma

This usually presents in children or young adults, usually male. The radiological appearance is variable and may show any combination or mixture of lysis and sclerosis, typically with periosteal reaction. Scintigraphically there is usually local increase of uptake, often intense. Scintigraphy at presentation may detect additional sites, particularly where plain film radiological confirmation is difficult to obtain. Scintigraphy may have an additional role to confirm or exclude recurrence after treatment (page 291).

Further reading:
Journal of Nuclear Medicine 1985; ***26****: 349–52*

Langerhans cell histiocytosis (histiocytosis X)

This has a wide spectrum from virtually benign to rapidly fatal. Scintigraphic appearances are variable. Many lesions are clearly visible as areas of increased uptake but some, probably the more malignant, provoke no osteoblastic response and are therefore not visible scintigraphically. Because of this variability and the prevalence of false negatives in the more malignant form, scintigraphy has little role to play in the routine investigation or follow up of this disease.

Further reading:
Journal of Nuclear Medicine 1996; ***37****: 1456–60*

Chondrosarcoma and fibrosarcoma

Both provoke an osteoblastic reaction and are therefore detectable scintigraphically. However the diagnosis is normally made on plain film and CT findings. There is no role for scintigraphy in the diagnosis or routine management of either of these conditions.

Benign bone tumours

Diagnosis of primary bone tumours depends on their radiographic appearance supplemented when necessary by histological confirmation. Scintigraphy plays little or no part in diagnosis or management of the majority and there have been few systematic studies. Benign abnormalities of bone are commonly an incidental finding and a knowledge of their scintigraphic appearance is important to avoid mis-diagnosis.

Osteoid osteoma

Osteoid osteoma is the only benign bone tumour in which scintigraphy may be a primary diagnostic modality. The typical history is an adolescent or young adult, more commonly male, with a history of well localized bone pain of several months duration, worse at night and relieved by aspirin. The typical radiological appearance is a sclerotic area containing a well defined radiolucent nidus. When present this is diagnostic but in many sites, for example vertebrae, carpal or tarsal bones and when the nidus is intramedullary rather than intracortical, the classical clinical and radiological features may be absent. The lesion is usually visible scintigraphically as a well defined area of increased uptake, present in all parts of the three-phase study and thus easily mistaken for infection. Most show intense uptake but there is a continuous spectrum and some are only faintly visible. On very high resolution or pinhole images the nidus is sometimes identified as a well demarcated photon-rich area with a rim intermediate in intensity between normal bone and the nidus. Radiographic confirmation may require CT. Excision biopsy both confirms and cures. Recurrence is documented only if excision is incomplete. Following block excision bone strength may take some time to recover; late pain is more commonly due to a stress fracture at the excision site than to recurrence. The use of an intra-operative probe during surgery allows the nidus to be identified with greater precision. The smaller block of bone resected is associated with more rapid return to normal mobility.

Further reading:

European Journal of Nuclear Medicine 1996; ***23:*** *1003–11*

Fibrous cortical defect and non-ossifying fibroma

These related conditions are usually an incidental finding. Seen most commonly in late childhood and adolescence, some persist into adulthood. The former are usually not associated with any abnormality on standard scintigraphic projections, but pinhole imaging of fibrous cortical defects may show faintly increased uptake in the rim. Uptake in non-ossifying fibroma is most often uniform and intense.

Enchondroma

This presents a varied picture. There is often a local increase in uptake, not always marked. In multiple enchondromatosis the intensity of uptake varies in different sites. Changes in intensity of activity on follow-up, especially increased uptake in the absence of any history of trauma, should alert to the possibility of malignant change. It is not possible to differentiate benign from malignant enchondroma on the basis of the scintigraphic appearance alone.

Osteochondroma

Osteochondroma show increased uptake which in general parallels that in the epiphyseal plate from which it was derived. There is therefore usually no increased accumulation of tracer in adults unless there has been recent trauma. In the absence of a history of injury, increased uptake should raise a suspicion of malignant change.

Haemangioma

Haemangioma is usually associated with slightly or moderately increased uptake but may be photon-deficient. Early appearance is similar to that in monostotic Paget's disease. An early blood-pool image may show increased blood volume but a dynamic study is rarely useful as the lesion usually occurs in a vertebra and is concealed during the early phases by activity in the great vessels.

Many other benign bone tumours also show increased uptake but scintigraphy plays no part in their routine diagnosis, which is usually made on radiographic findings.

Non-neoplastic conditions

Fibrous dysplasia

Both mono-ostotic and poly-ostotic fibrous dysplasia are usually associated with intensely increased uptake in affected areas, although 10–15% of cystic or 'ground glass' lesions are not associated with increased uptake. High uptake of gallium may also be observed. Common sites of involvement are the base of skull, mandible, facial bones (Fig. 1.19) and long bones of the lower limbs. The diagnosis can usually be confirmed radiographically, although in some cases the radiographic appearance is difficult to distinguish from Paget's disease. The poly-ostotic form may be unilateral (Fig. 1.20) and associated with precocious puberty and cutaneous 'cafe-au-lait' spots (McCune–Albright syndrome).

Further reading:
Journal of Nuclear Medicine 1990; ***31****: 1474–8*

Sickle cell disease

Bone marrow infarction is a common complication of sickle cell crises and is sometimes associated with fat embolism. Any part of the skeleton may be affected. The marrow space is usually expanded into the long bones of adults and is more extensive than normal in children. Asymmetries due to previous infarcts are commonly present; regeneration of marrow is variable and defects do not always persist. There are commonly multiple episodes at multiple sites. Imaging with either bone or marrow radiopharmaceuticals (page 210) in the initial one to three days shows absence of uptake of both tracers in the region of the infarct. By one week there is increased uptake of bone agents, ultimately reverting to normal over several months. The time scale depends on the site and extent of the infarct, previous history, age and whether or not the infarct is complicated by infection. Restoration of marrow may take longer and is less predictable.

Further reading:
Journal of Nuclear Medicine 1991; ***32****: 1617–8*

Paget's disease

This is a common asymptomatic incidental finding in populations of north European origin. About two-thirds of cases are male. The incidence varies across Europe, being higher in the north than the south. It is rare in Africa and Asia and under the age of 40 but rises in incidence with increasing age. Paramyxovirus particles, possibly measles or respiratory syncytial virus, have been observed by electron microscopy in Pagetoid osteoclasts, although a causative relationship has not been established. Uncommonly it is associated with pain, which may be due to stress

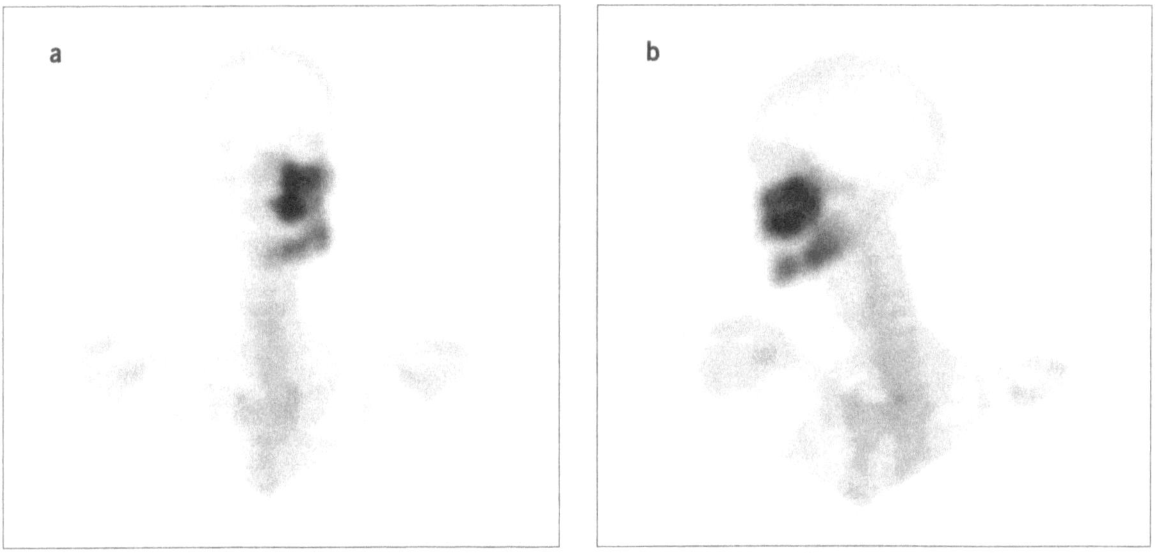

Fig. 1.19 Fibrous dysplasia of maxilla and mandible. (a) Anterior, (b) lateral.

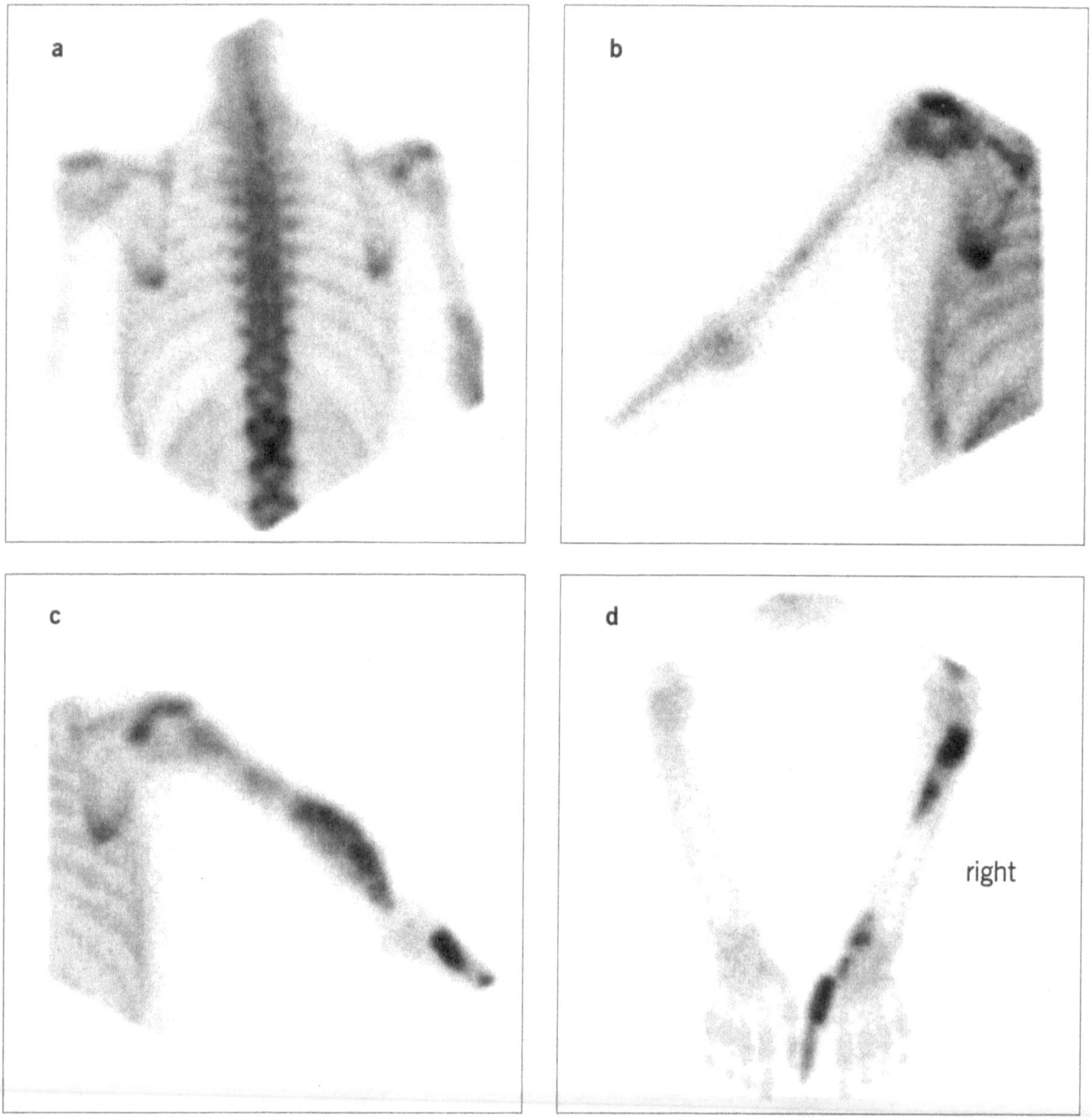

Fig. 1.20 Poly-ostotic fibrous dysplasia. Note that involvement is unilateral and of uneven intensity (compare Fig. 1.22). (a) Posterior thorax, (b) left ar,; (c) right arm, (d) both forearms.

or overt fracture, and with local hyperaemia. The majority of lesions are associated with greatly increased uptake of bone imaging agents but there is a continuous spectrum and rarely scintigraphy is normal despite unequivocal radiological changes of Paget's disease, namely expansion of cortical bone, coarsening and disturbance of trabecular architecture. There is no correlation between scintigraphic and radiographic appearance. In many cases radiological changes are minimal, but can be confirmed if adequate quality radiographs are carefully examined. Increased uptake in the absence of a radiological abnormality is also described but diagnosis must then be confirmed by biopsy. Affected bones appear photon-deficient on marrow and white cell imaging due to replacement of normal reticulo-endothelial marrow. There is high uptake of gallium into Pagetoid osteoclasts.

Characteristically Paget's disease starts adjacent to one end of a rib or an epiphysis in a long bone and progresses for a variable distance along the metaphysis, often involving the entire bone. The junction between normal and abnormal bone is identified radiologically by the 'flame' sign, which corresponds to the junction between normal and increased scintigraphic activity. Either calvarium or base of skull may be affected (Fig. 1.21). It is exceptional for Paget's disease to be multifocal in a single bone or to affect only part of the mid-shaft. Increased uptake typically extends continuously from a joint for a variable distance along the shaft. In contrast it is rare for metastatic disease to progress uniformly in this way. Paget's disease should therefore be suspected from the scintigraphic distribution (Fig. 1.22) but radiographic confirmation is nevertheless essential. In smaller bones and vertebral bodies this systematic progression cannot be defined and the whole bone usually appears involved. In these sites diagnosis cannot be made without radiographs; uptake due to Paget's disease is sometimes more intense than is commonly found in other conditions, possibly because of the high blood flow through Pagetoid bone. Paget's disease of the pelvis or scapula can often be suspected from its characteristic distribution, uniformly involving the whole of a bone or ossification centre. Metastatic disease is usually more patchy in its distribution. Other common sites include the tibiae and tarsal bones, but there may be monostotic involvement at any site or any combination of bones may be involved. Paget's disease of the skull is common. When the base of skull is involved the appearance may be of osteoporosis circumscripta (Fig. 1.21). The facial bones and mandible can also be affected. When scintigraphy is undertaken because of an unexplained raised alkaline phosphatase, it is essential to image the entire skeleton to identify peripheral as well as central Paget's disease.

The majority of untreated patients show no change on repeat imaging over many years. Occasionally additional lesions become evident and less commonly lesions may become less active. Sarcomatous change is a rare but well documented complication of Paget's disease. Increasing uptake, particularly when associated with pain, is suggestive of malignant change with the development of osteosarcoma but can also be due to stress fractures resembling the Looser zones of osteomalacia. In the absence of a

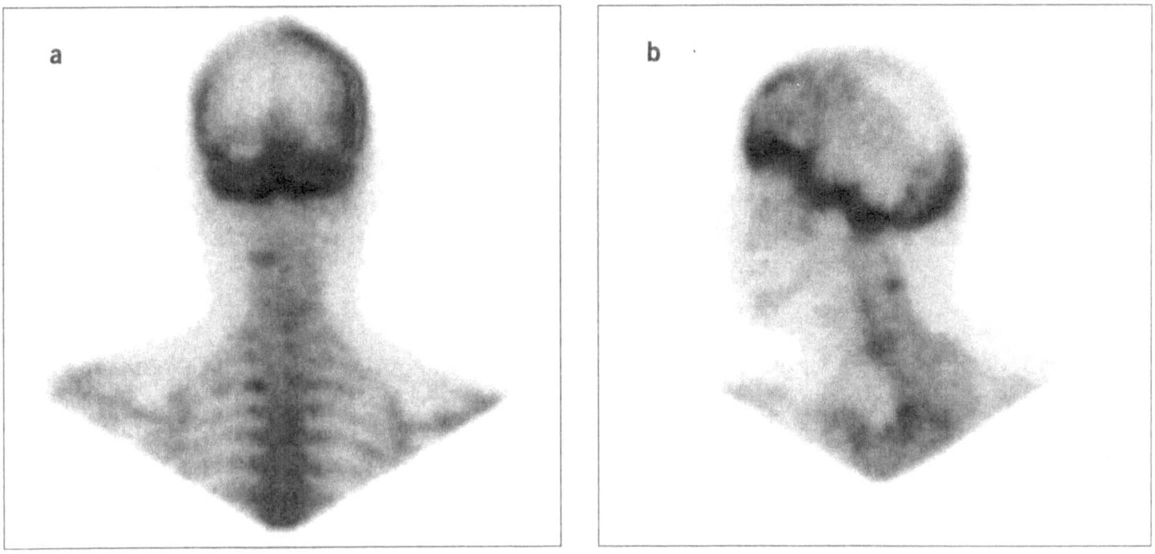

Fig. 1.21 Paget's disease of base of skull ('osteoporosis circumscripta'). (a) Posterior, (b) left lateral.

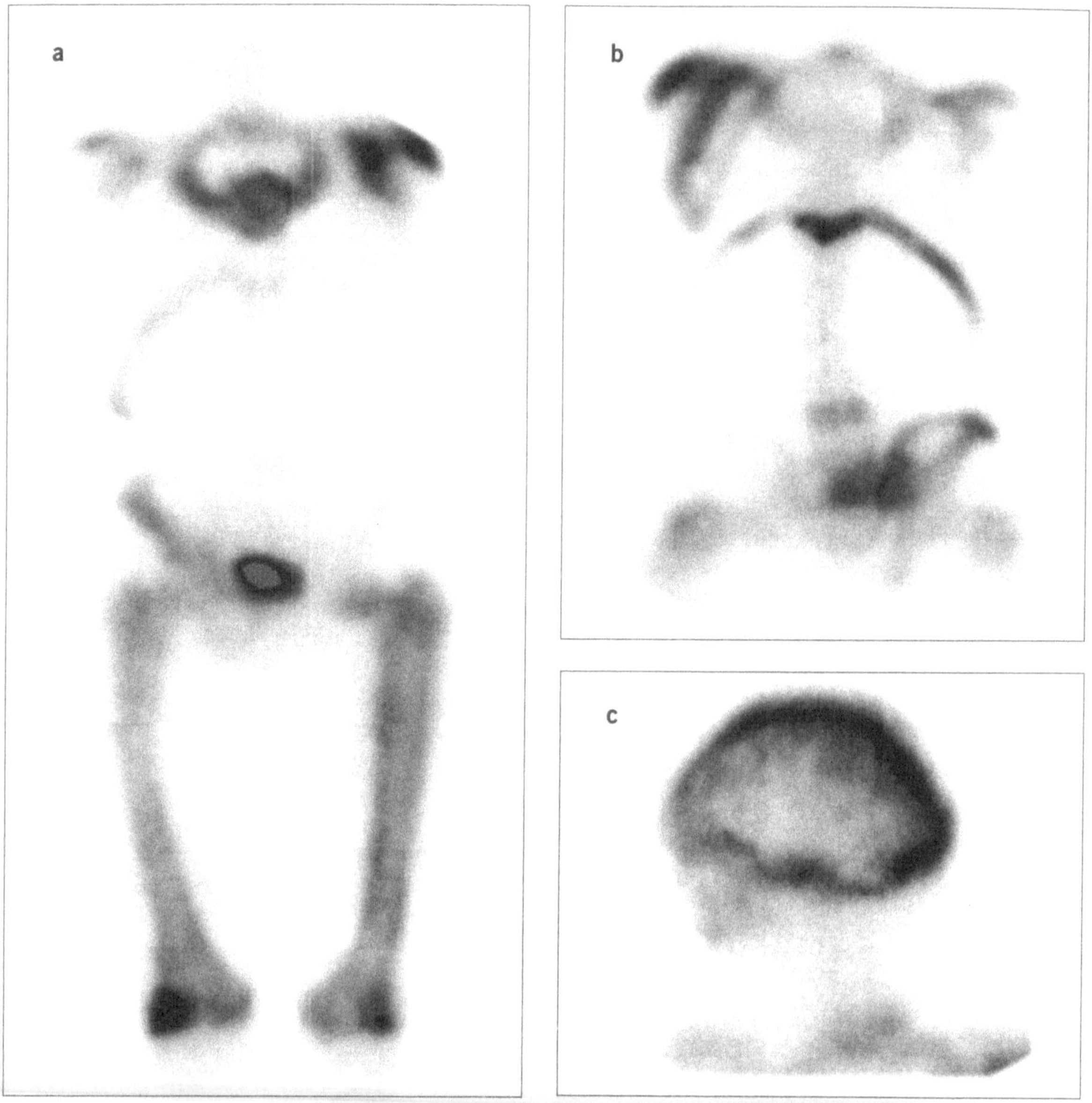

Fig. 1.22 Widespread Paget's disease with high uptake at many sites. Uninvolved bones are only seen if affected ones are set over-range so that detail is lost. This was an incidental finding in an asymptomatic patient with carcinoma of prostate. (a) Anterior, (b) posterior, (c) lateral skull.

previous scintigraphic examination this can rarely be distinguished from the high uptake commonly observed in uncomplicated Paget's disease. Metastasis from soft tissue primaries to Pagetoid bone does occur but is uncommon. Neither the radiographic nor the scintigraphic appearance is of any help in diagnosing or excluding dual pathology of this nature. Treatment of symptomatic Paget's disease with calcitonin or bisphosphonates is initially associated with reduction in uptake of bone imaging agents, which usually reverts to the previous pattern within 6–12 months of ceasing therapy.

Further reading:

Clinical Radiology 1985; ***36****: 169–74*

Journal of Nuclear Medicine 1995; ***36****: 121–6*

Hypertrophic osteoarthropathy

This may be associated with bronchogenic carcinoma, or less commonly with other conditions such as inflammatory bowel disease. A rare hereditary condition, pachydermoperiostosis, is associated with a similar appearance. Scintigraphically there is intense uptake along the shafts of affected long bones, giving a tubular appearance (Fig. 1.23). Scintigraphic changes tend to be more extensive than radiographic and when associated with cancer may resolve following radiotherapy of the primary tumour.

Further reading:

Journal of Nuclear Medicine 1991; ***32****: 1907–9*

Polycystic lipomembranous osteodysplasia

This is a rare, ultimately lethal, autosomal recessive condition usually manifest before the age of 30 with pain and tenderness in ankles and wrists, fracture after minor trauma and cyst-like lesions in the distal small bones of the extremities, especially the talus. The scintigraphic appearance is variable with increased uptake at affected sites. The main role of scintigraphy is to demonstrate the extent and distribution of skeletal involvement.

Further reading:

Nuclear Medicine Communications 1988; ***9****: 1005–11*

Tietze's disease

Costochondritis is a benign self-limiting non-specific inflammatory condition of an upper costo-chondral junction, presenting with swelling, local tenderness and, on occasion, pain. Pinhole imaging in the acute phase shows homogeneous, expanded, approximately spherical uptake at the rib ends. In chronic cases the pattern is crescentic.

Further reading:

European Journal of Nuclear Medicine 1994; ***21****: 947–52*

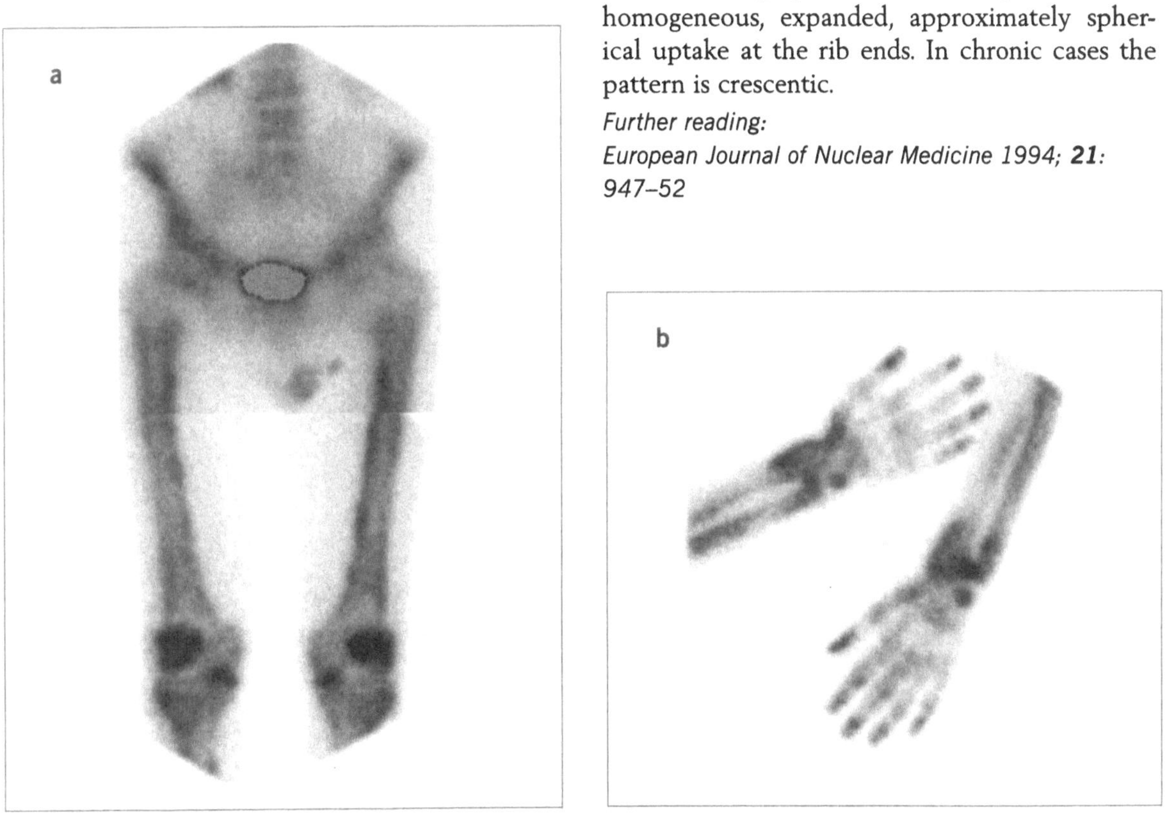

Fig. 1.23 Hypertrophic osteoarthropathy in patient with bronchogenic carcinoma. (a) Femoral, (b) radial and ulnar shafts appear tubulated due to increased periosteal uptake. Note also increased uptake in terminal phalanges, which were clubbed.

Infection

Acute osteomyelitis

Osteomyelitis may follow local trauma, particularly compound fractures, or may be haematogenous in origin. If the former, the effects of injury dominate the scintigraphic picture in the early stages; later the relative contributions of trauma, healing and infection are difficult to disentangle. Acutely, haematogenous infection is associated with oedema which occludes some intra-medullary blood vessels with bone infarction. Skeletal scintigraphy within 12–24 h of the onset of symptoms may appear normal or there may be a photon-deficient area. Osteoclasts remove devitalized bone and subsequently an osteoblastic reaction develops.

A two- (or three)-phase study (page 8) should be performed if the differential diagnosis includes infection. Increased activity in the two initial phases without any focal abnormality in the third (Fig. 1.24) indicates soft tissue inflammation whereas increased count rate in all three phases (Fig. 1.25) is suggestive of osteomyelitis. The dynamic phase is never useful in the trunk, where activity in soft tissues and major vessels usually obscures any asymmetry in bone blood flow. Distally the picture may be complicated by peripheral vascular disease. In patients with diabetes, asymmetry of flow is affected by the state of proximal arteries and may not reflect changes due to inflammation in distal tissues. Soft tissue infection is associated with hyperaemia but without increased uptake in bone; thus the

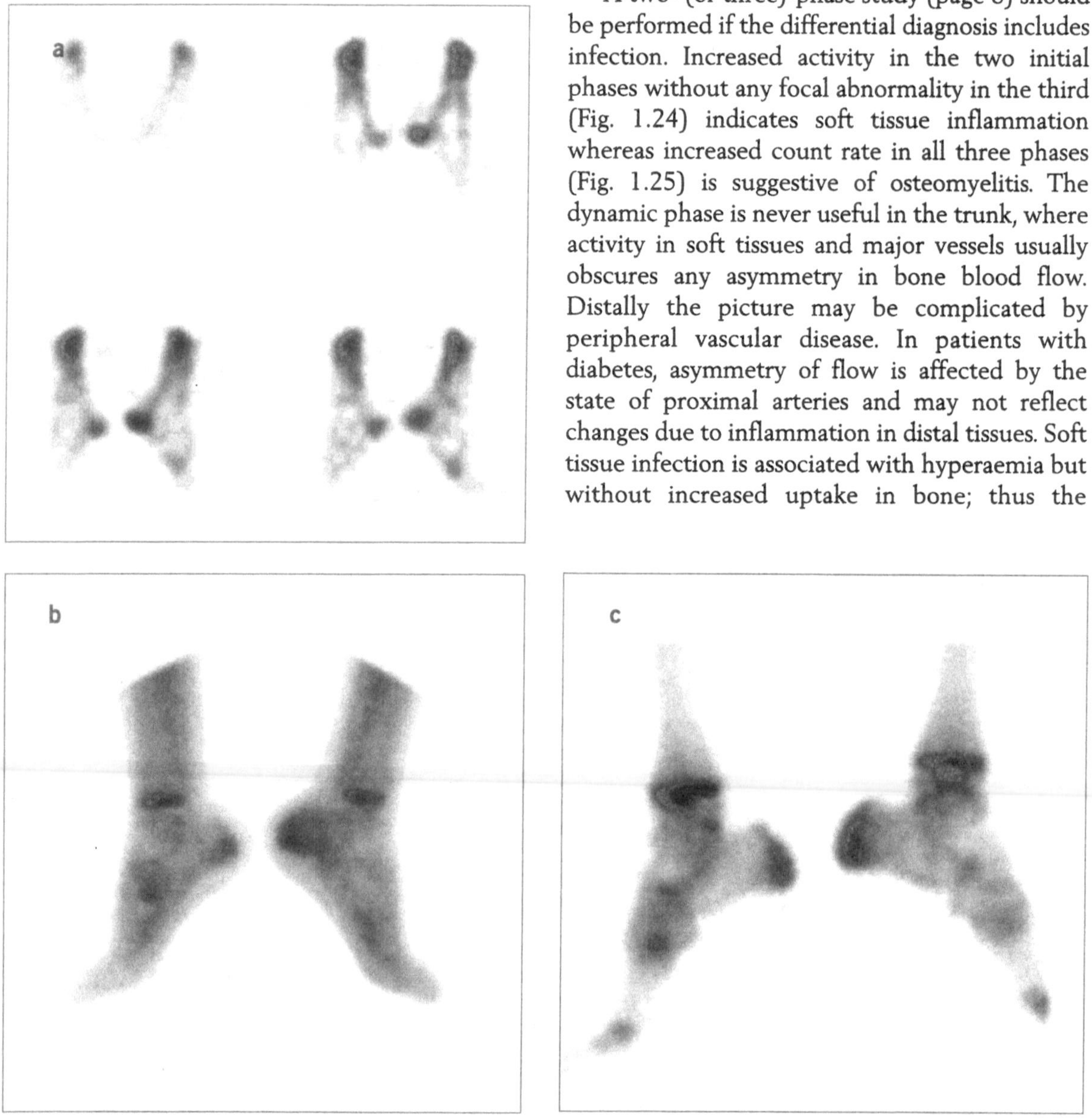

Fig. 1.24 Soft tissue infection. There is a high count rate in the left heel during (a) dynamic and (b) early phases but (c) only a minimal diffuse increase in the late phase.

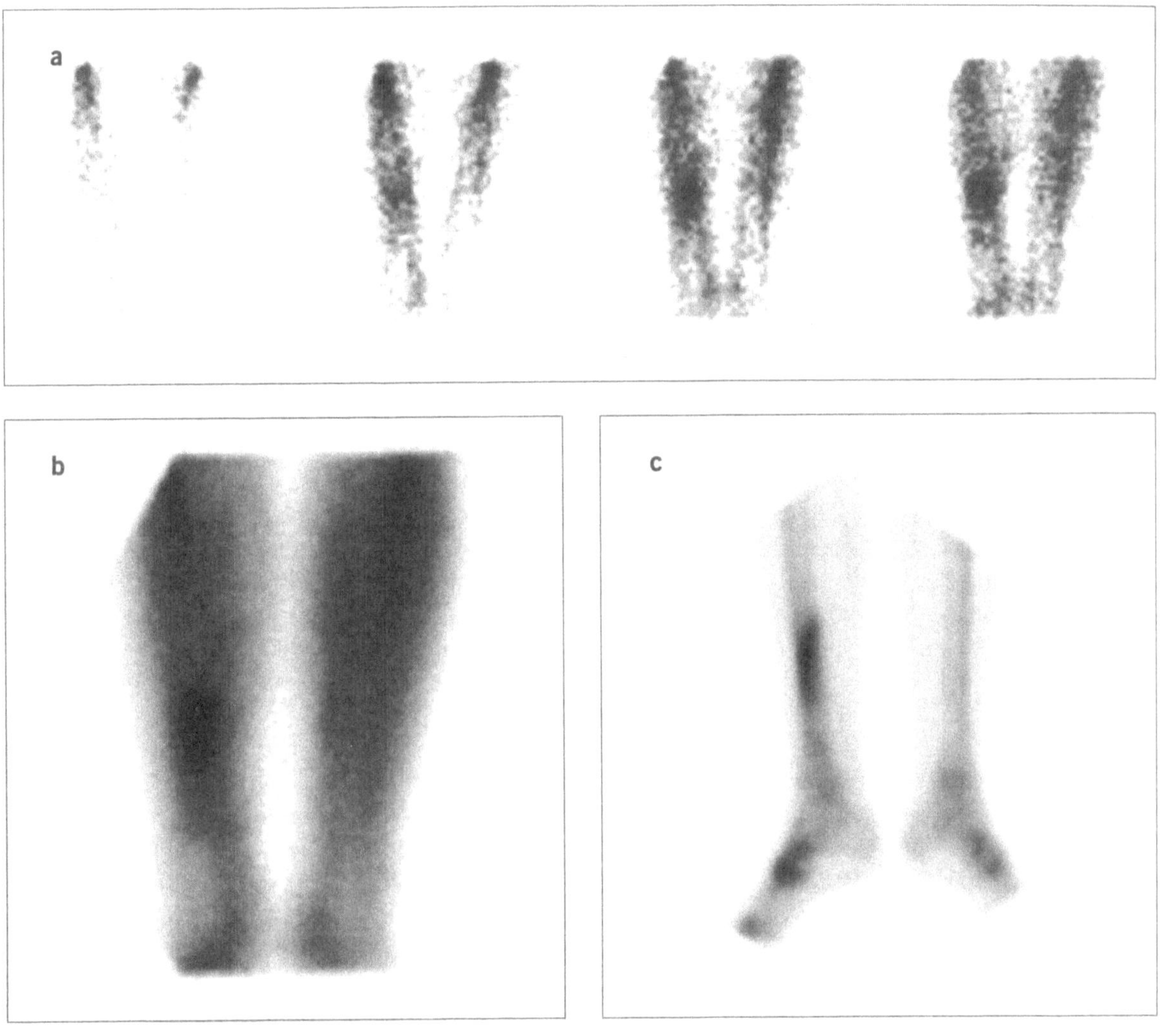

Fig. 1.25 Three-phase scintigraphy: osteomyelitis of the lower tibia. (a) Dynamic phase shows increased perfusion of affected region, (b) after 2 min there is substantial uptake into bone, (c) 3 h later there is a focal region of high uptake in the lower tibial shaft.

second phase is abnormal but not the third. Gallium uptake, although most commonly employed for diagnosing chronic infection is usually detectable by 24–48 h and becomes negative within six weeks of successful treatment. It has been superseded by the more accurate combination of three-phase bone scintigraphy supplemented by white cell imaging.

The value of bone scintigraphy in acute infection depends on the selection of patients, the duration of clinical history and treatment received prior to investigation. The best accuracy is obtained if the examination is performed acutely in untreated patients, when the second phase shows well localized hyperaemia and the third a photon-deficient area if there has not been time for a reaction to develop, or focal high count rate if it has. Once treatment has been initiated it is not possible to distinguish reliably between increased uptake in the third phase due to infection and that due to healing. However the early phase hyperaemia usually resolves over days rather than weeks. Thus increased uptake in bone without uptake in the early phase image is more suggestive of healing than active infection. Negative scintigraphy during or after antibiotic therapy does not exclude residual infection. Increased uptake in the third phase may take longer to develop in neonates, in whom there is a greater risk of false negative findings.

In uncomplicated osteomyelitis due to a pyogenic organism, scintigraphic changes precede radiographic. Radiographic evidence of infection may not be present at any time if adequate therapy is instituted sufficiently early. Scintigraphy gives no direct evidence as to the identity of the organism but it is sometimes possible to make an informed guess if all of the available evidence is considered. The length of the history is the best clue to the organism. A history of pain lasting several months is more suggestive of tuberculosis than of a pyogenic organism. The differences in prevalence and susceptibility between different populations to various organisms must also be kept in mind, for example the increased risk of salmonella osteomyelitis in subjects with sickle cell disease. Tuberculosis provokes greatly increased uptake, often with minimal radiological changes. Involvement of two bones on either side of a joint, particularly if this is symmetrical, is suggestive of septic arthritis whereas much higher uptake on one side than the other is more suggestive of osteomyelitis in the bone showing the higher uptake. In the spine, involvement of two adjacent vertebrae is suggestive of infection if the disc space is reduced, as tumour almost never involves or crosses a joint or intervertebral disc.

Whole-body scintigraphy is essential when investigating a patient with haematogenous osteomyelitis, as occult lesions are commonly present in addition to those which are evident clinically. Haematogenous infection may, like metastatic disease, involve any site. The distribution is essentially similar to that of neoplastic metastases as both are spread by the same pathway, namely the bloodstream. The distribution is therefore determined by the blood supply of bone, which in turn depends upon age. In children (except neonates) the greatest blood supply is to the epiphyses, where infection is commoner and may extend to cause septic arthritis. In the adult, nutrient arteries usually supply the mid-shaft region of long bones, which is the commonest site for both osteomyelitis and metastases. No such distinction can be made in the case of vertebral bodies or other small bones which are nevertheless involved by both processes more commonly than are long bones. In some sites reflux via venous plexuses is important.

Further reading:

American Journal of Roentgenology 1992: ***158****: 9–18*

European Journal of Nuclear Medicine 1995; ***22****: 1043–63*

Chronic osteomyelitis

Chronic osteomyelitis is associated with increased uptake of bone imaging agents. There is relatively high uptake of gallium (Fig. 1.26) but this is also seen in some acute inflammatory conditions and is thus of limited value in differential diagnosis. Sequestrae are commonly not visible on conventional scintigraphy but may be detectable as photon defects by pinhole imaging or SPECT if sufficiently large. Interpretation is often complicated by previous surgery. It is rarely possible to assess whether or not there is continuing active infection on the basis of a single investigation. Follow-up studies may be helpful provided that adequate clinical information, particularly on therapeutic interventions, is available.

In syphilis there is increased uptake in those bones which show periostitis on radiography. Gummae are photon-deficient.

Further reading:

European Journal of Nuclear Medicine 1996; ***23****: 792–7*

Diabetes

Diabetics have increased susceptibility to infection, are prone to peripheral sensory and autonomic neuropathy and to vasculopathy. Interpretation of scintigraphic findings is thus complicated by the possible co-existence of ischaemia, unrecognized trauma, autonomic disturbance and infection. When investigating the possibility of bone involvement by peripheral trophic ulcers, the initial dynamic phase of triple-phase bone scintigraphy is commonly confused by asymmetry of perfusion due to central and peripheral vascular disease. Uptake in the second phase indicating inflammation is usually present but does not differentiate whether it is due to unrecognized trauma or infection. This ambiguity persists into the third phase. Definite white cell accumulation is a reliable sign of infection, but a slight local increase is non-specific, especially in the partially treated patient. Malignant otitis externa (Fig. 1.27) is a potentially lethal complication usually associated with *Pseudomonas* infection of the external ear. If undertreated it may progress to thrombosis of intracranial venous sinuses. Bone changes are not usually visible on CT. Localized uptake of bone-seeking agents on SPECT of the skull is sensitive. In view of the danger of undertreatment some false positives can be accepted.

Further reading:

Journal of Nuclear Medicine 1994; ***35****: 411–5*

Journal of Nuclear Medicine 1988; ***29****: 1651–8*

Disease of joints

Osteoarthritis

Wear and tear of cartilaginous articular surfaces alters the distribution of stresses transmitted to underlying bone. High pressure which develops in joints during movement forces synovial fluid through infractions in hyaline cartilage into underlying bone. This provokes a reaction leading to the formation of sub-articular cavities. Damaged bone is removed and new bone laid down in response to these stresses, to spread and

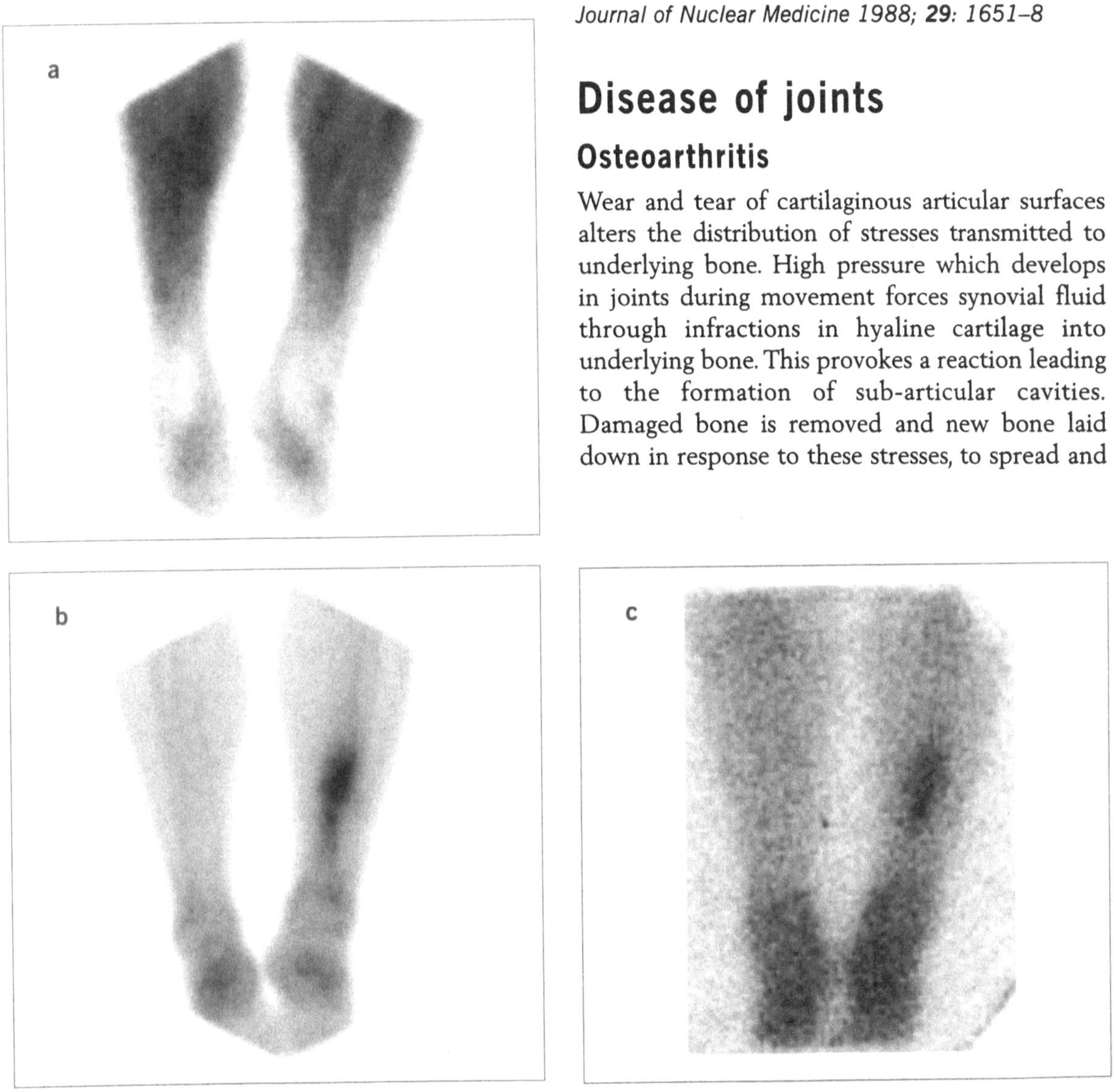

Fig. 1.26 Chronic osteomyelitis. (a) There is no increase of uptake in the tibial shaft in blood-pool phase but (b) high uptake in the late phase. This could be due either to continuing infection or to healing. (c) Uptake of gallium suggests chronic infection.

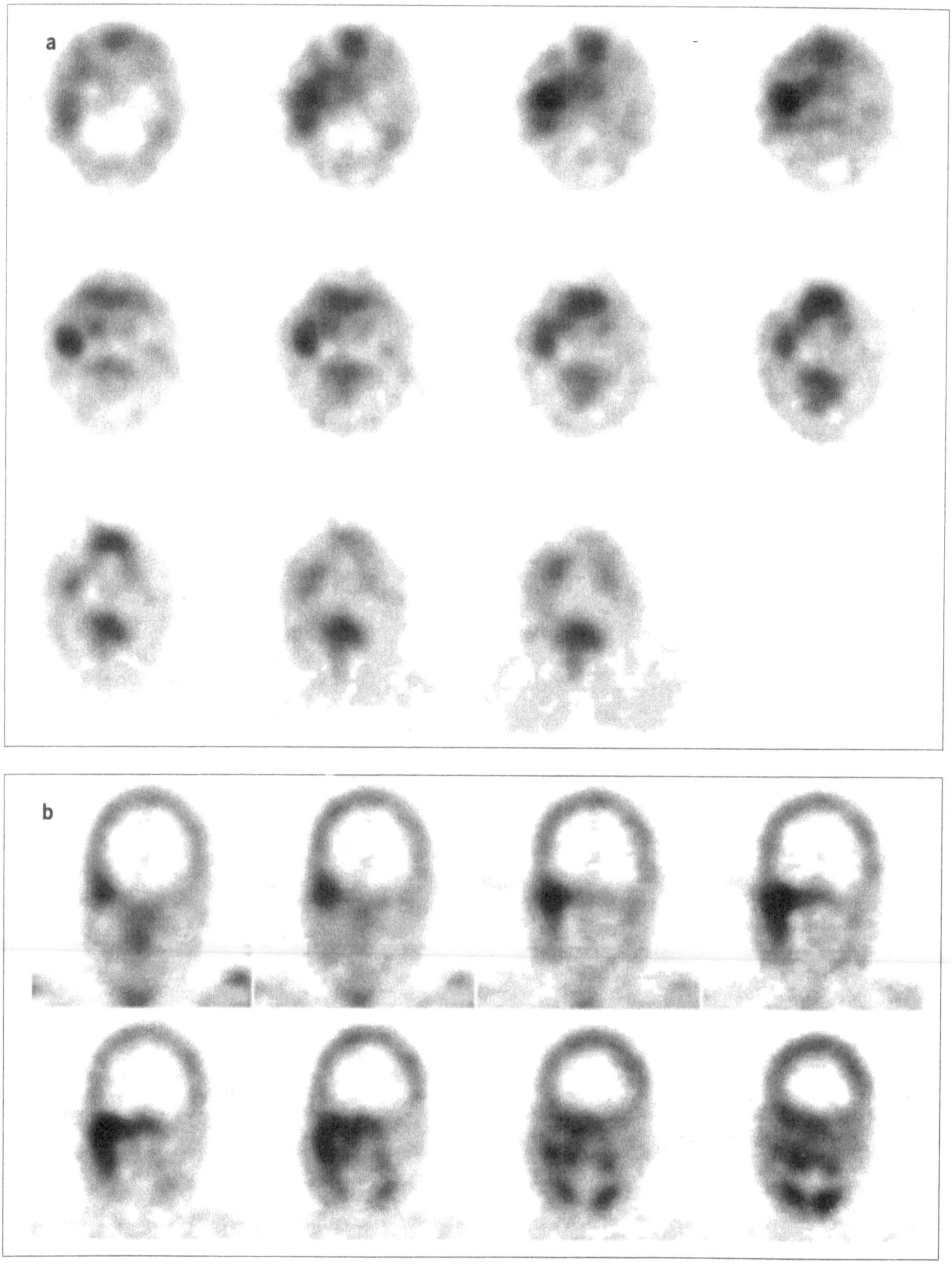

Fig. 1.27 Malignant otitis externa. (a) Transverse, (b) coronal and (c) sagittal SPECT showing localization confined to petrous temporal bone.

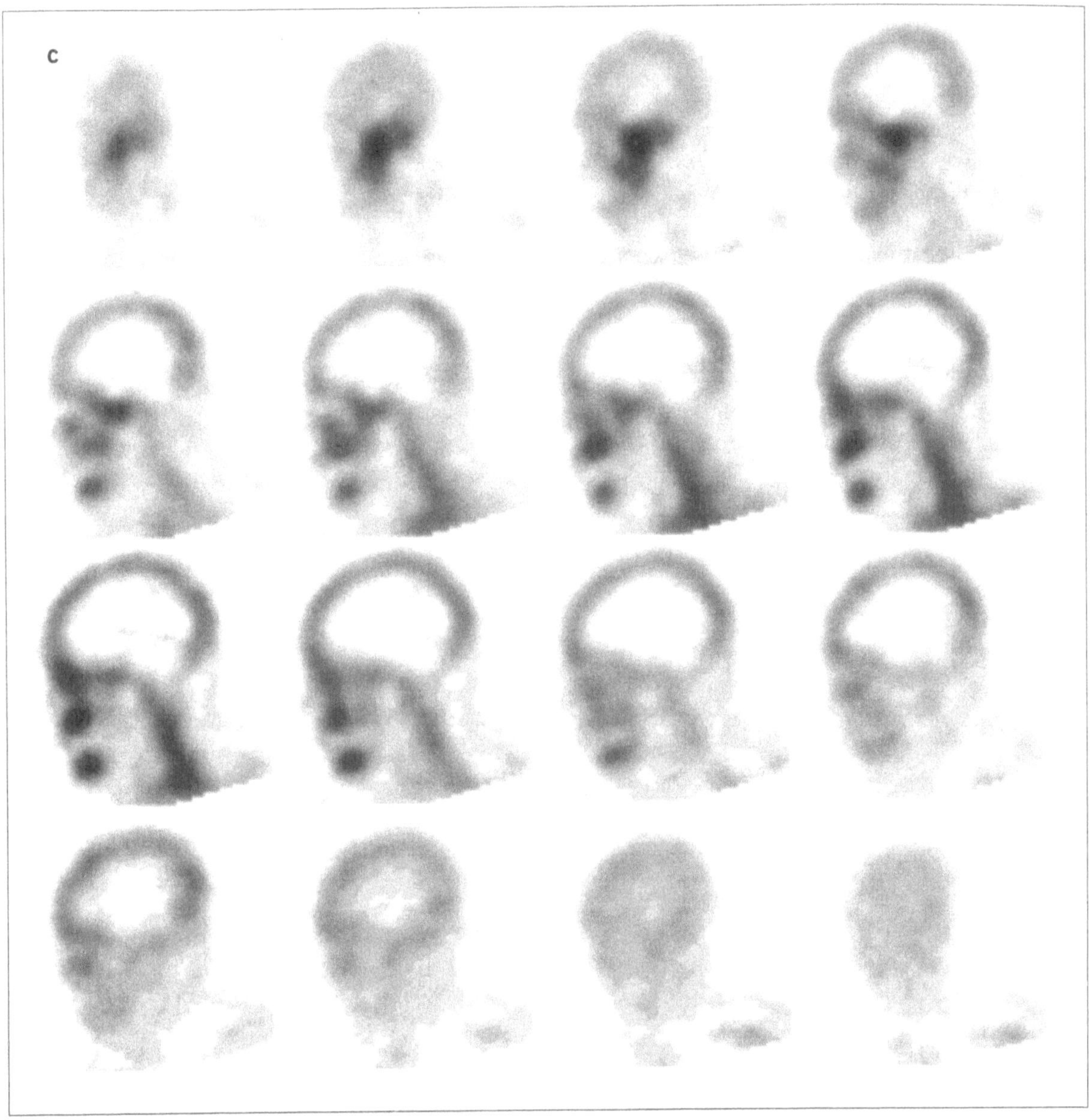

Fig. 1.27 *continued*

equalize the transmitted load. Osteophytes are one form of adaptive response, increasing the load-bearing surface and therefore reducing the load per unit area. New bone formation therefore precedes development of osteophytes or subarticular cavities and scintigraphic abnormalities precede radiographic changes. Persistence of scintigraphic abnormalities is an indication of incomplete adaptation associated with, or likely to give rise to, pain or stiffness.

Once osteophytes have formed a successful adaptation, the rate of bone turnover and the scintigraphic appearance both revert to normal. If the local stresses to which the bone is subjected remain excessive, the increased rate of bone turnover continues. There is therefore no correlation between the radiographic evidence of osteoarthritis and the scintigraphic appearance, which tends to correlate better with symptoms. Increased uptake in the immediate vicinity of the joint must always be interpreted with caution as this is commonly due to changes which precede overt radiological evidence of arthritis. Metastatic malignancy can occur in bone adjacent to a joint, but this is an uncommon cause of juxta-articular scintigraphic abnormality.

Scintigraphy is not often performed to assess osteoarthritis; rather osteoarthritis is a common incidental finding in patients being imaged for other reasons. In osteoarthritis of the knee, patients without abnormal focal scintigraphic accumulations are unlikely to have progressive deterioration in function within a five-year follow-up whereas those with abnormalities, especially an abnormal early phase indicating an inflammatory reaction, are at high risk of deterioration to an extent justifying operation (Fig. 1.28). The risk is greater if more compartments are involved.

Further reading:

*European Journal of Nuclear Medicine 1992;**19**: 849–901*

*Annals of the Rheumatic Diseases 1993; **52**: 557–63*

Chondromalacia patellae

There are two forms of this condition, which is an important cause of knee pain in adolescents and in older subjects. Pinhole scintigraphy of older subjects shows a characteristic pattern of spotty

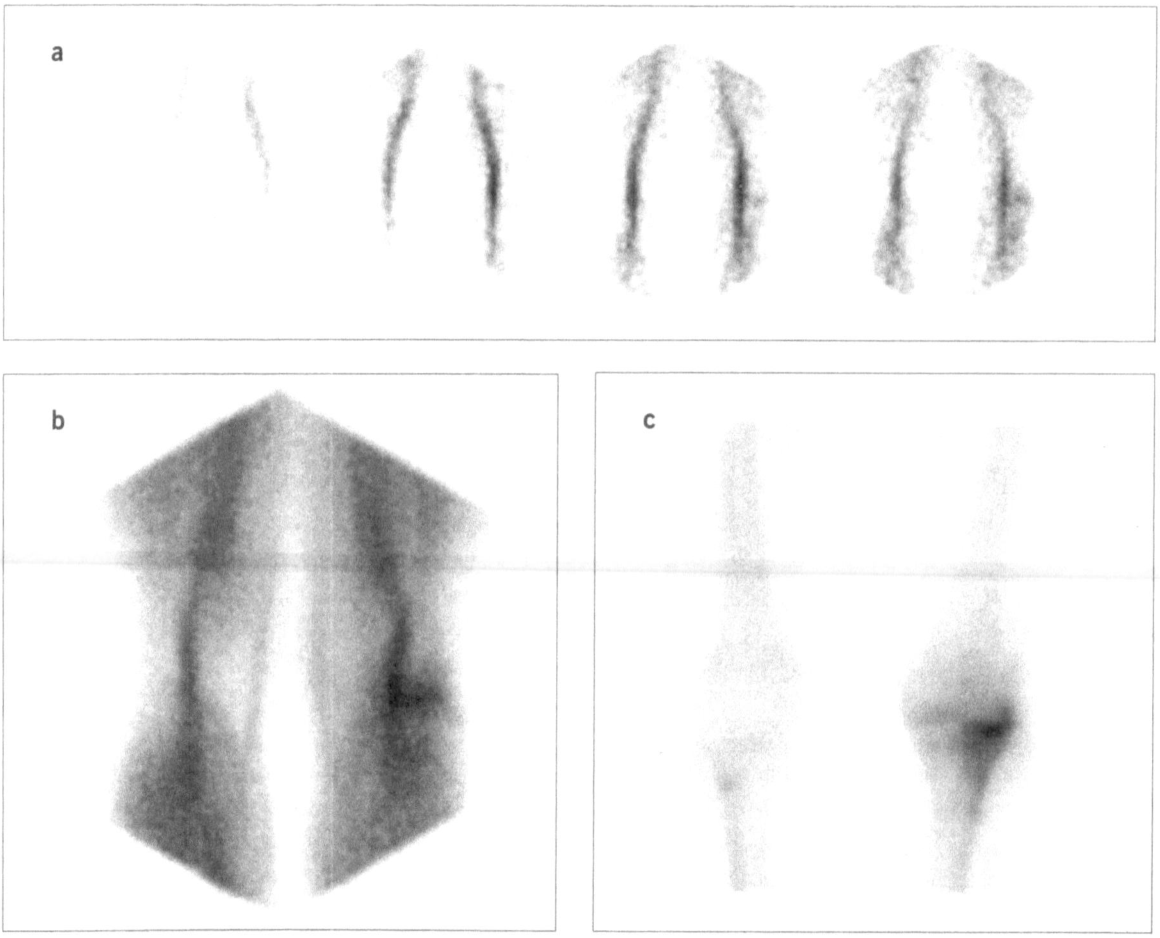

Fig. 1.28 High uptake confined to lateral tibial plateau in all three phases indicating high risk of progressive joint deterioration. (a) Dynamic, (b) immediate, (c) delayed.

uptake confined to the central retropatellar facet without other abnormalities in the knee joint.

Further reading:

Journal of Nuclear Medicine 1994; ***35****: 855–62*

Septic arthritis

There is increased uptake of bone-seeking radiopharmaceuticals on both sides of the affected joint, associated with increased blood flow and hyperaemia during the vascular and early blood-pool phases (Fig. 1.29). Commonly there is increased uptake in the entire limb distal to the affected joint due to hyperaemia. Infection is the only common cause of an abnormality affecting bones on both sides of a joint. However, as with osteomyelitis, the examination is most accurate in those cases where there is little clinical doubt and is often equivocal in clinically difficult cases, especially if partially treated. White cell imaging is more useful than in osteomyelitis as marrow is not present in the joint or peri-articular soft tissues. It is indicated if bone scintigraphy is equivocal.

Indications

It is more difficult to define indications for scintigraphy in osteomyelitis than in many other diseases. In a patient with typical signs and symptoms including localized pain, tenderness, inflammation, pyrexia and a neutrophil leucocytosis, it is arguable that no imaging is necessary but a hole drilled into the cortex at the appropriate site will relieve intramedullary pressure, provide pus for culture, relieve pain and reduce the risk of permanent disability. Such a patient, in whom the diagnosis is clear clynically, will usually also give an unequivocal scintigraphic

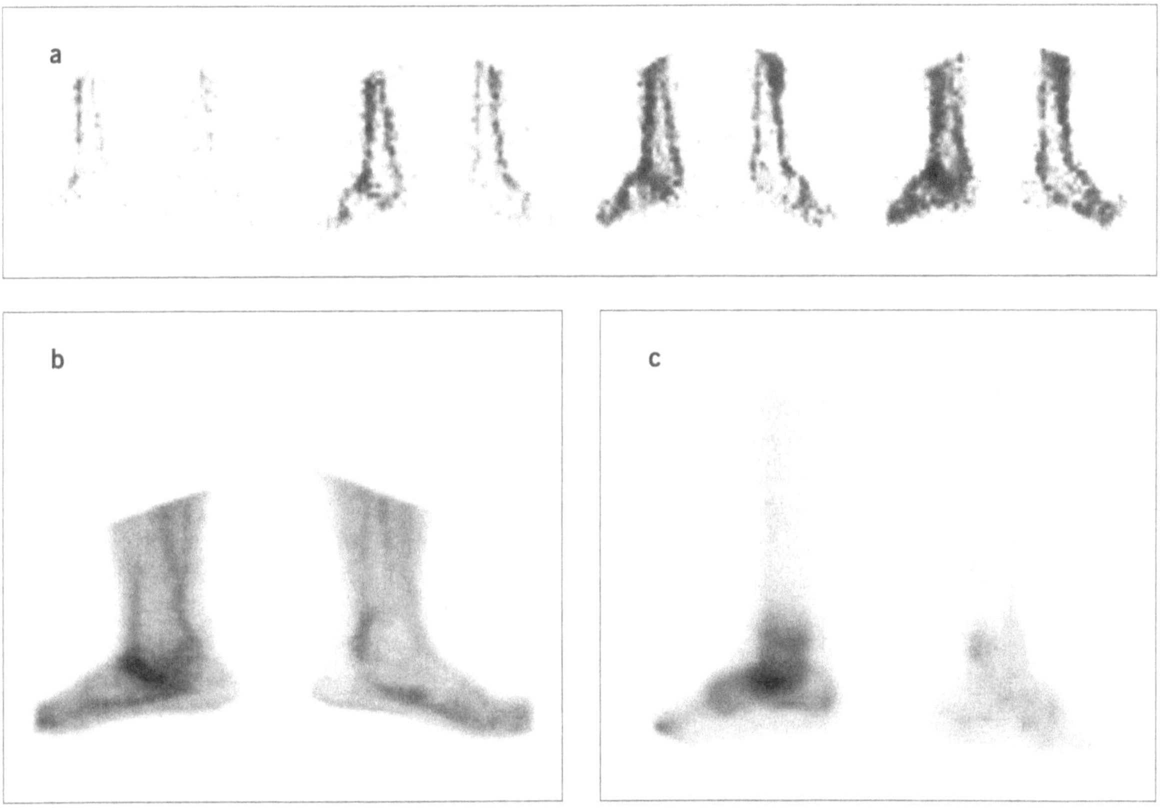

Fig. 1.29 Infective arthritis. (a) Dynamic, (b) immediate, (c) delayed. There is high uptake in the region of sub-talar joints in all three phases but not localized to one bone. The picture is similar irrespective of the organism. In this case the history was of several months of pain and tuberculosis was confirmed.

diagnosis. On the other hand when the clinical picture is blurred by partial therapy, previous surgery, trauma or any other reason scintigraphy is likely also to be equivocal. The role of scintigraphy thus depends largely on the availability of definitive treatment and the readiness of the surgeon to operate on clinical findings alone.

Rheumatoid and other erosive arthritides

In the acute phase, polyarthropathies show increased uptake in all affected joints, often symmetrically. Multiple joints, especially of hands and feet, are involved. Asymmetrical involvement of apophyseal joints in the spine can be misinterpreted as metastatic disease. The scintigraphic appearance correlates better with distribution of symptoms than with radiographic changes and is sufficiently characteristic to suggest the diagnosis, which must however be confirmed by standard radiological and serological criteria. Late in the course of the disease uptake may be due to secondary osteoarthritis. Anti-CD4⁻ and non-specific monoclonal antibodies and labels of extracellular water such as pertechnetate have been used as a measure of inflammation but are of little practical value. They have also been used to demonstrate epiphenomena such as Baker's cyst but have been largely superseded by ultrasound and contrast arthography. Gallium (pages 210, 262) is taken up strongly in active chronic inflammatory tissue, including rheumatoid. This has little practical application but may be observed as an incidental finding.

Further reading:
Radiology 1990; ***177****: 601–8*

Therapy

Non-absorbed β-emitting tracers are used to treat synovial hypertrophy, which contributes to the damage caused by rheumatoid arthritis. Given by direct intra-articular injection, the objective is to achieve a high local radiation dose determined by penetration of β-particles and limited to the thickened synovium. Radioisotopes employed include ^{32}P, ^{90}Y, ^{165}Dy, ^{169}Er, ^{186}Re and ^{198}Au. With all these, the limiting factor is minimizing systemic absorption from the inflamed joint.

Further reading:
European Journal of Nuclear Medicine 1995; ***22****: 970–6*

Ankylosing spondylitis

This is commoner in young adult males and is associated with expression of the HLA-B27 antigen. It most often presents with low back pain which may precede radiological changes, but at this stage the sacro-iliac ratio is usually elevated. The sacro-iliac ratio is that of count rate in both parts (synovial and fibrous) of the sacro-iliac joint to a low stress area of bone, preferably the upper sacrum clear of confounding structures such as the bladder.

The upper limit of normal is 1.5:1 in adults but is higher in adolescents before closure of the scale epiphyses of the ilium, in whom a normal range has not been established. In acute sacroiliitis the ratio may exceed 2:1. There is some overlap between normal and abnormal but in the acute phase considerably elevated ratios are usual. Borderline or normal values occur later, when there are radiological changes and if the patient is taking non-steroidal anti-inflammatory drugs. The measurement of the sacro-iliac ratio is thus a useful early indicator of an abnormality at the sacro-iliac joints and an objective criterion of response to anti-inflammatory drugs, but does not distinguish between the various causes of pain at this site. The test is not specific for ankylosing spondylitis, as increased ratios may occur following simple trauma and following unaccustomed activity, for example in sport.

Ankylosing spondylitis affects many other sites including sterno-clavicular joints and apophyseal joints of the spine. During the acute phase these may be involved asymmetrically to give a pattern indistinguishable from multiple metastases, unless the clinical picture is taken into account. Later, once fusion has occurred, disc spaces may be obliterated even in the lumbar region. Fixed deformity may cause difficulty in obtaining optimum projections. Fracture in an ankylosed spine is a fairly common serious complication, readily seen as a linear increase in uptake.

Further reading:
Nuclear Medicine Communications 1993; ***14****: 719–20*

Prosthetic joints

Prostheses are available for many joints but the only ones to have been extensively investigated scintigraphically are the hip and knee. There are many varieties of prosthesis for these and other sites, with a range of scintigraphic appearances even in normal subjects. Healthy prostheses are visible as photon-deficient areas. Most show some local accumulation of bone-seeking radiopharmaceuticals around the implant, but the intensity and distribution depend on the prosthesis (Fig. 1.30). Pain persisting after insertion of a prosthetic joint or recurring after an interval may be due to a number of causes. Two of the most common are infection and loosening of one or more components. Loosening of the femoral component of a hip prosthesis is most commonly associated with two areas of increased uptake in the late phase only, one at the proximal end medially in the trochanteric region and a second at the tip (Fig. 1.31).

Infection is associated with increased uptake in both early and late phases around the whole of the implanted component, but up to one-third of cases with the appearance of loosening are infected and one-third of infected cases do not have these characteristic signs. Analogous features have not been described for the acetabular component of a hip prosthesis or for knee prostheses. Infection usually manifests late and

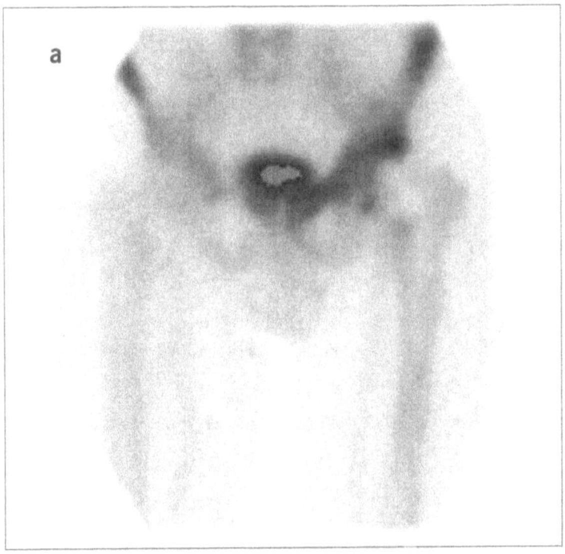

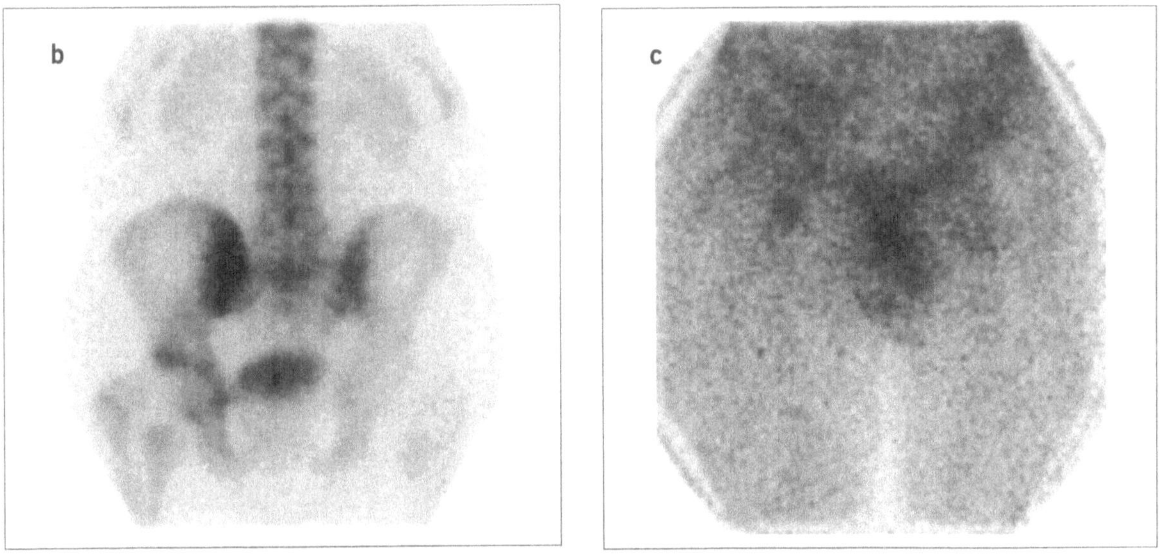

Fig. 1.30 Normal hip prosthesis. (a) Anterior, (b) posterior. High uptake in left side of pelvis was due to Paget's disease. (c) There is no local concentration of gallium.

is associated with high uptake of gallium (Fig. 1.32), whereas loosening usually shows uptake of the bone-imaging agent but not of gallium. Uptake of gallium disproportionate to intensity of bone agent uptake or uptake of technetium- or indium-labelled white cells improves accuracy compared with two-phase bone scintigraphy alone, and scintigraphy occasionally indicates that

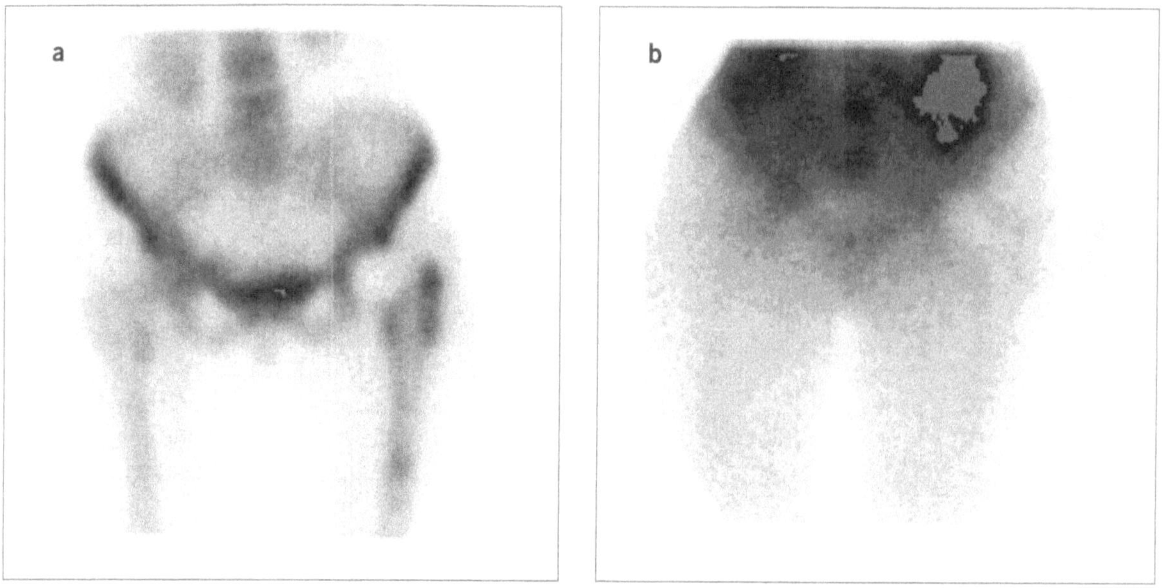

Fig. 1.31 Loose femoral hip prosthesis. (a) Uptake is increased at the points of leverage. (b) Gallium uptake is not increased.

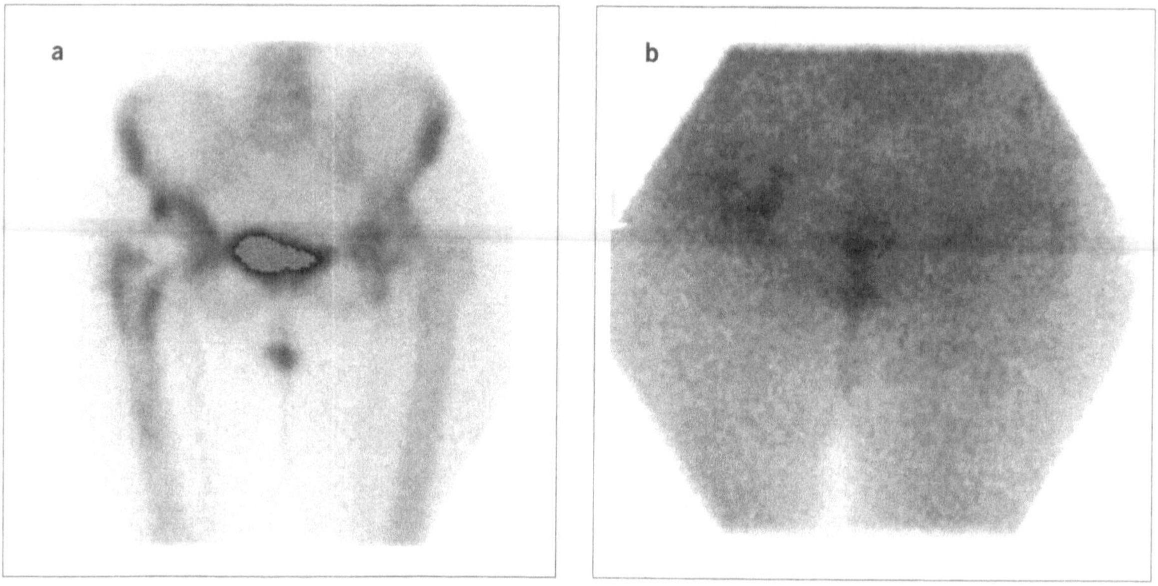

Fig. 1.32 (a) Increased uptake of Tc-MDP around both components of hip prosthesis. (b) Gallium uptake is localized only around acetabular component. There is thus a healthy femoral component and infection around the acetabular component.

the abnormality is not in the suspect component. No techniques can match the accuracy and speed of diagnosis of joint aspiration and contrast arthrography.

Plantar fasciitis

Plantar fasciitis is associated with focal uptake at the calcaneal insertion of the plantar fascia which varies in intensity with disease activity (Fig. 1.33). This is a common cause of heel pain which may on occasion require local steriod injection. The site of maximal inflammatory activity, which is the most effective site for local steroid injection, can often be most accurately localized by bone scintigraphy. High-resolution or pinhole plantar and lateral projections of the affected foot are required.

Further reading:

Lancet 1995; ***346****: 1400–1*

Aseptic (avascular) necrosis

This is of greatest importance in the sub-articular region, when it may be followed by infarction or collapse of a weight-bearing surface, with subsequent disability. It occurs locally following trauma, systemically in sickle cell crises, as a complication of some drugs, especially steroids and non-steroidal anti-inflammatories, idiopathically (for example Perthe's disease, page 292) or related to occupations such as diving or use of vibrating percussive tools. Imaging immediately after the event reveals a photon-deficient area corresponding to the infarct. Subsequently there is increased uptake, initially at the periphery and

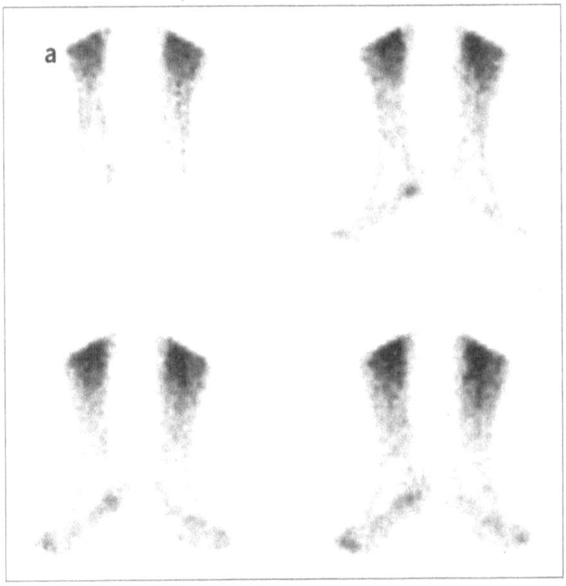

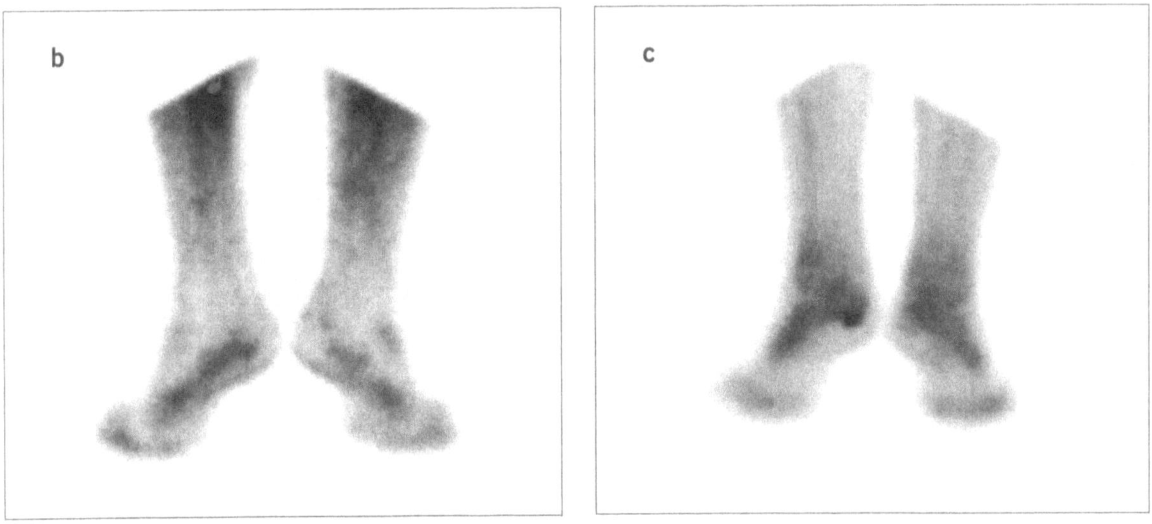

Fig. 1.33 During an acute exacerbation of plantar fasciitis (a) blood pool and (b) immediate phases show hyperaemia in plantar fascia. (c) Late phase; increased uptake confined to calcaneal spur.

subsequently throughout the affected region as dead bone is replaced. In adults symptoms are infrequent during the avascular phase.

Bone marrow oedema syndrome (algodystrophy of the hip, transient osteoporosis) may be related. Diagnosis is usually confirmed by MRI but should be considered in an adult presenting with acute hip pain with no obvious precipitating cause if scintigraphy shows increased uptake in the intertrochanteric region as well as in the femoral head in films at 3 h.

Dysbaric osteonecrosis

This occurs in divers who have descended deeper than 30 metres or other subjects who have been subjected to decompression. The incidence rises with depth reached, the duration and number of dives and the number of episodes of decompression sickness. It is however also seen in divers with no history of decompression sickness. The incidence of radiographically diagnosed bone infarcts is over 20% in those who have dived deeper than 300 metres. The commonest clinically significant sites are around the knees, the lower femur and upper tibia. Proximal humeral infarcts are rather less common but more frequently progress to symptomatic arthritis. Lesions in the femoral heads are comparatively rare. Scintigraphic lesions usually show increased uptake, first detectable about three weeks after the event. In the absence of complications they heal and disappear after three to four months. The earliest radiographic evidence is usually seen at about two to three months. Scintigraphic evidence of bone infarction precedes irreversible changes and should be the signal for appropriate prophylactic measures.

Back pain

Bone scintigraphy is indicated in patients with any proven malignancy who develop back pain not adequately explained on plain radiographs. In them, and more commonly in patients without proven malignancy, abnormalities are commonly due to degenerative disease of the spinal joints. Confirmation or exclusion of more serious pathology is of substantial practical importance. More focal abnormalities are detected by SPECT or pinhole than by planar imaging and the more precise localization of the abnormality can be essential in establishing the diagnosis. Following spinal fusion, persistent uptake at the operated level may indicate failure or non-union, whereas appearance of increased uptake at an adjacent level indicates transfer of stresses. The sensitivity and accuracy of these scintigraphic criteria have not been established. In younger subjects, increased uptake at the site of a radiologically confirmed spondylolysis (Fig. 1.34) suggests recent stress fracture, whereas uptake of normal intensity suggests established non-union. Uptake may be seen on the contralateral side when lysis is unilateral, suggesting that further fracture is impending. Absence of increased uptake in a collapsed vertebra is good evidence that it is of long standing, but continued increase in uptake is not an accurate means of determining the age of the injury.

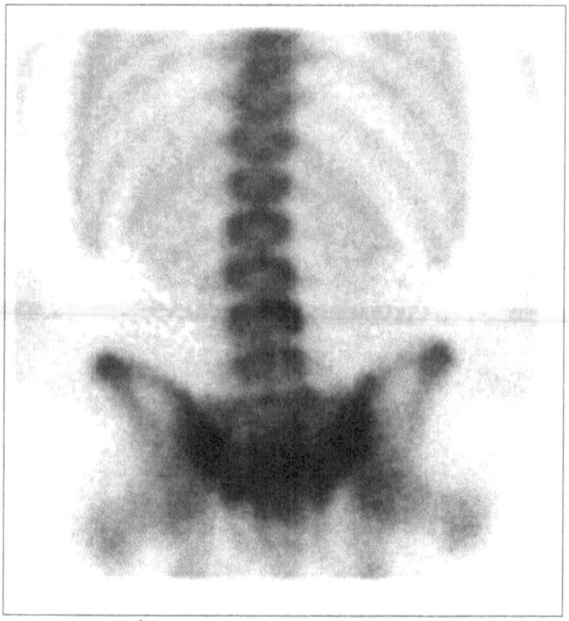

Fig. 1.34 Radiologically there are bilateral spondylolyses at L4. Only the right is active scintigraphically, indicating an established non-union on the left but some continuing attempt at healing on the right.

Further reading:
European Journal of Nuclear Medicine 1997; ***24****: 363–7*
Journal of Nuclear Medicine 1994; ***35****: 422–4*

Trauma

The majority of fractures are confirmed by clinical and radiological criteria, and scintigraphy is only occasionally employed to establish the diagnosis. More commonly, fracture is an incidental finding when scintigraphy is performed for unexplained pain. Immediately after injury, scintigraphy shows a photon-deficient defect due to blood clot and avascular bone between the viable bone ends, but this may be concealed by the complex shape of the fracture, if it is impacted, or by uptake and hyperaemia in adjacent soft tissues. Bone imaging radiopharmaceuticals are taken up in bruised and infarcted tissues, including devitalized muscle around a fracture. Subsequently granulation tissue is formed and differentiates into callus. Dead bone is removed, new calcified tissue laid down and converted into bone. Scintigraphy during this phase shows increased uptake. A few fractures have uptake detectable by 4 h. By 24 h 80% are visible but 5% still do not show increased uptake at 72 h. Most fractures are linear and the presence of a linear area of increased uptake, particularly in a long bone, should always give rise to the suspicion of fracture. Rib fractures are often multiple and reflect the cause (Fig. 1.35). Thus steering-wheel injuries form an arc of a circle whilst doors and tables usually have straight edges.

Following meniscal tears, uptake is seen in the corresponding tibial plateau. The extent is best demonstrated by SPECT. Scintigraphy may be performed without removing a cast, as absorption of 140 keV γ rays, even by plaster of Paris of normal thickness, is negligible. Increased uptake persists for a variable duration, depending on severity of the force which caused the fracture, whether or not there is infection, and stresses to which healing bone is subjected. A fracture caused by minimal trauma in an osteoporotic vertebra may no longer be visible scintigraphically after six months. A similar fracture in a normal vertebral body resulting from major trauma may remain visible for 18 months to two years. Alignment also affects the persistence of increased uptake. Where a fracture has healed in poor alignment, bone may be subjected to increased stresses which in turn provoke remodelling. Locally increased turnover may persist for many years. Following a clean osteotomy or simple transverse fracture of the tibia it takes several weeks before regions of increased uptake at each end of the fracture coalesce. Radiological union occurs some weeks later, but this reflects only the poorer spatial resolution of scintigraphy. When present, persistence of the defect between bone ends is evidence of pending non-union. In many patients this sign cannot be elicited, either because the broken ends are too close together initially to be resolved or because of the complex shape of the fracture, for example oblique or spiral.

Further reading:
Journal of Nuclear Medicine 1993; ***34****: 1403–9.*
1995; ***36****: 48–50*

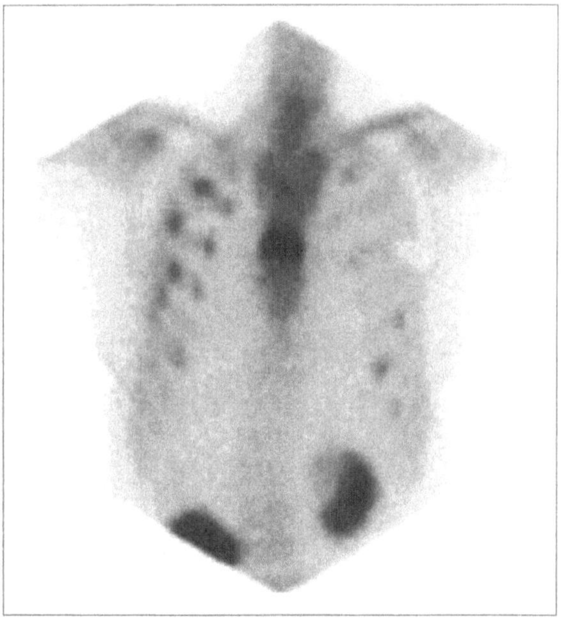

Fig. 1.35 Fractures to ribs and sternum due to cardiopulmonary resuscitation two weeks previously. Note pacemaker.

Indications

The principal clinical applications are detection of radiologically occult fractures in sites which may be difficult to visualize adequately by plain film radiography, including wrist, femoral neck, knee and dome of the talus, and to determine whether vertebral collapse is recent or of long standing. Whole-body imaging one to two weeks after injury is useful to confirm that no fracture has been missed on clinical examination of patients with multiple injuries. Although the majority of fractures not visible on initial radiographs are seen on repeat films at 7–14 days, there remain a number of patients in whom, despite a strong clinical suspicion, a fracture cannot be confirmed radiologically. Scintigraphy at this time will confirm or refute this suspicion (Fig. 1.36). Some of the fractures found prove to be clinically insignificant, for example of the triquetral, but many are clinically important. Scintigraphy is more sensitive than MRI for confirming scaphoid fractures.

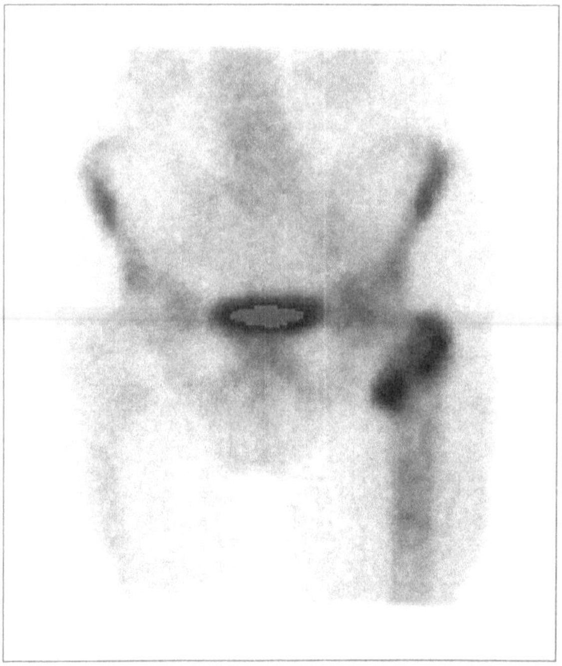

Fig. 1.36 Impacted inter-tronchanteric fracture of femoral neck one week previously not detected on plain radiographs.

Sources of error

Scintigraphy may not show any abnormality if performed within 72 h of injury or if fracture occurs in a previously irradiated area of bone. False positives are commonly due to minor bruising without fracture. This is particularly common in the ribs. Other causes of false positives include the presence of other non-traumatic lesions or artefacts such as spilt radioactivity or urine.

Stress-related changes

Localized bone pain is a common complaint in athletes, dancers and others who undertake unaccustomed or severe exercise. The lower limbs are usually involved but similar changes also occur in the spine. The commonest sites are calcaneum and bones of the foot, tibia and femur. Radiologically there may be evidence of periosteal new bone or incomplete fractures, but scintigraphy detects more abnormalities, often before any radiological abnormality is evident. If the part is rested resolution may occur without radiological changes becoming apparent at any time. A sequence of scintigraphic changes has been described in military recruits, some in the absence of symptoms. The mildest consist of a small, slight, ill-defined increase in cortical uptake, the next a larger, well defined elongated increase in the cortical region (Fig. 1.37), then a wide fusiform accumulation also involving the medulla and the last an extensive area of very high uptake affecting both cortex and medulla. All subjects in the last category and three-quarters of those in the third had radiological confirmation of stress fractures, but only 20% of those in the second and 4% in the first. Pain precedes radiological evidence of fracture and the scintigraphic abnormality may precede pain. Resolution may take weeks or months depending on the nature of the stress and how well the region is rested.

Prolonged and severe running or foot flexion on a hard surface is associated with periosteal elevation, possibly with avulsion of muscle insertions, and linear increased uptake at the insertion of the foot flexor muscles into the tibia (shin splints). An exaggerated length of stride in short military recruits gives rise to a similar appearance on the medial aspect of the proximal third of the femur at the insertion of the adductores brevis and longus (thigh splints).

Further reading:

Nuclear Medicine Communications 1994; ***15****: 341–60*

Indications

Scintigraphy is a better indicator of the optimal therapeutic regime than clinical history or radiographic appearance and is indicated in patients with back or lower limb pain following unaccustomed physical exercise.

Reflex sympathetic dystrophy

The syndrome, also known as Sudek's atrophy and the shoulder-hand syndrome, comprises prolonged pain and tenderness of all or part of a limb, vasomotor instability, trophic soft tissue changes and diffuse patchy rarefaction of affected bones indistinguishable from that seen in severe disuse. It most commonly follows injury, but a variety of causative factors have been implicated. No specific precipitating factor is identified in one-third of cases. The diagnosis is largely clinical, the principal features being pain of duration and severity disproportionate to the initiating injury. In the initial stage blood flow through the affected limb is increased and there may be a good response to vasoconstrictors such as calcitonin. This hyperaemia either resolves or passes into a late atrophic stage characterized by continuing pain, a cold atrophic cyanosed skin and joint stiffness, in which perfusion is decreased and vasodilators such as guanethidine are indicated.

Hyperaemia is associated with a local increase in uptake of skeletal imaging agents in the limb at the level of and distal to an affected joint or region; soft tissue inflammation often causes diffuse rather than focal increase in uptake in adjacent bone. The scintigraphic features of the initial stage of reflex sympathetic dystrophy are increased blood flow and diffuse soft tissue uptake in the early phase of the three-phase

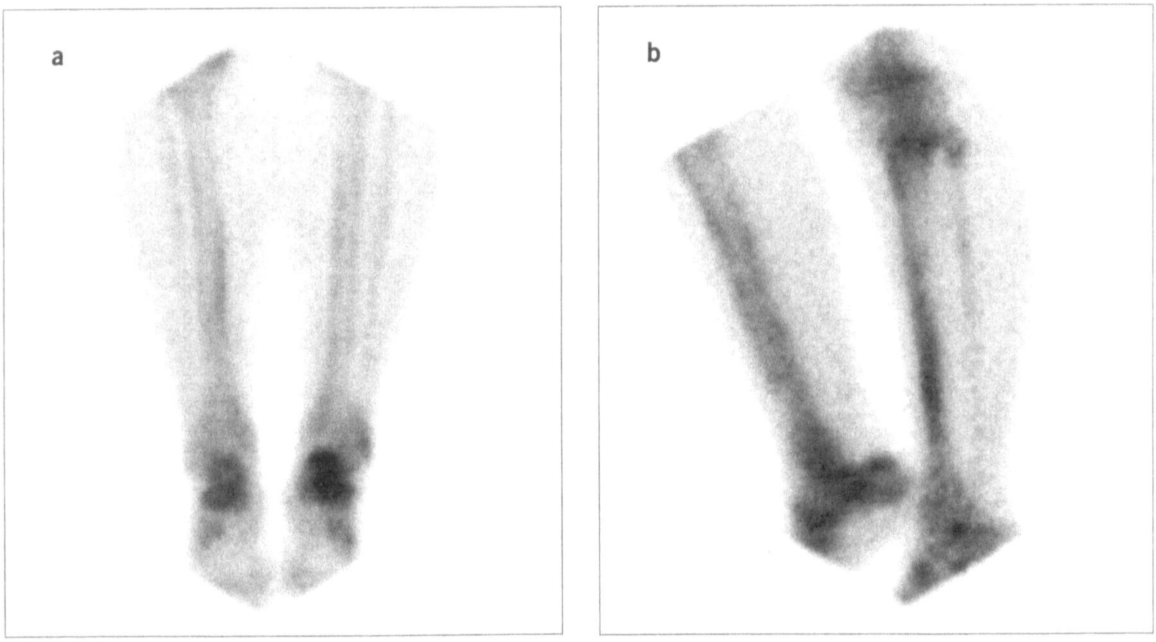

Fig. 1.37 Fusiform uptake in tibia due to stress in a professional dancer. (a) Anterior, (b) lateral. Uptake is localized mainly to the antero-medial aspect.

study, with a diffuse increase in all bones of the affected limb or part in the third phase (Fig. 1.38). The normal peri-articular accumulation may be emphasized. These changes remit if the condition resolves or become less evident in those which progress to the late stage. A focal form which does not affect the limb distal to the affected joint may be a rare variant.

Further reading:

European Journal of Nuclear Medicine 1995; ***22:*** *1187–93. 1996; 23: 256–62*

Symphasitis pubis

This painful condition may be associated with instrumentation of or trauma to the urethra. Erosion of the symphasis may be visible radiographically, but commonly radiographic changes are minimal. The condition is difficult to differentiate scintigraphically from residual activity in the bladder. Imaging at 18–24 h is therefore preferable. A pelvic inlet view is necessary to confirm that the bladder is indeed empty (Fig. 1.39).

Metabolic bone disease

Osteomalacia, hyperparathyroidism and renal osteodystrophy, as well as acromegaly and thyrotoxicosis, are associated with accelerated bone turnover and increased skeletal uptake of bone-seeking agents, which may be demonstrated scintigraphically or by measuring the 24-h whole-body retention.

Intoxication by vitamin A or its analogue isotretinoin may cause calcification in ligamen-

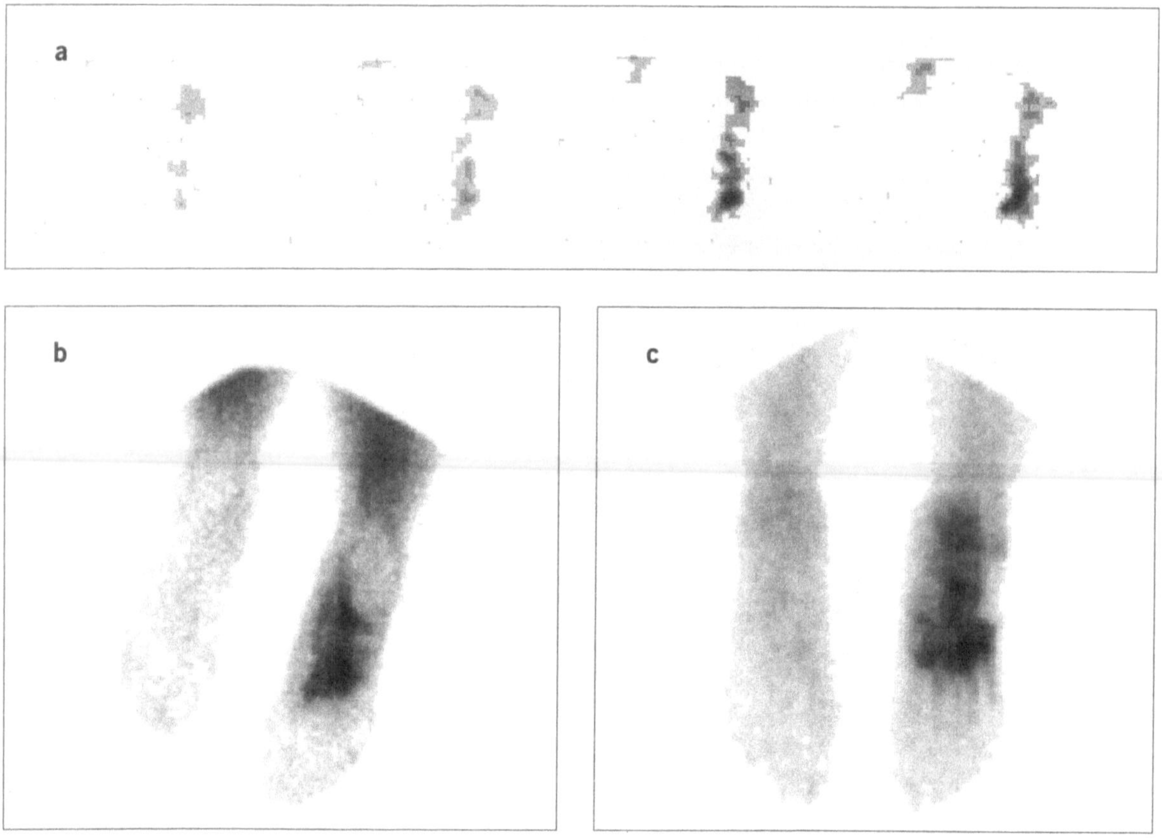

Fig. 1.38 Reflex sympathetic dystrophy. (a) Dynamic, (b) blood pool, (c) 3 h film showing diffusely increased uptake in all phases.

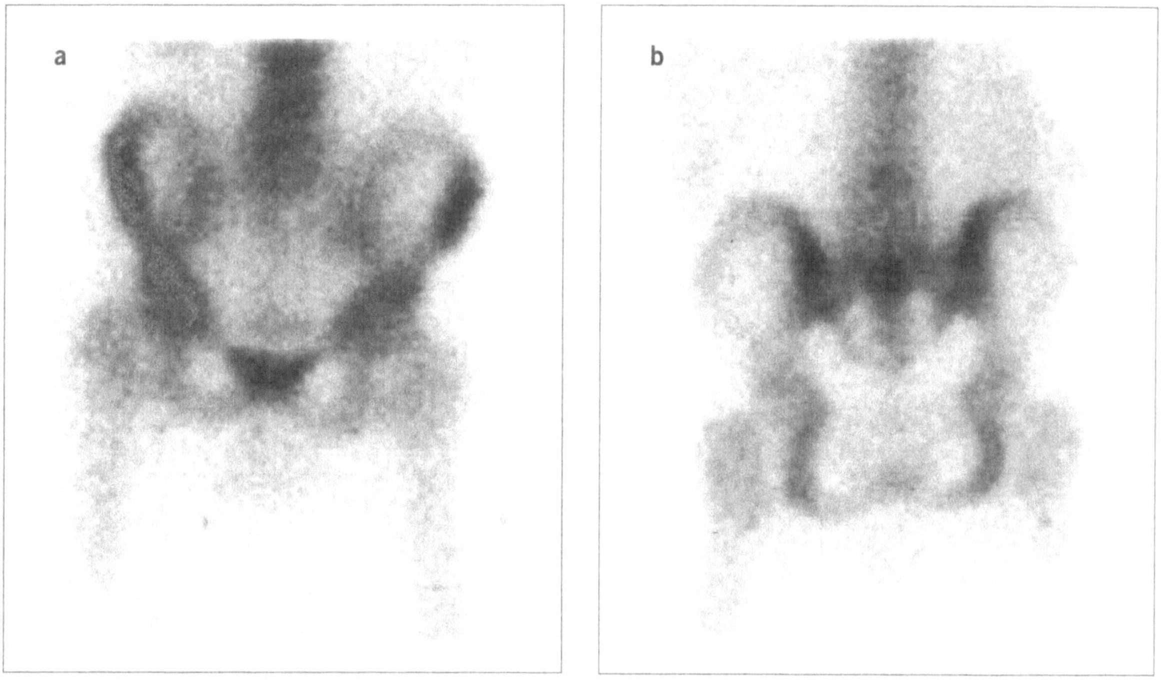

Fig. 1.39 Osteitis (symphasitis) pubis. On the anterior projection at 18 h (a) it is not possible to be certain whether activity is in symphasis pubis or bladder. (b) Pelvic outlet projection confirms that bladder is empty and uptake is therefore in pubis.

tous and tendonous insertions which can be visible scintigraphically. All of these conditions are best diagnosed by conventional biochemical parameters.

Fluorosis is associated with generalized high uptake producing a pattern resembling other 'superscans'.

Hyperparathyroidism

The appearance on imaging is variable, ranging from normal to the uniform high uptake of the 'super-scan' (Fig. 1.40). Brown cysts if present in hyperparathyroidism show high uptake and extraskeletal accumulations may be seen in ectopic calcification and chondrocalcinosis. Generalized high skeletal uptake of bone-imaging agents in carcinoma of bronchus may be associated with ectopic parathormone production.

Renal osteodystrophy

Delayed puberty due to renal failure in childhood may be associated with persistent uptake at epiphyseal sites long after the normal age of fusion and late manifestation of conditions usually associated with adolescence such as slipped femoral epiphysis (Fig. 1.41). The evaluation of renal osteodystrophy is complicated by slow soft-tissue clearance due to renal failure. It is preferable to administer radiopharmaceutical an hour or more before haemodialysis and count or image afterwards to distinguish bound from unbound activity.

Osteomalacia

The Looser zones of osteomalacia are incomplete fractures. Scintigraphy commonly reveals more than are visible radiologically and if the diagnosis is not suspected they may be easily misinterpreted as metastases (Fig. 1.43). Scintigraphic features common to all forms of metabolic bone disease include high uptake in the axial skeleton, long bones, calvarium, mandible and sternum, beading of the costo-chondral junctions (Fig. 1.43), low soft-tissue background and faint or

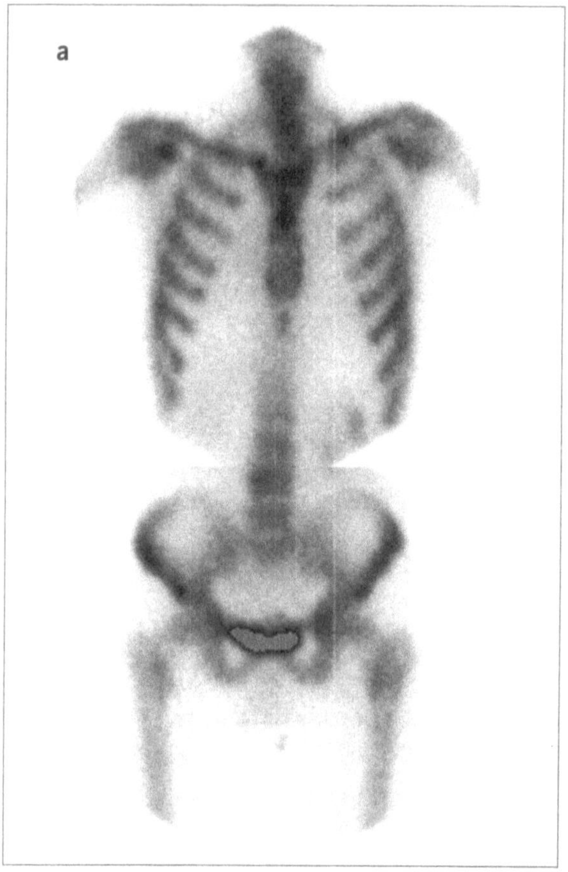

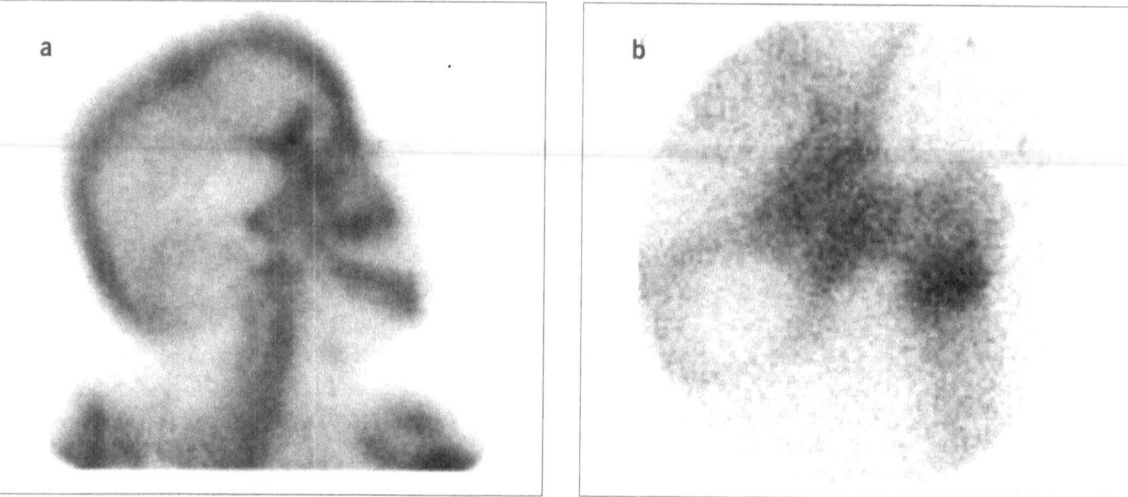

Fig. 1.40 Hyperparathyroidism. (a) Anterior, (b) posterior. Compare Fig. 1.14.

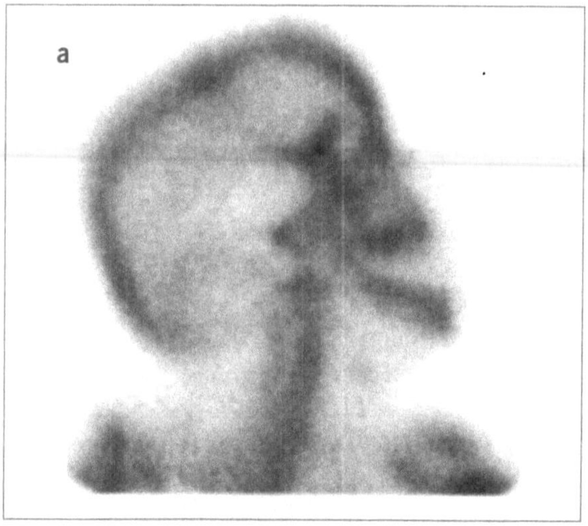

Fig. 1.41 A 21-year-old male with chronic renal failure, renal osteodystrophy and delayed puberty. (a) Abnormally high uptake in cranial vault, multiple unfused cranial sutures, maxilla and mandible, (b) stress fracture of left femoral neck, (c) abnormal pattern in head of the right femur due to a slipped epiphysis (pinhole images).

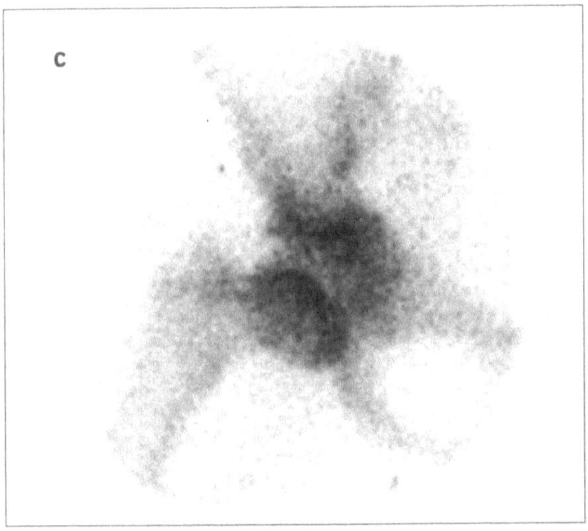

Fig. 1.41 *continued*

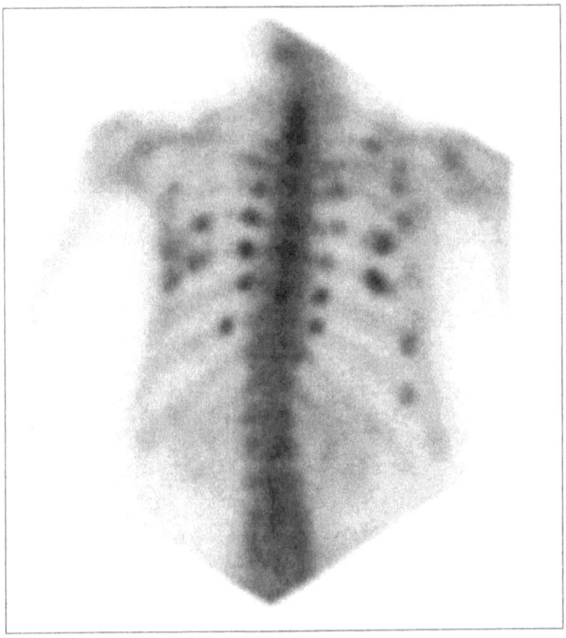

Fig. 1.42 Multiple rib fractures due to osteomalacia. This linear pattern is often seen. Few were visible radiographically.

absent images of the kidneys. No single feature is diagnostic. A scoring system marking each feature as normal, abnormal or grossly abnormal somewhat reduces subjectivity of interpretation. Scintigraphy is rarely used as a diagnostic test or for assessing progress of metabolic bone disease but unsuspected metabolic bone disease may be discovered when the examination is performed for other indications. Unexpectedly high retention should always alert to the possibility of underlying metabolic disease.

Further reading:

European Journal of Nuclear Medicine 1991; 18: 839–55

Bone metabolism

Total body calcium may be measured by neutron activation analysis. A reactor, cyclotron or radioisotope such as ^{252}Cf may be employed as the neutron source. An absorbed dose of 0.5–2.0 Gy permits total calcium to be measured with sufficient precision for longitudinal studies to be performed, or for comparisons to be made between well defined groups. There are however significant differences between populations and any centre performing this investigation must first establish its own normal range. Neutron activation of stable calcium (^{48}Ca) results in the formation of the 8.8 min half-life isotope ^{49}Ca. This emits a γ ray of 3.1 MeV. A gamma camera is of no value for detecting γ rays of this energy. A shielded whole-body counter equipped with one or more sodium iodide crystals 10–15 cm thick and not less than 15 cm in diameter is required. Measurement of calcium content of individual bones is more difficult. Part-body neutron activation with part-body counting, using an appropriately designed detecting system, is well established for limbs, but because of the difficulty of collimating a thermal neutron beam the limits of the irradiated volume are ill-defined. It is possible with this technique to follow changes in calcium content of a structure such as the forearm, but because the volume measured is ambiguous it is not possible to establish a normal range for a single measurement.

Photon absorptiometry measures bone mineral rather than bone calcium. There is in most circumstances a good correlation between the two methods. Photon absorptiometry is now usually performed using a collimated x-ray rather than an isotope source. Measurements are made

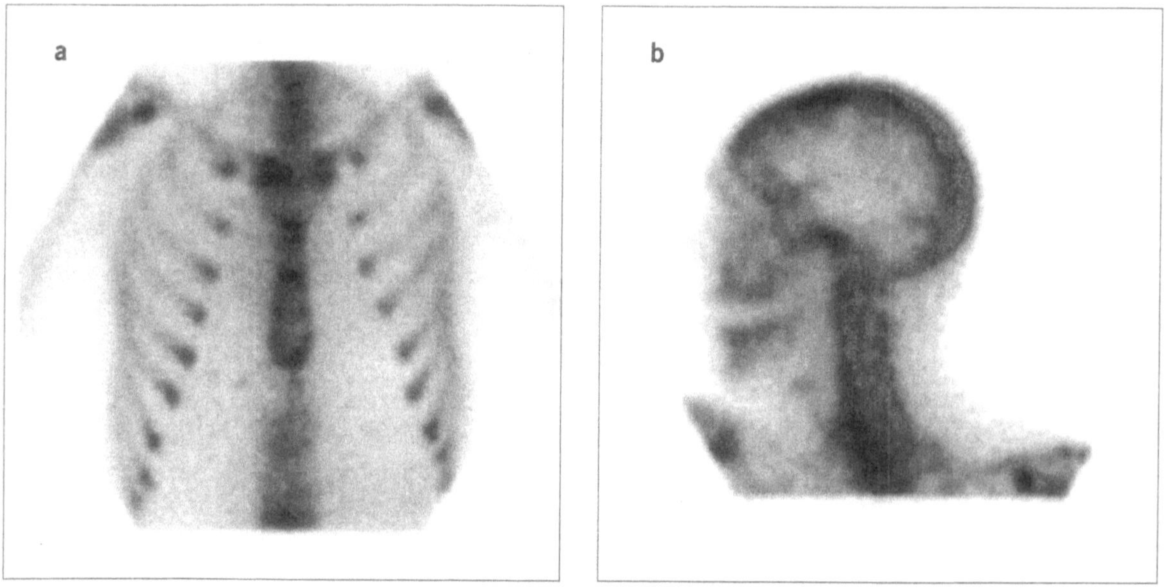

Fig. 1.43 Anterior thorax showing rib beading and 'tie' sternum in patient with renal osteodystrophy. (b) Lateral skull showing increased uptake in mandible, maxilla and over vault.

at two energies, a lower where there is predominately photo-electric absorption, giving a large difference between bone and soft tissue, and a second where absorption is mostly in the Compton range and attenuation is a function of total tissue mass, relatively unaffected by the chemical composition. The method is capable of good reproducibility and precision. X-ray sources have largely replaced isotopes as they have a higher output and thus reduce the duration of measurement. Clinical usefulness depends on a number of assumptions inherent in the physics of absorption measurement and on the premise that bone mineral density directly affects bone strength. The former set of assumptions give rise to systematic discrepancies between equipment from different suppliers, which make direct comparisons difficult. Software changes can also invalidate follow-up studies on the same equipment. The latter is valid only if bone blood flow and mineral composition are identical in the reference and test populations. Reducing blood supply to bone increases mineral density but decreases strength, wheres altering mineral composition, for example by replacing hydroxyapatite by fluroapatite, also increases bone density but decreases strength. Clinical interpretation of measurements should therefore made with caution.

Further reading:

Journal of Nuclear Medicine 1994; ***35****: 1159–61*

Calcium absorption

Calcium absorption can be measured by administering a trace amount of ^{45}Ca or ^{47}Ca orally and measuring faecal recovery over 5–7 days. ^{47}Ca, being a high energy γ emitter, can be measured without sample preparation. Its decay product ^{47}Sc is not absorbed and thus acts as an internal marker for completeness of faecal collection. A more elegant technique is to measure plasma levels after simultaneous administration of two radioisotopes of calcium, one orally and one intravenously. The quality of the information depends upon the number of plasma samples taken. Some information can be obtained from a single sample but the results from multiple samples taken over 3–6 h are more reliable. This technique permits correction for calcium secretion into the gut as well as absorption from it. This is principally a research technique with no established clinical application.

Calcium turnover

Turnover studies require either a γ-emitting radionuclide which can be measured by whole-body counting or total urinary and faecal collections. The latter are clearly impractical over an extended period, while the 4.5-day half-life of ^{47}Ca limits the duration of studies which can be performed to 2 weeks or less. The nearest analogue with a convenient radionuclide is ^{85}Sr. This is a quite good, but not perfect, analogue as it is absorbed less readily from the gut than calcium and is excreted more readily in urine. Calcium metabolism is complex and there is no entirely satisfactory mathematical model to account for either partial or whole-body retention curves.

2 Lung

Lung perfusion scintigraphy

Principles

The commonest indication for this investigation is to detect suspected pulmonary emboli by displaying regional defects within the lungs of the distribution of right ventricular output. The total pulmonary blood volume is slightly less than 0.5 l; under resting conditions at any instant about 10% is in capillaries. Capillary blood volume increases two- to three-fold on exercise. Regional distribution is demonstrated by labelling blood entering the right atrium with particles large enough to be trapped in any capillary they encounter, which in practice means the pulmonary bed. There are two implicit assumptions, namely that all particles are trapped in the lungs on their first transit from the right ventricle and secondly that intravenously administered particles are completely mixed with blood coming from all other regions as they pass through the right atrium, right ventricle and pulmonary outflow tract, so that site of injection does not affect distribution within the lungs. Both of these assumptions are valid under most circumstances. The particles employed must be physiologically inert and should not be so large that they are trapped proximal to the capillary bed. Any larger than 150 μm (micrometres) impact in arterioles and thus affect relatively large drainage regions. Particles smaller than 10 μm are likely to pass through the capillary bed, to be extracted by the reticulo-endothelial system, particularly in liver and spleen, instead of being trapped in lung. Any soluble tracer not bound to particles initially mixes with the total plasma volume. Its subsequent fate depends on its chemical form.

Radiopharmaceuticals

Two types of particle are employed:

Microspheres

These are spherical particles with a comparatively small size range. Dimensions depend on the preparation but are usually between 50 and 100 μm. Those used clinically are manufactured from heat-denatured human serum albumin and are digested by phagocytic cells within a few hours. Non-biodegradable particles including polystyrene and carbon are available for experimental use but should not be employed clinically in humans.

Macroaggregates

These are also prepared from denatured human serum albumin (macroaggregated human serum albumin, MAA). They are irregular in shape and usually have a wider range of particle size within any preparation, although their mean diameter is similar to that of microsphere preparations. Their rate of digestion tends to be somewhat faster than that of microspheres. There are substantial differences in rate of dissolution between preparations from different suppliers, some releasing sufficient free pertechnetate for thyroid and gastric mucosa to be visualized within 30 min of injection. It is therefore advisable to delay injection until immediately before imaging.

To be detectable scintigraphically, both microspheres and macroaggregates must be labelled before use with ^{99m}Tc. Kits, usually freeze-dried to prolong their shelf-life, are available commercially which allow this to be achieved by the addition of sodium pertechnetate at room temperature. In addition to preformed particles the kits contain a reducing agent, usually a tin salt such as stannous chloride. Free pertechnetate should not be present in sufficient amounts to be detectable in the final preparation at the time of injection. Any present is taken up by thyroid and gastric mucosa, which are consequently visualized. Pertechnetate is also released later when particulate preparations are digested after impaction in the pulmonary vasculature. The extent to which thyroid and stomach are visualized therefore increases with the interval between injection and imaging.

Technique

In use there are no practical or clinical differences between microspheres and macroaggregates and choice may normally be made on the basis of cost. The number of particles to be administered should be the minimum compatible with obtaining adequate diagnostic information and must be small relative to the number of capillaries in the pulmonary vascular bed (more than half a billion). If fewer than 50000 particles are injected the images are difficult to interpret because of random inhomogeneity in distribution of the particles associated with poor statistical quality. On the other hand there is no benefit to be gained by injecting more than 500000 particles. It is therefore desirable to aim for a preparation in which the required activity is contained by between 100000 and 250000 particles. Even in patients with advanced lung disease this has no measurable effect on the diffusing capacity for carbon monoxide. Some patients with chronic lung disease show a small (2–3%) fall in oxygen saturation, which is not discernible clinically and returns to the baseline value within an hour. The risk of adverse reactions is extremely small, even in patients with severe pulmonary hypertension or a right to left shunt, unless more than 100 times the recommended number of particles is administered.

The usual administered activity for adults is between 50 MBq and 75 MBq. Administered activity may sometimes be reduced to 25 MBq without loss of diagnostic accuracy in young patients well enough to tolerate an extended imaging time, for example in pregnancy or lactation. At least 200000 and preferably 400000 counts should be acquired per projection using a general purpose collimator. There is no advantage in employing an high-resolution collimator, as resolution is limited by respiratory movement. The practical lower limit of administered activity is determined by how long a patient can remain still without unacceptable discomfort or distress. Six projections should be obtained: anterior, posterior, both posterior oblique and both anterior oblique. Some departments prefer lateral to anterior oblique, but in the lateral projection a non-perfused segment may be partially concealed by 'shine through' from the contralateral lung. More abnormalities are observed on posterior and posterior oblique projections than anterior and anterior oblique. Omitting some projections increases the risk of missing significant abnormalities and may decrease the level of confidence when deciding whether or not a defect is segmental, but increasing from a six-projection to an eight-projection protocol has a minimal effect on sensitivity or accuracy.

Hazards, adverse effects and precautions

Aggregation of particles may occur during injection if there is a delay after blood is withdrawn into the syringe and before the full activity is administered. This may result in uneven distribution in the lungs of the resultant smaller number of larger particles, which can render the study non-diagnostic. If uneven distribution with a number of small randomly distributed high count rate regions is observed there is no alternative but to delay imaging until the unsatisfactory tracer dose has dispersed. This takes at least 4 h and sometimes more than 12 h. Extravasated particles are digested at the site of the injection, with the release of free pertechnetate. There are no toxic effects other than slight local soreness at the site of the extravasation but the study is non-diagnostic unless sufficient of the original dose is intravascular or a separate injection is given intravenously.

In patients with a right to left shunt there is likely to be some systemic embolism of injected particles. This has been employed to measure shunt size (page 146). Severe pre-existing pulmonary arterial hypertension is sometimes cited as a contra-indication to injection of particulate radiopharmaceuticals. There is a significant hazard only if excessively large particles are employed or if the number of particles administered exceeds the maximum advised, 500 000. At recommended levels there are rarely even minor side effects and the risk of significant toxicity has not been shown to differ from subjects with no shunt or pre-existing lung disease. Indeed this number of particles has been deliberately injected into the aortic root without ill effects to demonstrate and measure distribution of systemic arterial output in normal volunteers. If the number or size of particles is not strictly controlled, pulmonary hypertension may be exacerbated in patients with severe vascular pruning due, for example, to chronic lung disease. The concentration of particles in a preparation is usually stated in the package insert. Other reports of adverse reactions refer to discontinued preparations which employed non-catabolized aggregates such as iron hydroxide. These are no longer available.

Older preparations

Other particles have been used in the past, including iron hydroxide colloid aggregates. These particles are not biodegradable and may not be physiologically inert. The risk of an adverse reaction, possibly associated with serotonin release from lung histiocytes, is much higher. They are no longer commercially available and should not be used.

Lung ventilation scintigraphy

Ventilation imaging displays the behaviour within lung of an inspired gas. It may be used to demonstrate regional distribution of ventilation, regional lung volume, air trapping or to measure regional lung function. True ventilation imaging requires a radioactive gas, but under certain circumstances an approximation to the regional distribution of ventilation can be obtained with an aerosol. Other parameters can be studied only with gases. Commonly, when employing longer half-life radioactive gases such as xenon, the study is limited to a single projection, but if a dual-headed gamma camera is available or a limited single-breath technique is used for some projections there is no reason why several projections should not be obtained. Indeed this is to be recommended. There is little difference in complexity of use, convenience or availability between xenon and aerosols. When taking account of the total cost, including disposables, the cost difference is comparatively small. Some departments are reluctant to use radioactive gases because of a perceived difficulty in handling them safely, but with proper training the difficulty, hazard and radiation risk to staff is no greater than with aerosol preparations.

Further reading:

Journal of Nuclear Medicine 1996; ***37****: 239–44*

Radiopharmaceuticals

Gases

A suitable gas must be non-toxic, non-irritant, non-metabolized and with a convenient radioactive isotope. In practice the only one commonly available is xenon-133 (^{133}Xe). Xenon-127 (^{127}Xe) has more useful gamma ray energies (Appendix 1), higher than that of technetium but is not currently available for medical use. The principal disadvantage of ^{133}Xe is that, because of the relatively low energy of its principal gamma ray, 10 cm of inflated lung causes 45% attenuation of counts. In any projection, detected counts come from an ill-defined volume but principally the nearest half of lung thickness to the camera. Moreover its photopeak falls within the Compton scatter plateau of ^{99m}Tc. As count rate from the usual activity of xenon is lower than that from the usual administered activity of ^{99m}Tc-labelled particles, a ventilation study is impractical for several hours after performing a perfusion study. The decision to ventilate must therefore be taken before performing the perfusion study. This problem does not arise with ^{127}Xe. A further advantage of ^{127}Xe is that the perfusion study can be used to select the optimum projection for the ventilation study. The substantial differences in attenuation between technetium and either xenon isotope makes it impossible to calculate true ventilation/perfusion ratios from a dual-isotope study.

Technique

Equipment

To obtain a ventilation study with xenon the patient breathes from a closed one-way circuit containing air or oxygen and a carbon dioxide absorber such as soda-lime. Complete disposable circuits are available, or a shielded spirometer may be employed. The cost of disposables for the latter is less than that of disposable circuits but, depending on through-put, the capital cost of an adequately shielded spirometer may take many years to recover from the lower cost of non-reusable items. Either a face-mask or a mouthpiece may be used. It is difficult to obtain a good gas-tight seal around a face mask. A 'snorkel'-type mouth piece in conjunction with a nose clip is thus to be preferred. Appreciable patient co-operation is necessary whichever method is employed. The circuit must include a chamber containing soda-lime to absorb carbon dioxide and prevent hypercapnoea. Immediately before use a reservoir in the circuit must be charged with sufficient oxygen (5–10 l) to prevent hypoxia during the duration of the examination.

Procedures

Xenon is injected into the circuit immediately before the start of an inspiration. The first stage in a typical protocol is acquisition of 5 s frames for 60 s. This allows sufficient time for the computer to be started shortly before the gas is introduced into the mouthpiece, as close as possible to the airway and just before inspiration, so that one or two frames are obtained showing the true distribution of ventilation while giving some leeway in timing the administration. A normal breath is preferable, as the distribution of ventilation may change on forced inspiration, but it may be easier to achieve co-operation if the patient is instructed to first expire and then to inspire, the injection being given at end expiration, just before the instruction to inspire. It is advisable to rehearse the patient two or three times before administering the xenon. Only frames collected during the first breath represent the true distribution of ventilation. If analogue images alone are to be obtained, up to 700 MBq may be required to obtain an adequate number of counts in the time available, normally less than 10 s. Prolonging this by breath holding may alter the distribution of inspired tracer. Moreover many patients are unable to hold their breath for any significant period. Digital imaging with filtration to reduce the noise content of the images permits the activity necessary to be limited to 70 MBq.

Frames corresponding to the first breath are identified as the first two containing radioactivity in the lungs. These are summed to form the ventilation image, in which the count-rate distribution is proportional to regional ventilation. The examination continues with the collection of further frames at two per minute while the patient continues to rebreathe the xenon in the closed circuit. Except in severe obstructive airways disease, inhaled gas comes into equilibrium with all gas in the lung within 4 min. Extending the period for equilibration beyond this increases the amount of xenon absorbed in fat elsewhere in the body, thereby slowing wash-out from extrapulmonary sites such as brain, subcutaneous and abdominal fat and liver, and increasing the radiation dose without improving diagnostic accuracy. At equilibrium, count rate is proportional to regional lung volume. The circuit is then broken so that the patient inhales room air but continues to exhale into the bag or spirometer, to collect as much as possible of the xenon as it is washed out. Further images are collected at two frames per minute until the count rate approaches background (Fig. 2.1). In the normal subject, wash-out takes less than 90 s and occurs at a similar rate from all parts of the lung. Areas of air trapping are

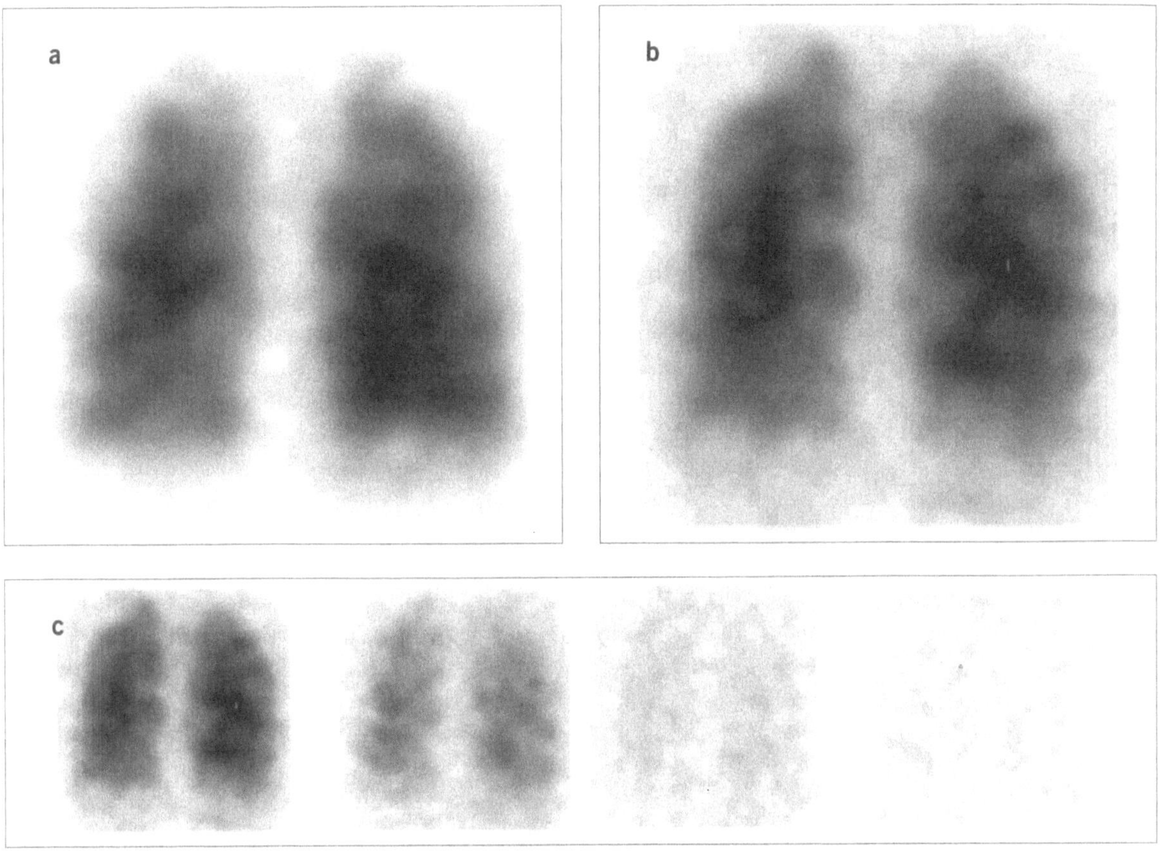

Fig. 2.1 Ventilation series (80 MBq ^{133}Xe, adult with no lung disease). Posterior projection. (a) First breath, (b) equilibrium, (c) consecutive 30 s frames during wash-out. The distribution of radioactivity in the first frame resembles that in all subsequent frames. Wash-out is rapid and symmetrical.

revealed by their slow wash-out. The true ventilation study thus has three phases: ventilation, lung volume and wash-out.

Interpretation

Because of the low energy of ^{133}Xe, counts detected in any one projection come predominately from the half of lung nearest to the collimator. No single projection visualizes all lung segments. In the posterior projection only posterior segments are adequately seen. Segmental defects anterior to the hilum tend to be better visualized on lateral projections, those dorsal to the hilum on posterior oblique projections. The medial basal segment of the right lower lobe may not be visualized in any projection. The size of lower lobe and basal defects tends to be underestimated with the result that segmental defects are misdiagnosed as sub-segmental. If only a single view ventilation study is performed the six-view perfusion scan may reveal defects in segments not examined in the ventilation study, the significance of which is therefore difficult to evaluate. A single view ventilation study fails to obtain clinically important information in up to 20% of patients. For this reason first breath images in both the left and right posterior oblique or lateral projections should be obtained routinely before a full examination in the posterior projection. The additional radiation dose from these single-breath studies, without the rebreathing component, is small (about 10 μSv per view) and is justified by the substantial improvement in diagnostic accuracy. The principal costs are increased examination times and utilization of xenon. If a dual-headed camera is available either both posterior obliques or both laterals should be obtained routinely.

Further reading:
*Journal of Nuclear Medicine 1993; **34**: 370–4*

Hazards, side effects and precautions

Toxic effects from xenon are unknown at the concentrations used for radioisotope lung ventilation studies. They are observed only when the xenon concentration exceeds 35% of the inspired gas pressure, for example when cerebral blood flow is measured by computed tomography using stable xenon. This concentration is many orders of magnitude greater than can be achieved when using radioxenon. Environmental contamination and radiation exposure of staff may result if patients are unable to co-operate sufficiently in use of the face-mask or mouthpiece. The room in which the examination is performed must thus have adequate air-flow and ventilation.

Steady-state ventilation

Ultra-short-lived gases, in particular krypton-81m (^{81m}Kr, half-life 13 s), have also been employed. The use of the latter is restricted by the limited availability of the generator system from which it is obtained, which has a useful working life of only one day. The cost is high compared to xenon and access is limited, but because the gamma ray energy (190 kV) is sufficiently above that of technetium it is practical to ventilate after the perfusion study. A first-breath image can be obtained as with xenon, to give the true distribution of ventilation. It is not possible to obtain wash-out data because of the short half-life. However a steady-state technique is more commonly employed. For this the patient breathes from a reservoir with an overflow to which ^{81m}Kr is added at a constant rate by passing the air through a dry rubidium/krypton generator, to give a relatively constant inspired concentration of ^{81m}Kr. Equilibrium is reached within four half-lives, by which time the rate of arrival is equal to the combined rates of wash-out and of radioactive decay. In patients with a normal rate of ventilation (up to 2 l/min) the count rate varies linearly with regional ventilation and radioactive decay is responsible for two-thirds or more of the krypton loss. Strictly speaking the images record inspired volume per unit time, not ventilation per unit volume and are not wholly independent of lung volume, but in practice the error is negligible if this is ignored. Because there is a steady state, the duration of acquisition for each view is not limited by the ability of the patient to suspend respiration. It is thus possible to collect high count density images in multiple projections. In the normal subject these are indistinguishable from perfusion images in the same projections.

The important limitation of this technique is that as respiratory rate increases, count rate becomes more closely related to lung volume than to ventilation. At high respiratory rates it represents only lung volume (page 284). Thus subjects with tachypnoea, especially children, are difficult to evaluate. There should always be doubt in these circumstances as to whether a true ventilation image has been obtained. Various other short-lived gases including oxygen-15 have also been used but are available only where there is an on-site cyclotron. Because of their longer half-lives the images are more closely representative of lung volume than of regional ventilation.

Further reading:
*British Journal of Radiology 1979; **52**: 353–70*

Aerosol scintigraphy

Types of aerosol

An aerosol is a stable suspension in air of any mixture of solid or liquid droplets or particles. These are widely used as an alternative to radioactive gases, but do not provide identical information. The nature of information obtained depends on properties of the aerosol, in particular range of particle sizes present and characteristics of the particles themselves. Aerosols may be classified as wet or dry and as soluble or insoluble. The distribution of an aerosol depends to a considerable extent on particle size. The unit used to describe this is the mean aerodynamic diameter (MAD), defined as the diameter of a spherical particle with a density of 1 g/ml which settles at the same rate as the particle under consideration. The largest particles are deposited in the mouth and upper airways. Aerosol particles with a MAD greater than 2 μm are almost entirely deposited in the airways and do not reach the lung periphery to any useful extent. An higher percentage of small particle than of large particle aerosols reach the alveoli, but very small particles may be exhaled without being deposited. Particles less than 1 μm MAD have significant peripheral distribution which, in the normal subject, closely resembles in appearance the ventilation image.

Large particles are more likely to be deposited on the walls of airways than in alveoli and result in a more uneven pattern of deposition. This is more marked in patients with chronic lung disease whose air-flow is more turbulent (Fig. 2.2). Particles of wet aerosols are liable to coalesce and deposit more proximally in the respiratory passages; peripheral penetration may be obtained more easily with dry aerosols. However good peripheral penetration can be obtained with small particle wet aerosols. The advantage of aerosols is that because of the relatively low rate of absorption it is possible to obtain a set of projections comparable to the perfusion study. The disadvantages of aerosols are that there is commonly uneven central deposition in airways rather than in alveoli, especially in patients with obstructive airways disease and that a larger activity of MAA is required than would otherwise be the case, because the same isotope is employed both for the aerosol and the perfusion study. A delay is often necessary between aerosol ventilation and perfusion examinations. This can be several hours depending on the rate of clearance of whichever technetium preparation is given first and the relative activities administered. A further disadvantage of aerosols is the low (<10%) efficiency of delivery systems, necessitating 200 MBq or more to be dispensed in order to achieve a deposition of 20 MBq in lung.

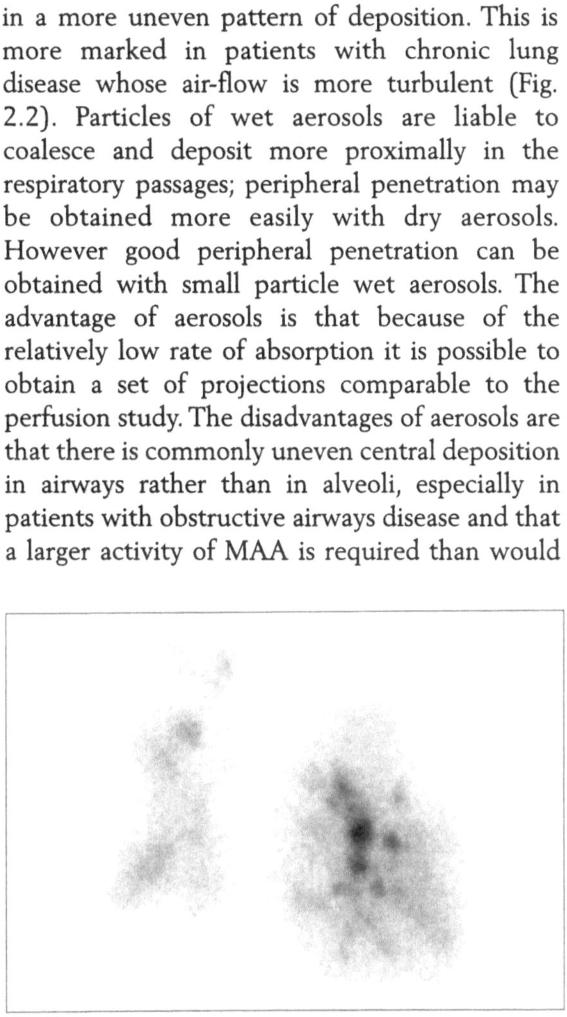

Fig. 2.2 Uneven deposition of aerosol in subject with cystic fibrosis. Posterior projection.

Further reading:

*Journal of Nuclear Medicine 1996; **37**: 239–44*

Equipment

A number of aerosol-producing devices are available commercially. The aerosol is usually labelled with ^{99m}Tc. The majority are either air (gas) jet or ultrasonic nebulizers, many of which produce wet aerosols. In most there is a settling chamber between nebulizer and mouthpiece to remove as many as possible of the larger particles. There are some differences in their performance, especially size distribution of particles. It is advisable to spend some time comparing available systems clinically before deciding which to purchase. For diagnosis of pulmonary embolism the smallest particles, which pass as far peripherally as possible without impacting, are desirable. However with most preparations there is a certain amount of central deposition in trachea and larger airways. The systems producing the smallest particles (0.14 μm MAD) make a 'smoke' by heating a graphite crucible containing pertechnetate to 2500°C in an inert atmosphere.

Technique

If the aerosol study is to be combined with a perfusion study, some consideration must be given to the order and timing of radiopharmaceuticals and relative administered activities for aerosol and perfusion images, in order to minimize cross-talk between the ventilation and perfusion components of the examination when technetium is used for both. With all aerosol-producing devices the patient inhales aerosol through a mouthpiece during normal tidal breathing until an adequate count rate is achieved over the lungs. The distribution of deposition is affected by the pattern of breathing. Several minutes may be required to deposit approximately 20 MBq. To achieve this, more than 200 MBq has to be introduced into the system. Before any disposable components can be discarded the equipment must be stored in a shielded container within an authorized store until there has been sufficient radioactive decay for disposal to be permissible. Exhaled aerosol is collected on a filter to minimize atmospheric contamination, which also has to be stored before it can be discarded. Environmental contamination can nevertheless occur if the patient is not able to co-operate.

Dynamics

Soluble aerosols, in particular pertechnetate and DTPA, are fairly rapidly absorbed from lung by passive diffusion across the alveolar lining, but there is considerable variation in the rate at which this occurs. The rate of diffusion varies inversely with molecular weight. Pertechnetate is absorbed at about twice the rate of DTPA (page 82). Particles deposited in airways are cleared upwards by ciliary action and ultimately swallowed. Because of the relatively slow wash-out even of soluble compounds, it is possible to obtain multiple projections comparable to those obtained with perfusion imaging. It may be practical to perform an aerosol study with 20 MBq pertechnetate and an hour or so later a perfusion study with 80 MBq of technetium macroaggregates with acceptably low residual contribution from the aerosol study.

Normal appearance

Perfusion

The distribution of pulmonary blood flow in the lungs is gravity-responsive and depends on posture. In the normal erect subject at rest the

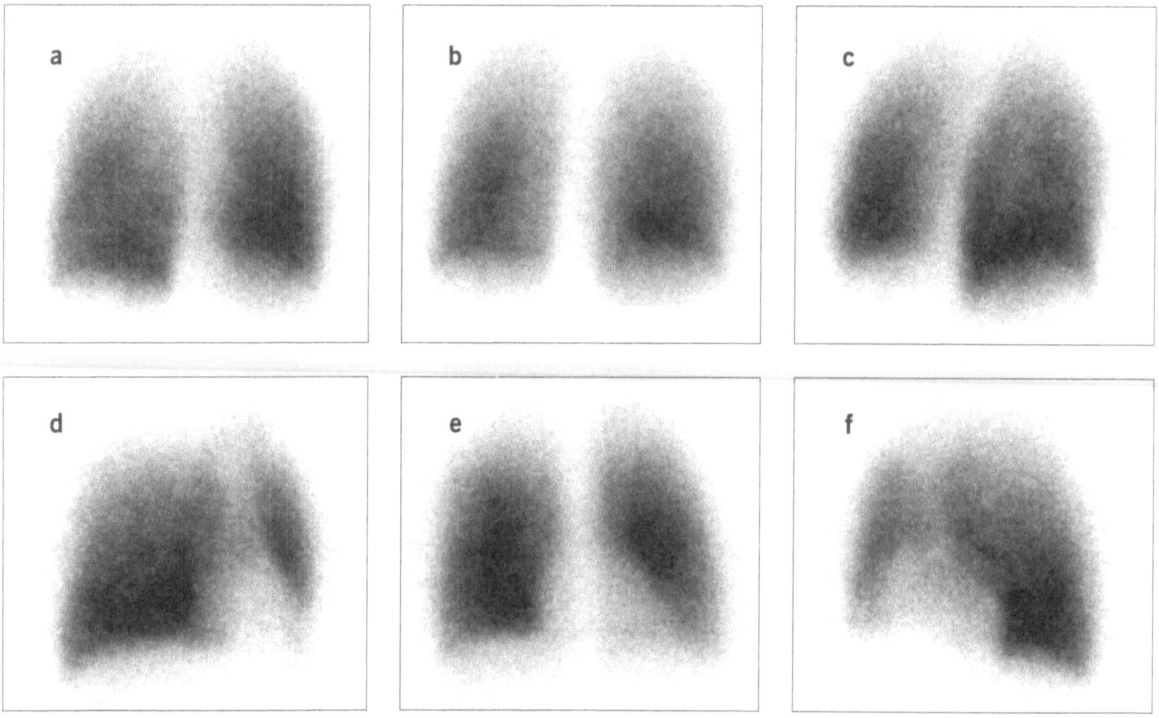

Fig. 2.3 Normal six-projection lung perfusion study. Patient was erect when injected. The normal gradient of perfusion, decreasing from base to apex, is best seen in the oblique projections. (a) Left posterior oblique, (b) posterior, (c) right posterior oblique, (d) right anterior oblique, (e) anterior, (f) left anterior oblique.

lower zones carry the bulk of both ventilation and perfusion. On changing from the erect to the supine position distribution alters, with greater perfusion of more dependent posterior segments. On exercise there is progressive recruitment of alveoli, with increasing perfusion and ventilation of upper segments. Within segments the perfusion pattern appears as a smooth gradient from superior to inferior, although this is in part a reflection of the limited resolution of the gamma camera. At least six projections are required to display all broncho-pulmonary segments (Figs 2.3, 2.4). These display distribution of perfusion at the time of injection.

A rise in pulmonary venous pressure, either as a result of exercise or of disease, reduces the vertical gradient, making perfusion more nearly uniform from apex to base. When interpreting a perfusion study it is important to know the posture of the patient at the time of injection as there is no subsequent redistribution. It is desirable that posture should be specified in departmental policy. The erect posture is usually preferred as most patients coming for lung scintigraphy are able to walk or sit. It provides important physiological information about the basal condition of the patient which is clinically relevant. If a patient is unable to stand or sit upright it is acceptable to administer the particles in any posture, but this must be recorded and the information available when interpreting the images.

Ventilation

The distribution of ventilation normally closely resembles that of perfusion and is also posture-dependent. For comparability, particles for the perfusion study must be administered with the patient in the same posture as the ventilation study. In chronic lung disease non- or under-

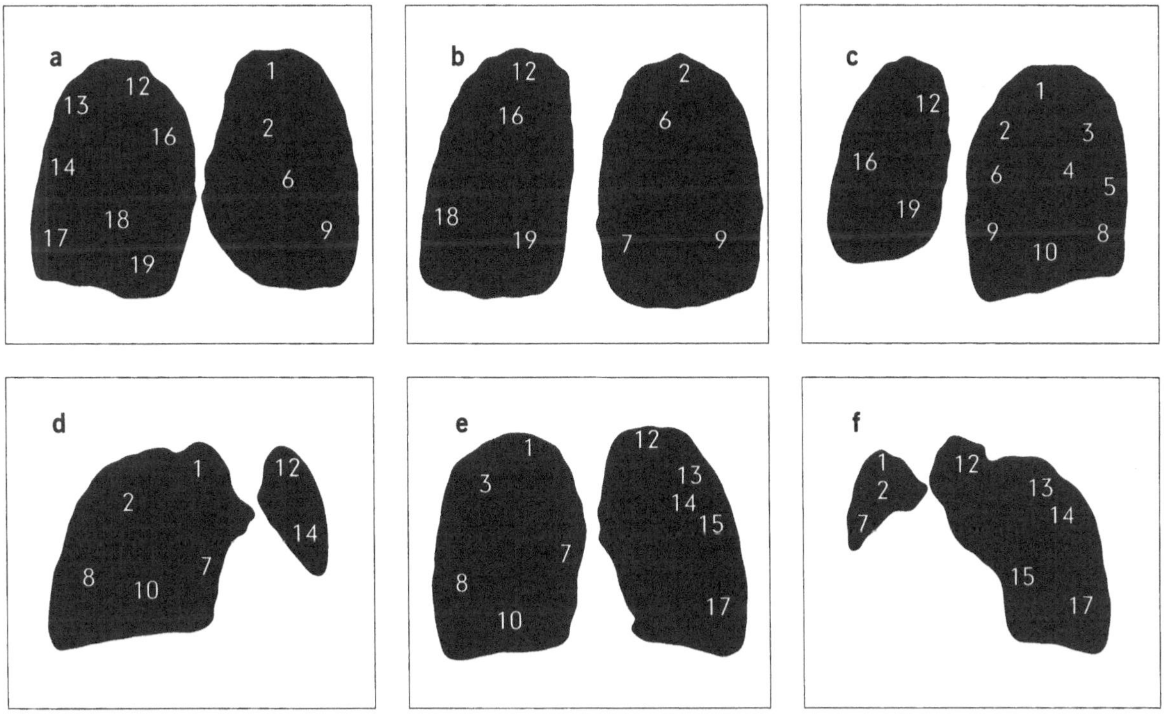

Fig. 2.4 Schematic representation of lung segmental anatomy. Projections correspond to Fig 2.3. Segments are numbered as follows. **Right upper lobe**: 1 apical, 2 posterior, 3 anterior. **Middle lobe**: 4 lateral, 5 medial. **Lower lobe**: 6 superior, 7 medial basal, 8 anterior basal: 9 posterior basal, 10 lateral basal. **Left upper lobe**: 12 apico-posterior, 13 anterior. **Lingular**: 14 superior, 15 inferior. **Lower lobe**: 16 superior, 17 medial anterior basal, 18 lateral basal, 19 posterior basal.

perfused areas are also hypoventilated. Ventilation is reduced where there is air trapping, in some cases despite normal perfusion, but is normal in regions which are non-perfused in consequence of a recent pulmonary embolism. The characteristic 'V/Q' mismatch in pulmonary embolism thus consists of normally ventilated (V) segments with reduced or absent perfusion (Q) and must not be confused with the reverse mismatch (poor ventilation of regions with normal or only slightly reduced perfusion) of chronic lung disease.

The regional distribution of ventilation may be abnormal in asymptomatic smokers with normal chest radiographs, possibly due to airways narrowing. They often show minor focal defects of ventilation which may persist for many years. Lung disease may cause extensive anomalies on the ventilation image despite little or no radiographic abnormality (Fig. 2.5). Although in the normal subject all phases of the ventilation study appear similar, this is not true when lung function is abnormal. A matched defect of both ventilation and perfusion with no matching radiographic defect is regarded as indeterminate for pulmonary embolism. If it corresponds to a region of air trapping, the probability of pulmonary embolism is reduced. For this reason it is essential to obtain all phases of the ventilation study in all patients.

Pulmonary thrombo-embolism

Clinical features

The commonest indication for lung perfusion and ventilation scintigraphy is investigation of suspected pulmonary embolism. Clinical signs of pulmonary embolism are insensitive and non-specific. The classical triad of haemoptysis, pleuritic chest pain and dyspnoea is found in less than one-third of patients with pulmonary embolism, whereas electrocardiographic changes of right heart strain or abnormalities of lung function both have poor sensitivity and low specificity. On the other hand there is a strong correlation between presence of certain *a priori* risk factors and incidence of pulmonary embolism. These risk factors are:

- clinical evidence suggesting a deep vein thrombosis
- history of previous venous thrombosis or thrombo-embolism
- recent surgery
- cancer
- pregnancy
- puerperium
- high-dose oestrogen therapy (over 50 μg per day)
- exacerbation of inflammatory bowel disease
- any haematological abnormality associated with hypercoagulability or hyper-viscosity
- recent reduction in mobility especially if associated with heart failure, myocardial infarction, stroke, infection, trauma, other prolonged illness, a long car or plane journey or any other prolonged immobilization
- increasing age.

Pulmonary embolism must not be confused with pulmonary infarction. The two may occur together but the terms are not synonymous. Embolism can and usually does occur without infarction because of the dual (pulmonary and bronchial) arterial supply of lung. Pulmonary embolism occurs when an object, most commonly a thrombus dislodged from a peripheral or pelvic vein or, infrequently, the right side of the heart, enters the pulmonary artery or one of its branches and becomes impacted. Rarely, pulmonary embolism may be due to tumour or foreign body. Pulmonary infarction occurs only when the bronchial arterial supply is compromised, leading to ischaemia of lung tissue. In many cases pulmonary embolism does not affect viability as the bronchial arterial supply remains patent and is adequate for survival of lung tissue. Pulmonary embolism without infarction produces no changes on the chest radiograph. Such changes as have been described, for example reduction in number or size of pulmonary arterial branches, are unreliable 'soft' signs which do not stand up to objective scrutiny. The only reliable radiological signs (pleural effusion, consolidation and/or collapse) are non-specific and, when they do occur, indicate

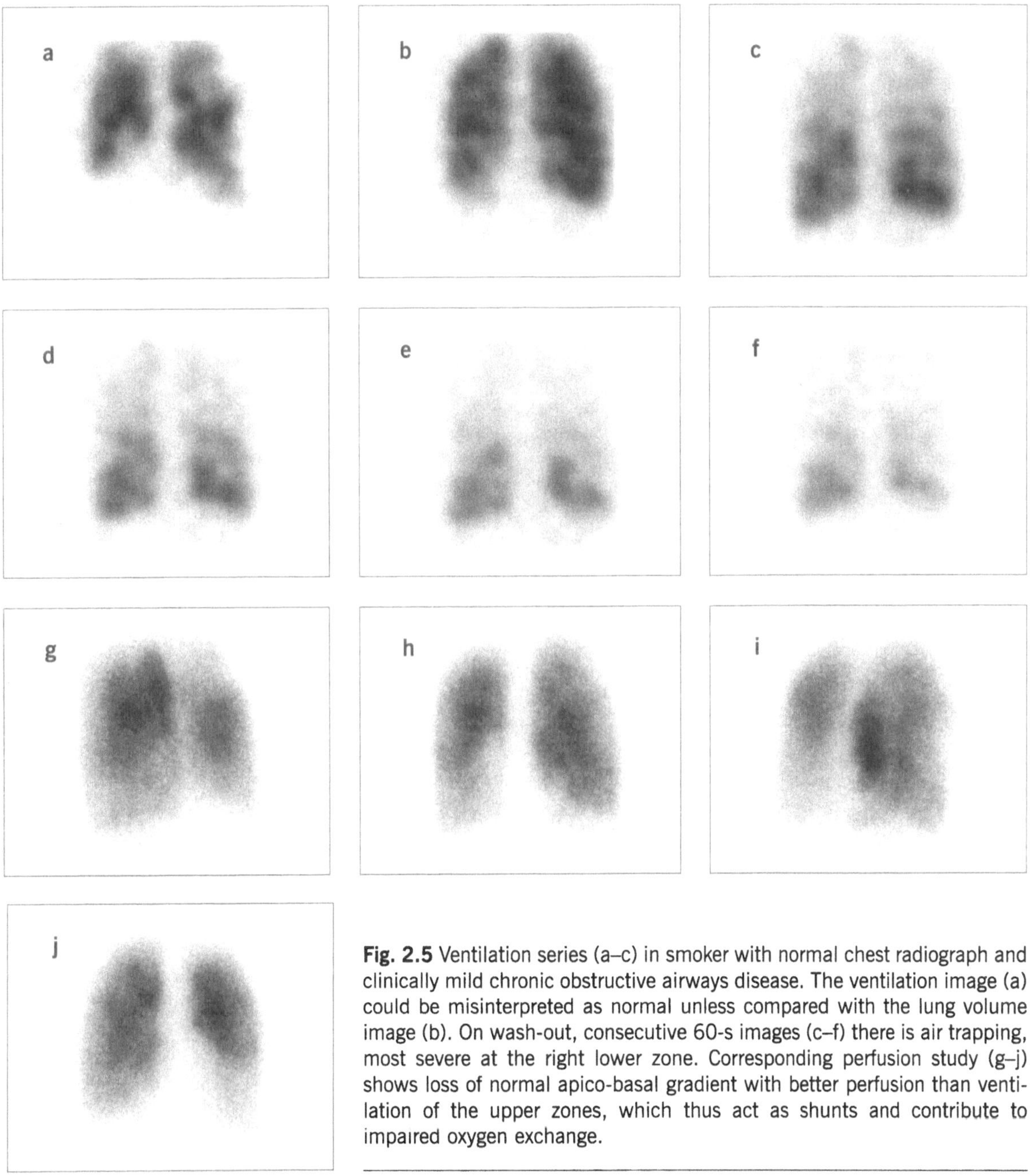

Fig. 2.5 Ventilation series (a–c) in smoker with normal chest radiograph and clinically mild chronic obstructive airways disease. The ventilation image (a) could be misinterpreted as normal unless compared with the lung volume image (b). On wash-out, consecutive 60-s images (c–f) there is air trapping, most severe at the right lower zone. Corresponding perfusion study (g–j) shows loss of normal apico-basal gradient with better perfusion than ventilation of the upper zones, which thus act as shunts and contribute to impaired oxygen exchange.

pulmonary infarction. Pulmonary embolism can be diagnosed by scintigraphy, spiral CT angiography, MRI or pulmonary angiography.

Scintigraphic changes

The characteristic findings on perfusion scintigraphy in a patient who has had a pulmonary embolism within the previous 24 h and whose lungs were previously normal are the presence of perfusion defects affecting two or more whole segments without corresponding ventilation or radiographic abnormalities (Fig. 2.6). A segmental perfusion defect is an area of absent perfusion corresponding in size and position to an identifiable broncho-pulmonary segment. Thus the defect should be well defined and wedged-shaped in at least one projection, with the base of the wedge on the pleural surface. It is commonly possible to identify affected segments by name, even allowing for variability of broncho-pulmonary architecture.

In addition, a number of sub-segmental defects are commonly present. It is often difficult to be certain whether or not these are associated with a ventilation mismatch, particularly if only a single projection ventilation image has been obtained, because of the poorer statistical quality of the first-breath image and because there is no single projection which displays all segments (Fig 2.7).

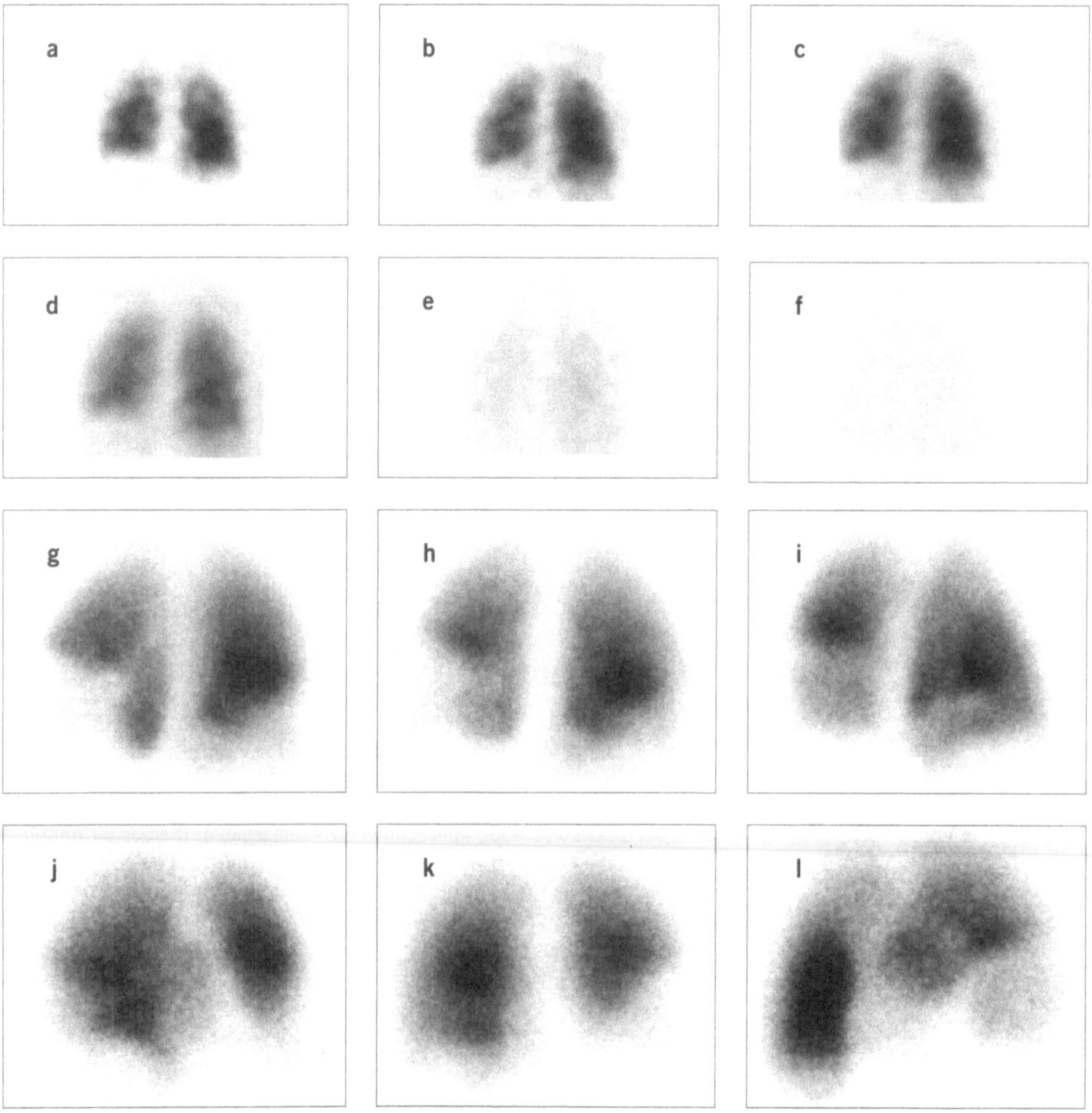

Fig. 2.6 (a) First breath (ventilation, posterior), (b) lung volume frame, (c–f) consecutive 30-s wash-out frames, (g) left posterior oblique, (h) posterior, (i) right posterior oblique, (j) right anterior oblique, (k) anterior, (l) left anterior oblique perfusion. The pattern of ventilation and wash-out is normal apart from an insignificant discrepancy between ventilation and lung volume at the right upper zone. There is a perfusion defect at the left base corresponding to at least two broncho-pulmonary segments. Interpretation: high probability of pulmonary embolism.

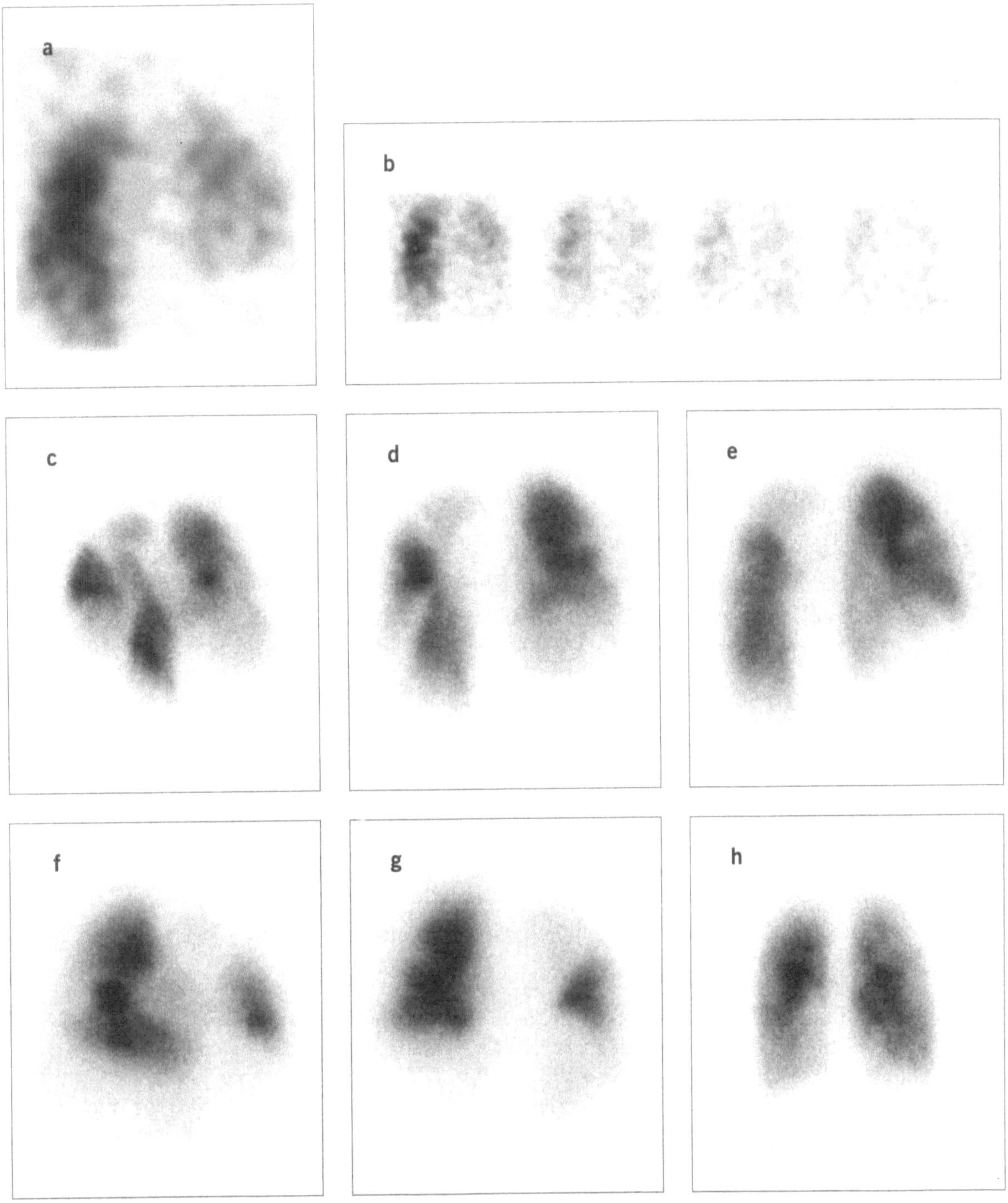

Fig. 2.7 (a) First breath (ventilation, posterior), (b) lung volume and consecutive 30-s wash-out frames, (c) left posterior oblique, (d) posterior, (e) right posterior oblique, (f) right anterior oblique, (g) anterior, (h) left anterior oblique perfusion. The pattern of ventilation and wash-out from the left lung is normal. Ventilation and volume of the right lung are reduced but there is no air trapping. There are multiple bilateral segmental perfusion defects in areas with normal ventilation (compare Fig. 2.3). There is also a matched reduction of ventilation and perfusion at the right base. Radiologically there was some consolidation and collapse at the right lung base. Interpretation: high probability of pulmonary embolism. The radiological changes are compatible with, but do not prove, pulmonary infarction.

Following pulmonary embolism, ventilation is retained in non-perfused segments for a variable period. The characteristic perfusion changes become less well defined over a period of days or weeks as a collateral supply develops from adjacent segments. In addition clot tends to break up and move more distally. In consequence perfusion defects often alter to involve sub-segments rather than segments. These may be difficult to distinguish from non-segmental defects occurring in chronic lung disease. Lung scintigraphy should therefore be performed early, at a low index of suspicion. Delaying investigation increases the probability of inconclusive findings.

The pattern may return to normal within a week, especially in younger patients without previous lung disease, and most patients show some improvement by this time. However perfusion defects and less often ventilation perfusion mismatches sometimes persist for months or years. Depending on the population studied, some series report that in up to 20% of cases ventilation perfusion mismatch persists after a pulmonary embolism for long enough to cause uncertainty in detection of subsequent events. There are no reliable scintigraphic criteria for determining the age of pulmonary emboli.

The distribution of emboli reflects that of pulmonary blood flow at the time of embolization. Posterior emboli tend to be more common if the event occurred when the patient was supine and more basal segments are involved if the patient was erect or semi-erect. It is relatively uncommon for one segment only to be affected. In most patients there are several segmental and sub-segmental defects. Pre-existing lung disease or pulmonary hypertension with upper lobe blood diversion alters the distribution pattern both of perfusion and of emboli.

Defects corresponding to an entire lobe are less common and occlusion of a main pulmonary artery, leading to hypoperfusion or non-perfusion of a whole lung, is rarely due to pulmonary embolism. When seen this is more commonly due to other conditions such as congenital absence of the pulmonary artery or a large central tumour (Fig 2.8).

Perfusion defects may be caused by many conditions other than pulmonary emboli. Consolidation visible on the chest radiograph results in a matched perfusion and ventilation defect which may persist for a variable time after resolution of the radiographic abnormality. In patients with chronic obstructive airways disease segmental anatomy is often considerably distorted, with resulting uncertainty as to whether or not defects are segmental. In some studies the term 'non-segmental defect' is restricted to heart, mediastinum and effusions. A perfusion study without a ventilation study is rarely interpretable when lung anatomy is distorted.

Other causes of ventilation perfusion mismatch are rare. It may be found in patients with significantly greater obstruction to a pulmonary artery than to the corresponding bronchus. This may be due to:

- a bronchogenic carcinoma obstructing a pulmonary artery branch to a greater extent than the corresponding bronchus
- more rarely, secondary tumours
- large inflammatory gland masses in the mediastinum, for example tuberculosis
- mediastinal fibrosis (whether idiopathic or due to radiation)
- compression by vascular structures such as aortic aneurysm
- congenital anomalies of the pulmonary vasculature
- Takayashu's disease
- schistosomiasis
- cryptogenic fibrosing alveolitis.

In obstructive airways disease the pattern is the inverse of that seen in pulmonary embolism, namely affected regions are perfused but are under-ventilated. On rebreathing, under-ventilated areas fill in and there is delayed wash-out (Fig. 2.5). In patients on positive pressure ventilatory support with lobar atalectasis, areas both of mismatch and of reverse mismatch may be observed. A full ventilation study thus displays not only pulmonary embolism but also the presence and extent of airways obstruction.

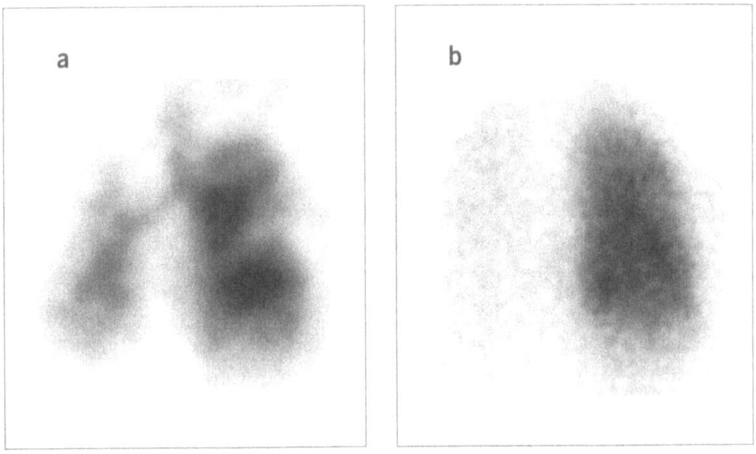

Fig. 2.8 (a) First breath (ventilation, posterior), (b) lung volume, (c) left posterior oblique, (d) posterior, (e) right posterior oblique, (f) right anterior oblique, (g) anterior, (h) left anterior oblique perfusion. Mismatch of an entire lung is common due to causes other than lung cancer. In this patient the mismatch is due to mediastinal spread of a bronchogenic carcinoma.

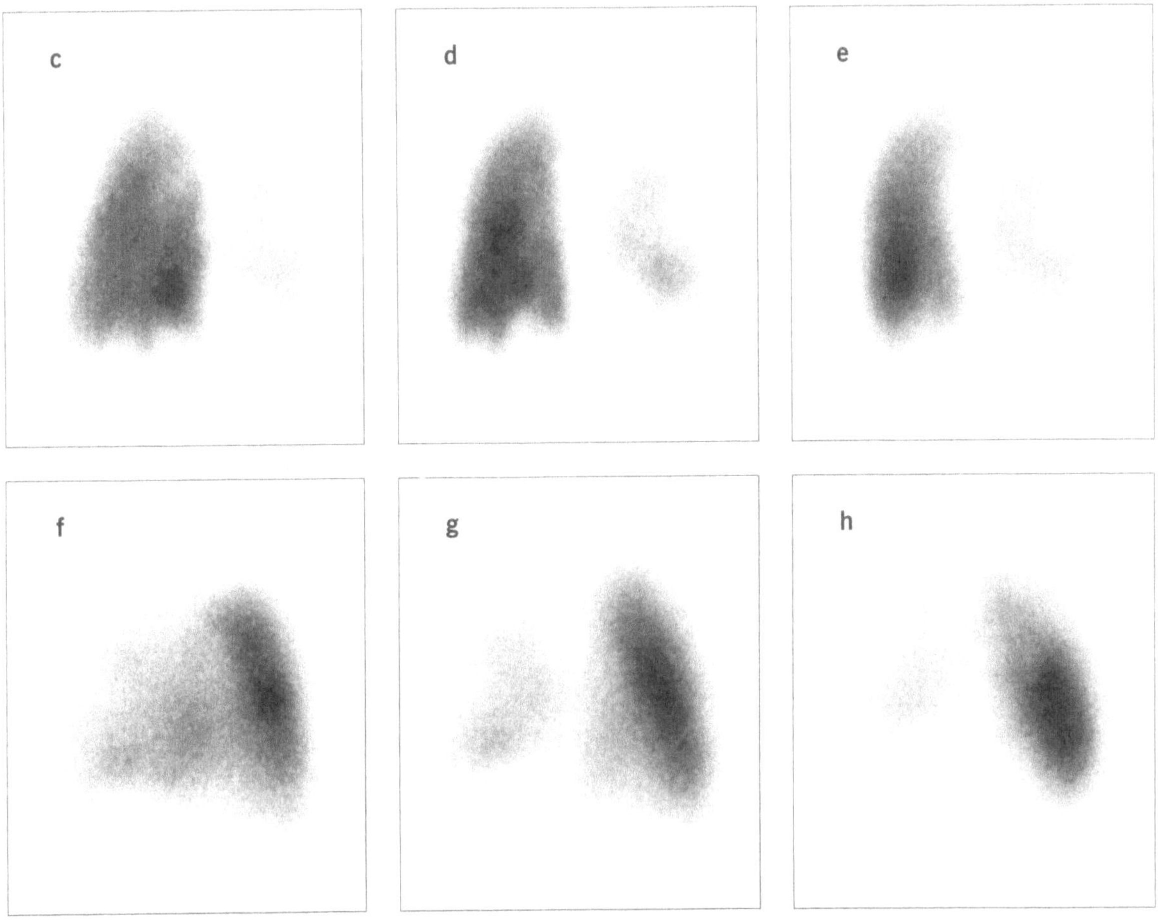

Strategy of investigation

When using ^{133}Xe the ventilation study must be performed before the perfusion study. Two alternative strategies are commonly employed, choice depending on locally available resources. If a radiologist is available to read the chest radiograph before every perfusion study, it may be permissible to omit the ventilation study in non-smoking, non-asthmatic patients with a completely normal chest radiograph taken within 12 h of the examination. If this level of supervision is not practicable a ventilation study should precede every perfusion study unless ^{81m}Kr or ^{127}Xe is available, as the gamma energy of ^{133}Xe falls within the Compton plateau of technetium, with consequent degradation of the image unless unacceptably high activities of xenon are used. Strict criteria of normality must be employed. It is not sufficient merely to note the absence of radiographic consolidation, collapse or line shadows. If there is any radiographic abnormality or history of obstructive airways disease, asthma or smoking, a ventilation study must always be performed. Pulmonary oedema is associated with matched defects which are often rounded rather than wedge-shaped and may correspond to radiographic abnormalities (Fig. 2.9), Irrespective of the chest radiographic findings, ventilation studies must always be obtained in patients with asthma, as matched ventilation perfusion defects due to bronchial mucus plugs are common. These defects are matched but can be misdiagnosed as pulmonary embolism in the absence of a simultaneous ventilation study. A recent chest radiograph is always essential for interpretation of the ventilation and perfusion study, irrespective of whether both ventilation and perfusion patterns are normal or abnormal. Pulmonary infarcts which have collapsed to leave only line shadows on the chest radiographs are below the resolution limit of isotope imaging devices. In the presence of linear shadows on the chest radiograph a normal ventilation and perfusion study does not exclude a pulmonary infarct, although it is generally considered that the radiographic changes take several days to develop, so that the event must have occurred some days previously. Follow-up studies (in a relatively small number of patients) suggest that such patients are not at risk if treatment is withheld.

Interpretation

The typical features of recent pulmonary embolism are two or more broncho-pulmonary segments in which there is normal ventilation but absent perfusion, often with one or more additional affected sub-segments, in a patient with a normal chest radiograph. When the chest radiograph is abnormal, two or more areas of mismatch that do not correspond to radiographic abnormalities and are segmental or sub-segmental are also associated with a high probability of pulmonary embolism. At the other end of the spectrum an entirely normal pattern of both ventilation and perfusion in a patient with a normal chest radiograph effectively excludes recent pulmonary embolism. Unfortunately many patients fall into neither of these categories. The evidence provided by ventilation perfusion scintigraphy is indirect as there is no direct thrombus label available. In these patients the confidence of diagnosis depends on other criteria such as number and extent of defects and the *a priori* estimate of clinical risk. Areas of consolidation cannot be ventilated and although in most cases there is a matched deficit of perfusion, in some cases sufficient pulmonary arterial flow persists to contribute to arterial oxygen desaturation, i.e. the region acts as a right to left shunt. A similar phenomenon is occasionally seen in regions of lung which are non-ventilated due to partial bronchial obstruction. This is similar to the reverse mismatch seen in chronic obstructive airways disease.

A number of sets of criteria have been proposed to categorize patients as high, intermediate or low probability of pulmonary embolism. The most complete were developed from the Prospective Investigation of Pulmonary Embolism Diagnosis study (PIOPED), a large prospective study comparing ventilation and perfusion scintigraphy with pulmonary angiography. The fundamental premise underlying the categories is that, as the findings in most patients neither definitively confirm nor totally exclude the possibility of pulmonary embolism, it is possible only to indicate a probability. PIOPED originally classified patients into five categories: high, intermediate, low or very low probability of pulmonary embolism and normal. Some other classifications use only three or four categories. The criteria used in the original PIOPED study and analysis have been subject to much criticism, including their definition of some categories and that an inadequate number of patients with normal V/Q scintigraphy had angiographic confirmation.

There is good evidence that combining *a priori* probability of pulmonary embolism with scintigraphic findings further refines interpre-

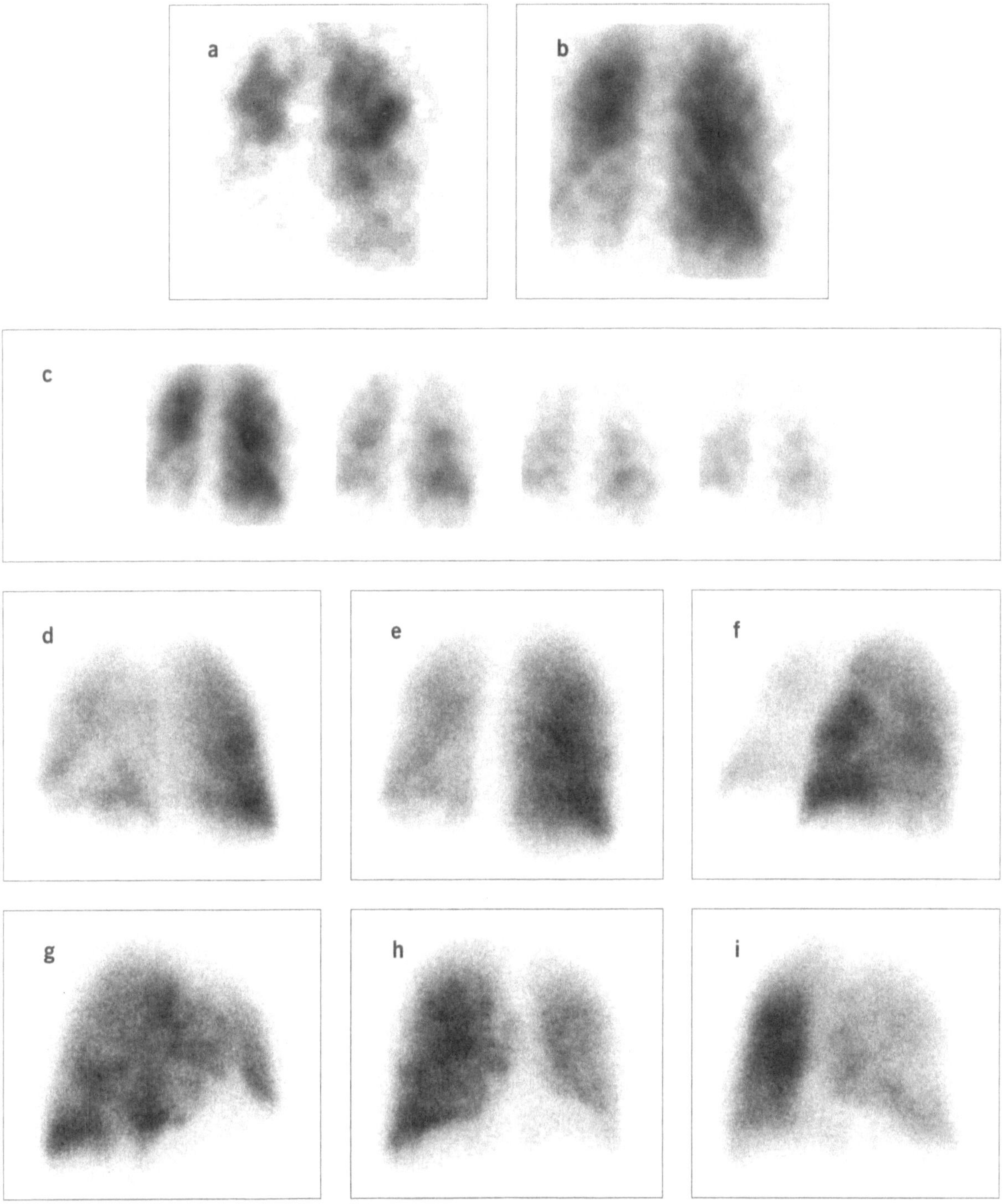

Fig. 2.9 (a) First breath (ventilation, posterior), (b) lung volume.(c) consecutive 60-s wash-out frames, (d left posterior oblique, (e) posterior, (f) right posterior oblique, (g) right anterior oblique, (h) anterior, (i) left anterior oblique perfusion. The ventilation pattern is unremarkable but the wash-out is slow. The gradient of perfusion from base to apex is less marked than normal, there is reduced perfusion of the entire left lung, the perfusion pattern in both lungs is uneven but with no clearly segmental defects. There are no segmental ventilation perfusion mismatches. This pattern is non-specifically abnormal and, depending on the clinical and radiological picture, is associated with an intermediate to low probability of pulmonary embolism.

tation. Subsequent review of the PIOPED data has led to the following revised criteria. Ventilation perfusion mis-match of 75% or more of at least two segments indicates an high (<80%) probability of recent pulmonary embolism, even in the absence of identifiable risk factors. In patients with two or more risk factors the probability increases to almost 100%. There is a very small incidence of false positives attributable to central tumours, gland masses or because the defects are residues of previous episodes of pulmonary thrombo-embolism. The extent of segmental defects estimated by comparison with normative maps tends to systematically underestimate the percentage of the lobe or segment affected, so that there is an in-built tendency to under-diagnose high probability of pulmonary thrombo-embolism.

At the other end of the scale a normal ventilation and perfusion study with lung outlines corresponding to those seen on the chest radiograph effectively rules out pulmonary embolism. Defects corresponding to <25% of a segment or which appear smaller than the corresponding radiographic abnormality, non-segmental matched ventilation and perfusion defects regardless of size with or without matching radiographic abnormalities such as an enlarged heart or mediastinum, elevated diaphragm or enlarged hilum and defects corresponding to a large pleural effusion (Fig 2.10) are associated with a <10% incidence of pulmonary embolism if there are no associated risk factors, rising to >20% if two or more risk factors are present. A single perfusion defect associated with a small pleural effusion, obliterating the costo-phrenic angle, is associated with an higher risk, up to 25%, irrespective of other risk factors. Other combinations of scintigraphic findings in patients with no risk

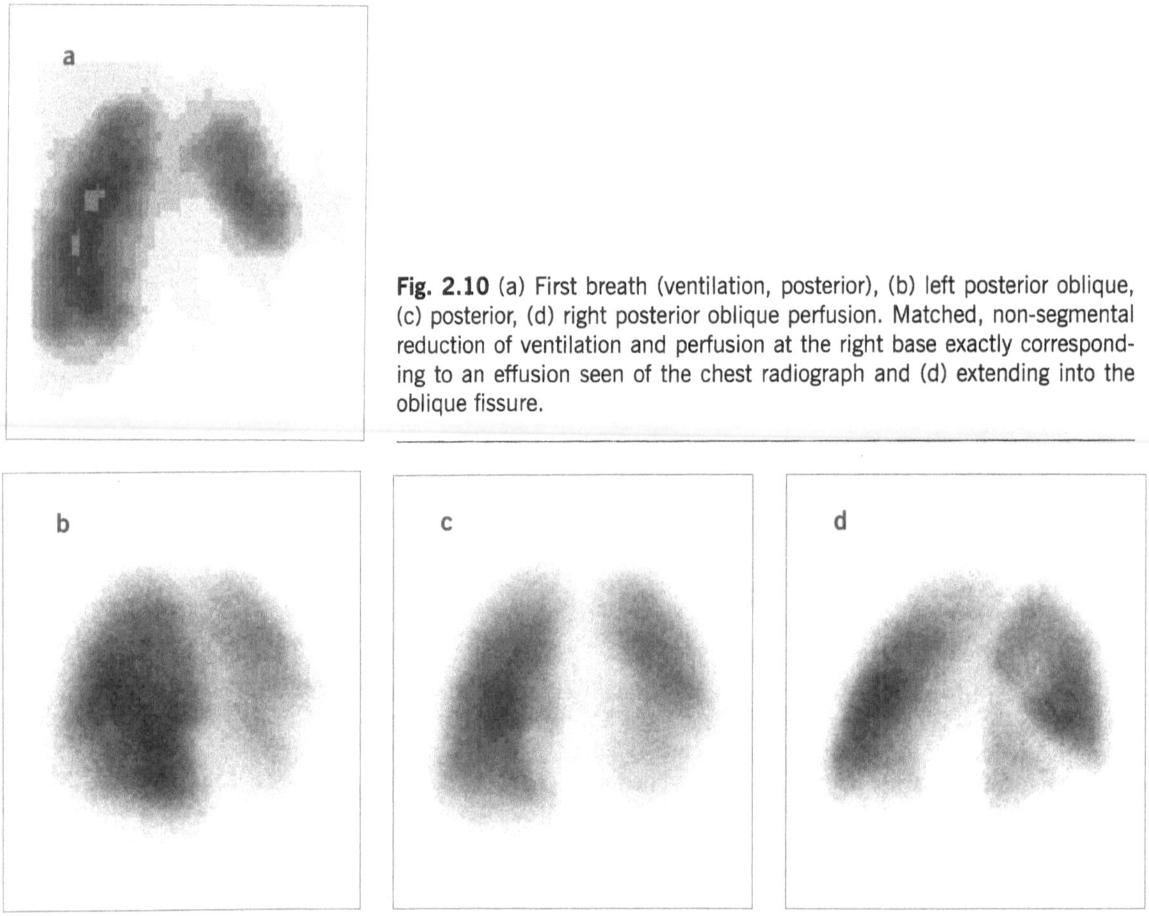

Fig. 2.10 (a) First breath (ventilation, posterior), (b) left posterior oblique, (c) posterior, (d) right posterior oblique perfusion. Matched, non-segmental reduction of ventilation and perfusion at the right base exactly corresponding to an effusion seen of the chest radiograph and (d) extending into the oblique fissure.

factors are associated with a 25% incidence of pulmonary embolism. This increases to 45% if two or more risk factors are present. When the clinical, radiographic and scintigraphic findings in combination are inconclusive spiral CT or pulmonary angiography may be indicated. Labelled monoclonal antibodies and peptides have been used experimentally to image clot directly, but experience with them is limited and no preparation is available for routine clinical use.

Further reading:
Journal of Nuclear Medicine 1993; ***34****: 370–4. 1995;* ***36****: 2380–87. 1996;* ***37****: 577–84, 1310–1316*
European Journal of Nuclear Medicine 1995; ***22****: 1099–1100. 1996;* ***23****: 1547–53*

Bronchogenic carcinoma

Although usually obvious radiologically, occasional cases are encountered in which a small radiographically occult central tumour obstructs the pulmonary artery without occluding the bronchus, giving rise to a mis-match between ventilation and perfusion. The presence of a ventilation and perfusion defect larger than would be expected from the chest radiograph has been cited as evidence of extensive mediastinal spread, implying inoperability, but this is not used widely enough for the reliability of this criteria to have been properly assessed. A number of tracers accumulate in primary and metastatic lung tumours (page 270).

Further reading:
European Journal of Nuclear Medicine 1992; ***19****: 355–68*

Obstructive airways disease and lung function

Ventilation scintigraphy can be used in the assessment of regional lung function. Effective half-life (page 323) of a tracer gas in any region of lung is related to its mean residence time in that region. The time constant (equal to 1.44 times effective half-life) reflects both airways resistance and compliance. Normal lung regions have a time constant of 40–50 s. Severe obstructive airways disease can extend this to 200 s. The distribution of ventilation is influenced by regional compliance, airways resistance and minute ventilation rate. Regional wash-out times therefore cannot be compared unless the total minute ventilation is standardized. Even this assumes that regional flow does not change with ventilation. Although probably true in normal subjects, this is unlikely to be a valid assumption in patients with lung disease.

Ventilation, like perfusion, shows a gravity-dependent gradient in the normal subject. The normal resting tidal volume is 1.5–2 l/min, with the greater part of the volume at the base in the erect subject. The distribution of ventilation is affected both by lung volume and the rate of inspiration. Meticulous attention to detail is essential if reproducible quantitative results are to be obtained. However inhomogeneities of regional wash-out rate persist even when the rates themselves are altered. A qualitative demonstration of regional lung function can readily be obtained during the final (wash-out) phase of a xenon ventilation study. This information is not available with aerosols. This is a sensitive and specific method of detecting regional obstructive airways disease. Quantitative interpretation is of little value unless the necessary controls are exercised during collection of data. Although it is straightforward to demonstrate the existence of irregularities of ventilation it is very difficult to measure them accurately.

In asthma and obstructive airways disease there is a spatially matched reduction in both ventilation and perfusion, but there may not be exact quantitative matching of the severity of the reduction. This can be an important factor contributing to arterial desaturation. Reverse mismatch, that is perfusion of poorly ventilated regions with delayed wash-in and wash-out, can give rise to confusion. The wash-out phase of the ventilation study is the most sensitive test for the presence of obstructive airways disease. In up to 4% of patients with evidence of air trapping, the chest radiograph is normal (Fig. 2.5). Abnormalities are relatively uncommonly seen on lung volume images, which are however an essential precursor of wash-out. One parameter which can readily be calculated, and for which the lung volume image is required, is regional ventilation per unit volume of lung. This should be approximately uniform throughout the lung fields. In patients with cystic fibrosis (and other lung diseases) this may be used to demonstrate effectiveness of interventions such as physiotherapy.

The ratio of ventilation to perfusion can be measured only by comparing the distribution of ventilation with that of perfusion using a solution of the same radioisotope. The difference in depth and volume of lung examined, resulting from the incongruity of gamma ray energy between technetium and either xenon or krypton is too great for valid ratios to be otherwise obtained. If a solution of xenon in saline is employed 90% of the xenon is lost into the alveoli on the first pass through the lungs because of the low solubility of xenon in aqueous solutions. Subjective qualitative comparison of technetium perfusion and xenon ventilation is clinically useful for follow-up of some patients with cystic fibrosis or chronic lung disease.

^{81m}Kr can be used to demonstrate regional ventilation in adults with a normal respiratory rate. Because of the short half-life it is possible to obtain multiple projections at a low radiation dose. However, as minute volume increases regional count rate progressively reflects regional lung volume rather than ventilation and is correspondingly less a simple picture of ventilation. This is particularly a problem in young children (page 284). For this reason ^{81m}Kr is more useful for imaging than for quantitative studies.

Air trapping cannot be demonstrated by ^{81m}Kr because of the rapid rate of radioactive decay, but the high photon flux permits the reduced ventilation into these areas to be visualized more readily than with a single breath of ^{133}Xe. For most purposes ^{133}Xe is the preferred agent for the study of obstructive airways disease, particularly because the wash-out phase is more sensitive than the wash-in phase. SPECT is not practical with gaseous agents because of the rapid changes which can occur in their distribution compared to the duration of a SPECT acquisition.

Pulmonary hypertension

Three perfusion patterns have been described:

1. large asymmetrical perfusion defects suggestive of previous pulmonary embolism
2. multiple, small, ill-defined defects thought, in view of associated abnormalities observed on pulmonary angiography, to be due to occlusion of smaller arterial branches
3. a normal perfusion pattern.

Detection of new pulmonary emboli is thus difficult unless a previous study is available for comparison.

Further reading:
Journal of Nuclear Medicine 1994; ***35****: 793–6*

Muco-ciliary function

Aerosols containing either insoluble particles such as colloid, macroaggregate or carbon particles or large molecular weight soluble tracers such as albumin are cleared slowly from alveoli and therefore leave a high background for the perfusion study. Non-diffusible tracers reaching alveoli are removed slowly by macrophages. Clearance of deposited particles from airways is the result of muco-ciliary action; cilia are not found in alveoli. Muco-ciliary function can be studied using a non-absorbable aerosol with particles large enough to deposit in larger airways. If particle size is too large there is a greater deposition in the oro-pharynx or on the carina. Wash-out may be measured in sequential planar or SPECT images. By measuring the ratio of central to peripheral clearance in consecutive images over an hour or so it is possible to obtain an assessment of muco-ciliary function. The rate of wash-out from the central region, corresponding to the larger airways, should be compared with that peripherally to distinguish muco-ciliary clearance from absorption. If an insoluble or non-absorbable preparation is used there should be no diffusion, but in practice there may be some elution of tracer. In chronic lung disease the wash-out pattern may be irregular or discontinuous due to stasis, reflux of mucus and coughing. This may be the principal mechanism of clearance in subjects with chronic airways disease, which is associated with substantial impairment of ciliary function. There is no reliable technique of achieving maximal central deposition as turbulence of air flow is an important but uncontrollable variable.

Further reading:

Journal of Nuclear Medicine 1995; ***36****: 1355–62*

Lung permeability

Alveolar permeability

The literature is confusing, as this term has been employed to describe two unrelated phenomena, permeability of alveolar endothelium to influx of small solutes such as DTPA or permeability of capillary vascular endothelium to efflux of plasma proteins from lung capillaries. The former may be estimated by the rate of wash-out of an aerosol containing a soluble, diffusible tracer from the peripheral (alveolar) region of lung. The rate of absorption from alveoli is inversely proportional to molecular size and is therefore faster for the smaller molecular weight pertechnetate (163 daltons) than for Tc-DTPA (493 daltons).

The rate of absorption is slower in subjects over 50 years than in younger adults. Absorbed DTPA behaves like intravenously administered DTPA: that is, it is distributed in the extracellular volume and excreted by the kidneys. Technetium DTPA is most commonly employed, although there do not appear to have been any systematic comparisons between it and pertechnetate. Quantitative studies may require correction for rising extra-pulmonary background resulting from absorbed activity. If two exponentials can be fitted to a 30 min wash-out curve, the lung is likely to be abnormal with two populations of alveoli.

A global measure of the rate of absorption may also be made by measuring renal clearance, as renal clearance of DTPA is much faster than absorption from lung, so that absorption is the rate-limiting step which determines clearance rate. The mean normal clearance rate of Tc-DTPA is approximately 1% per minute in non-smokers, with an upper limit of 2% per minute. The normal mean in smokers is over 3% per minute.

Clinical applications

Disease increases permeability and thus increases the rate of absorption, which is consequently faster in smokers than in non-smokers, in workers exposed to silica dust and in *Pneumocystis* pneumonia. Increased permeability precedes radiological abnormalities. It has been suggested that this test is useful in early diagnosis of *Pneumocystis carinii* pneumonia in patients with AIDS. It is probably a non-specific finding, the apparent specificity being a consequence of prevalence in the group studied. There is overlap between the wash-out rates of smokers and of patients with pneumocystis. Unless baseline values for an individual are available the test is thus of limited value in smokers.

Further reading:

European Journal of Nuclear Medicine 1997; ***24****: 81–7*

Capillary permeability

Pulmonary accumulation of indium-transferrin can be measured, to provide an index of lung capillary permeability to plasma proteins. Earlier workers used radio-iodinated (^{131}I) human serum albumin. There is also increased interstitial accumulation of diffusible intravenous tracers such as Tc-DTPA. Transferrin is a large molecular weight β_1-globulin which forms stable complexes with a number of trivalent metals, including indium. These normally remain intravascular but leak into the extravascular space when capillary endothelium is damaged. ^{113m}In has been used hitherto but is no longer available. The biological properties of ^{111}In are essentially identical, although the radiation dose is substantially higher. To correct for changes in blood volume, technetium-labelled red cells are administered simultaneously. At each time point, the ratio of count rate over the lung to that over the heart is calculated for both isotopes and the ratio of the two ratios (indium ratio to technetium ratio) plotted. The slope is a measure of the rate of extravascular accumulation of plasma proteins. This is increased in adult respiratory distress syndrome and other conditions associated with interstitial oedema, for example following drainage of a pleural effusion or pneumothorax.

Further reading:

European Journal of Nuclear Medicine 1997; ***24:*** *449–61*

Trauma

Following closed chest injury, whether from blunt trauma or blast, perfusion defects precede development of overt radiological changes. These may appear segmental and are easily confused with the defects of pulmonary embolism, which however usually occurs later in the post-traumatic period. Fat embolism also occurs in the early post-traumatic period and results in small non-segmental perfusion defects or a mottled appearance on perfusion imaging. Inhalation of smoke or toxic gases initially produces oedema of the airways mucosa and alveolar air trapping, usually with little or no clinical signs. Two to seven days later progressive airways obstruction becomes associated with areas of collapse and the development of infection. Xenon wash-out demonstrates air trapping at an early stage, before radiographic changes are evident. A perfusion study rarely adds useful information and is not usually necessary. Inhalation injury, especially from hot gases, usually gives a pattern of uneven air trapping, despite access to the entire lung by the noxious agent.

Aspiration

Aspiration of a foreign body may cause bronchial obstruction. Although usually readily evident if chest radiographs are taken in inspiration and expiration, ventilation scintigraphy, especially during the wash-out phase, may provide definitive evidence of airways obstruction when radiographs are equivocal.

Radiation

The characteristic change of radiation pneumonitis is a greater depression of perfusion than of ventilation in the irradiated area. Radiation fields rarely correspond to broncho-pulmonary segments. There is usually little difficulty in distinguishing the ventilation-perfusion imbalance of radiation pneumonitis from that of pulmonary embolism (Fig 2.11). Capillary permeability is increased in the irradiated area, which therefore shows increased accumulation of intravenously administered diffusible tracers such as Tc-DTPA.

Further reading:

European Journal of Nuclear Medicine 1993; ***20:*** *515–46*

Cryptogenic fibrosing alveolitis

A wide variety of appearances may be observed including an asymmetrical distribution of ventilation, delayed wash-out and mismatches of ventilation and perfusion. Those latter may be mild or severe, may mimic the mismatch of pulmonary embolism, or the reversed mismatch of obstructive airways disease. Assessment of asymmetry of lung function may be particularly useful when evaluating candidates for lung transplantation.

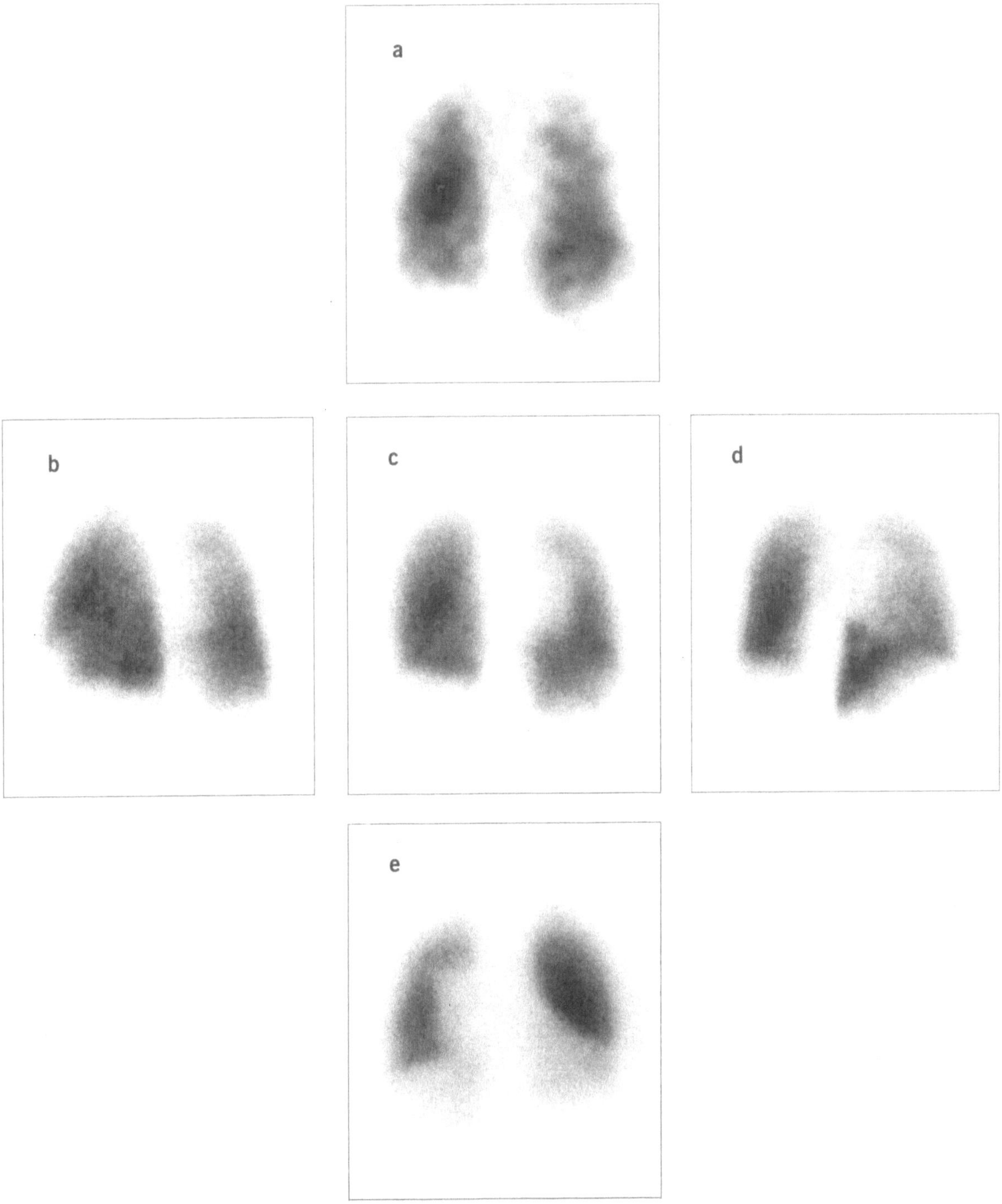

Fig. 2.11 (a) First breath (ventilation, posterior), (b) left posterior oblique, (c) posterior, (d) right posterior oblique, (e) anterior perfusion. The rectangular perfusion defect with no equivalent ventilation abnormality is due to previous radiotherapy for Hodgkin's disease. Characteristically, radiation fields are geometrical and do not correspond to anatomical boundaries.

3 Urinary tract

Techniques employing radioisotopes form the basis of most clinical measurements of differential function in the renal tract. In many situations Nuclear Medicine also offers the most sensitive practical method of assessing global function. Processes which influence function can be assessed separately; these include renal blood flow, glomerular filtration and proximal tubular function or mass. In addition neuromuscular activity of the renal pelvis and ureter, vesico-ureteric reflux, bladder emptying and urinary residual volume can be evaluated.

Radiopharmaceuticals

Tracers can be categorized by the aspect of renal function they demonstrate and are most conveniently considered under four categories. These have differing although to some extent overlapping roles and comprise:

- substances excreted by tubular secretion
- those excreted by glomerular filtration
- those bound by cells, principally in the proximal tubules
- non-extracted substances used for measuring blood flow.

Tubular secretion

Only proximal tubular function can be measured. The clinically important radio-labelled compounds employed are technetium mercapto-acetyl-triglycine (MAG3) and sodium ortho-iodohippurate (Hippuran, OIH). OIH labelled with ^{131}I was one of the first radioactive compounds used for investigation of renal function and until recently retained a pre-eminent place. MAG3 has now largely replaced OIH in clinical use because MAG3 combines lower radiation dose with useful quantification and better quality images. OIH is also available in limited quantities labelled with ^{123}I. This is more expensive than MAG3 and has few if any advantages. Absorbed radiation dose is an important consideration. In the normal subject this is low with all of these compounds because effective half-life (Appendix 2) is principally determined by the rapidity of excretion. In renal failure dose to the kidney from ^{131}I may be increased more than 500-fold compared with the normal subject. With all compounds excreted in urine, absorbed dose to the bladder is increased in subjects with urinary retention but under com-parable conditions technetium compounds are always associated with an appreciably lower absorbed dose than ^{131}I-labelled equivalents. A number of other technetium-labelled compounds secreted by the renal tubules have been described. Some have an higher extraction efficiency than MAG3 but none is generally available.

Both MAG3 and OIH are secreted into the lumen by cells lining proximal tubules. This is an active process requiring expenditure of energy. The clearance rate of all secreted tubular labels is reduced by metabolic inhibitors such as deoxyglucose, by probenecid and if the plasma concentration is so high that the tubular transport mechanism is saturated and the maximum rate of tubular transport (Tm) is exceeded. Under any of these circumstances clearance is reduced and may fall to a level similar to glomerular filtration rate. Tubular transport can be saturated by pre-treatment with relatively large amounts of unlabelled OIH, MAG3 or PAH (sodium para-amino-hippurate, the reference non-radioactive compound used for measuring tubular secretion), as all share a common pathway. Saturation does not occur under normal clinical circumstances because the quantity of tracer administered is several orders of magnitude less than the limit which the kidneys can handle.

The rates of urinary excretion of both MAG3 and OIH are similar, even though the fraction of MAG3 activity present in renal arterial blood extracted on each passage through the kidney is lower than of OIH. The single-pass extraction efficiency of MAG3 is a little over 50% whereas that of OIH is about 65%. There is some inter-subject variation. The apparent anomaly is explained by protein binding of MAG3, which retains most of it in plasma whereas OIH diffuses both into the extravascular space and red cells, giving it a larger 'volume of distribution' and hence a lower plasma concentration. The ratio of OIH to MAG3 extraction does not remain constant as renal function deteriorates. There is a little hepatic excretion of both MAG3 and OIH. This rarely give rise to diagnostic difficulty, even in renal failure, provided that renal excretion curves are interpreted in conjunction with appropriate images. Some patients in renal failure continue to extract MAG3 and OIH from blood but are unable to excrete it into the tubular lumen. This gives a rising curve with a persistent and increasing nephrogram in the absence of obstruction.

Further reading:
*Journal of Nuclear Medicine 1995; **36**: 603–6.*

Glomerular filtration

Numerous substances are excreted by glomerular filtration, many of which can be labelled with convenient radioisotopes. The general requirements are a molecule which carries no net charge, is associated with little or weak protein binding and has a molecular weight of less than approximately 600 daltons. Many metal chelates fall into this category, as do most radiological contrast media and a number of other substances including the polysaccharide inulin (the original reference compound) and vitamin B_{12}. When measurement of glomerular filtration rate (GFR) is combined with imaging, ^{99m}Tc-DTPA (diethylene-triamine-pentaacetic acid) is most commonly used. ^{111}In-DTPA is also suitable, especially as a second tracer when simultaneous dual isotope measurements are required. It is more expensive than Tc-DTPA and its two gamma ray energies require a medium energy collimator with corresponding loss of sensitivity for any given resolution. Both chromium (^{51}Cr-EDTA) and ^{125}I-sodium iothalamate are widely employed for measurement of total GFR by blood clearance but not for imaging or single kidney measurements.

In contrast to secreted agents, the single-pass extraction efficiency for all filtered agents is approximately 20%. There are small systematic differences between the various agents, due principally to differences in extent to which they are able to diffuse out of the vascular 'compartment'. Under most circumstances these are not of sufficient magnitude to be clinically significant. ^{99m}Tc-DTPA is usually employed for imaging and can be used for quantitative studies. Its principal disadvantage is that although some commercial formulations consistently give labelling yields of 98% or greater, the labelling efficiency of others is variable. Some preparations cannot reliably achieve an adequate radio-pharmaceutical purity and are therefore unsuitable for measurement of GFR, although they can be used for qualitative imaging. Radio-pharmaceutical quality must be assayed in each preparation before it is used for quantitative tests.

Further reading:
*European Journal of Nuclear Medicine 1992; **19**: 30–35*

Fixed tubular labels

A number of substances, including toxic heavy metals such as mercury and cadmium are fixed in the kidneys, most commonly by cells of the proximal tubules. A number have been used for renal scintigraphy. The compound in this group currently most widely employed as a diagnostic agent is technetium dimercapto-succinate (DMSA). Technetium monomercapto-succinate (MMSA) has similar properties but is not generally available. Uptake of DMSA by proximal tubules is blocked by pre-treatment with large doses of unlabelled PAH or DMSA but not by probenecid and may be related to pathways of trace metal conservation. Blood clearance is relatively slow. DMSA is preferably administered by clean intravenous injection as absorption from an extravasation is poor and erratic. There is somewhat better absorption from the intramuscular site. Renal uptake does not reach a plateau for some hours after injection and the background level continues to fall. The highest contrast between kidney and background occurs at 15 h. Imaging is possible for at least 24 h, limited only by radioactive decay. Late imaging is sometimes helpful to distinguish bound tracer from excreted activity retained in the renal pelvis.

Tc-glucoheptonate (GH) behaves somewhat differently. It is partially excreted by filtration and partially bound by a mechanism distinct from that responsible for DMSA uptake, as uptake is not blocked by PAH or DMSA but is reduced by probenecid. Because of the incomplete excretion it is not suitable for GFR measurement, whereas interpretation of images or measurement of renal uptake is complicated by the difficulty in distinguishing excreted from fixed activity in the renal areas. A delay of at least 4 h between injection and imaging is required for clearance of excreted activity. As an agent for imaging renal morphology Tc-GH is inferior to Tc-DMSA, although it may give different information. No separate clinical role for this has been established.

Further reading:
Seminars in Nuclear Medicine 1982; ***12****: 330–44.*

Blood flow

Renal perfusion (as distinct from 'effective renal plasma flow', page 102) can be calculated by measuring uptake of microspheres injected into the left atrium or aortic root, wash-in or wash-out of an inert tracer such as xenon or single pass of an intravascular tracer such as labelled red cells. The theory and methodology are similar to measurement of blood flow in other organs (page 221). No clinical role has been established for renal blood flow measurements.

Further reading:
Nuclear Medicine Annual 1994 Raven Press, New York pp. 251–284

Choice of radiopharmaceutical

^{99m}Tc-MAG3 is the agent of choice for gamma camera renography (page 84) and indirect reflux studies (page 104) in patients with unimpaired renal function because it is excreted so quickly. ^{99m}Tc-DTPA must be used for measurement of glomerular filtration and is preferable for renography in some patients in renal failure (page 108). In most patients higher contrast when compared with adjacent non-renal structures is achieved with MAG3 than Tc-DTPA. Because of the fast excretion and high extraction 15 MBq ^{99m}Tc-MAG3 is adequate in most adults unless time-condensed images of ureteric function or complex mathematical manipulations of the curves are required, when 50–100 MBq may be needed. A dose of 150–200 MBq ^{99m}Tc-DTPA is necessary for imaging, especially in renal failure, because of the lower contrast. As the absorbed dose per unit of activity is similar for both compounds there is a useful saving in absorbed radiation dose when using MAG3. 10–20 MBq ^{99m}Tc-DTPA is sufficient if global GFR is to be measured using blood clearance only, without imaging.

For imaging renal cortex, to obtain a more accurate estimate of single kidney function or to measure absolute uptake, 75 MBq ^{99m}Tc-DMSA is given by intravenous injection. Imaging should be performed not less than 90 min after injection, to allow time for clearance of excreted activity. In most subjects the relative function figures obtained at 90 min are similar to those obtained at later times. In the presence of a dilated system, partial obstruction or for quantitative studies the interval may be extended to 15–18 h. Uptake does not reach a plateau for 15–18 h. This delay is therefore required for reproducible absolute uptake measurements.

Renography

Technique

The technique is similar whether employing ^{99m}Tc-MAG3, ^{99m}Tc-DTPA, ^{131}I-OIH or ^{123}I-OIH. Probe renography with a pair of collimated detectors, one over each kidney, usually supplemented by a third viewing a blood-pool region such as the heart and sometimes a fourth over the bladder, is now obsolete. It is impossible to position renal detectors so as to be certain that each is receiving counts from the whole of one kidney and none from the other. Such errors, although not common in experienced hands, render this technique too unreliable for further clinical use. Renography should therefore always be performed with a gamma camera. Few cameras have adequate sensitivity for ^{131}I, and the absorbed dose associated with it, although acceptable in subjects with normal renal function, becomes unacceptably high in ureteric obstruction and in renal failure. ^{131}I-OIH should thus no longer be used routinely because agents which give as good or better clinical information and a lower absorbed radiation dose, in particular MAG3, are now available. The cost and limited availability of ^{123}I-OIH rule it out for routine use, although it is still used in a few centres. Whichever radiopharmaceutical is chosen, it must be administered by clean intravenous injection. This can be confirmed by imaging the injection site. Extravasated activity can significantly distort renal curves.

There is no consensus on the optimal acquisition protocol, despite various dogmatic pronouncements. Most centres acquire a rapid sequence, 12–30 frames per minute for the initial 1–5 min of the study, followed by longer frame times, typically two frames per minute for up to a further 40 min. A general purpose or high sensitivity low energy collimator should be employed with technetium radiopharmaceuticals. Some centres acquire one frame per second for the first 2 min, although the evidence that this provides reliable additional information, in particular for detection of renal artery stenosis, is contentious. In normal subjects the time–activity curves from both kidneys are similar, although they are rarely identical. The count rate for both kidneys should fall to less than 50% of the peak count rate within 15 min of reaching the peak, in which case the study may be terminated. If renal function or drainage is impaired a 30–45 min study is required. It is rarely necessary to record continuously for longer than 45 min but additional single frames, if necessary up to 24 h, are often helpful to determine the level of an obstruction, to differentiate between the various causes of a prolonged rising or delayed falling phase of the renogram and to differentiate a prolonged nephrogram from a prolonged pyelogram. In very late frames, for example the following morning, excreted activity is occasionally visualized in bowel.

There is little to be gained from a matrix finer than 64 × 64, corresponding to a pixel size of 6–7 mm, unless separate cortical regions of interest are required, when a smaller pixel size is essential to minimize partial volume effects. The apparently higher resolution of smaller pixels is however partially counterbalanced by increased statistical noise as available counts are distributed amongst more pixels. The examination can be performed either with the patient supine above an under-couch camera or seated with their back against the collimator face. An anteriorly placed camera is preferable for pelvic or transplanted kidneys and is sometimes helpful in horseshoe or ectopic kidneys. The supine posture is preferable for a number of reasons. If either kidney is mobile, the two are usually more symmetrical when supine than when erect, thus reducing errors in differential function calculation. Many patients are nervous and vaso-vagal reactions associated with hypotension or syncope are not uncommon, often before any injection has been administered. Hypotension, by reducing perfusion pressure and consequently GFR, invalidates the investigation. Syncope is less likely when supine as gravitational pooling in dependent regions is reduced. Orthostatic hypotension is correspondingly less severe. A similar problem occurs in patients who have received an intravenous diuretic, which not infrequently induces hypovolaemia.

A further advantage of performing the examination supine is that there is usually less patient movement. If acquisition time is extended most patients find the supine posture less uncomfortable. Gravity plays no part in drainage of urine from the renal pelvis in the normal subject. It becomes important only when the pelvis is dilated and akinetic. The rate of drainage should thus not be affected by posture in the normal kidney. If the collecting system is dilated it is advisable to allow the patient to stand for several minutes at the end of the study and if

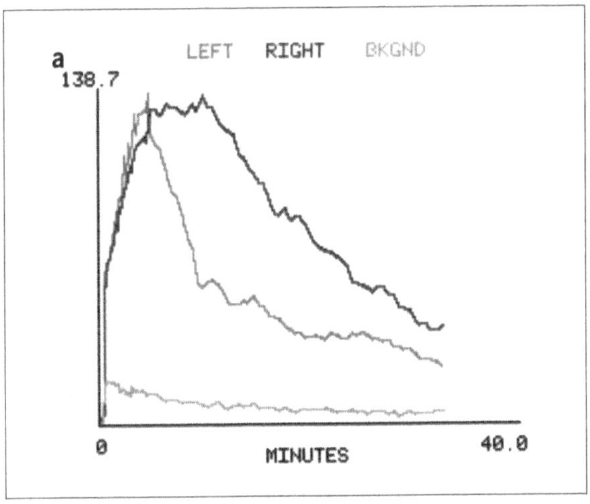

Fig. 3.1 (a) Normal unsubtracted renogram and background curves. Adult given 15 MBq ^{99m}Tc-MAG3. (b) Summed frames for initial 90 s from arrival of tracer at the kidney, demonstrating nephrographic phase during which all renal activity is in parenchyma. Note that hepatic activity, mostly in blood pool, overlies the upper pole of the right kidney to a variable extent. (c) Frame corresponding to peak of curve from left kidney. (d) Frame corresponding to peak of curve from right kidney. The concentration of activity in renal pelvis is such that parenchymal activity is difficult to distinguish. (e) At 30 min most residual activity is in renal pelvis. A faint nephrogram is usually visible.

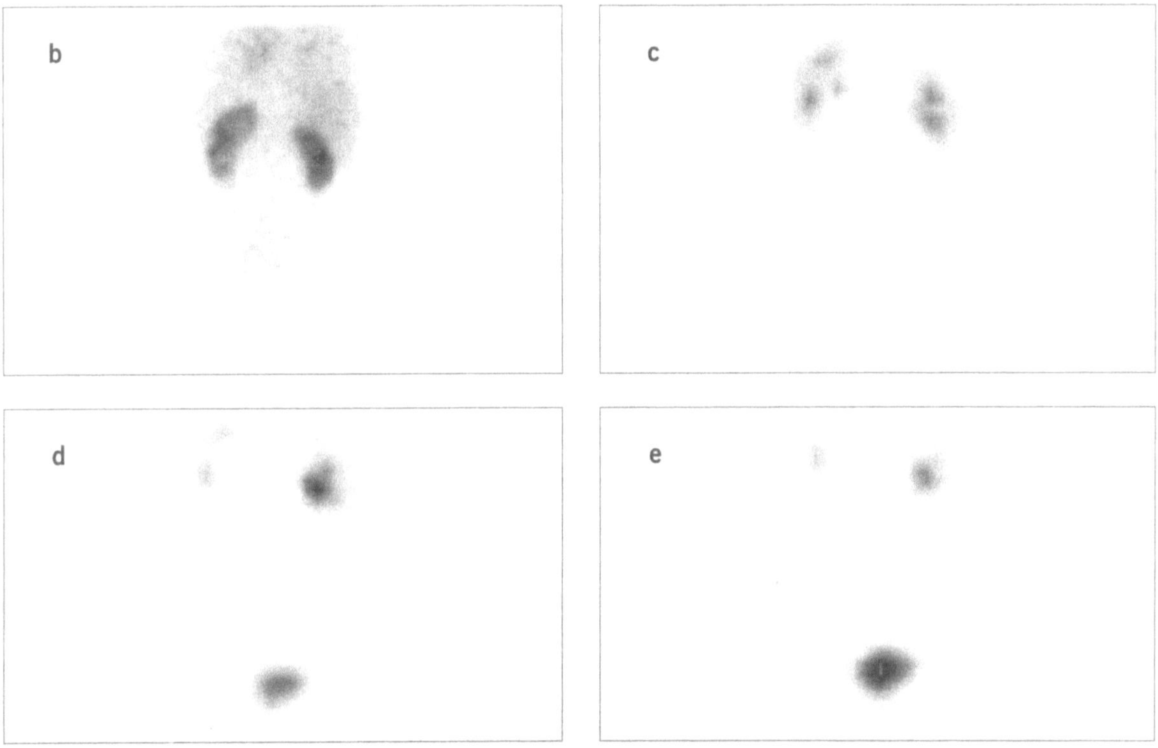

possible to void. Emptying of the collecting system in a post-voiding film excludes obstruction. If a rising curve is encountered during the study, the patient should in the first instance be given a sip of water to drink. If this fails to provoke contraction of the renal pelvis within 30 s a diuretic may be given. The rapidity of response to swallowing indicates that it is not due to an increased rate of urine production, which takes at least 3 min to be evident after an intravenous diuretic.

Activity

A dose of 15 MBq ^{99m}Tc-MAG3 is adequate for most purposes unless time-condensed images (page 179) of ureters or complex mathematical manipulations such as (de)convolution are required, in which case 75–100 MBq should be given. A dose of 150–250 MBq ^{99m}Tc-DTPA is necessary. There is no data to support the common contention that 100–150 MBq ^{99m}Tc-MAG3 is necessary in other situations.

Data analysis and interpretation

Time–activity curves (Fig. 3.1a) should be obtained from regions of interest (ROI) enclosing each kidney and renal pelvis and from a vascular non-renal area, for example between the kidneys to include aorta and inferior vena cava. If the field of view of the camera is large enough a region of interest drawn over heart may be employed. Cortical regions excluding renal pelvis may be preferable to whole kidney regions but are difficult to define when cortex is thin. They require a finer matrix for acquisition. The function of the vascular region is primarily to assess quality of the injected bolus and if deconvolution analysis is to be employed (see below) and not for 'background correction' (page 88). It is sometimes useful to have a time–activity curve from the bladder. It is rarely helpful to view all the frames in the study, although any anomalies in the curves should be evaluated by checking appropriate frames to exclude movement or other artefacts. Physiological causes of discontinuities are common, especially erratic mass contractions of the renal pelvis and uretero-pelvic or vesico-pelvic reflux. As a minimum, images corresponding to the initial 90–120 s of the study (Fig. 3.1b) to identify anomalies in the nephrogram, peak(s) of renal time-activity curves (Figs 3.1c, d) to see the general size and configuration of the calyces and pelvis and the final frames of the examination (Fig. 3.1e) should be inspected to determine where residual activity is localized. If the kidneys peak at different times, images should be viewed at both.

The normal renogram curve is affected by contributions from many sources in addition to renal extraction and renal excretion. These include blood flow, blood volume, diffusion of tracer in both directions between plasma and an ill-defined and non-homogeneous 'extravascular space', hepatic extraction and excretion, the size of the renal pelvis and its neuromuscular activity. So far it has defied attempts at comprehensive mathematical analysis. The curve initially rises steeply to a maximum, usually within 5 min when using MAG3 or OIH and slightly later with DTPA. The magnitude and time of maximum count-rate is largely determined by renal function but is also appreciably influenced by volume of

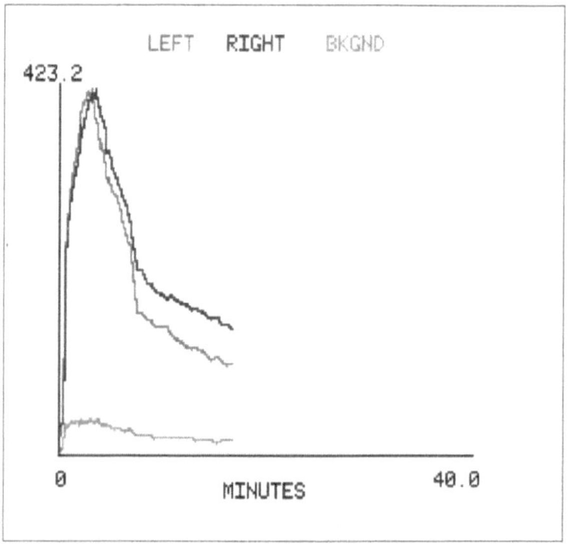

Fig. 3.2 Same subject as Fig. 3.1a pretreated 15 min before MAG3 injection with 20 mg frusemide and oral hydration.

the collecting system. The peak represents the instant at which rate of tracer extraction by kidney from blood equals the rate of loss down ureter. Peak height is significantly influenced by volume of the renal pelvis. If the same output flows into a bigger bag it takes longer before starting to overflow. When two kidneys with renal pelves of different sizes have similar urine output, the peak occurs later and is higher in the kidney with the larger pelvis. Increasing urine flow decreases any difference (Fig. 3.2). The peak is followed by a non-linear descending phase, which falls from the peak to less than 50% of the maximum within 15 min, the rate of fall being slower with DTPA than with MAG3. The rate of accumulation, estimated either by the slope during the initial 120 s from arrival of tracer at the kidney or by the total amount of tracer accumulated in each kidney during this time, is determined mainly by renal blood flow and extraction efficiency. Although fractional clearance is constant once mixing is complete, rate of uptake falls as the concentration of tracer remaining in the circulation decreases. In the initial frames tracer is in renal parenchyma, whereas at later times more of the residual activity is in the collecting system. Some activity

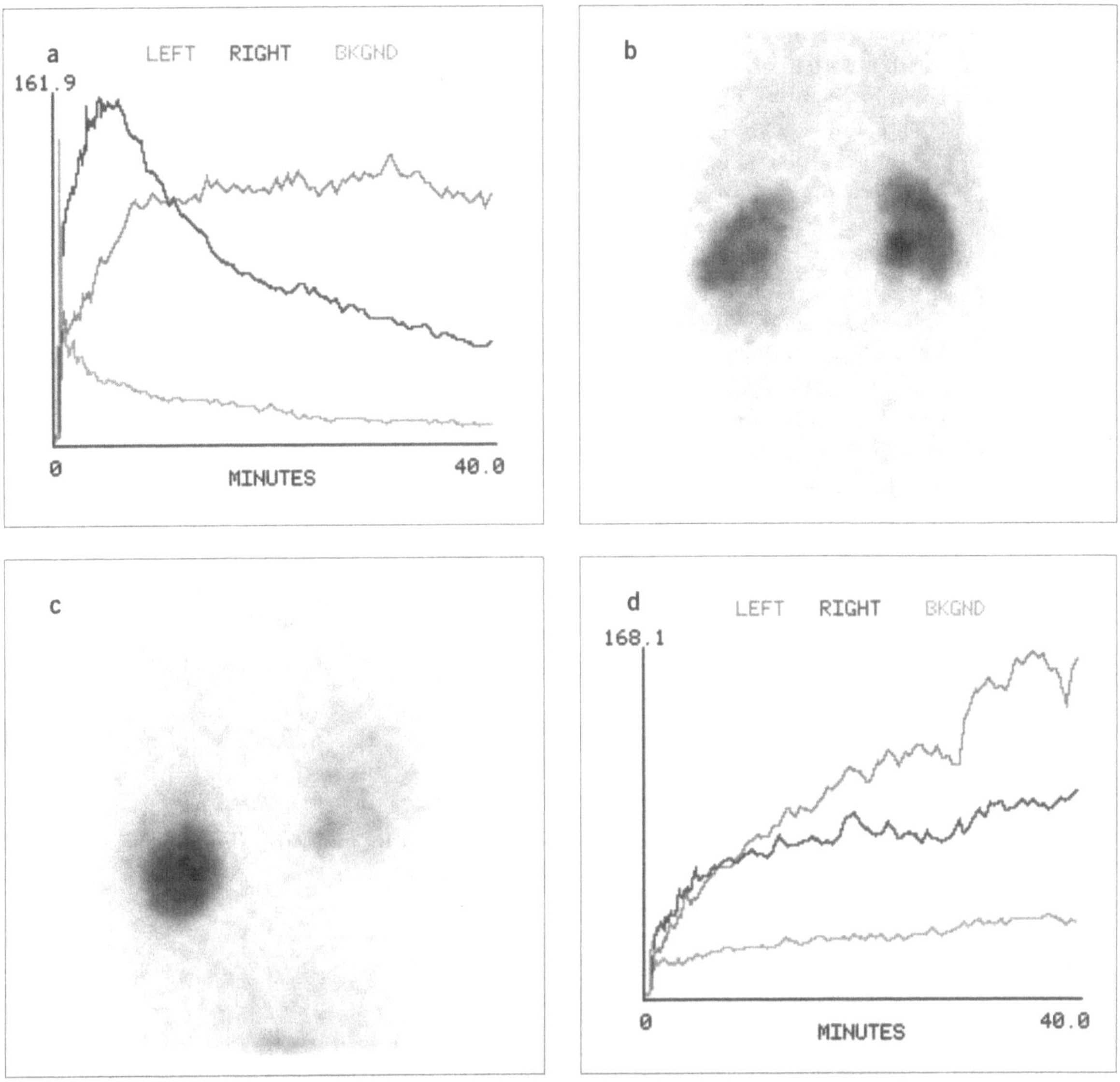

Fig. 3.3 (a) Normal right kidney but flat curve on left despite pretreatment with diuretic and oral hydration. (b) Frame corresponding to peak of right renal curve. (c) Final minute showing activity in distended left extra-renal pelvis. The flat curve is due to this and not to obstruction. (d) The same subject the previous day following an extravasated injection. Both renal curves rise throughout the examination, as does the background curve as tracer is absorbed from the extravasation. This could be misinterpreted as obstruction with impaired function if the background curve is not inspected in every patient. A small partial extravasation is more difficult to recognize but can also significantly distort the curves.

remains visible in cortex throughout the examination. Substantial persistence of parenchymal activity, equivalent to the persistent nephrogram of excretion urography, may be due to obstruction, hypertension, hypotension, acute tubular necrosis or chronic renal failure. Films at 2 h and later, sometimes up to 24 h, may be required to determine whether residual radioactivity is in parenchyma or pelvis (Fig. 3.3a–c) and if the latter the level of any obstruction.

It is usual also to obtain a time–activity curve from a vascular non-renal area adjacent to the kidneys synchronously with the renogram curves. This serves several functions. This curve should reach its maximum within the first minute and fall steadily thereafter. A flat or rising vascular curve indicates either renal failure or, more commonly, a partially extravasated injection. This can distort renogram curves (Fig. 3.3d), in extreme cases giving a rising curve suggestive of obstruction in a normal kidney. As MAG3 is relatively painless when extravasated a partial extravasation is easily overlooked unless the background curve is always examined in conjunction with the renogram. Any extravasated radiopharmaceutical is absorbed from the injection site more slowly than after intravenous injection and at an unpredictable rate; it is subsequently excreted by the kidneys. A large extravasation is easily identified, even if the injection site is not within the field of view, by the bilaterally delayed peaks of the renal curves and slowly rising background count rate, in contrast to the normal falling background curve. A small partial extravasation is not always easily recognizable. A flat or rising background curve should always raise the suspicion of a partial extravasation.

The largest error is estimation of the contribution to detected counts from non-renal 'background'. It is often stated that subtraction of a background curve from the renogram gives a net renal curve representing only renal function. Although true in principle, there are in practice many pitfalls as the time–activity curve of any 'background' area is unlikely to be identical to that of non-renal counts detected within renal regions of interest. Depending on distance between kidney and background regions a substantial but variable fraction of counts detected in any peri- or juxta-renal region is scatter arising from the kidney, while a significant fraction of counts detected in renal areas are scatter originating from non-renal tissues. It is arguable whether the proper model for assessing accuracy of renal background regions of interest is the contralateral side to a solitary kidney, when renal bed is occupied by other vascular structures such as gut which do not necessarily have a similar blood volume, or whether a unilateral hydronephrosis (Fig. 3.4, but see also Fig. 3.21) is a better model. Renal regions may be freehand, rectangular or elliptical. Separate cortical and pelvic regions may be drawn either manually or automatically by comparing early and late frames. 'Good' background areas are ones where by chance the errors cancel. A short crescent adjacent to the infero-lateral aspect of each kidney is commonly chosen, but this underestimates the contribution to background from overlying liver in the crucial early frames. Alternatively a crescentic region almost encircling the kidney but excluding ureter may be used. Interpolative background subtraction from an elliptical circumferential region is also sometimes employed to calculate relative function from the initial nephrogram before any activity is present in ureter, but cannot be used once tracer has entered the ureter. A number of more complex mathematical approaches have also been proposed but have not gained widespread acceptance. All methods of background correction are relatively crude approximations. Unless applied with extreme caution and a clear understanding of possible sources of error, background correction can as easily introduce inaccuracy as reveal occult information. In general 'background subtraction' should be employed with circumspection.

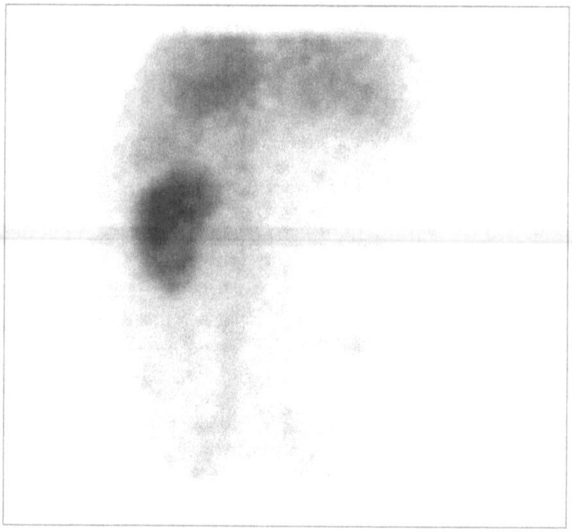

Fig. 3.4 Initial (nephrogram) phase in subject with right hydronephrosis. The count rate around the hydronephrosis is higher than that in the region corresponding to kidney. Any 'background' region would over-subtract.

Further reading:
Journal of Nuclear Medicine 1990; ***31****: 1710–16*
Nuclear Medicine Communications 1994; ***15****: 636–42*

Diuretics

The administration of a diuretic to promote maximal urine flow is useful in certain patients, particularly to establish a diagnosis of pelvi-ureteric junction obstruction (PUJO) and to distinguish between obstruction and stasis in a dilated collecting system and/or ureter. There is little to choose between available intravenous diuretics. Bumetanide (1 mg) has a shorter duration of action than frusemide (20 mg), but the difference is marginal and in practice it is reasonable to use whichever is available locally at lower cost. In the normal subject, curves obtained with and without a diuretic are similar, apart from a slightly steeper down slope in the former case. Two techniques have been employed. Some confusion has been engendered by a failure to appreciate that they give different information and have different indications.

A diuretic is essential if pelvi-ureteric junction obstruction (page 106) is suspected, as this diagnosis can be confirmed only at high urine flow rates. If the differential diagnosis includes pelvi-ureteric junction obstruction the patient should be orally hydrated with 400–500 ml of water and given a diuretic (20 mg frusemide or 1 mg bumetanide intravenously) 15 min before starting the gamma camera renogram, to ensure maximal diuresis at the time of study. A higher dose (40 mg frusemide) is often advocated but is associated with an higher incidence of symptomatic hypovolaemia. The rate of administration should not exceed 4 mg/min. The patient must be well hydrated to achieve maximum flow-rate of urine and should preferably remain supine for the duration of the examination to eliminate the risk of orthostatic hypotension. Although urine flow is increased substantially even in dehydrated subjects by diuretics such as frusemide or bumetanide, the effect is less than can be achieved with adequate hydration. It is not usually necessary to resort to intravenous hydration, but sufficient time must be allowed for orally administered fluids to be absorbed and the diuretic to take effect. In hot weather a larger volume of oral fluids may be required. The patient must be advised to maintain a fluid intake greater than 0.5 l per hour for 3–4 h, until the diuretic has worn off.

The alternative procedure, if a flat or rising curve is encountered more than 15 min after radiopharmaceutical injection, is to administer diuretic during data acquisition. For this to be practical the software should be capable of displaying renogram curves while the study is in progress. Many commercially available nuclear medicine computer systems are deficient in this respect. The response normally commences 3 min after administration of diuretic. A 50% fall in count rate in the affected kidney within 15 min of diuretic injection, associated with a concave curve (Fig. 3.5) excludes obstruction with an high level of confidence. The principal advantage gained clinically by this manoeuvre is reduction in the number of equivocal examinations, especially those with non-specific or indeterminate flat curves.

Further reading:
British Journal of Urology 1992; ***69****: 113–20*
Urological Radiology 1992; ***14****: 79–84*

A prompt fall in amount of activity in renal pelvis can be elicited in many subjects by swallowing a liquid or solid bolus, which provokes propagated peristalsis (Fig. 3.6), as well as by threat. This reflex indicates that the collecting system is capable of emptying and is therefore not obstructed. Stimulation of the swallowing reflex should always be attempted before resorting to a diuretic.

Further reading:
European Journal of Nuclear Medicine 1994; ***21****: 521–4*

Precautions

Administration of intravenous diuretics to patients taking angiotensin-converting enzyme (ACE) inhibitors can precipitate severe hypotension. Pre-existing electrolyte disturbances, including those caused by some anti-malarials, are associated with an increased risk of ventricular

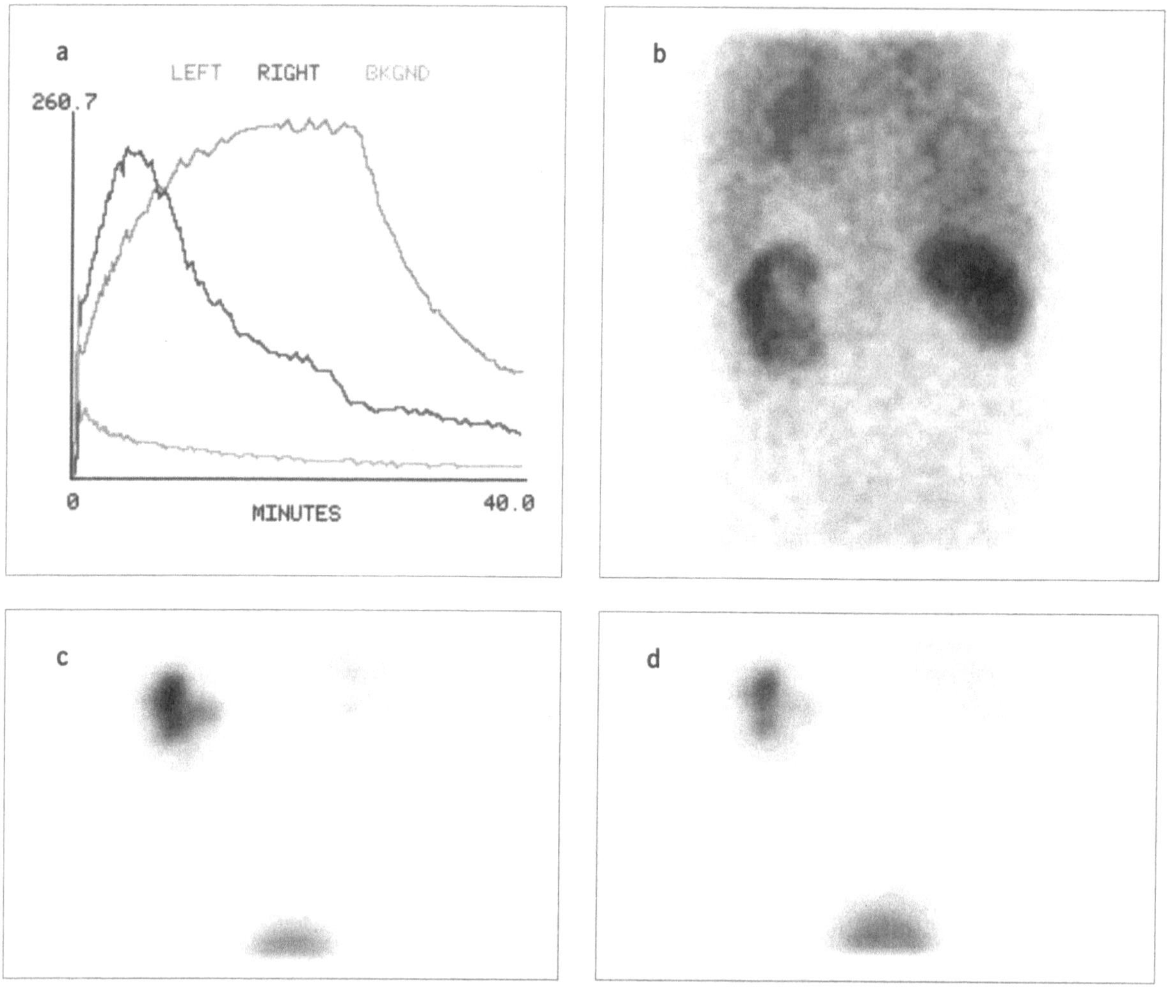

Fig. 3.5 Response of non-obstructed left kidney with good function to diuretic. The fall in count rate starts 3 min after injection. A response is also visible from the right kidney even though it is draining. (a) Initial 90 s, (b) just before administration of diuretic, (c) 7 min later.

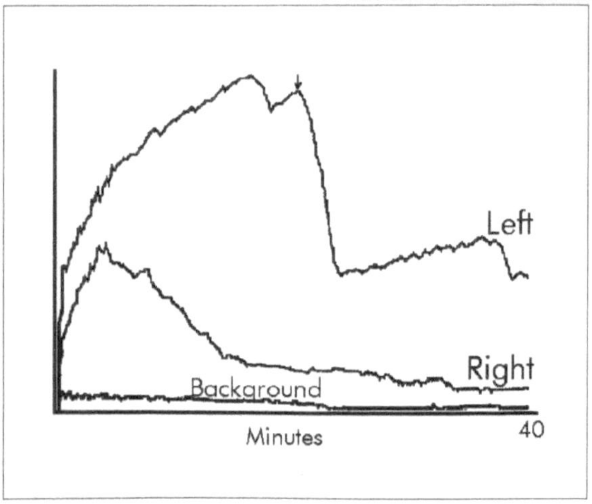

Fig. 3.6 Response to swallowing. The initial dip is due to a contraction of the renal pelvis which failed to clear the ureter, with rapid reflux of its contents back into the pelvis. The second swallow (arrow) provoked a propagated peristaltic wave which emptied the ureter. The third dip was spontaneous.

arrythmias. Acute anaphylactic reactions, although rare, are described. The possibility of cross-reactivity with sulphonamides has been raised. Hypovolaemia leading to symptomatic hypotension is likely unless an adequate fluid intake is maintained until effects of the diuretic have worn off.

Interpretation of abnormal curves

A rising curve which falls in response to swallowing a sip of water (Fig. 3.6) indicates a clinically insignificant disturbance in initiation of peristaltic contraction in the renal pelvis, without obstruction. A rising curve which does not respond to swallowing, but which starts to fall within 3 min following a diuretic and which decreases to half or less within 20 min, particularly if the descending phase is concave, indicates a non-obstructed kidney with an hypokinetic or akinetic pelvis (Fig. 3.5). A rising curve which does not respond to diuretic suggests obstruction, although a poorly functioning kidney with a grossly dilated collecting system may fail to respond. Other causes of a persistently rising curve, or one which remains on a plateau, include renal artery stenosis, hypotension, low output into a dilated collecting system and an extravasated injection. The latter is usually bilateral. Obstruction should not be diagnosed on the basis of a rising curve alone. Other evidence should be sought, including a prolonged nephrogram, reduced function and a collecting system which retains activity to the level of the obstruction (page 106).

When renal function is substantially impaired a kidney may no longer be capable of increasing its output in response to a diuretic sufficiently to flush out activity which has accumulated in the dilated system. Raised blood urea acts as an osmotic diuretic and the kidney may not be capable of further increasing its output. Thus failure by a poorly functioning kidney with a slowly rising curve to respond to diuretic is not necessarily evidence of obstruction, although a response does exclude obstruction. The use of a diuretic in this way reduces the number of indeterminate flat curves, especially in patients with a dilated collecting system.

Single kidney function

There are a number of methods for measuring the contribution from each kidney to overall function. These estimate total amount of tracer extracted separately by each kidney before any has escaped down the ureter, compare initial slopes during the uptake phase, employ the principle of 'stop flow' to measure increase in count-rate following application of sufficient pressure to the abdomen temporarily to obstruct both ureters, or measure uptake at equilibrium of a fixed tracer such as DMSA. Accuracy of all methods is limited by difficulties in distinguishing between perfusion and extraction, determination of counts which originate not from kidney but from adjacent soft tissues (background correction) and allowance for differences in renal depth. Stop-flow is little used because it is difficult to sustain complete occlusion of both ureters for an adequate period by abdominal compression.

In dynamic studies the maximum interval over which split function can be calculated without abdominal compression is limited by the minimum time for a molecule to pass from renal artery to ureter. This may be longer than 90 s, especially in obstruction or renal failure, but there is no reliable and simple method of determining the duration in any individual patient. If an appreciably longer time is selected the error leads to under-estimation of the contribution from the kidney with the smaller renal pelvis because more excreted activity is likely to have escaped from the field of view. Split function may be calculated from summed renogram frames or from time–activity curves by comparing net counts in each renal area for 90 s or 120 s after arrival of activity at the kidney, during which time none can have left. Ratio of slopes of uptake curves can also be used but is subject to greater error than ratio of cumulative counts over the same interval. Although when renal function is normal non-renal ('background') counts contribute a relatively small percentage of the total, their contribution becomes progressively larger as renal function deteriorates and the number of counts originating from kidney decreases. The problem is compounded at early times, when there is substantial vascular activity in both the liver and spleen; these organs overlap the kidney by an inconstant and unpredictable amount, which varies from one patient to another and alters with posture. Background correction, for example by using an elliptical circumferential region with Goris interpolative subtraction (page 140) is therefore essential for calculation of split

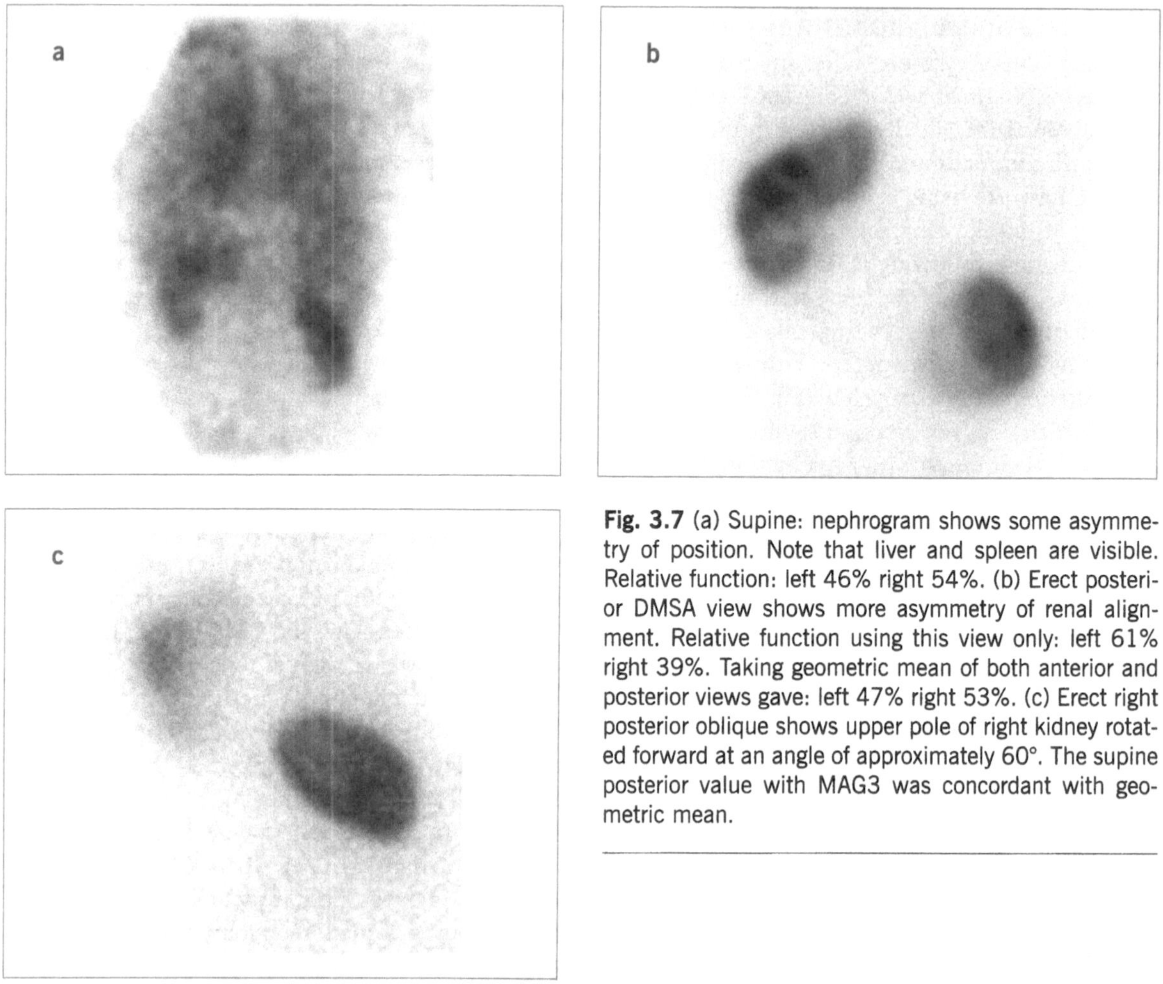

Fig. 3.7 (a) Supine: nephrogram shows some asymmetry of position. Note that liver and spleen are visible. Relative function: left 46% right 54%. (b) Erect posterior DMSA view shows more asymmetry of renal alignment. Relative function using this view only: left 61% right 39%. Taking geometric mean of both anterior and posterior views gave: left 47% right 53%. (c) Erect right posterior oblique shows upper pole of right kidney rotated forward at an angle of approximately 60°. The supine posterior value with MAG3 was concordant with geometric mean.

renal function. A crescentic region just lateral to the lower pole of each kidney does not take the varying hepatic activity into account.

Estimates of split renal function obtained during the renogram in a single projection, usually posterior, are reproducible but not accurate. Depth correction is required for accurate relative function measurement as kidneys are often asymmetrically situated. This is not always obvious on visual inspection of posterior frames. As gravity plays no part in drainage of urine from the normal renal pelvis this does not affect the rate of drainage. In some subjects one or both kidneys are mobile and move caudally on change of posture. A kidney does not descend with a simple vertical movement. It rotates about the lower pole, so that the upper pole comes to lie more ventrally when erect than in the supine position (Fig. 3.7). In most subjects errors due to difference in renal depth can be reduced by performing the examination supine to minimize asymmetry. Dilatation of the renal pelvis is usually associated with ventral displacement of the renal parenchyma which is not posture dependent (Fig. 3.8) When using a technetium compound, either can lead to a serious under-estimate of the contribution that kidney is making to overall function because the thickness of soft tissue required to halve detected count rate from any given quantity of technetium is less than 5 cm. When a kidney rotates ventrally or is displaced by a dilated pelvis its mean distance from the posterior wall and thus from the gamma camera or detector may increase by more than this amount, resulting in a substantial reduction in observed count rate without any change in radioisotope content of the kidney.

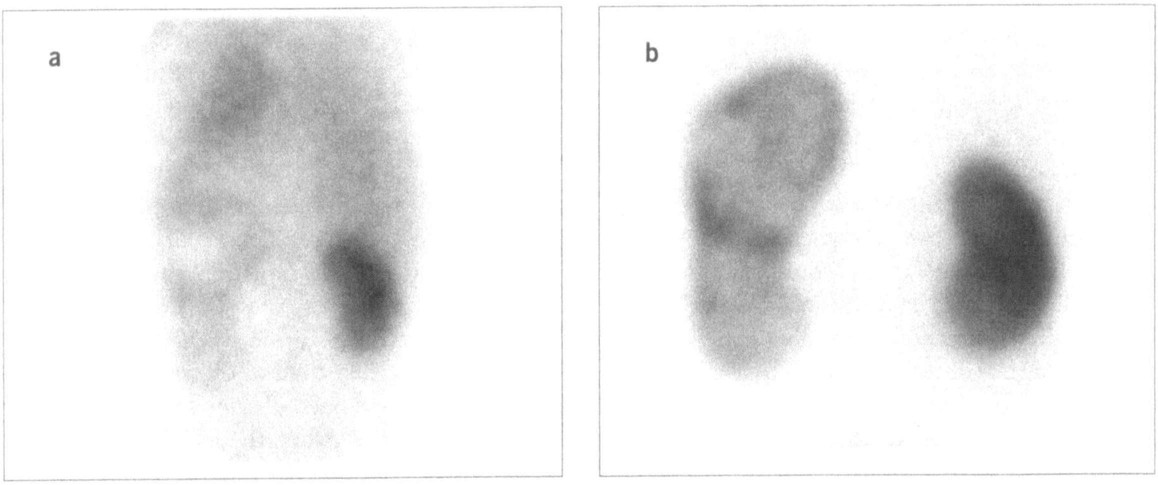

Fig. 3.8 (a) MAG3 nephrogram in patient with distended left intra-renal pelvis. Relative function (MAG3 or posterior DMSA): left 27% right 73%. (b) Posterior DMSA image; geometric mean of anterior and posterior gave relative function of: left 45% right 55%. Function of the left kidney is seriously underestimated if posterior projection only is considered.

The geometrical mean (GM) of count rate in two opposed projections (square root of the product of background-subtracted count rate of each kidney in each view) is independent of depth but during the initial phase of a renogram the right kidney is usually obscured in the anterior projection by blood pool in liver. Less satisfactory alternatives such as measuring mean depth of kidney in a lateral projection or ultrasound must therefore be employed. In normal subjects background is proportionately smaller for DMSA than for excreted compounds and hepatic uptake does not overlie kidney in the anterior projection. DMSA is thus usually more accurate, except in renal failure when background is high (Fig. 3.9). Both methods give similar

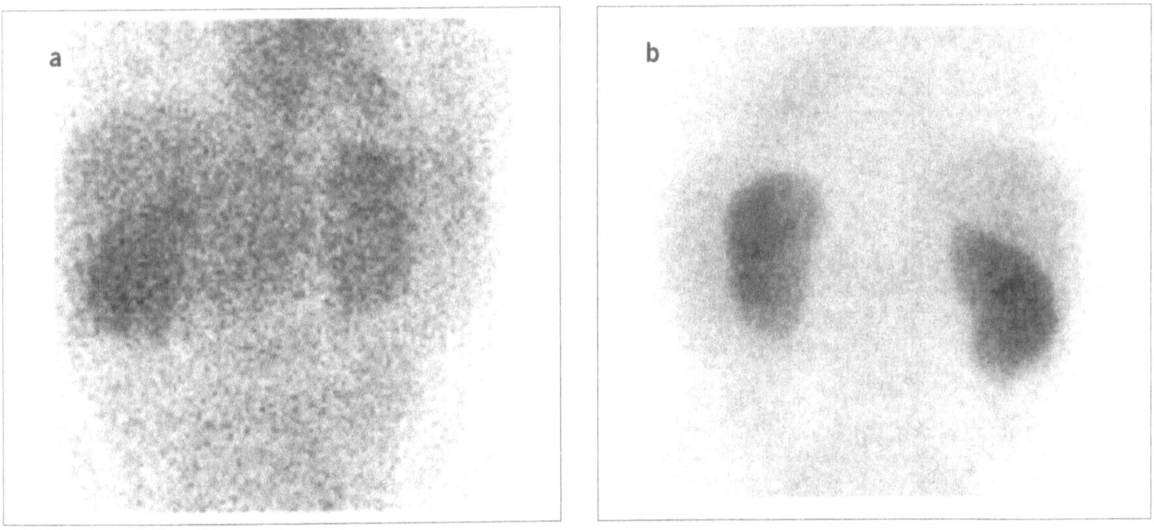

Fig. 3.9 DMSA anterior (a) and posterior (b) projections in patient with moderately impaired renal function. Note high extra-renal uptake, especially in liver, spleen and marrow.

results under most circumstances, although DMSA can over-estimate contribution of an obstructed kidney if uptake is measured earlier than 15 h after injection, due to variable retention of excreted activity. On the other hand if function is depressed background correction tends to be over-estimated when MAG3, OIH or DTPA are used, with consequent under-estimation of the contribution of the poorer kidney. This method can also be used to determine absolute uptake, by comparing calculated GM count rate with that from a phantom containing a known fraction of the administered activity immersed in a water-bath of the same thickness as the patient, or by making a calculated attenuation correction. Absolute uptake of DMSA correlates well with GFR determined by blood clearance.

Some of these limitations can be overcome if stop-flow is employed with an excreted agent, although this is not common practice. Ureters can be occluded by external pressure on the abdomen, as is usual during excretion urography. The technique depends on the observation that clearance is unaltered for the first 20 min after occlusion of a ureter. The increase in net renal count-rate over a timed interval, usually 10 min, is proportional to clearance of tracer. Thus measurement is not limited to 2 min and may be made 10–15 min into the examination, when vascular pool in liver and spleen make a relatively small contribution to detected counts.

Deconvolution

There have been numerous attempts at mathematical analysis of renogram curves to obtain objective numerical information, the most widely advocated approach being (de)convolution of renal time–activity curves with a blood pool curve, with the objective of estimating the curve which might have been obtained had the original injection been made directly into the renal artery rather than a peripheral vein, in proportion to renal blood flow and with no recirculation (Fig. 3.10). The mean transit time of tracer through the kidney can then be calculated. The nuclear medicine literature usually refers to deconvolution, although mathematically the term convolution is commonly employed irrespective of the direction of the process. This provides a less subjective means of comparing differences in the same patient than unaided visual analysis. Claims that separate transit times can be calculated for cortex and pelvis should be treated with reserve because of overlap and cross-talk from activity in either area into the other. Obstruction is associated with more prolonged transit times than other causes of a rising curve. Disadvantages of this approach include:

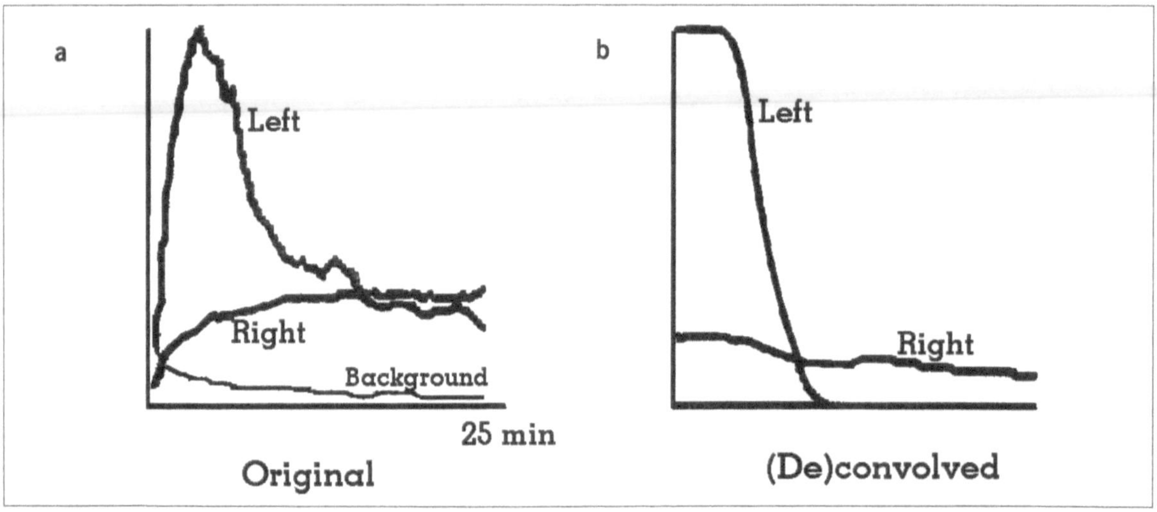

Fig. 3.10 (a) Normal left renogram with dysplastic right kidney. (b) Calculated deconvolved curves.

- absence of a single unambiguous solution to the mathematical analysis
- difficulty in identifying a pure 'cortical' region, especially when the cortex is thin
- discrepancy between observed blood pool and true renal arterial curves
- uncertainties in estimating non-renal contribution to renal and non-vascular components of the blood-pool curves
- statistical noise in the data.

All of these tend to amplify statistical noise in the raw data, which is further magnified in the course of convolution. Convolution is an appropriate technique only when low noise data is being analysed. Higher administered activities are thus required than if simpler methods of analysis are employed.

Emptying of renal pelvis is due to regular, relatively slow, peristaltic contractions of low amplitude, too small to be detected on the conventional renogram. In some patients large irregular contractions are sometimes observed as discontinuities in the renogram curve (Fig. 3.15a). Irregularities in clinical curves due to random statistical fluctuations in count-rate can be sufficiently reduced by smoothing to be discounted. Other irregularities, due to irregular mass contractions of the renal pelvis, ureter or less often to vesico-ureteric or uretero-pelvic reflux, are of clinical significance. Convolution routines requiring heavy smoothing of such data destroy real patho-physiological information, creating curves without physical or clinical meaning. The choice of blood-pool and background regions is also critical. Caution should therefore be exercised when interpreting convolution, as many factors can influence the validity of results obtained.

Convolution is advocated to distinguish recoverable from non-recoverable impairment of renal function, differential diagnosis of the dilated renal pelvis and renovascular hypertension. Although convolution does distinguish between those obstructed and those non-obstructed kidneys in whom the diagnosis was also readily obtained by visual inspection, clinical examination and history, it may not help to differentiate when diagnosis is in doubt, especially when there is low urine output into an hypotonic dilated collecting system. On balance this procedure infrequently gives useful information which cannot be obtained as reliably from simple parameters such as time to peak and rate of fall, provided that numerical indices are interpreted in conjunction with visual inspection of curves and images. Failing routinely to evaluate all available data reduces the accuracy and usefulness of renography.

Further reading:
Journal of Nuclear Medicine 1995; ***36****: 147–52*

Sources of error

All radiopharmaceuticals must be given by clean intravenous injection. The effect of an extravasation on renal curves depends on its size and rate of absorption. In some cases no effect is visible to naked eye inspection. In others there is delay in time to peak and reduction or abolition of the down slope. The rate of absorption of an extravasation of MAG3 is variable and unpredictable. In some cases absorption is sufficiently rapid to give a rising background curve. In others it has no visible effect. The rate of absorption of extravasated DTPA is less variable. It usually produces a flat or rising curve in the absence of renal pathology. Arm movement which results in projection of an extravasation over a renal area gives rise to random discontinuities in curves which are easily recognized if the frames are examined. If the extravasation is projected over the region chosen as background this is invalidated. When GFR is measured by plasma clearance, extravasation gives a falsely low value. It is not possible to make a correction and the examination must be repeated at a later date. A full bladder may be associated with a rising renal curve but the renal pelvis drains after voiding. Patient movement can cause gross distortion of the curves.

Scintigraphy

Technique

Imaging with either DMSA or GH demonstrates distribution of functioning proximal tubules. In

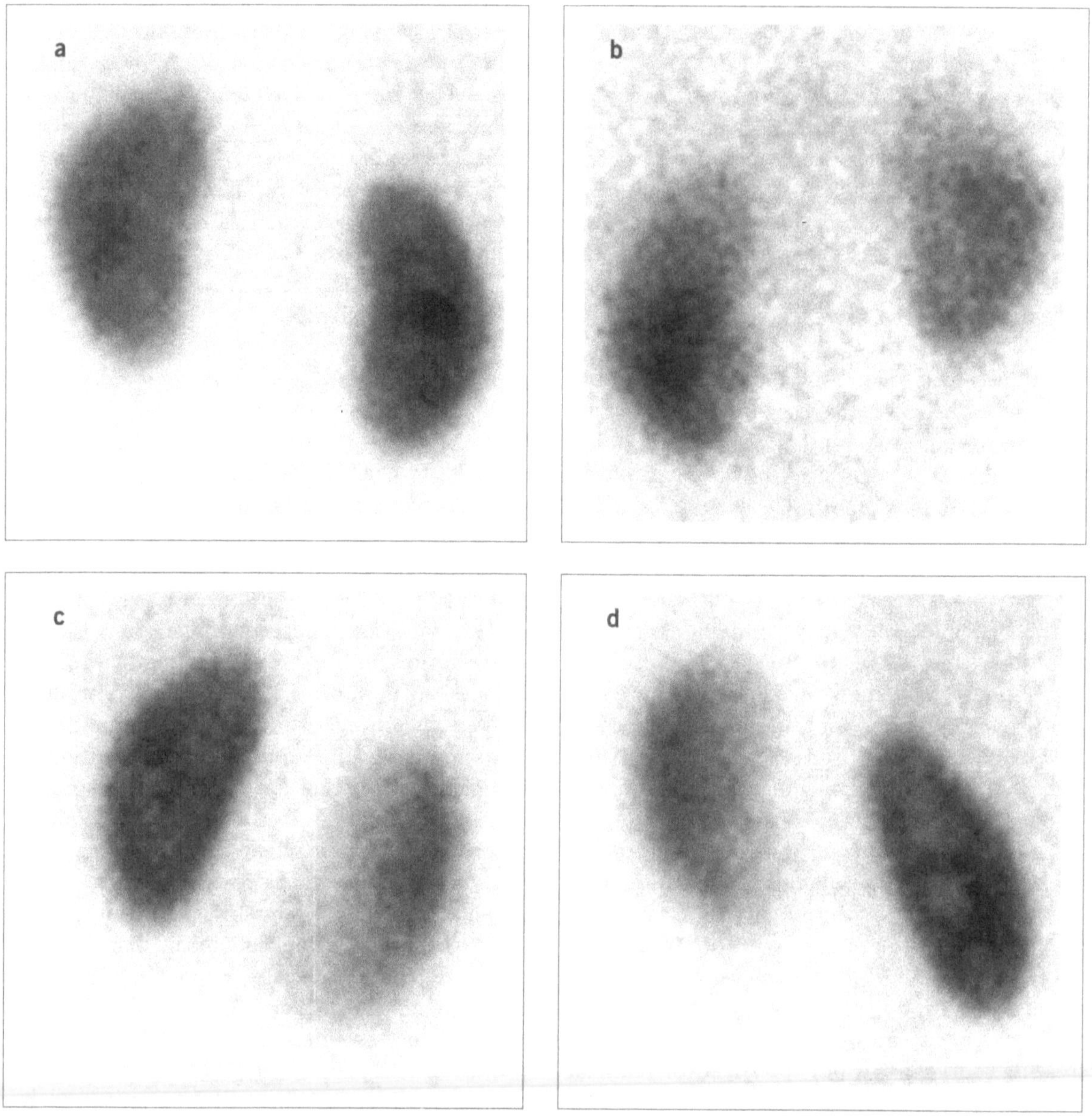

Fig. 3.11 Normal DMSA: (a) posterior, (b) anterior. (c) left posterior oblique, (d) right posterior oblique, Pyramids are most clearly seen in the oblique projections.

the normal subject uptake in cortex is uniform. The pyramids, often more easily seen in oblique than posterior projections, are identifiable as fairly symmetrical regions of lower count rate but are not totally photon-deficient, as in all projections they are overlain by some thickness of cortex (Fig. 3.11). The clarity with which they are seen is variable. They are conspicuous in some subjects and indistinguishable in others. Identifiable detail is largely determined by whether respiration is diaphragmatic or costal. When respiration is predominantly costal there is usually little movement of the kidneys. Diaphragmatic respiration is associated with considerable degradation of image quality. It may be difficult to distinguish movement blur due to respiratory motion from other, more easily eliminated, sources of unsharpness. The upper and lower poles are ill-defined when there is respiratory movement, whereas patient movement causes loss of definition of all borders. This is of considerable clinical importance, as the potential to detect small regions of cortical loss is impaired by movement.

Four projections should be obtained, anterior, posterior and both posterior obliques, using a gamma camera equipped with a general purpose low energy collimator. In the case of ectopic (Fig. 3.12a–c) or horseshoe (Fig. 3.12e, f) kidneys additional anterior obliques may be required. Data should be presented as good quality monochrome images in which the entire range of counts is visible with no over-range of pixels. Relative function should be calculated using the geometric mean of background-subtracted counts in posterior and anterior views (page 93).

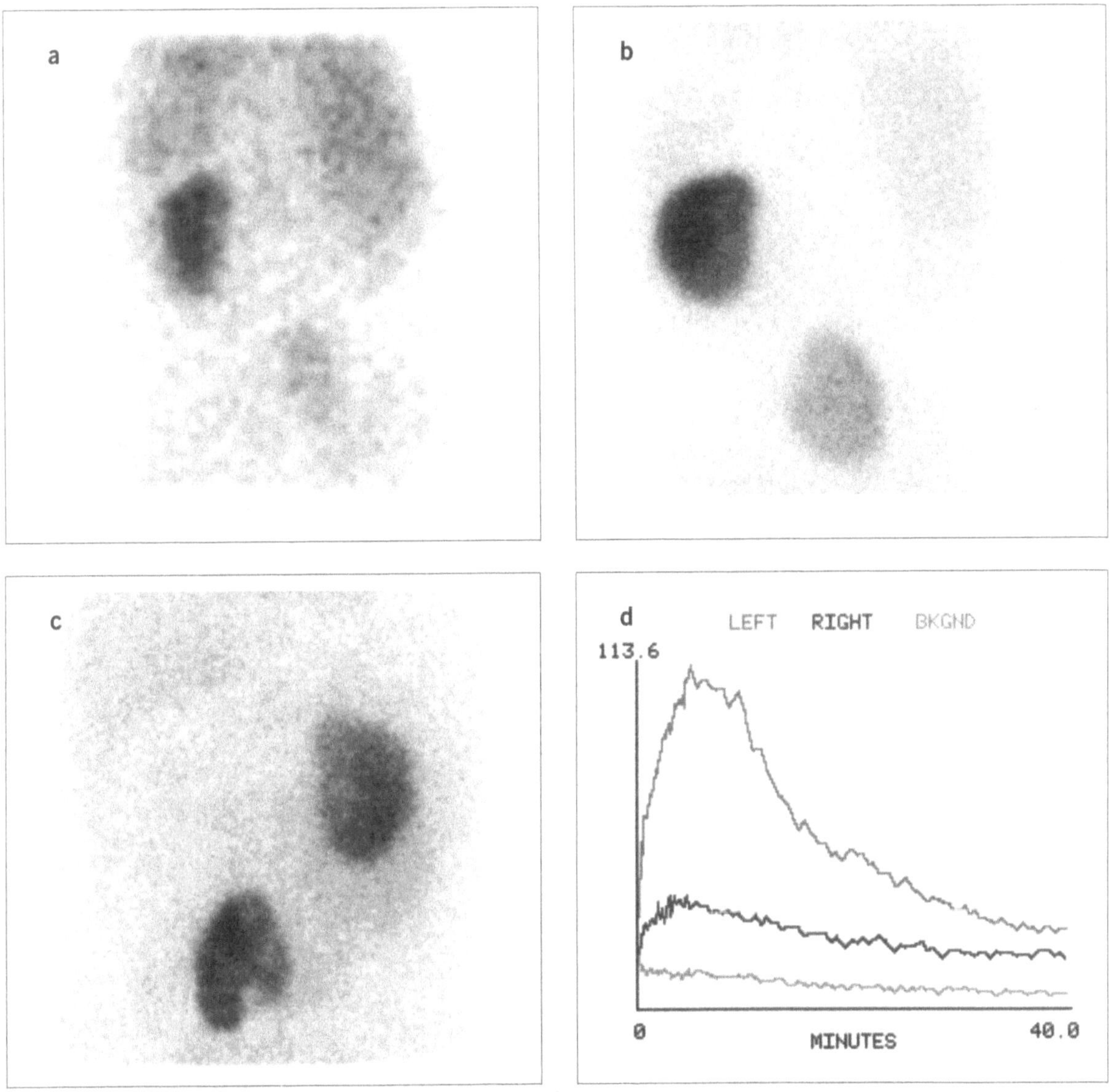

Fig. 3.12 (a) Posterior MAG3 nephrogram, (b) posterior DMSA, (c) anterior DMSA of patient with right pelvic kidney; diagnosed by ultrasound as 'absent right kidney'! (d) Renogram curves underestimate contribution of pelvic kidney. Relative function, posterior MAG3: left 86% right 14%. Geometric mean DMSA: left 58% right 42%. (e) Horseshoe kidney, anterior, (f) posterior. Reproducibility of relative function is poor as distinction between moieties is arbitrary. The anterior projection (e) is essential to evaluate a horseshoe kidney. (**e,f** *overleaf*)

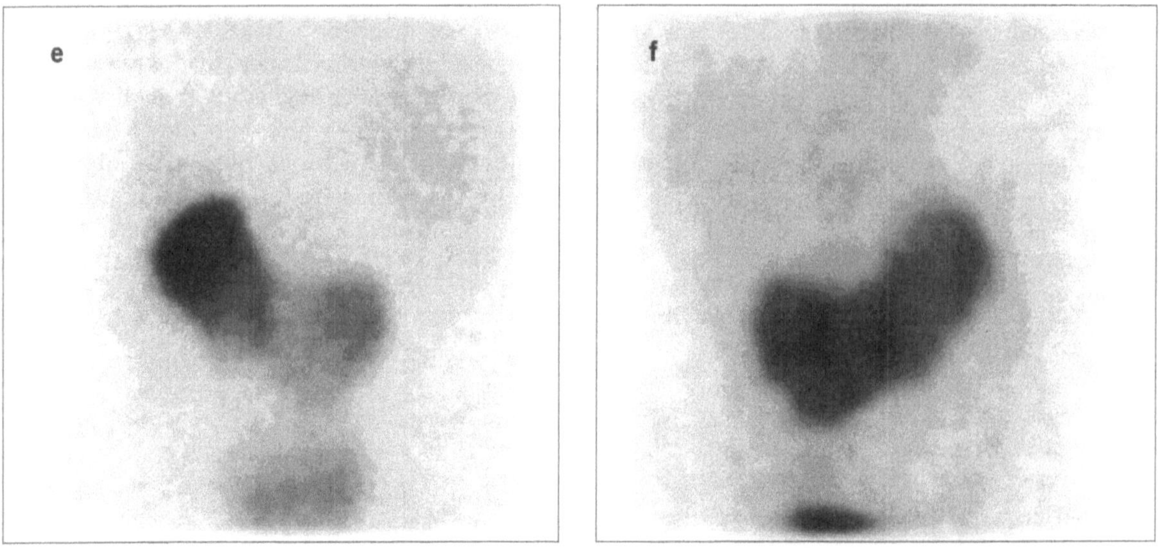

Fig. 3.12 *continued*

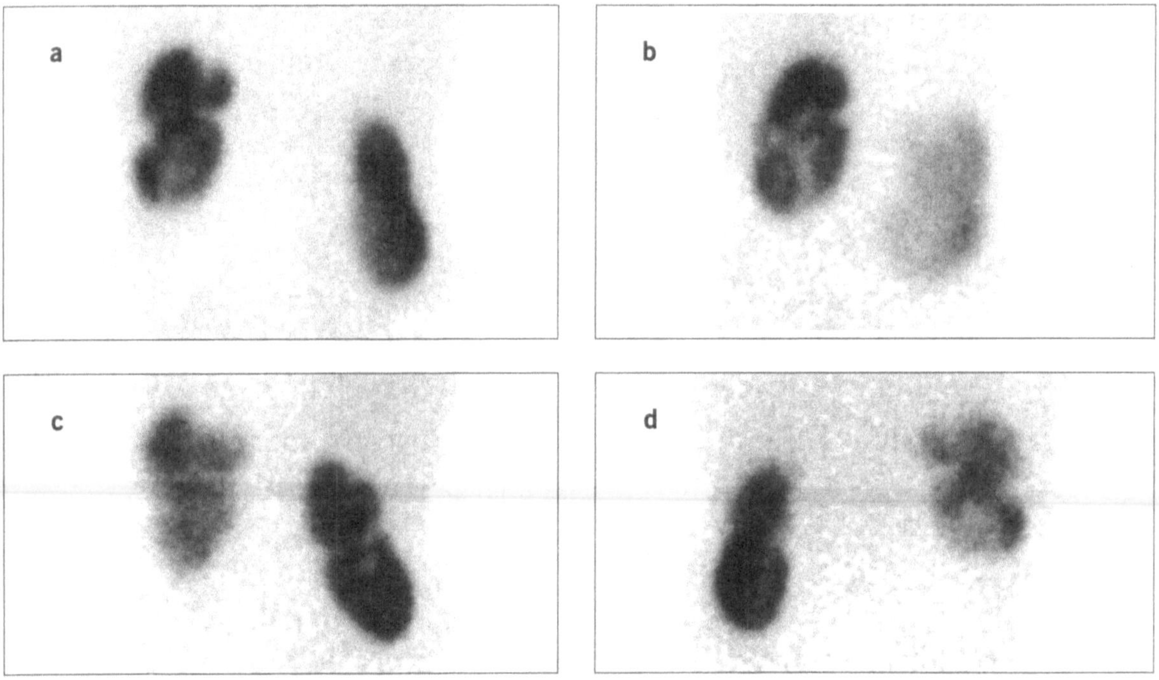

Fig. 3.13 (a) Posterior, (b, c) both oblique, (d) anterior projections of DMSA study in patient with reflux nephropathy and scarring of both kidneys.

Interpretation

In the normal subject photon deficiencies correspond to renal pyramids, but the surrounding cortex should be complete. Abnormalities appear as absolute or relative reduced uptake or as focal or segmental photon-poor defects. Defects which breach the integrity of the cortex surrounding pyramids may be reversible if associated with a recent episode of pyelonephritis or may be due to scarring (Fig. 3.13). An irregular outline can also be due to persistent foetal lobulation, but may be differentiated from scarring as cortex outlining pyramids remains intact. Especially at the poles, pathological defects are commonly wedge-shaped and may be seen more readily in oblique projections. Rounded defects, especially in the mid-part, may be cysts (Fig. 3.14). This must be confirmed by ultrasound. Concentric thinning of cortex over one pole is less common than segmental defects but is difficult to detect scintigraphically and may be better seen by ultrasound. A substantial difference in size between kidneys is also abnormal although it is not possible to distinguish long-standing from recent causes.

Acute obstruction can cause global reduced uptake of all tracers in the affected kidney (Fig. 3.15) which may however be enlarged, although with GH and to a lesser extent DMSA this can be

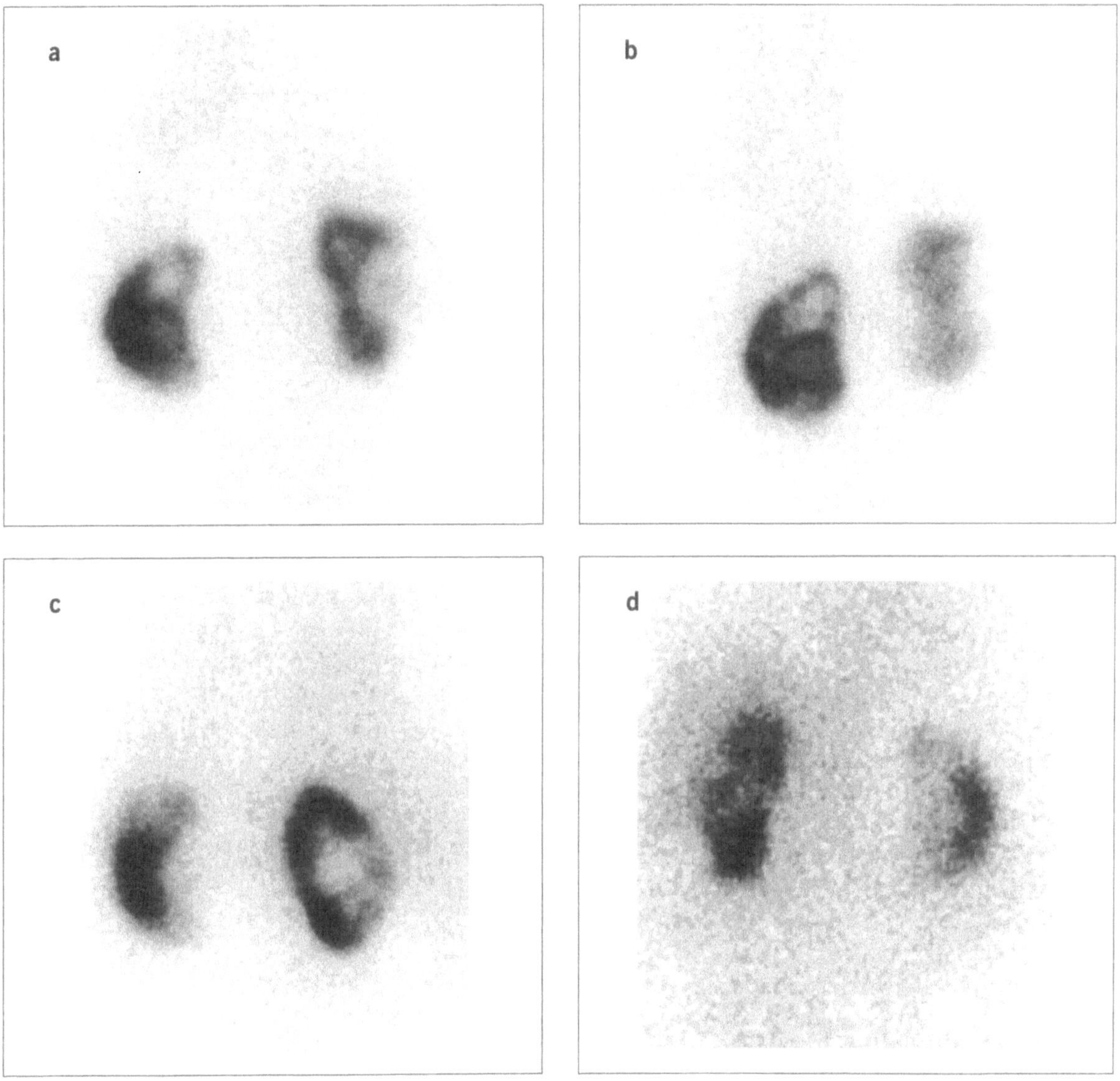

Fig. 3.14 (a) Posterior, (b, c) both oblique, (d) anterior projections of DMSA study in patient with polycystic kidneys.

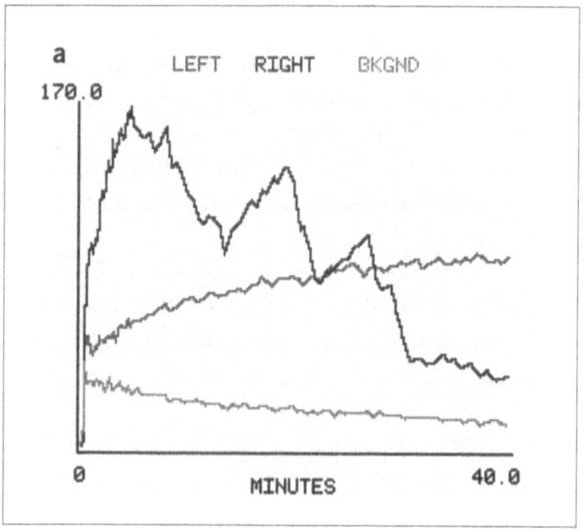

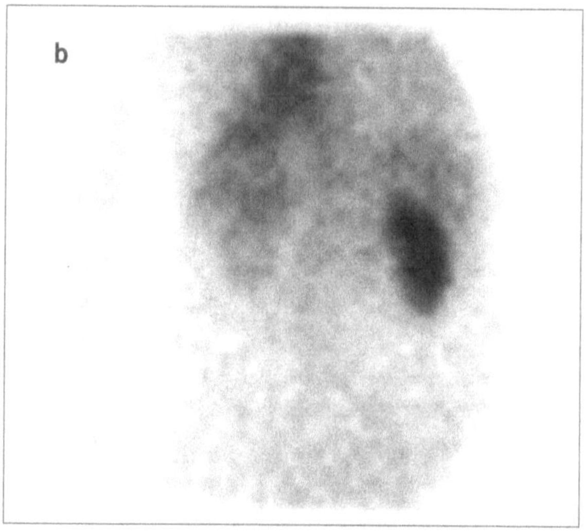

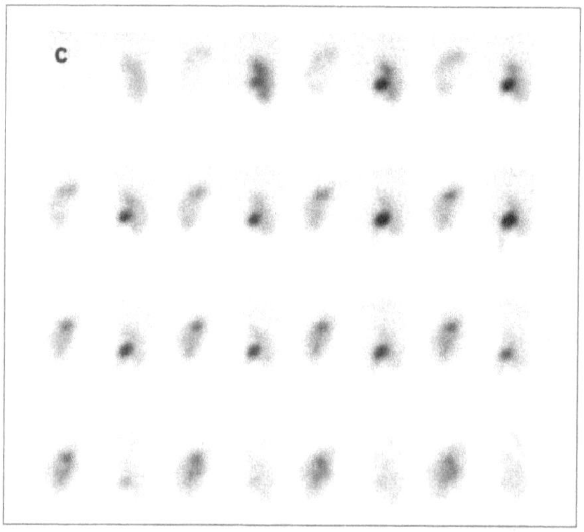

Fig. 3.15 Recurrent renal colic with passage of calculi: (a) severely depressed left renal function; current episode with left ureteric calculus impacted for 72 h. (b) Nephrogram, relative function: left 16% right 84%. The left improved to 35% after relief of the obstruction. (c) Note also abnormal but asymptomatic pattern of emptying of right renal pelvis.

partially masked by retained activity in the collecting system unless images are repeated 15–18 h after injection. Low uptake relative to the size of the kidney (the 'pale' kidney) indicates depression of function which may be reversible, but there is no reliable criterion to determine with certainty at a single examination which kidneys are likely to improve. Absence of any improvement within a week or so of initiating treatment of the underlying disease suggests that significant recovery is unlikely. However slow recovery once started may continue for many months.

Infection more commonly causes segmental defects at one or more renal poles, less commonly in the mid-part. Imaging during or immediately following an episode of acute pyelonephritis may reveal large defects which often decrease in size (Fig. 3.16) or resolve completely on follow-up after treatment of the acute episode. It is not possible to distinguish reversible from irreversible defects on a single examination or to estimate the age of any defect. Defects seen when the patient is afebrile and asymptomatic are usually permanent, although in children they commonly become less conspicuous with time, presumably due to eccentric growth of adjacent normal tissues.

Clearance measurements

Many methods have been described for measuring renal function. The majority derive from the basic formula for clearance:

$$\text{Clearance} = \frac{\text{(concentration in urine)(urine volume per min)}}{\text{(plasma concentration)}}$$

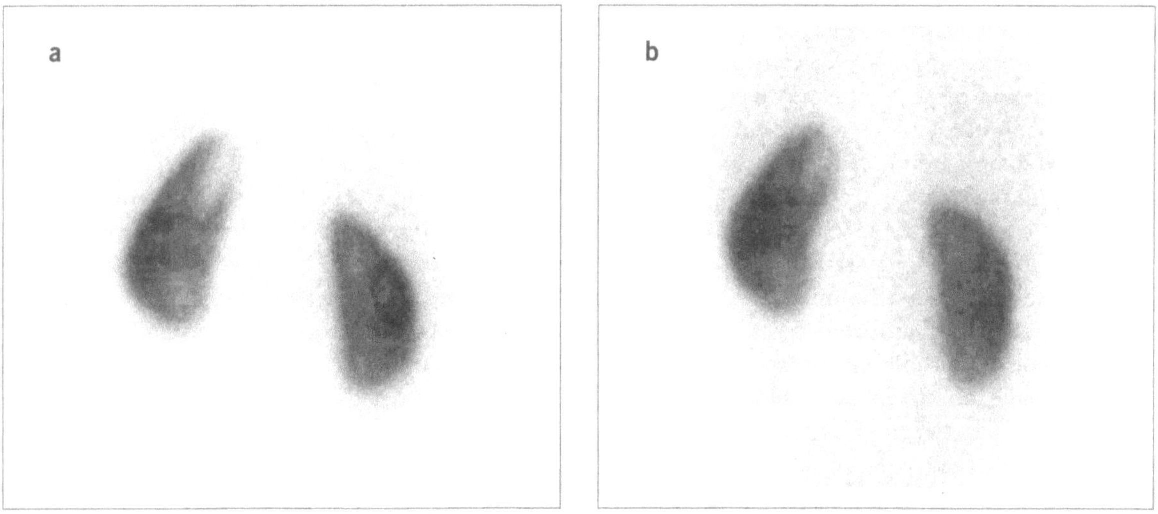

Fig. 3.16 (a) Segmental defect at left upper pole due to pyelonephritis. (b) Partial reversal three weeks later. Many defects observed during an acute episode disappear completely within a few weeks.

Clearance is thus the volume of plasma containing the quantity of tracer that is excreted per minute in urine. If the agent employed is excreted exclusively by glomerular filtration, the parameter measured is glomerular filtration rate. A substance completely cleared from the blood on a single passage measures renal blood flow. Only particulate tracers such as microspheres fulfil this requirement and even then there are reservations (page 222). The clearance of any other substance is a measure of the clearance of that substance and nothing else. It may reflect tubular function, blood flow, some other feature or any combination. It may be reproducible and clinically useful. It should however be described unambiguously.

Absolute clearance depends on body size. Clearance is therefore usually expressed normalized to a standard body surface area of 1.73 m^2 to facilitate inter-comparison. Normograms are available which allow surface area to be calculated from height and weight. At the extremes of body size, and especially in neonates, these normograms are inaccurate. Alternatives such as normalizing to extracellular volume or initial volume of distribution have therefore been proposed. Although physiologically more appropriate these have not yet achieved general acceptance.

Further reading:
European Journal of Nuclear Medicine 1991; ***18****:* *385–90*

Urinary clearance

To calculate urinary clearance both plasma and urine concentrations and urinary output must be measured (Appendix 2). The plasma concentration is maintained at a constant level, confirmed by multiple blood samples, by constant rate or feedback-controlled continuous infusion. There are a number of practical problems which render this method unsuitable for routine clinical use. It is difficult to achieve an unvarying plasma concentration of most tracers for an adequate duration, the more generally available assays of classical non-radioactive tracers such as inulin or PAH are subject to appreciable error and further significant uncertainties are introduced by residual urine in bladder either at start or end of the collection period. To achieve a complete urine collection it is necessary to ensure that the patient has an empty bladder both at the beginning and at the end of every timed interval. In practice, even in patients who are catheterized, it is extremely difficult to ensure that the bladder is completely empty. A difference in residual volume between the beginning and the end of the period can result in either under- or overestimation of clearance. To eliminate the need for the difficult and often inaccurate chemical assays of PAH and inulin and because gamma emitters can easily be measured in plasma or whole blood, appropriate radioactive tracers (page 81) are often employed for clearance measurements.

As an alternative to constant infusion, sodium iothalamate (^{125}I) can be administered in a small volume (<0.5 ml) by subcutaneous injection. The most practical method of measuring GFR by urinary clearance is to measure blood and urine concentrations between 30 min and 120 min after subcutaneous ^{125}I- or, if available ^{131}I-iothalamate, taking two or more blood samples per 30 min urine collection and using a counter over the bladder to estimate residual urine volume. Reproducibility is improved if the subject remains under controlled basal conditions throughout the study, fasting except for a specified quantity of fluid. GFR is expressed normalized to a body surface area of 1.73 m^2, the normal range being 80–140 ml/min/1.73 m^2. It is not possible to achieve a sufficiently constant blood concentration of substances excreted by the tubules following subcutaneous injection.

Some confusion has been caused by the term 'effective renal plasma flow' (ERPF) originally introduced to describe PAH clearance. This name was chosen because any compound which is completely extracted on a single pass through an organ can potentially be employed to measure blood flow through that organ. When renal function is normal, extraction efficiency of PAH is >85%, but <100%. It was postulated that the difference between actual extraction of PAH and 100% extraction was due to shunting, some blood by-passing tubules, so that PAH clearance could be equated with tubular perfusion. This assumption is now known to be incorrect and the term is therefore misleading. To compound the confusion the same expression is also used to refer to clearance of OIH, which is less than PAH clearance, whereas others apply the term indiscriminately to clearance of either substance and a fourth group to MAG3 clearance. The single-pass extraction efficiency of all these compounds is substantially less than 100%, is less than that of PAH and there is appreciable individual variation. OIH extraction efficiency is about 65% in healthy subjects, which is about 85% of the clearance of PAH. The ratio of PAH to OIH clearance does not remain constant as function deteriorates. It is thus both more accurate and less misleading to specify unambiguously which clearance has been measured, avoiding altogether the debased term 'ERPF'. There are small differences in measured clearances of the various filtered agents, so that even with GFR it is preferable to specify the tracer used.

Blood clearance

A number of simplifications have been described, some of which attempt to overcome the need for urine collection and its associated errors by using only blood samples. These in effect calculate apparent 'volume of distribution' of the injected tracer at some time after injection, or more commonly the change in volume of distribution over a measured time interval. The blood clearance curve of a substance excreted by glomerular filtration over the first 4 h after a bolus intravenous injection can be approximated by the sum of two exponential functions, the faster of which is ascribed principally to mixing of tracer with extravascular fluid and the slower to excretion. The alteration in volume of distribution or plasma half-time can be correlated with clearance (Appendix 2). This obviates the need to measure urinary output. The main sources of error arise from analysis of experimental curves into two components and in the assumption that plasma clearance is equivalent to urinary clearance. Compartmental analysis never gives an unique solution and when renal function is impaired the components may not differ greatly in their half-times. There is always some hepatic clearance of substances excreted by the kidneys, but this is small when renal function is normal. When however renal function is impaired the relative contribution of hepatic clearance is proportionately greater. Plasma clearance may thus not reflect urinary clearance accurately in renal failure, although under most other circumstances the error is negligible.

Adequate delineation of the blood clearance

curve requires eight or more blood samples spaced progressively over the 4 h following injection for GFR measurement and 1.5 h for tubular clearance, with more frequent sampling at earlier times. Typical sampling times are 2, 5, 10, 20, 30, 40 and 90 min for tubular clearance and 5, 10, 20, 35, 60, 90, 120 and 240 min for GFR. Accurate recording of actual sampling times is essential. Simplifications have been proposed which estimate slope of the slow component by taking only two relatively late blood samples, assuming that the contribution of the faster component is negligible at that time, or which calculate the volume of distribution at a single time. The accuracy of the two-sample method, taking accurately timed samples at 120 min and 180 min, is adequate for clinical measurement of GFR but does not give sufficient precision for clinically useful measurement of clearance of secreted compounds. The single sample methods are less satisfactory. Despite claims that a single sample at 44 min is optimal there is in fact no one optimal time. Although in subjects with normal renal function the error is acceptably small, it becomes larger as renal function deteriorates. In consequence simplified methods using one plasma sample are subject to confidence limits too wide to be of clinical use in many patients and have no advantage over calculated creatinine clearance. It should be noted that most papers which compare two or more of these techniques appear satisfied if a high correlation coefficient is obtained. However inspection of the data reveals substantial scatter around the line of identity. Confidence intervals for calculation of the ordinate value at any given value of the abscissa (or vice versa) are as important as correlation coefficient.

Further reading:
Journal of Nuclear Medicine 1996; ***37****: 1883–90*

Renal uptake

An alternative method, applicable to both tubular and glomerular labels, is to express amount of tracer taken up by kidneys in the initial 90–120 s as a fraction of total administered activity. This measures single kidney function using yet another parameter, renal uptake, which is related to clearance although there is some uncertainty whether the confidence limits of the correlation are narrow enough to justify replacement of clearance by uptake. It is probably true to say that, except in renal failure, uptake measurements are sufficiently reproducible for sequential studies to be clinically useful, but it is preferable to distinguish between uptake and clearance measurements. Inter-conversions should be regarded with caution. The principal sources of error in these early uptake measurements are the estimation of non-renal 'background' counts and correction for asymmetry of renal depth. Most published methods incorrectly assume a constant relationship between height, weight and renal depth, rely on a lateral projection or employ ultrasound measurements. Slight obliquity in a lateral projection can affect estimates of renal depth. If ultrasound is performed with the patient prone and scintigraphy performed with the patient erect or supine there may be considerable differences in renal orientation and depth.

Residual bladder volume

A number of techniques have been described for measuring residual bladder volume either by ultrasound or by isotopes. Ultrasound methods are accurate when residual volume is relatively large and the bladder regular in outline. Isotope methods are accurate for determining residual volume at the end of an examination but cannot determine it at the beginning when there is no radioactivity in the bladder. The count rate from full bladder is measured, the patient voids in privacy and collects urine passed. The residual count rate in the bladder region and that from the voided specimen are measured. The difference in count rate (after background correction) is correlated with the volume voided. If the concentration of activity in urine is much greater than that in surrounding soft tissues it may be preferable to omit 'background correction' as most of the counts detected in the background area are scattered counts arising from bladder. When residual volume is small it may be justifiable to use a difference area as background.

Vesico-ureteric reflux

There are two isotopic techniques for diagnosing vesico-ureteric reflux, direct voiding radionuclide cystourethrography (DVRC) and indirect voiding radionuclide cystourethrography (IVRC).

Direct voiding radionuclide cystourethrography (DVRC)

The technique is essentially similar to radiographic micturating cystourethrography (MCU). It involves catheterization and installation into the bladder of activity dissolved in saline or water. The patient is catheterized, the bladder emptied and refilled with a solution of any soluble radiopharmaceutical. A dose of 20–50 MBq pertechnetate dissolved in a volume determined from a nomogram relating bladder volume to age, height and weight is commonly employed as the cheapest and most readily available. The entire filling phase is recorded on a gamma camera to detect 'low-pressure' reflux. When the bladder is so full that the patient is unable to tolerate more the position is changed from supine to erect, either standing or seated as appropriate, with their back against a large-field gamma camera fitted with a general purpose collimator. The patient then voids in front of the camera, much as in an MCU. The entire voiding process is recorded on the data processing system at a rate of 12 frames per minute. Although performed in some departments, many feel that if instrumentation is to be performed the better anatomical resolution of the MCU should be obtained.

Indirect voiding radionuclide cystourethrography (IVRC)

The indirect method is performed shortly after the end of a gamma camera renogram, when the bladder is full and the patient feels able to void but without adding additional activity. Catheterization is thus eliminated. MAG3 is preferred to DTPA because the higher extraction and more rapid excretion results in a lower background in the renal areas. The patient stands or sits with their back against a large-field camera which acquires at 12 frames per minute for as long as necessary. The camera should be started before the patient is instructed to void and if possible five frames or more should be acquired before voiding starts. The patient then voids into a suitable receptacle. The study should be filtered (page 319) and time–activity curves obtained from the renal areas. The study is viewed in cine mode or ROIs are superimposed on the frames to confirm that there is no patient movement. Reflux is diagnosed if the count rate in the renal region of interest increases by more than three standard deviations of the number of counts present in the first of the frames selected and in the absence of movement (Fig. 3.17). It is sometimes possible on filtered frames to identify lesser degrees of reflux into a ureter, but it can be difficult to distinguish between activity passing upwards from bladder and that passing downwards from kidney. Radiological grading schemes are not applicable to the IVRC. Reflux may be identified also on time-condensed images (page 179) of ureters. These are difficult to obtain in small children as much of the length of the ureter may be concealed by full bladder.

Interpretation

Interpretation of both investigations is similar. Reflux is detected as a rise in the time–activity curves from renal areas, confirmed by visual inspection of corresponding frames. Both must always be inspected. The ureters are not usually visualized on unprocessed frames, but time-condensed parametric image (page 179) may permit ureteric peristalsis to be visualized. There is as yet insufficient information to determine the role of these time-condensed images in detection of vesico-ureteric reflux. Initial reports in adults suggest they may prove useful.

Further reading:
Nuclear Medicine Communications 1991; ***12:*** *397–407*

Obstruction

An obstruction is defined as 'anything that stops or blocks a way or passage, or hinders or prevents progress'. When the passage is a tube along which fluid is flowing, obstruction is manifest as an alteration in one or both of the parameters which define flow. These parameters are pressure and rate of flow. If pressure remains constant the effect of an obstruction is to reduce rate of flow. Conversely if rate of flow is to remain constant pressure must rise; imaging investigations provide information about flow rate but give no information about pressure. Clinical urinary tract obstructions are almost always incomplete; the passage is only partially blocked. Many of the controversies which exist regarding the problem of clinical assessment of obstruction arise from failure to appreciate the existence of two independent components, only one of which is readily amenable to non-invasive study.

There is no non-invasive technique capable of estimating pressure and it must be accepted that this important parameter can be measured only by invasive methods. Renography however gives important information both about function and flow rate. Although in experimental animals there is evidence that filtered agents are more sensitive to ureteric obstruction than tubular ones, there is no clinical confirmation that either MAG3 or DTPA has any advantage over the other when investigating acute obstruction. There is a subjective impression that because of its faster excretion MAG3 gives a clearer distinction between normal and abnormal. However radiation dose is an important consideration and the lower activity of MAG3 needed confers upon it a substantial advantage in acute obstruction.

Renal colic

Acute renal obstruction is most commonly due to a calculus in a ureter and presents with a characteristic clinical picture, but may be due to many other causes. As the majority of calculi are radio-opaque, the first imaging investigation to be performed should be a plain radiograph of the abdomen, supplemented if necessary by angled or oblique projections or tomography to confirm the

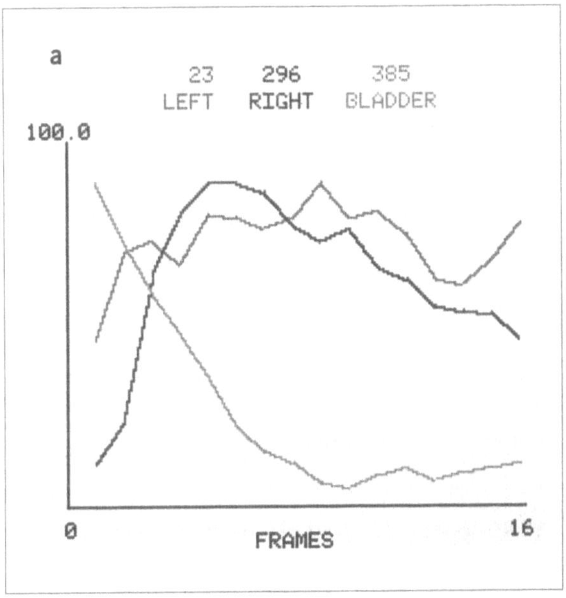

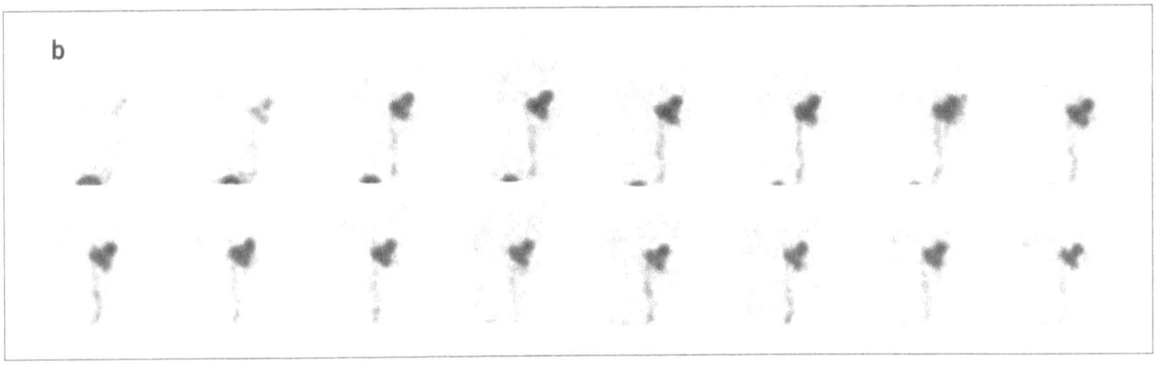

Fig 3.17 (a) Time–activity curves from both renal areas and bladder during voiding, showing gross reflux on right but also significant reflux on left. (b) Frames showing the reflux. The amount of activity refluxing into the renal pelvis is small compared to that in the bladder and is thus easily overlooked unless each curve is normalized to its own maximum and the bladder is over-range.

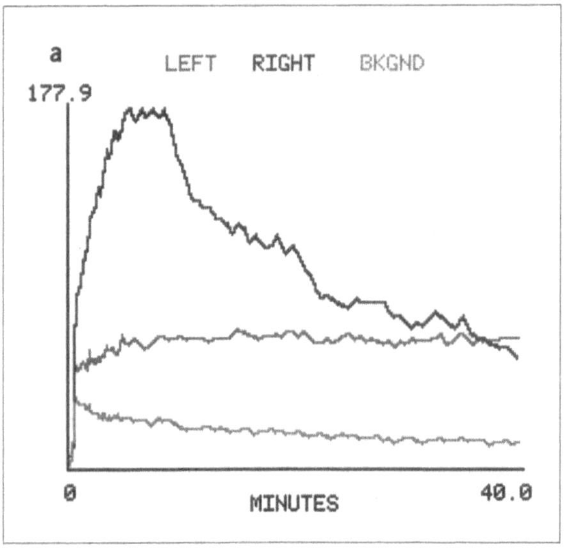

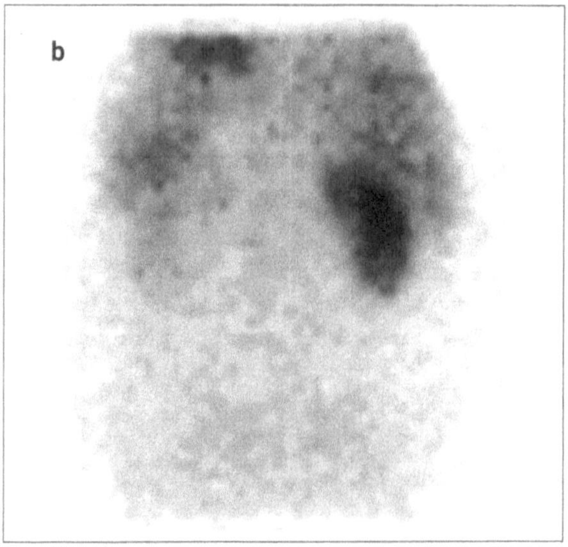

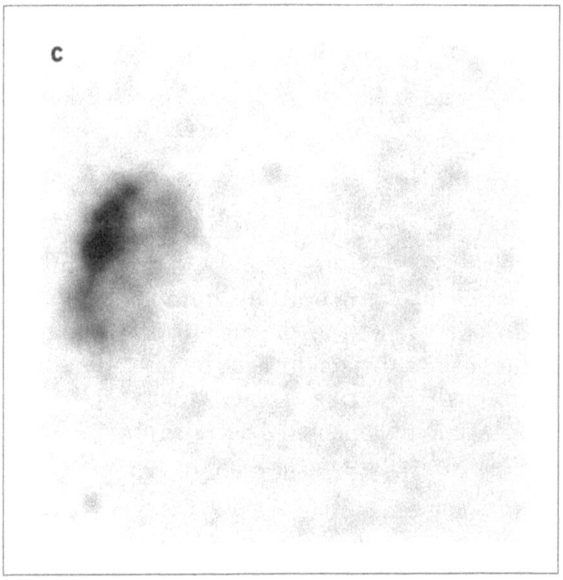

Fig. 3.18 (a) Obstructed renogram (MAG3) with depressed function, rising curve and no response to diuretic at 25 min. (b) Initial 90 s frame. (c) Nephrogram is still present on left 40 min after injection. Activity has cleared from the right, which is barely visible.

level of the calculus. Excretion urography is required to determine whether any opacity is in the line of a ureter but gives little information about renal function and can on occasion be misleading (page 107). The renographic features of obstruction are an enlarged kidney, a rising curve which does not respond to diuretic, a prolonged nephrographic phase followed by retention of tracer in the collecting system and ureter to the level of obstruction and, if obstruction has been present for more than 48 h, impaired function (Fig. 3.18). Renography is required initially to measure how well the obstructed kidney or kidneys are functioning and is the best method for follow-up, to determine whether there has been any change in function of the kidney or severity of obstruction.

Pelvi-ureteric junction obstruction (PUJO)

A congenital anomaly of shape of the renal pelvis may be associated with episodes of renal colic or may be asymptomatic but may nevertheless cause progressive renal damage. It is commonly diagnosed antenatally by ultrasound. In affected subjects the ureter forms a sharp angle with the renal pelvis rather than one tapering gently into the other. Sudden distension of the renal pelvis by a rapid increase in rate of urine output may convert the junction into an obstructing flap valve. PUJO can only be detected reliably under conditions of high urine output. This is achieved by administering 20–40 mg frusemide 15–20 min before the renogram is started. In the interval the patient should drink 400–500 ml of fluid to ensure adequate hydration and a maximum diuresis.

Interpretation

Information which may be obtained by renography includes confirmation of:

- the presence of renal function on the side of symptoms
- whether symptoms are due to obstruction to drainage of urine formed by that kidney
- the condition of the contralateral kidney and
- a quantitative estimate of function both kidneys.

The classical renographic appearance of an obstructed kidney is a curve which continues to rise for the entire duration of the examination, at least 30 min. A number of other conditions also produce a slowly rising curve, including an extravasated injection, renal artery stenosis, a kidney whose function has been impaired by previous disease and a dilated collecting system. Comparison of early and late frames from the study, supplemented as appropriate by additional later frames at intervals up to 24 h, facilitates differential diagnosis. Thus in renal artery stenosis delay occurs mainly in cortex (prolonged nephrogram) and there is no abnormal accumulation of activity in, or dilatation of, the collecting system. When the rise is due to an obstructed collecting system this is readily visible on the images, both by initial prolonged retention of tracer in the cortical region (nephrogram) and, if imaging is continued for long enough, visualization of activity accumulating in renal pelvis and ureter, which are commonly dilated to the level of obstruction. In the presence of marked hydronephrosis and hydroureter or renal impairment in association with a dilated collecting system, it is often impossible to differentiate obstruction from stasis.

The majority of obstructed kidneys initially exhibit no measurable impairment of function. Even if a single measurement reveals asymmetrical function there is no means of determining whether this is of long standing or due to the current event. Clearly, a shrivelled kidney is unlikely to improve its contribution significantly. However it is much more difficult to determine whether a kidney of normal or moderately reduced size, but with decreased uptake or excretion of tracer, might improve. Factors affecting prognosis include duration and severity of obstruction and whether or not infection supervenes. This can only be determined by consecutive investigations, the value of which depends on their reproducibility. Progressive deterioration is an indication for intervention. A difference in relative function of more than 4% on consecutive examinations is significant. Even some kidneys which initially show no detectable function sometimes improve, occasionally after a considerable time. At present there is no established criterion for predicting recoverability in an individual case. A kidney initially with low uptake relative to its size (the 'pale' kidney) should always be regarded as potentially recoverable.

Excretion urography

It is important to appreciate why this investigation can give a misleading impression of renal function, which is assessed by visual evaluation of density of the pyelogram. This is determined by the number of iodine atoms interposed between the x-ray tube and film and depends on a number of factors, but especially the amount of iodinated compounds filtered through glomeruli, volume of the collecting system and ureter and extent to which glomerular filtrate is concentrated during its passage along tubules. If renal function is poor or renal blood flow impaired, although the numbers of glomeruli and of tubules may be greatly reduced those tubules which do survive are often capable of concentrating the reduced volume of filtrate presented to them. They may indeed concentrate it to a greater extent than the contralateral healthier kidney with a greater solute load. Hence the comparatively small amount of contrast excreted is present in high concentration and is washed out slowly. As visible density depends on concentration of contrast medium, not total amount excreted, a small amount of contrast contained within a small volume excreted by a kidney with low output appears dense on the radiographs. The net result is that the denser pyelogram may be found on the side of the more poorly functioning kidney. Unfortunately this is not invariably the case and it is never possible by simple visual inspection to determine whether a dense pyelogram represents a healthy kidney or an unhealthy one. Excretion urography may thus be misleading for assessment of renal function. In contrast the initial part of the renogram is directly related to rate of excretion of tracer, and the area under the curve, at least for the first 90–120 s, is a function of the total amount excreted.

Renal failure

The principal indications for radionuclide investigations in patients with renal impairment or failure are to distinguish between obstructive and non-obstructive renal failure, to estimate relative renal function and determine whether the separate contribution of each kidney alters as the disease evolves. Total function can be assessed by standard biochemical parameters such as serum urea and creatinine, although their limitations are well documented. Measurement of glomerular filtration rate (GFR) is potentially more accurate; clearance of substances such as PAH, OIH or MAG3 secreted by proximal tubules less so, depending on the method employed. Differential function can be assessed non-invasively only by isotopic means. A number of factors influence choice of radiopharmaceutical. Experimental unilateral ureteric obstruction results in both an absolute and a relative decrease in glomerular function, more severe than the decrease in OIH clearance. The effect on DMSA uptake is intermediate between the two. MAG3 resembles OIH in this respect. Neither MAG3 nor OIH can be used to differentiate obstructive from non-obstructive renal failure as both compounds continue to accumulate in proximal tubules but are not excreted for some hours after injection in all patients with acute tubular necrosis (ATN) and some patients in chronic renal failure. They may thus give a rising curve in the absence of obstruction. The conditions under which this occurs have not been fully defined and it is not possible to predict the patients with chronic renal failure in whom this will occur. For this reason, although MAG3 is preferred under most circumstances, DTPA is more straightforward to interpret in chronic renal failure and acute tubular necrosis despite its slower excretion.

Because of this and contrary to widespread belief, ^{99m}Tc-DTPA is the agent of choice to differentiate between obstructive and non-obstructive renal failure, 200 MBq being the typical activity in adults. The high photon flux is necessary to obtain adequate images during the early vascular phase. An initial study lasting for 30 min, framing at the rate of six frames per minute for the first 2 min, in order to observe the vascular phase, and two frames per minute thereafter is usually adequate but may be supplemented by single images at progressively increasing intervals for up to 24 h. Although some residual renal activity is usually visible in the absence of any obstruction, this is usually clearly located in cortex and comparatively little above background. The normal ureter is never visualized scintigraphically on unprocessed images. If there is any clinically significant obstruction the kidney, pelvis or ureter is visualized as an area of characteristic shape. Obstruction is rarely complete and for this reason it is important to image as frequently as convenient during this 24 h period and not to rely on a single late image.

The role of DMSA in renal failure is less clear. It is commonly possible to identify kidneys with DMSA which are not visible on excretion urography. This is particularly useful in confirming the presence of small non-obstructed kidneys, but as DMSA gives no indication of drainage it is not the initial investigation of choice in this condition. DMSA uptake correlates with creatinine clearance but when function is poor, high background makes uptake measurements less accurate. Although DMSA has some role in renal failure for determination of relative function, it is in general subsidiary to DTPA.

A patient in renal failure should never be dehydrated and for diagnostic purposes the maximum safe level of hydration should always be employed. It sometimes helps to give a diuretic, although in the presence of uraemia the osmotic load which the patient is carrying is itself a powerful diuretic, the effect of which may be enhanced little if at all by other agents at normal dose levels.

Renography is best performed with the patient supine. These patients are often too ill to sit steadily for any length of time and the examination is invalidated if orthostatic hypotension occurs or if there is appreciable patient movement. Imaging in isolation is of limited value in differential diagnosis. Even the distinction between obstructive and non-obstructive renal failure cannot be made without quantitative information as to whether the amount of tracer in a dilated collecting system is increasing or decreasing. A falling curve excludes significant obstruction, although a rising one does not confirm this diagnosis, as low output into a dilated system, renal artery stenosis, hypotension and an extravasated injection may all give a rising curve in the absence of obstruction. Acute tubular necrosis is the only condition associated with a specific form of curve, namely an initial spike due to high blood flow in the absence of renal extraction. This is seen only with DTPA; MAG3 gives a rising curve.

Hypertension

The range of systolic and diastolic blood pressures in any population does not have a Gaussian distribution but is skewed to the right. Diastolic pressures greater than 90 mm are associated with an increased risk of cerebral and coronary vascular events and renal failure due to progressive renal damage. The more common associations of hypertension include chronic renal failure, diabetes, reflux nephropathy and glomerulonephritis. There are a number of uncommon endocrine causes including Cushing's disease (page 170), Conn's syndrome (page 170), thyrotoxicosis (page 157) and phaeochromocytoma (page 171). Renal artery stenosis is well documented but rare. In the majority of patients no cause can be found.

The principal indication for investigating renal function is to determine adequacy of steps to limit renal damage secondary to raised blood pressure or the underlying disease. A renovascular cause for hypertension, as distinct from effects secondary to the hypertension, is found in a small minority, 1% or less depending on the population selected.

Renal artery stenosis

There are no features in the renogram specific either to hypertension or to renal artery stenosis. A number of techniques have been proposed to identify renal artery stenosis causing renovascular hypertension. The most widely used is renography following ACE (angiotensin converting enzyme) inhibition. This blocks formation of angiotensin II, which maintains perfusion pressure distal to a stenosis by increasing tonus of the efferent arterioles, from angiotensin I. The fall in perfusion pressure causes a reduction in glomerular filtration rate and proximal tubular urine flow, accounting for the similar results observed with both glomerular and tubular agents. Renography during erect exercise and after a large dose (1.5 g) of aspirin have also been suggested. The latter technique has a low sensitivity and cannot be recommended. Experience with the exercise renography is restricted to one or two centres.

ACE inhibition renography is not appropriate for screening as the changes observed are relatively non-specific and the danger of severe hypotension high by comparison. The usefulness of this test depends on the population selected. It is of greatest value in patients whose *a priori* risk is approximately 30%. Clinical features associated with an increased risk of renovascular hypertension due to renal artery stenosis include:

- onset of hypertension under the age of 25 or over 45
- an abdominal bruit
- an accelerated rate of clinical deterioration
- worsening of control of blood pressure
- occlusive disease elsewhere in the circulation
- a diastolic pressure greater than 95 mm refractory to three or more drugs.

In older subjects in whom renal artery stenosis is most likely to be due to atheromatous disease, response to angioplasty is usually short lived. In these patients preservation of renal function may be more important than control of blood pressure. In younger subjects fibromuscular hyperplasia is more probable and angioplasty may provide a definitive cure.

Technique

25 mg captopril is given orally (or enalaprilat 0.04 mg/kg intravenously) after an overnight fast and the patient hydrated with 300–500 ml of water or juice. Blood pressure must be monitored for at least the next hour. It is advisable for the patient to recline comfortably with an in-dwelling cannula inserted into a convenient vein in case hypotension requires urgent volume expansion. Features which identify patients at highest risk of severe hypotension are a history of previous myocardial infarction or stoke, angina and transient ischaemic attacks. Renography is started 1 h after ACE inhibition and may be performed with either a tubular or a glomerular agent for not less than 20 min. Specificity may be improved by administration of 20 mg frusemide within 3 min of the radiopharmaceutical but this increases the risk of severe hypotension.

Interpretation

The characteristic features of unilateral renal artery stenosis following ACE inhibition are a difference of more than 5 min between the times of peak activity and a reduced rate of wash-out such that, with MAG3, the ratio of counts 20 min after the peak to peak counts is greater than 0.18. Cortical mean transit time is also increased. If the post-captopril renogram shows asymmetrical renal function the examination may be repeated on another day without ACE inhibition. Any asymmetry in differential function due to a renovascular cause is emphasized by ACE inhibition. Alternatively a baseline study may precede ACE inhibition in every patient, in which case both may be performed on the same day after an interval of 2 h and using a higher dose of radiopharmaceutical for the second study.

The author has been unable to reproduce the high sensitivity (65–91%) or specificity (93–100%) cited in the literature for this test. In his experience this technique has a sensitivity and specificity of at best 50%. A possible explanation for this discrepancy may lie in differences in referral criteria.

Sources of error

Similar, although usually bilateral, changes have been reported in many other renal diseases including lupus nephritis, polyarteritis, interstitial nephritis, diabetic nephropathy, obstructive uropathy and essential hypertension. False negatives have been reported in renal failure.

Further reading:

American Journal of Kidney Diseases 1994; ***24****: 665–73*

Journal of Nuclear Medicine 1996; ***37****: 1876–82*

Reflux nephropathy

(see also page 288)

The most common cause of preventable renal failure is thought to be reflux nephropathy. It commonly presents in infancy or childhood, often in the neonatal period and is frequently asymptomatic until renal damage is extensive. Any child found to have significant bacteriuria must therefore be further investigated. Cortical loss is commonly missed at excretion urography because of the difficulty in identifying the renal outline in children. Renal scintigraphy with DMSA is considerably more sensitive than excretion urography for detecting scars or other cortical defects (Figs 3.13, 3.16). Ultrasound is somewhat less sensitive than DMSA except for detecting concentric cortical loss. No single investigation detects all cases. Overall DMSA is more sensitive than ultrasound for detecting scarring in children. Ultrasound is however always required to determine the size of the collecting system and detect calcification. Relative function should be obtained with DMSA as a baseline for follow-up in every case.

Vesico-ureteric reflux associated with infection can lead to progressive renal damage. However reflux is a common phenomenon in children and may not by itself be significant. Reflux usually resolves with growth, but the suggestion is erroneous that absence of reflux on a single examination is evidence that no further investigation need be performed, as it has been shown that any single test for reflux has at best only approximately a two in three chance of detecting it. The optimal combination of investigations is regular urinalysis to detect occult infection, combined with an initial ultrasound, DMSA scintigraphy and IVRC in all children in whom infection or scarring is proven (page 289).

Post-operative changes

Some operative procedures for PUJO have the incidental effect of denervating the renal pelvis, thus eliminating the swallowing reflex. Following cystectomy with formation of an ileal loop it is usual to observe reflux into the renal pelvis (Fig. 3.19); indeed absence of reflux raises the

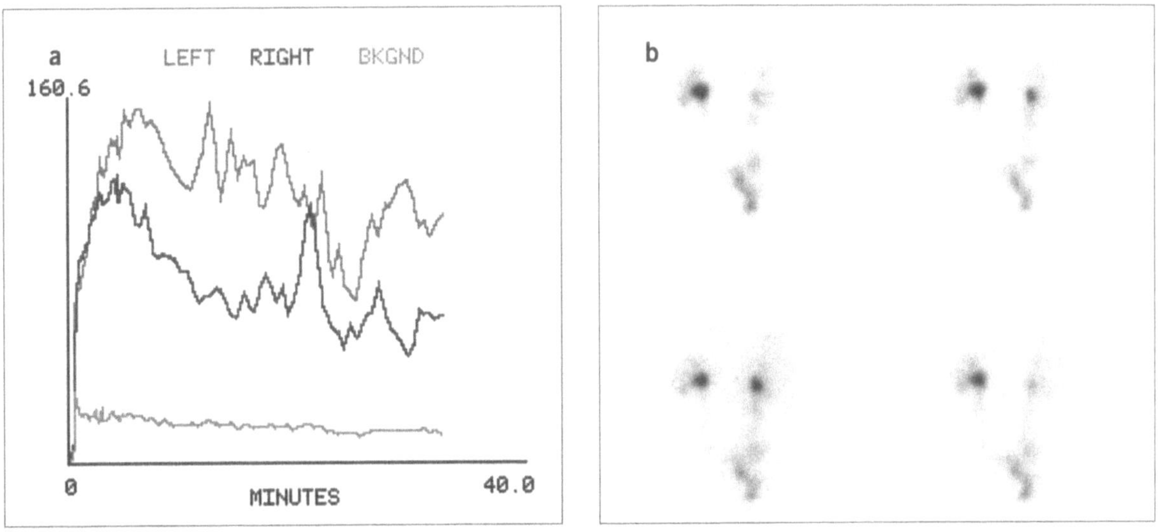

Fig. 3.19 (a) Curves showing spontaneous reflux from ileal loop bladder into renal pelvis; (b) frames corresponding to right side reflux. This is a usual finding in patients with an ileal loop and is evidence that the ureters are not obstructed.

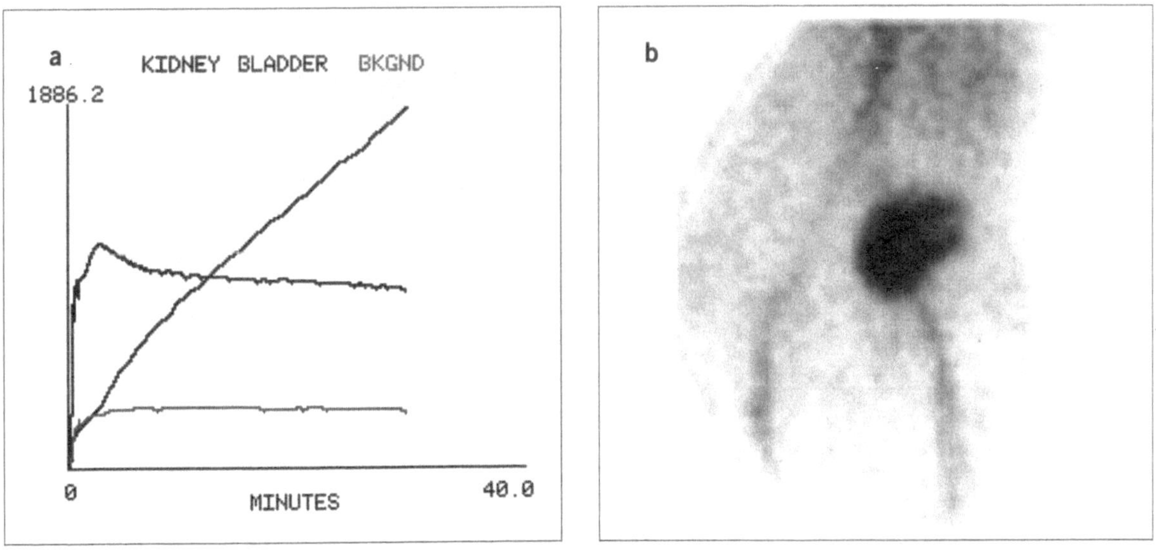

Fig. 3.20 (a) DTPA renogram from healthy transplant. (b) Nephrogram, (c) 30 min after injection. (**c** *overleaf*)

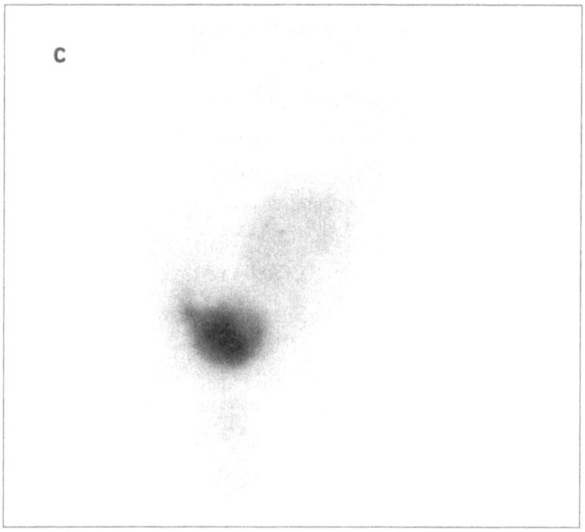

Fig. 3.20 *continued*

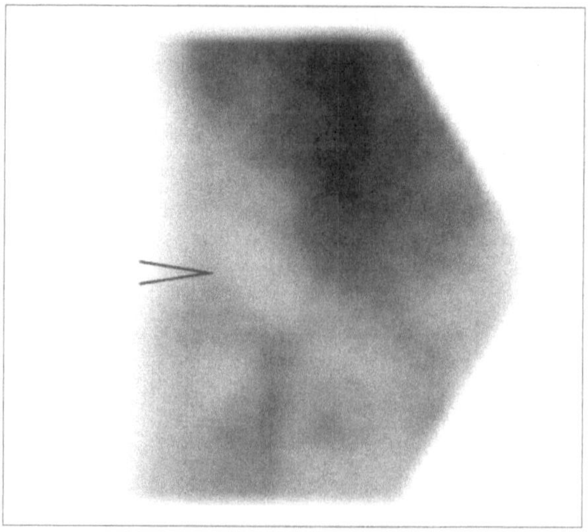

Fig. 3.21 Infarcted transplant (open arrow) is photon-deficient.

possibility of stenosis at the site of implantation.

Renal transplants

The role of nuclear medicine in management of renal transplants has diminished with the adoption of cyclosporin and other anti-rejection therapy and availability of Doppler ultrasound. A healthy transplant has a renogram similar to that from an orthotopic kidney (Fig. 3.20). In the acute phase renography with DTPA can differentiate an infarcted transplant (Fig. 3.21), which appears as a photon-deficient area

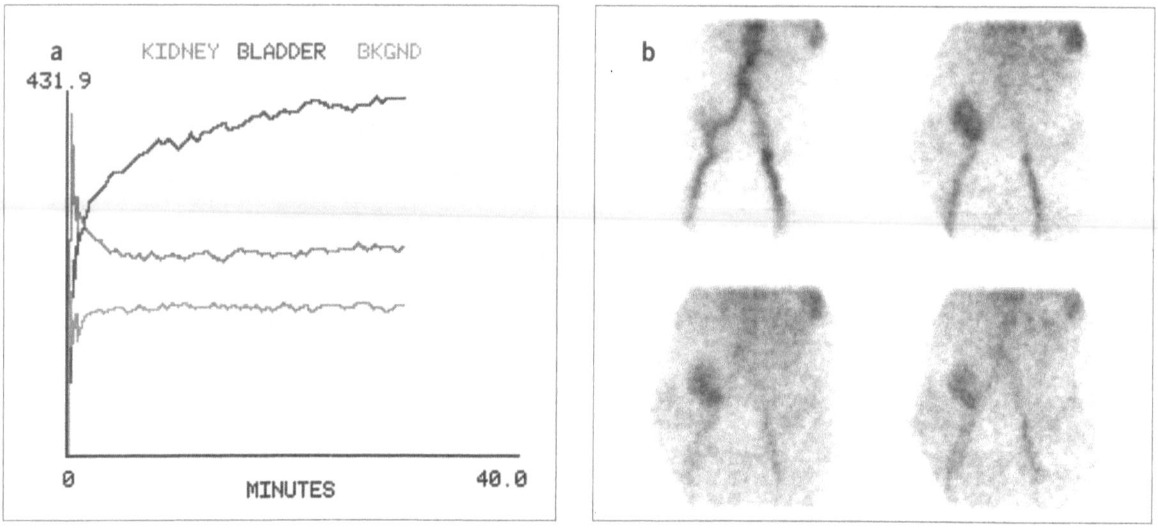

Fig. 3.22 (a) Typical DTPA curve of ATN in a transplanted kidney, with early vascular peak and subsequent slowly developing nephrogram. (b) Consecutive frames showing count-rate in the transplanted kidney which peaks early, at the same time as spleen (top right-hand corner of figure) and decreases thereafter, indicating a well vascularized or hyperaemic kidney which fails to extract DTPA.

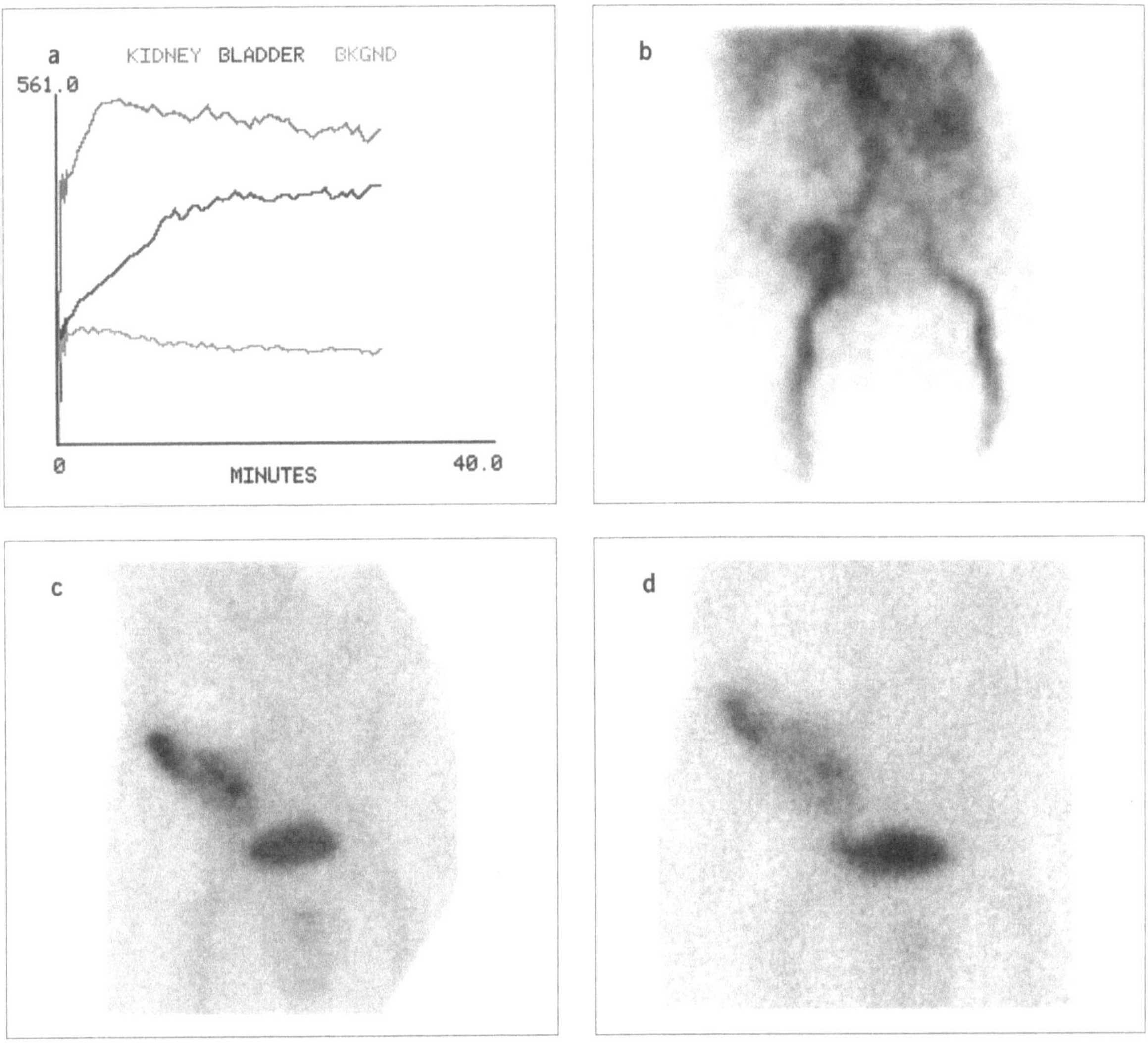

Fig. 3.23 Urine leak. (a) Curves from transplanted kidney, the contralateral side and leak (mislabelled bladder). (b) The transplant is well seen in the early nephrographic phase. (c) By 10 min a collection is becoming visible above and lateral to the transplant. (d) By 30 min concentration is higher in the collection than in the kidney.

sometimes with surrounding hyperaemia, from ATN which is hyperaemic but fails to extract and excrete DTPA (Fig. 3.22). Commonly there is a non-specific mixed pattern with a small initial spike and subsequently a flat featureless curve. Urine leaks are usually readily visible on later films as enlarging collections outside the boundaries of the renal tract (Fig. 3.23). There is no specific pattern of rejection in a single examination. Diagnosis is by exclusion. If patients are systematically followed up at regular intervals with DTPA renography, a fall in the perfusion index heralds rejection. Perfusion index is defined as:

(total counts per pixel in ROI drawn over the iliac artery, integrated to the peak of the first pass) / (total counts per pixel over the same period in the renal ROI) × 100

The normal value is <150. Non-systematic renography is rarely useful in follow-up of renal transplants.

Further reading:

European Journal of Nuclear Medicine 1991; ***18****: 227–8*

4 Heart

Imaging regional myocardial perfusion, to distinguish between jeopardized myocardium, infarct and healthy muscle, is currently the most important cardiac investigation employing radioisotopes. There are other techniques with less common or less well established clinical applications which display infarcts, wall motion, ventricular function, fatty acid metabolism and adrenoreceptor activity locally or globally. It is possible to demonstrate flow patterns in some congenital defects. There are useful but less often used non-imaging methods for measuring global ventricular function and shunt size.

Myocardial scintigraphy

Although myocardial perfusion imaging has been a routine clinical investigation with an established role for over 20 years, many uncertainties remain unresolved concerning interpretation, optimization of technique and even indications. Potassium analogues, in particular thallium, have been the mainstay for many years. A number of technetium-labelled agents introduced more recently have achieved considerable popularity because, despite certain disadvantages (see below), they provide improved image quality.

In most patients with coronary artery disease but who have not suffered a myocardial infarct, myocardial perfusion is homogeneous at rest. Heterogeneity can be demonstrated only under conditions of stress, which may be physical or pharmacological. Controlling the level of stress is one of the most important variables constraining reproducibility of myocardial perfusion imaging. An approach combining elements from a number of suggested protocols may be the most productive. For simplicity they are initially described separately.

Radiopharmaceuticals

Thallium-201 (thallous chloride, Tl^+)

The hydrated thallium ion (Tl^+) has a similar ionic radius to the hydrated potassium ion (K^+) and, when administered intravenously as the chloride at tracer concentrations, behaves *in vivo* as an analogue of potassium. It is cleared rapidly from the circulation after intravenous injection with a high (approximately 90%) first-pass extraction efficiency in muscle, as a result of the 'sodium pump' associated with depolarization following muscle contraction. Intracellular thallium is mainly diluted unchanged in the potassium pool and not fixed. Some may be converted to Tl^{3+} which is unlikely to diffuse readily. On the second pass and subsequently there is a bi-directional flux, determined by relative intracellular and extracellular concentrations. Plasma concentration falls more rapidly than intracellular, resulting ultimately in net wash-out from heart at a rate of approximately 10% of the residual radioactivity per hour. Initial distribution within myocardium is proportional to regional blood flow at the time of injection. Extra-cardiac localization of the residue is largely determined by the distribution of cardiac output at injection. In exercising subjects, uptake is high in active muscles but relatively low in splanchnic areas. Administration to resting subjects results in higher concentrations in liver and other viscera than when injected during exercise.

Redistribution starts immediately in all tissues. Factors determining blood levels are complex and no satisfactory mathematical model has been proposed. Blood levels of thallium fall more rapidly after a meal than in fasting subjects, possibly due endogenous insulin affecting activity of the sodium/potassium pump. This is associated with increased wash-out from myocardium and impaired redistribution. Patients should therefore be restricted to fluids and low-calorie light snacks in the interval between initial and redistribution thallium imaging. After some hours distribution within myocardium is proportional to amount of viable muscle present. In normal subjects this redistribution is largely complete within 3–4 h, although in the presence of severe ischaemia up to 24 h may be required for full equilibration, by which time count rate is much lower than initially, principally because of wash-out. Thallium is second only to FDG (page 267) as a marker of viability. When used in conjunction with a coronary vessel vasodilator (page 119) the sensitivity of thallium is only marginally lower than that of FDG but thallium is more readily available and less costly.

Further reading:
European Journal of Nuclear Medicine 1996; ***23****: 188–94*

Technetium-labelled fixed agents

The available compounds are sestamibi, tetrofosmin and furifosmin. They are non-polar, moderately lipophilic complexes which exhibit some wash-out but little or no redistribution. Separate injections must thus be administered, one at peak stress to demonstrate regional blood flow and a second at rest to show muscle mass. Unlike thallium the order is immaterial. Both parts of the examination may be performed on the same day using not less than three times the original activity for the second injection, or on separate days, in which case both may be of similar activity. There is minimal wash-out of tetrofosmin from myocardium. There is some wash-out of sestamibi but redistribution is negligible. The mechanism of uptake is unclear and may not be the same for all agents but involves an active process, being depressed or abolished by metabolic inhibitors. Statistical quality of the images is better than thallium and absorbed radiation dose lower. These agents provide a clearer representation of regional perfusion at the time of injection because there is effectively no redistribution and technetium has better imaging properties than thallium.

No difference in clinical efficacy has been demonstrated between the various preparations. Their main disadvantage is rapid hepatic excretion, giving rise initially to higher hepatobiliary than cardiac concentration of activity, the distribution of which alters as it passes from hepatic parenchyma into small intestine via the bile ducts and gall-bladder. There may be associated duodeno-gastric reflux (page 252). These can produce artefacts in SPECT reconstruction, which assumes distribution of radioactivity to be unchanging in the field of view during acquisition. A delay is therefore required before imaging while extra-cardiac activity is taken up and cleared through the liver. The technetium-based agents provide a clearer and more accurate map of regional blood flow than thallium because they do not redistribute, but the latter may be more accurate in distinguishing viable from non-viable myocardium.

Further reading:
Journal of Nuclear Medicine 1997; ***38****: 419–24*

Teboroxime

This is also labelled with technetium but differs in important respects from the compounds described above. The single-pass extraction efficiency is higher than that of thallium or other technetium agents and is linearly related to perfusion even at very high flow rates. Uptake is passive and appears to be dependent on blood flow alone. It thus gives no information on viability. It is rapidly washed out of myocardium so that imaging must be started immediately. Regional wash-out rate may be the most useful parameter to measure. The wash-out curve is biexponential, the fast component having an half-time of 5–6 min at rest and 2.5–3.0 min during exercise or adenosine infusion. Sublingual isosorbide dinitrate before injection at rest enhances ability to differentiate reversible perfusion defects. For practical purposes it can be used only for dynamic planar imaging or with fast multi-headed SPECT cameras.

Further reading:

*Journal of Nuclear Medicine 1994; **35**: 689–92, 1265–73*

Other agents

The most sensitive radiopharmaceutical for identifying viable myocardium is ^{18}F-fluorodeoxyglucose (FDG), usually employing positron emission tomography (PET). Although a positron emitter, ^{18}F can be used with a gamma camera if a sufficiently heavy collimator or coincidence counting are available (page 267). It is more accurate than the technetium agents alone in distinguishing between viable and non-recoverable myocardium and slightly more sensitive than thallium. Its main disadvantages are cost and limited availability because of the short half-life. No clear clinical role has been established.

Further reading:

*Journal of Nuclear Medicine 1997; **38**: 582–6*

Myocardial metabolism

Myocardial metabolism is reflected by uptake of fatty acids and has been investigated using a variety of ^{11}C and ^{123}I-labelled compounds, in particular analogues of medium chain-length pentadecanoic and hexadecanoic fatty acids such as 15-(*p*-iodophenyl)3*R*,*S*-methylpentadecanoic acid (BMIPP). Lower uptake of fatty acid analogues than of thallium has been observed in hypertrophic cardiomyopathy, but there is limited clinical experience with these compounds. Their cost and restricted availability preclude routine clinical use. ^{13}N-ammonia and ^{43}K are similarly restricted to the immediate proximity of a cyclotron by their short half-lives. Isotopes of rubidium and caesium are inferior to thallium as potassium analogues because of their larger ionic radii. A generator-produced positron-emitting rubidium isotope, ^{82}Rb, is however available and used in some PET facilities which do not have a local cyclotron. Experience with these compounds is limited. None of these agents has an established clinical role.

Further reading:

*Journal of Nuclear Medicine 1997; **38**: 559–63*

MIBG

Meta-iodo-benzyl-guanidine (MIBG) can be labelled with ^{123}I (page 172). It is taken up exclusively by noradrenaline re-uptake receptors. These are present in normal heart, absent from denervated transplanted hearts and present at reduced concentration in hypertrophic and dilated cardiomyopathy, ischaemic heart disease, hypertension, hypothyroidism, anthracycline toxicity and arrythmias. An increased wash-out rate is a non-specific but common feature of compromised myocardium.

Further reading:

*Journal of Nuclear Medicine 1997; **38**: 447–51*

Ischaemia

Ischaemia can be detected in myocardium or brain using an iodinated misonidazole derivative ^{99m}Tc-nitroimidazole. This compound is not trapped unless first reduced by enzymes present only in living tissue. It does not accumulate in infarcts. In the presence of adequate oxygen it is re-oxidized and washes out. Accumulation thus occurs only in viable hypoxic tissue. Little clinical experience has been reported but this or a later derivative clearly has clinical potential.

Further reading:

*European Journal of Nuclear Medicine 1995; **22**: 265–80*

Particulate tracers

Direct intracoronary injection of microspheres or MAA (page 57) is the only method of demonstrating the separate distribution of each coronary artery. Use of two isotopes permits both left and right coronary territories to be mapped simultaneously. The procedure has largely been superseded by thallium and its analogues, despite their inability to distinguish arterial territories with any precision. The technique was formerly widely employed and is safe provided there is rigorous control of the number of particles injected and their size.

Further reading:

*Seminars in Nuclear Medicine 1980; **10**: 178–86*

Administration and imaging schedules

Thallium

80 MBq ^{201}Tl (as the chloride) is given by clean intravenous injection at peak stress to demonstrate regional perfusion or at rest to demonstrate viability. Absorption from a perivenous site is slow compared with duration of stress effects. Unless the extravasation is partial and small the procedure must be repeated. There is also a high local radiation dose. Ulceration has been described at a site of extravasation of thallium although there is some dispute as to the relationship between injection and subsequent ulceration. When performing a stress/redistribution study with thallium the initial set of images, representing regional perfusion, must be started within 5 min and acquisition completed within half an hour, before redistribution is appreciable. The redistribution study is obtained 3–4 h later, by which time change is slow and duration of acquisition less critical. If a defect found in early images does not at least partially fill in on the late set, a second intravenous injection of thallium should be given on the same or a subsequent day. Some centres dispense with redistribution images and routinely give a second injection of thallium when the patient is resting after the initial images, but this results in all patients receiving a higher dose of thallium. Some may image within 15 min of the second injection but a longer wait, up to 4 h, is desirable if extent of irreversible damage is not to be overestimated. In most patients 30 min is adequate. Preceding re-injection with 20 mg isosorbide dinitrate sublingually improves accuracy by increasing the number of defects shown to be reversible. Other nitrates are likely to be equally effective but have not been formally evaluated for this application. If hibernating myocardium (page 127) is suspected, a late image should be obtained not earlier than 4 h after the resting injection. Prior administration of a nitrate may allow this interval to be shortened but data to confirm this is incomplete.

Technetium agents

Separate injections are given, one at peak stress and the other at rest. The order in which the rest and stress doses are administered is immaterial. If stress and resting studies are performed 24 h apart the fractions may be of equal size. Where a one-day protocol is employed, the second dose should be at least three times greater than the first, typically 250 MBq and 750 MBq. Total administered activity should not normally exceed 1000 MBq unless gated SPECT is required. The dose equivalent from the usual 80 MBq ^{201}Tl is similar to that from 2 GBq of a technetium agent. Reinjection thallium protocols are associated with a substantially higher adsorbed radiation dose. As with thallium, preceding rest injection with an oral or sublingual nitrate improves accuracy. It is advisable to wait at least 45 min for clearance of hepatic uptake after rest injection before starting SPECT. Half an hour is sometimes allowed after exercise because of lower splanchnic perfusion. A longer delay has the added advantage of reducing lung and blood background, giving higher contrast. A light meal before imaging encourages gall-bladder emptying, driving excreted activity away from the heart and minimizing duodeno-gastric reflux of excreted activity.

Exercise stress testing

The patient may be exercised erect or supine, on a treadmill or bicycle ergometer, using any standard incremental stress test, for example the modified Bruce protocol. Walking is the most familiar form of exercise for most subjects and is readily simulated on a treadmill. ECG, heart rate and blood pressure must be monitored and recorded continuously if the test is to be conducted safely and because knowledge of these parameters is essential for interpretation. Exercise is associated with diversion of perfusion away from the splanchnic circulation and thus gives lower hepatic and splanchnic uptake than pharmacological stress.

The criteria for terminating a test are any one of:

- the patient achieving the maximum appropriate heart rate (200 – age)
- development of runs of ectopic beats or a significant dysrythmia
- fall in blood pressure

- more than 2 mm horizontal depression of the ST segment
- angina
- exhaustion.

The doctor supervising the test must be familiar with cardio-pulmonary resuscitation and have appropriate emergency facilities immediately at hand. Automatic ECG interpreting devices have difficulty in distinguishing between flat and up-sloping ST segments and over-report ST depression. Reliance on automated interpretation results in substantial under-stressing of some patients. This is an important source of error.

Pharmacological stress

This is necessary whenever adequate exercise is not possible, for example because of arthritis or claudication, and is the procedure of choice in patients with left bundle branch block. Four drugs are available. Overall safety and effectiveness is similar for all provided relevant contra-indications are observed.

Dipyridamole

This blocks both reabsorption and metabolism of adenosine, increasing circulating level three- to four-fold. Adenosine is a vasodilator which can also cause atrioventricular block. Its actions are opposed by caffeine, theophylline and related compounds, which must be withheld for 24 h before the test. Vasodilatation decreases coronary vascular resistance and increases perfusion of normal myocardium. Stenosed vessels have little or no reserve capacity to dilate. Contrast between normal and abnormal areas is thus increased but there is potential to create a steal, leading to acute myocardial ischaemia. Dipyridamole is given by slow intravenous infusion at a rate of 0.14 mg/kg/min for 4 min. A longer infusion does not improve hyperaemic response and is associated with an higher incidence of side effects. Radiopharmaceutical is administered 2 min after completion of dipyridamole infusion.

Further reading:

Journal of Nuclear Medicine 1995; ***36****: 575–80, 2016–21*

Adenosine

This is administered at 140 μg/kg/min by constant infusion. Tracer is given after 3–4 min and infusion continued for a further 1–2 min. Both adenosine and dipyridamole have the same final route of action. There are rarely significant ST segment ECG changes in patients receiving adenosine or dipyridamole.

Further reading:

Journal of Nuclear Medicine 1995; ***36****: 276–7*

Toxicity and precautions

Caffeine and methylxanthines should be omitted for at least 12 h before the examination as they compete with adenosine for the same A2 receptors. Adverse effects are similar for both drugs. These include:

- a moderate increase in heart rate in most patients
- facial flushing and chest pain in over one-third of subjects, including those without coronary disease
- vascular headache
- anxiety
- dizziness
- tachycardia
- occasionally morbid fear
- dyspnoea
- throat or epigastric pain.

They may be more intense with adenosine infusion because higher plasma levels are achieved but resolve more rapidly because of the shorter biological half-life.

Both adenosine and dipyridamole are contra-indicated in patients with asthma or obstructive airways disease; there is a serious risk that bronchospasm may be provoked or exacerbated. Second or third degree heart block is a contra-indication unless a pacemaker is fitted but there is no evidence that pre-existing first degree heart block progresses to higher grades with adenosine. First degree block appearing during infusion may progress and the infusion should be terminated immediately. Patients with sinoatrial disease are particularly at risk of profound bradycardia, which may progress to sinus arrest. Aminophylline injection should always be at hand to

reverse bronchoconstrictor effects.
The dose of aminophylline should not exceed 5 mg/kg, given by slow intravenous injection over 20 min. Neurological symptoms resembling a transient ischaemic attack have been described. The bronchodilator effect of aminophylline may wear off before dipyridamole is inactivated. The patient must therefore be kept under observation for more than 1 h after aminophylline. In practice both dipyridamole and adenosine have a good safety record provided that patients with contraindications are recognized. Serious adverse effects occur in less than one examination in 10 000 and are reversible provided proper precautions are in place and staff are adequately trained.

Further reading:
Nuclear Medicine Communications 1994; ***15****: 578–85*

Dobutamine
This is a short-lived predominant β-1 agonist with a plasma half-life of 2 min. It increases heart rate and myocardial contractility. Its mild α-1 agonistic activity permits safe infusion into a peripheral vein and it has a relatively low propensity for provoking dysrythmias. At high doses it also increases blood pressure. It is administered by constant infusion at progressively increasing dose levels, starting at 2.5 μg/kg/min and increasing by 2.5 μg/kg/min every 5–8 min until one of the end-points is reached or a dose of 20 μg/kg/min has been given for 8 min. If there is persistent bradycardia, diagnostic accuracy may be improved by giving 0.25–1.0 mg atropine to provoke a tachycardia. Its action more closely simulates exercise than does that of adenosine as the overall effect is to increase myocardial oxygen demand, causing normal distal coronary vessels to dilate whereas stenosed arteries may not be able to allow perfusion to increase; indeed flow resistance may increase at a stenosis. It is the pharmacological agent of choice in asthma. In most other patients no convincing difference in efficacy has been demonstrated but it is the only one of the four which permits both an incremental dose schedule and recognized end-points. It should not be employed in patients with left bundle branch block, in whom spurious (false positive) reversible antero-septal perfusion defects are commonly observed.

Further reading:
Journal of Nuclear Medicine 1994; ***35****: 737–9. 1997;* ***38****: 424–7*

Toxicity and precautions
Dobutamine is contraindicated in patients with hypertension, which may be exacerbated. Flushing, nausea, headache, tingling and light-headedness are related to stimulation of the sympathetic nervous system. Ventricular tachycardia occurs in about 4% of patients but is usually not sustained. Atrial or ventricular premature beats occur in 10–15% of patients. Supra-ventricular tachycardia, atrial fibrillation and dyspnoea also occur. Hypotension is occasionally observed.

Atropine blocks parasympathetic (vagal) activity and thus causes tachycardia. It may precipitate urinary retention, especially in older male subjects. It causes dry mouth, difficulty in swallowing, dilatation of the pupils and difficulty in visual accommodation. This may affect the ability of the patient to drive safely for several hours. Rare side effects include confusional states and precipitation of acute glaucoma.

Arbutamine
This is also a potent short-acting sympathomimetic designed as a diagnostic agent with both chronotropic and inotropic effects, although its plasma half-life is longer than that of dobutamine. It is given by constant infusion using a pump controlled by feedback from heart-rate and blood pressure monitors attached to the patient. The pump can thus be set to achieve a preset heart rate or double product [(heart rate) × (systolic blood pressure)] and to maintain this for a predetermined time. This is a relatively new agent and experience with it is limited.

Further reading:
Journal of the American College of Cardiology 1995; ***26****: 1159–67*

Combined exercise and pharmacological stress
In normal subjects myocardial uptake of tracers is highest after dobutamine, marginally lower with adenosine or dipyridamole and substantially

lower with exercise. Lung and liver uptake are however also higher after pharmaceutical agents than exercise. Clearance from all tissues is slower. Exercise reduces the incidence or severity of some adverse effects of dipyridamole, especially dizziness, and reduces hepatic uptake of cardiac tracers by redistributing cardiac output away from viscera and into muscles. It has therefore been proposed that patients should exercise during or after stress but before administration of radiopharmaceutical to divert perfusion away from splanchnic areas. The combination of dipyridamole and exercise probably gives the best contrast between normal and abnormal myocardium but no difference in accuracy has been observed in the relatively small series reported. Restricted and symptom-limited exercise, for example walking on the level, have both been proposed. It has not been established which exercise protocol is best. Any difference is clearly small and a large series would be required to establish precedence. The duration of action of dipyridamole in man is uncertain, but is at least 15 min and possibly up to 1 h. Because of their short duration of action, adenosine and dobutamine infusions have to be continued until the radiopharmaceutical has been given and are thus less convenient for a combined protocol. Rates of uptake and wash-out and the extracardiac distribution of tracer differ depending which drug is used and whether exercise is employed. Quantitative criteria must be validated separately depending which protocol is employed.

Further reading:

Journal of Nuclear Medicine 1994; ***35****: 535–41*

Imaging

Protocols

The principal protocol variations are summarized in Table 4.1.

Planar

Good clinical results can be obtained with planar imaging using a general purpose low energy collimator to give an adequate count density in three or four projections within 30 min, from anterior to left lateral in equal angular increments (Fig. 4.1).

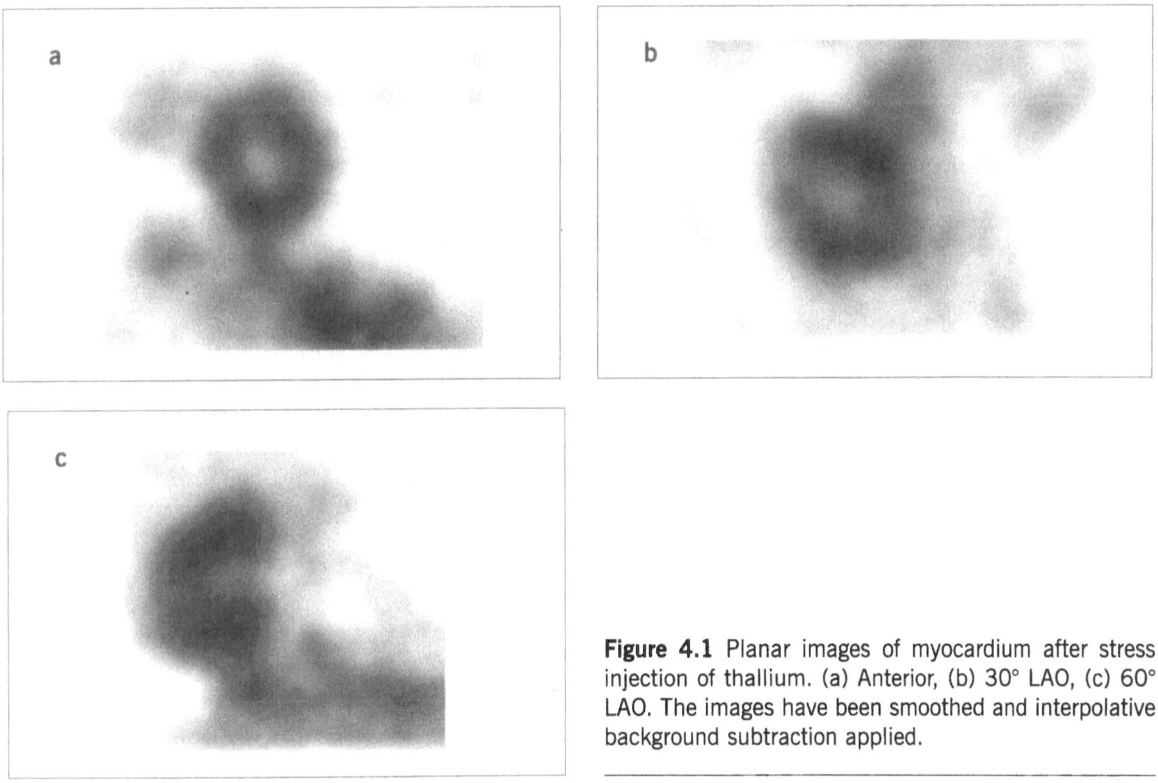

Figure 4.1 Planar images of myocardium after stress injection of thallium. (a) Anterior, (b) 30° LAO, (c) 60° LAO. The images have been smoothed and interpolative background subtraction applied.

Table 4.1 Summary of protocols suggested for myocardial perfusion imaging.

	Step 1	*Step 2*	*Step 3*	*Step 4*
1	Stress	80 MBq Tl	SPECT	3–4 h wait
2	Stress	80 MBq Tl	SPECT	3–4 h wait
3	Stress	80 MBq Tl	SPECT	3–4 h wait
4	Stress	80 MBq Tl	SPECT	40 MBq Tl
5	Stress	80 MBq Tl	SPECT	>18 h wait
6	Stress	250 MBq Tc	0.5–1 h wait	SPECT
7	Stress	250–500 MBq Tc	0.5–1 h wait	SPECT
8	250 MBq Tc	0.5–1 h wait	SPECT	Stress
9	250–500 MBq Tc	0.5–1 h wait	SPECT	>18 h wait
10	**0.3–1.0 mg sublingual GTN**	**5–10 min wait**	**80 MBq Tl**	**15–60 min wait**
11*	**0.3–1.0 mg sublingual GTN**	**5–10 min wait**	**250 MBq Tc**	**0.5–1 h wait**

	Step 5	*Step 6*	*Step 7*	*Step 8*	*Step 9*
1	SPECT				
2	SPECT	18–24 h wait	SPECT		
3	SPECT	>18 h wait	80 MBq Tl	0.5–4 h wait	SPECT
4	2–4 h wait	SPECT			
5	80 MBq Tl	0.5–4 h wait	SPECT		
6	750 MBq Tc	0.5–1 h wait	SPECT		
7	>18 h wait	250–500 MBq Tc	0.5–1 h	SPECT	
8	750 MBq Tc	0.5–1 h wait	SPECT		
9	Stress	250–500 MBq Tc	0.5–1 h wait	SPECT	
10	**SPECT**	**Stress**	**250–500 MBq Tc**	**0.5–1 h wait**	**SPECT**
11*	**SPECT**	**Stress**	**750 MBq Tc**	**0.5–1 h wait**	**SPECT**

Recommended protocols are highlighted in bold.
* A 2-day protocol as in 7 can be substituted, in which case 250–500 MBq Tc is given on each occasion. The order of stress and rest studies can be reversed.
'Tc' may be any of the technetium-labelled myocardial agents.

Accurate repositioning is important for comparability between early and late films. This is best achieved by careful angulation of the camera head while the patient remains supine. It is more difficult to reproduce accurately the positioning of an erect patient. Smoothing or filtration (page 319) may enhance planar images but must be employed with circumspection as contrast is altered and artefacts may be introduced. Approximate arterial territories corresponding to scintigraphic regions in planar images are shown in Fig. 4.2.

Tomography

SPECT is generally considered preferable to planar imaging as it enables better separation of structures and visualization of the posterior wall of the left ventricle. It provides higher contrast images. Phantom studies indicate that a high resolution or long-bore collimator optimized for tomography is preferable to general purpose or high sensitivity (page 305). The examination may be performed with the patient either supine or prone. The latter minimizes artefacts often observed in the inferior wall due to superimposition of liver, but many older patients have difficulty in maintaining this posture. It is necessary for patients to keep their arms above the head in order to allow the camera close to the thorax in lateral projections; many find this difficult for the extended time required with a single-head system. A 180° rotation (or 90° if two heads at right-angles are available) from 45° RAO to 45° LPO, acquiring 64 projections in a non-circular body contouring orbit is usually employed. The matrix chosen should give a pixel size between one-half and one-third of system resolution (page 306) at the depth of the heart. Most comparisons have found no advantage for a 360° acquisition even when using technetium. The count rate of thallium may be improved by including the higher energy (168 keV) peak in addition to the more commonly employed 80 keV emissions. The patient should not move their arms during acquisition as this changes the pattern of scattered radiation, thus introducing potentially serious errors into the reconstruction.

Gated SPECT acquisition is possible with more modern equipment. A higher administered activity is required than for planar imaging and even so acquisition time is substantially extended. Its clinical role is unclear.

Further reading:

Quarterly Journal of Nuclear Medicine 1997; ***41****: 1–9*

Data presentation

Equivalent stress and redistribution or re-injection short axis, horizontal long axis and vertical long axis images corrected for obliquity should be displayed side by side (Fig. 4.3). The choice of whether these are presented in monochrome or colour depends on the

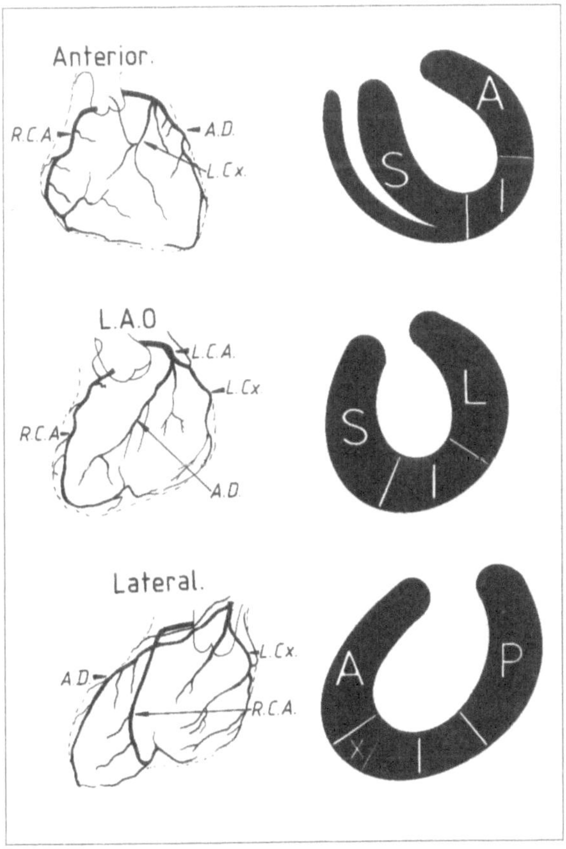

Figure 4.2 Diagrammatic representation of distribution of vascular territories. A, anterior or antero-lateral wall; I, inferior wall, superimposed on apex in anterior projection; P, posterior wall; X, apex; S, septum; RCA, right coronary artery; LCA, left coronary artery; Lcx, circumflex branch of the left coronary artery; AD, anterior descending branch of the left coronary artery.

characteristics of the recording medium (page 315) and personal preferences of the observer, remembering that 5% of the male population is red/green colour blind.

Data processing

As an initial quality check SPECT images should be replayed as a cine loop to detect patient movement during acquisition. Other methods of detecting movement artefact include visual inspection of a summed image of all the frames (Fig. 4.4), superimposing a line over selected frames or inspecting the sinogram (Fig. 4.5). Depending on the plane in which movement has occurred, it may be possible to correct by shifting affected frames. If this is not possible, for example if movement is not confined to a single plane, the acquisition must be repeated. Both pre-filtration and filtration are usually required for reconstruction (page 319), although some systems combine these. Optimum settings depend on characteristics of the image and available software. A number of protocols appear to give equivalent information but different implementations of nominally similar protocols may differ substantially. Caution must therefore be exercised when comparing examinations performed on different equipment or from different departments.

A number of techniques of quantifying both

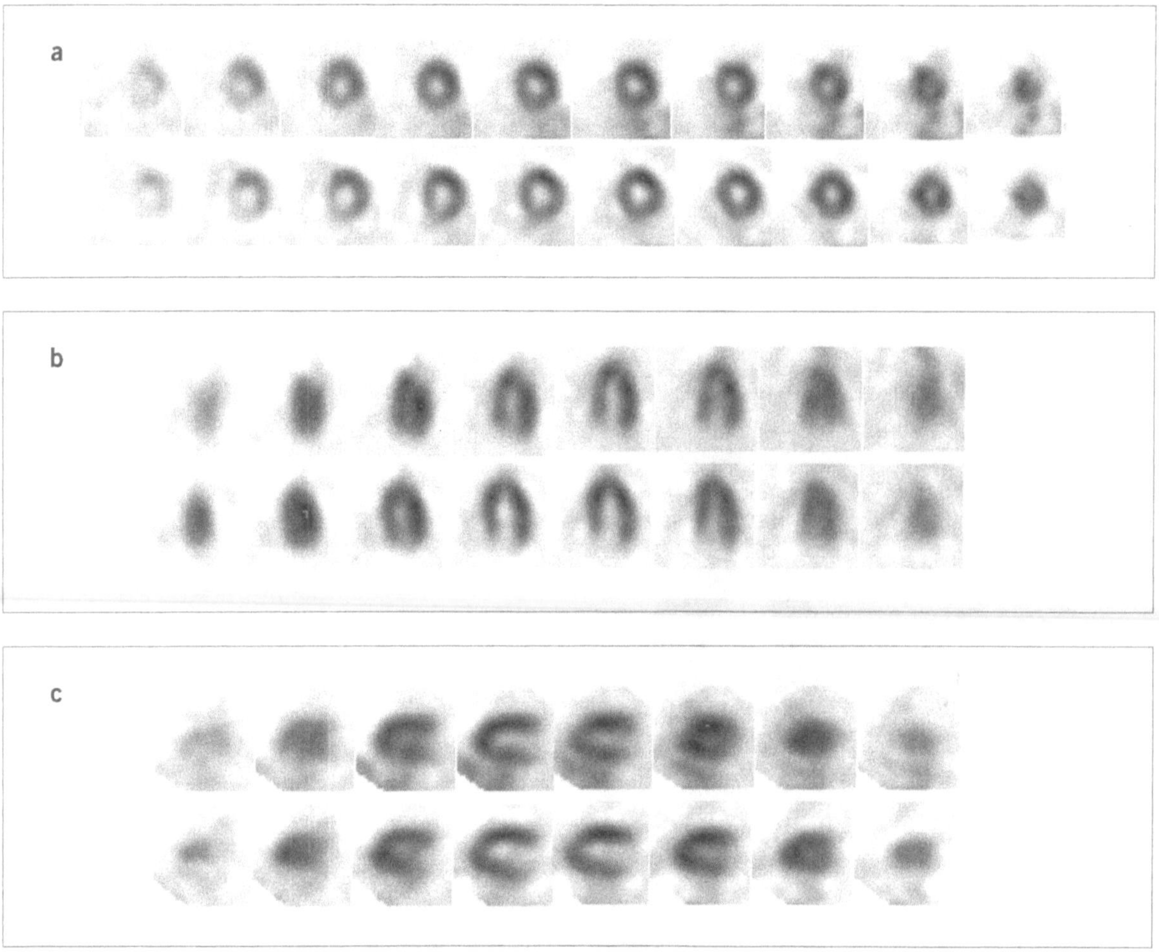

Figure 4.3 Pharmacological stress with adenosine infusion. Normal short axis (a), horizontal long axis (b) and vertical long axis (c) sections. Note physiological apical thinning, unchanged on redistribution. In each set the upper row is at rest and the lower row is a stress study.

planar and SPECT images by comparing normalized count rates in segments have been proposed. Semi-quantitative analysis of planar images reduces subjectivity of comparisons between early and late images provided that repositioning is accurate. It facilitates identification of deviations from a normal range, provided this has been validated locally. In one widely used method for planar images the heart is enclosed in a circular region of interest. After point by point interpolated background subtraction (page 140) a number of equal angular sectors, 16 or more radiating from the centre, are defined and the number of counts in each calculated. These are displayed graphically normalized to the maximum number of counts in any sector (Fig. 4.6). Superimposed plots of radial profiles in stress, redistribution and if available re-injection images, with limits derived from populations of normals, may help to highlight significant abnormalities.

More elaborate processing such as polar analysis of SPECT images is sometimes helpful. These are obtained by calculating radial profiles for consecutive short axis slices, identifying either peak or mean counts and displaying the slices as a series of concentric circles, apex in the centre and each slice in turn forming a series of concentric rings. Thus the polar image in effect shows the heart as though the viewer is looking into the left ventricle towards the apex from the base and the ventricle has been spread flat

Figure 4.4 Sum of all frames in tomographic acquisition.

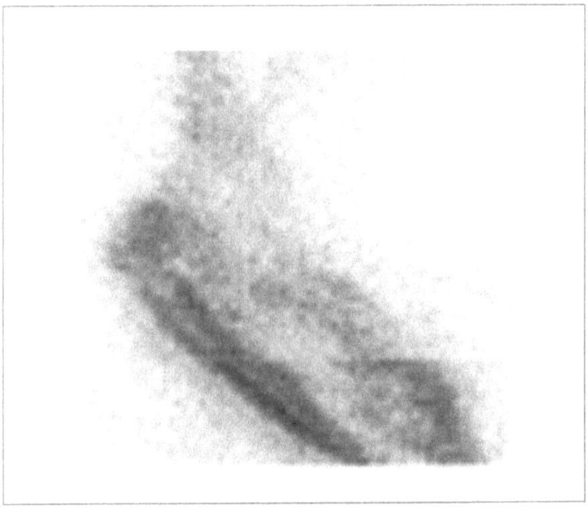

Figure 4.5 Sinogram obtained by 'stacking' SPECT frames to make a three-dimensional matrix which is rotated 90° in the z dimension. This new frame represents the same single line of pixels in each of the original projections. Discontinuity in lower third is evidence of patient movement.

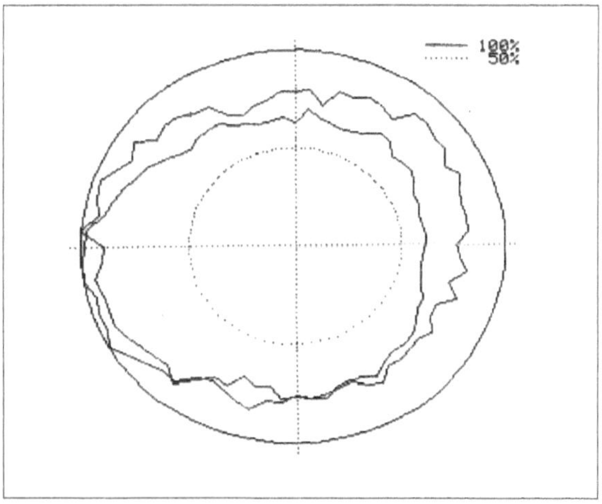

Figure 4.6 Radial profiles of count rate from a pair (rest and stress) of planar short axis projections (Fig. 4.3a) normalized to highest count rate in any sector.

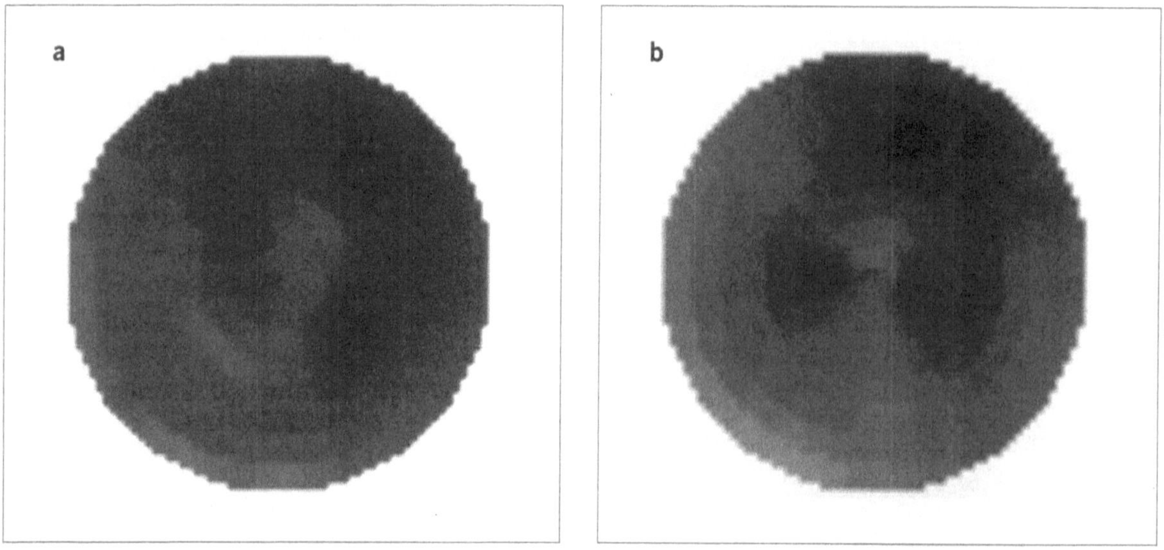

Figure 4.7 Polar images derived from (a) rest and (b) stress sectional data in Fig. 4.3a.

(Fig. 4.7). When using semi-quantitative criteria of analysis the normal range is dependent on all steps in processing. Where comparison with a normative database is employed recommended processing protocols must be strictly followed. Any deviation may invalidate the comparison. Normative databases are gender dependent and also assume comparability of physical dimensions between the reference and test populations. Commercially available databases are mostly derived and validated in the USA and may not be applicable to other populations. Attenuation correction using a transmission source is preferable.

Further reading:
European Journal of Nuclear Medicine 1994; ***21****: 154–7*

Interpretation

Planar images

In the normal subject planar images show the left ventricle as a U- or C-shape of uptake clearly visible above lung activity (Fig. 4.1). Papillary muscles are sometimes identified as small areas of slightly higher count rate on the lateral wall. The right ventricle is seen faintly immediately following stress but is usually not visible on redistribution or re-injection unless there is right ventricular hypertrophy. Uptake in lung is homogeneous and higher in cardiac failure and in smokers than healthy non-smokers. The ratio of lung to heart count rate is significantly higher in patients with coronary artery disease who have been adequately stressed than in normal subjects; a ratio >0.52 is a strong predictor of subsequent cardiac events and is commonly associated with regional abnormalities of wall motion. This is not a reliable criterion if exercise is sub-maximal and may be less reliable in smokers than non-smokers. Nevertheless this information is sufficiently useful to justify an anterior planar image routinely prior to SPECT acquisition.

Small sub-endocardial areas of ischaemia or infarction may not be visible, but these are usually due to disease in vessels too small to be amenable to angioplasty. A defect present only on stress, or larger on stress than at rest, is evidence of ischaemia. It is possible to make some estimate of the vascular territory involved (Fig. 4.2) but the range of variation is such that angiographic distribution of disease is correctly predicted in little over half of all cases. Angiography is always required in patients with a reversible defect to define anatomy and differentiate small vessel from large vessel disease. Pathological defects occur most commonly in the inferior and anterior walls or at the apex. The severity of count deficit is prognostically significant. There is a correlation between extent to which thallium uptake is reduced and amount of muscle damage. More than half of defects in which count rate in the stress study is reduced by less than 50% show improved contractility after revascularization, even though they do not fill in on 24 h redistribution images.

Further reading:
Journal of Nuclear Medicine 1995; ***36****: 944–51*

The criterion of reversibility on same-day (redistribution) thallium images or after resting injection of any of the technetium agents overestimates infarction and fails to identify a quarter or more of treatable ischaemic areas. For this reason, if a defect persists a further injection of thallium should be given with the patient at rest. Routinely preceding thallium injection with a nitrate improves accuracy of detection of viable myocardium. If a perfusion defect is present following a rest injection, imaging may be repeated at 4 h to identify hibernating myocardium, although depending on population the detection rate is likely to be low. A defect which fills in late indicates hibernating myocardium which is likely to recover if revascularized. A fixed defect has less than a 5% chance of improving after surgery. Defects which fill in only on re-injection or very late redistribution (24 h) usually indicate more severe stenosis. Thallium reinjection is more sensitive than the technetium agents for detecting viable myocardium, although when re-injection is preceded by a nitrate there appears to be little if any difference between the various agents.

Further reading:
Journal of Nuclear Medicine 1994; ***35****: 674–80.*

SPECT

The criteria for SPECT (single photon emission computed tomography) are similar to those for planar images. The higher contrast and clearer delineation of anatomical relationships facilitate interpretation and localization. In the short axis SPECT reconstruction of a normal subject there is uniform activity in myocardium, which appears as a series of rings of similar count rate. The vertical and horizontal long axis images are approximately U-shaped with uniform uptake throughout (Fig. 4.3). The extent of territory supplied by each coronary artery is variable. Fig. 4.8 gives an idealized approximation but it is not possible to identify individual vascular territories with any degree of confidence. There may be a slightly lower count rate at the apex due to normal thinning of the myocardium. This can be distinguished from an apical abnormality if gated SPECT is performed. The normal physiological defect at the apex is less prominent in the systolic frame than in end diastole. Left ventricular wall thickness is comparable with the resolution of SPECT imaging and changes in wall thickness thus cannot be measured directly. Wall thickening is evidence of contraction and thus important evidence of functional integrity. Although not directly measurable by SPECT, increased peak count density in systole compared to diastole is good inferential evidence of wall thickening.

Ischaemia is identified as an area of reduced count rate in the stress image which at least partially fills in on redistribution (Fig. 4.9, 4.10). A filling defect present in both stress and redistribution images suggests infarction (Fig. 4.11).

In severe ischaemia redistribution may not be adequate to fill in a defect within 3–4 h. Any patient with a defect persisting in the redistribution image should therefore have a further imaging after 24 h or re-injection of thallium. Re-injection is preferable as count-rate is low at 24 h, principally because of wash-out and excretion. If a defect persists after re-injection but hibernating myocardium is considered a possibility, a further set of images should be obtained 4 h later. A defect present on the early re-injection image but which fills in after some hours is indicative of hibernating myocardium. Hibernation is a term used to describe myocardium which has reduced its function (contraction) to match a reduced blood supply, typically a stenosed vessel, but which can improve its contractile function if an adequate blood supply is restored. Detection of hibernating myocardium may be enhanced if thallium is administered during a glucose clamp (infusion of thallium over 30 min in 250 ml 10% glucose containing 10 mEq potassium chloride and 5 IU insulin) to enhance potassium turnover.

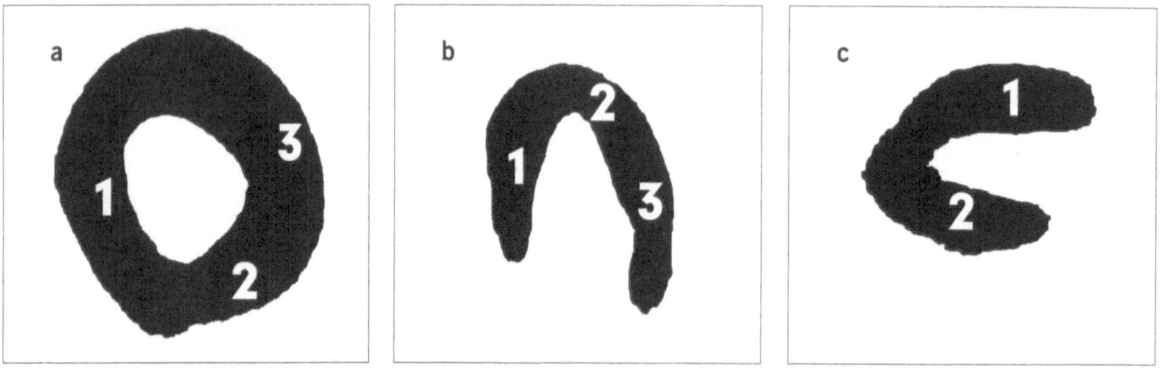

Figure 4.8 Diagram indicating typical arterial territories superimposed on Fig. 4.3. 1, Left anterior descending (LAD) artery; 2, right coronary artery (RCA); 3, circumflex branch of left coronary (LCx) artery.

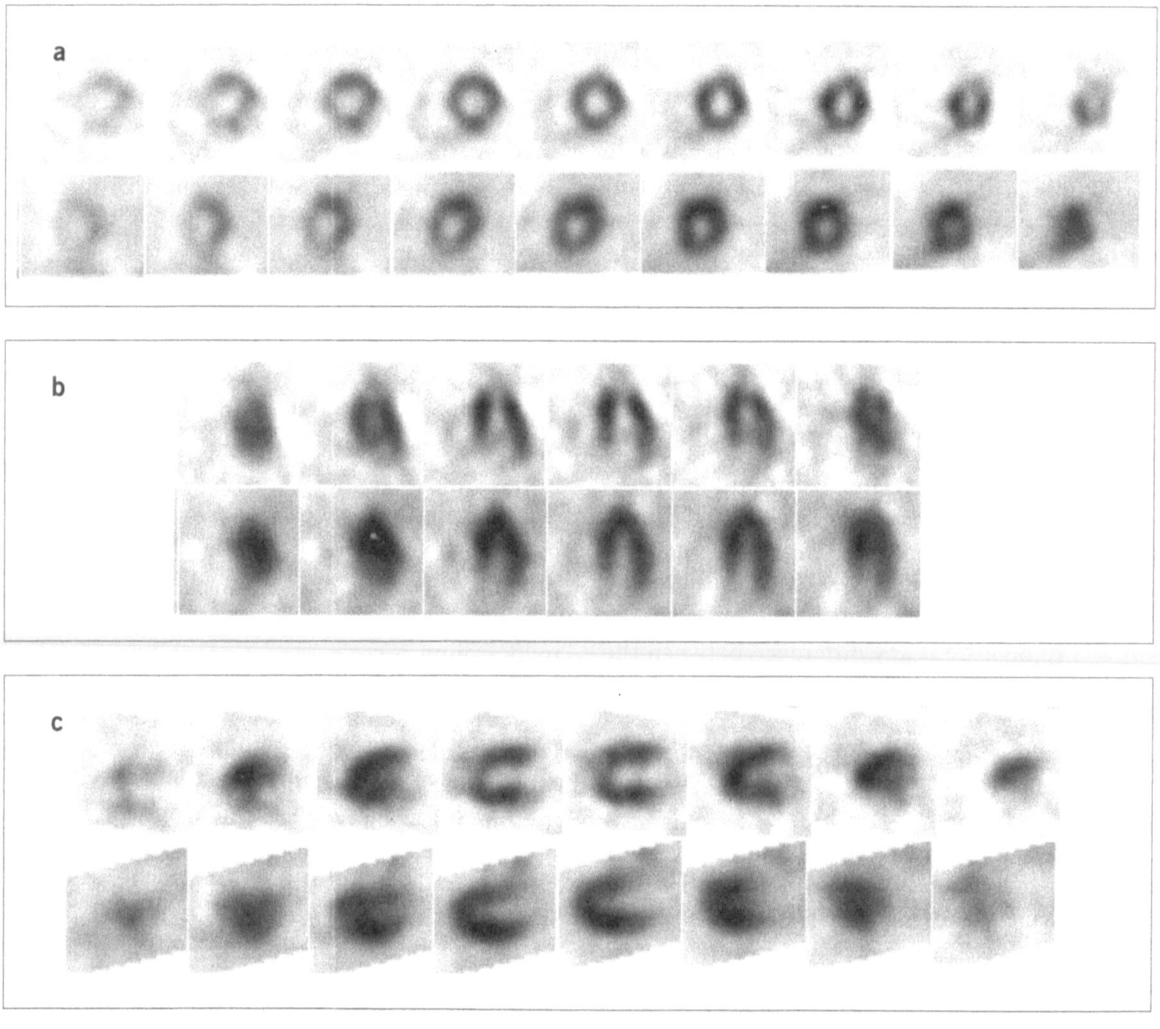

Figure 4.9 Anterior defect on tetrofosmin stress study (upper rows) which fills in after thallium rest injection (lower rows) indicating ischaemia. (a) Short axis, (b) horizontal long axis, (c) vertical long axis.

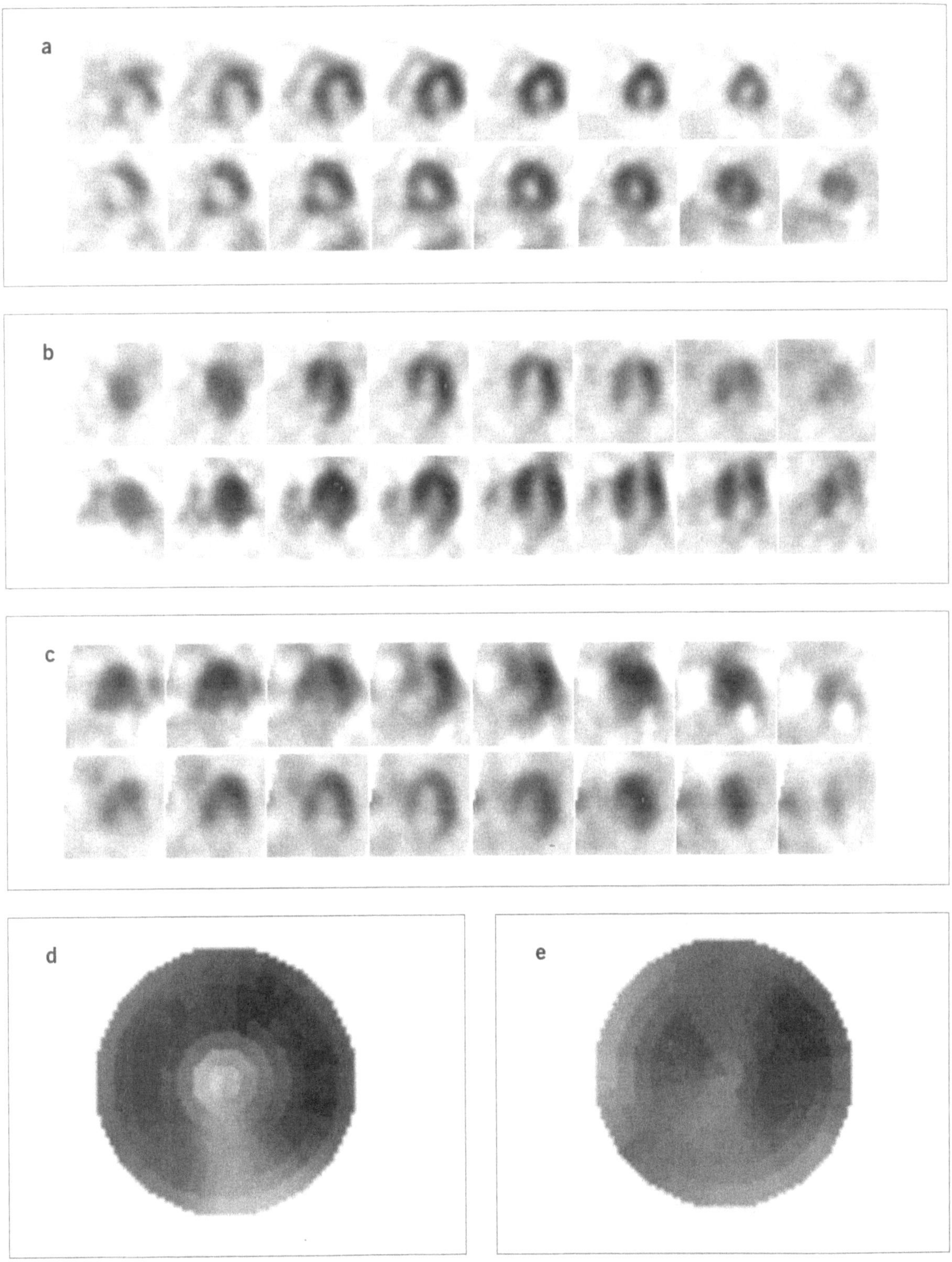

Figure 4.10 Same protocol as Fig. 4.8 showing reversing inferior (RCA territory) defect due to ischaemia. (d) Rest and (e) stress polar images.

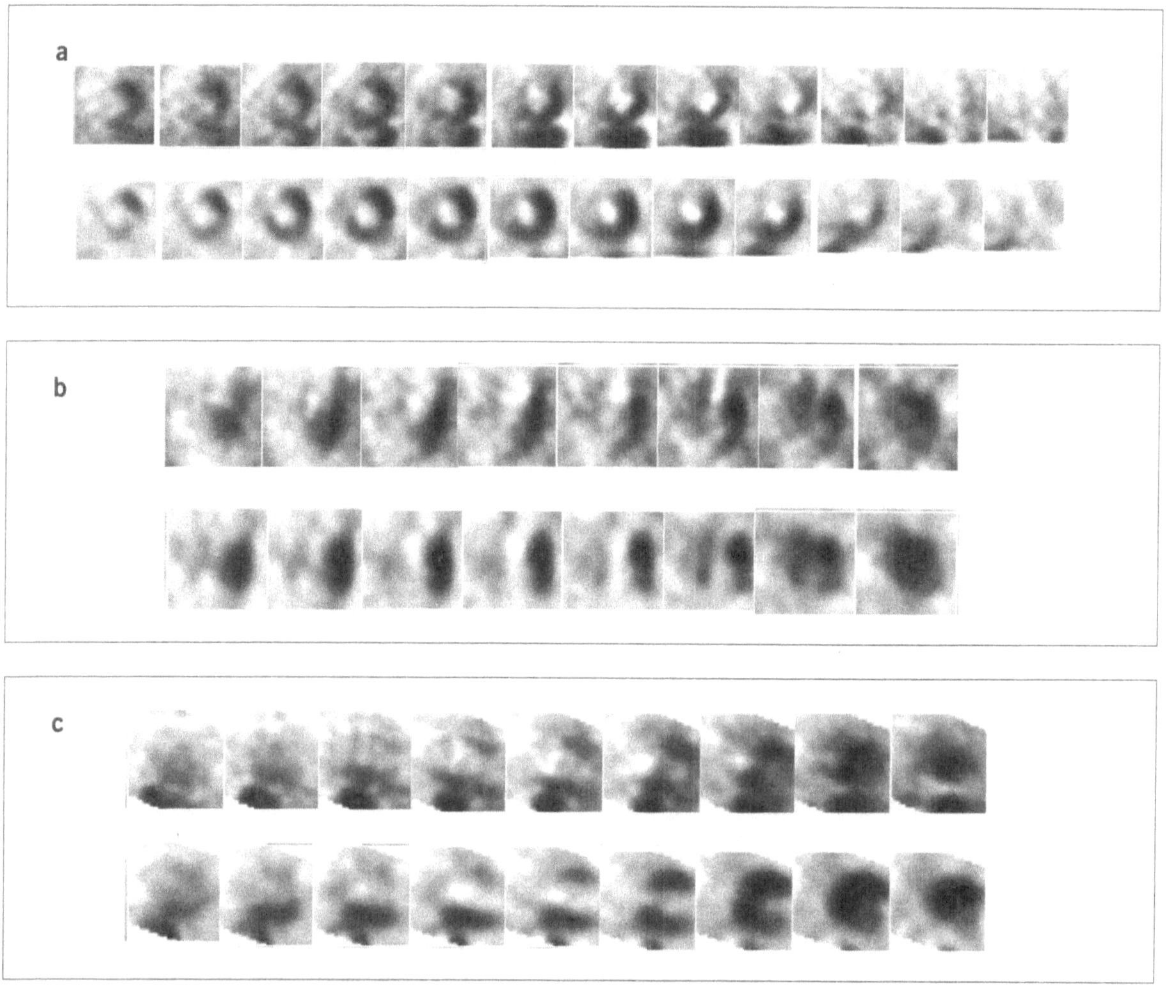

Figure 4.11 Fixed defect due to previous large anterior infarct. (a) Short axis, (b) horizontal long axis, (c) vertical long axis. Upper row rest injection of thallium preceded by GTN, lower ^{99m}Tc-tetrofosmin during pharmacological stress (adenosine).

The alternative and preferable technique is to precede injection of thallium by a nitrate such as 300–1000 μg glyceryl trinitrate spray (GTN) or 5–10 mg isosorbide dinitrate tablet. Either should be given sublingually. GTN has the faster onset of action. There is a strong case for preceding thallium reinjection with a nitrate in all cases.

Further reading:

Journal of Nuclear Medicine 1995; ***26****: 1377–83, 1945–52*

Reverse redistribution, defects not present after stress but which appear in the redistribution or re-injection images, is occasionally encountered. This should not be confused with differential attenuation of thallium and technetium commonly observed when a dual isotope protocol is employed. Reverse attenuation is most commonly observed following thrombolytic therapy of coronary arterial thrombosis, when it indicates perfusion of stunned myocardium which is however likely to recover. Stunning occurs when blood supply to a region of muscle is restored after a period of ischaemia, for example following thrombolysis. It is identified as a region of abnormal reduced wall motion with normal or near-normal perfusion. In practice the distinction between hibernation and repeated stunning due to repeated ischaemic events is not clear cut. When reversed redistribution is observed in patients with a low *a priori* probability of coronary artery disease it is not associated with an unfavourable prognosis and may be due to normal variability of regional thallium clearance or to artefact, whereas in patients with known coronary artery disease it tends to be associated with abnormal wall motion and may indicate a mixture of scar and normal, stunned, or hibernating myocardium. In many cases no explanation is found. When paradoxical redistribution is observed a separate thallium injection should be given at rest, imaging at 30 min and if in doubt again at 4 h.

Further reading:
Journal of Nuclear Medicine 1995; ***36****: 1019–21. 1996;* ***37****: 742–3*

Sources of error

The commonest cause of a false negative is premature termination of stress, before a recognized end-point has been achieved. Extravasation of radiopharmaceutical invalidates the investigation, which may have to be repeated on another day unless a duplicate isotope dose is available. With thallium and to a lesser extent technetium agents overlying soft tissues, especially large breasts, breast prostheses or liver may cause artefactual filling defects, usually in the inferior wall of the left ventricle. Breast artefacts may also be seen laterally. Their location and severity depend on the size of the patient and type of support worn. Prostheses give rise to defects which may differ in location and intensity from normal breast tissue. The problem can be alleviated if the patient can tolerate lying prone for imaging or if a transmission source is employed for regional attenuation measurement.

SPECT images may be invalidated if the patient moves during the study. In some cases defects become less obvious or may be concealed altogether; in others artefactual defects can be created if a region containing myocardium in some projections is substituted by a non-cardiac volume in others. As technetium agents do not redistribute it should not be necessary to repeat the stress procedure but only repeat imaging. If movement occurs during a thallium stress acquisition there may be no alternative to repeating the stress procedure. Upward 'creep', a continuous craniad shift of the heart during the period of acquisition, may be related to a transient increase in total lung volume during exercise.

This is most commonly seen with thallium stress acquisitions because imaging must be started before the patient has fully recovered from the stress procedure. It can be detected by viewing the cine display or placing selected images, for example every eighth, in a row and superimposing two parallel lines which indicate the limits of the first projection. The heart should be contained within the same lines in every projection. It is sometimes possible to correct for movement by shifting affected projections before reconstruction, but in many cases there is no satisfactory alternative to repeating the examination.

Failure to detect any perfusion defect in a patient who has not achieved a recognized end-point is unreliable. If a patient is not able to exercise adequately for non-cardiac reasons, pharmacological or dual stress should be used. Failure to image sufficiently late after re-injection precludes the possibility of detecting hibernating myocardium. The technetium agents tend to underestimate the extent of recoverable myocardium and to overestimate extent of irreversible disease.

Further reading:
Journal of Nuclear Medicine 1993; ***34****: 1355–6. 1994;* ***35****: 674–80. 1995;* ***36****: 905–6, 921–31*

Indications

Myocardial imaging is most useful in patients with or suspected of coronary artery disease in whom the *a priori* probability of myocardial ischaemia is neither very high nor very low. The pick-up rate does not justify its use as a screening test in asymptomatic low-risk subjects. If there is typical angina pectoris or there are unequivocal electrocardiographic signs of ischaemia, coronary angiography is indicated. Scintigraphic imaging is indicated for:

- evaluation of atypical chest pain

- when ECG interpretation is compromised, e.g. by bundle branch block
- detection of regional myocardial ischaemia, e.g. in association with previous infarction
- evaluation of extent of viable but ischaemic myocardium
- evaluation of amount of viable myocardium, e.g. when considering revascularization
- evaluation of functional significance of an angiographic stenosis
- detection of hibernating or stunned myocardium
- risk stratification in hypertension or following myocardial infarction [Three features have strong predictive value for further cardiac events following myocardial infarction: extent of ischaemic regions, their number, and increased lung uptake of thallium. The latter correlates with reduction in ejection fraction (page 146).]
- identification of ischaemia in hypertrophic cardiomyopathy [Defects on the resting study precede clinical signs of disease. Increasing number or size of defects in a clinically stable patient carries a serious prognosis..]

Further reading:
*Journal of Nuclear Medicine 1993; **34**: 1013–9*

Infarct imaging

Radiopharmaceuticals

A number of technetium-labelled agents may be used to visualize severely ischaemic or recently infarcted myocardium. The highest contrast is seen with ^{99m}Tc-pyrophosphate but uptake may be observed with most, if not all, of the phosphate derivatives used for bone scintigraphy, with ^{99m}Tc-glucoheptonate as well as many other compounds, most now only of historical interest. Quality control of pyrophosphate is important as activity persisting in the blood pool can either mimic or disguise an infarct. Freshly prepared preparations are most reliable. The tracer cannot reach regions which are totally avascular, uptake being confined to the periphery and areas which have some residual blood supply but where myocardium is severely ischaemic or dead. An anti-myosin monoclonal antibody labelled with ^{111}In has also been employed.

Timing

Pyrophosphate imaging has a high false negative rate if performed within 24 h of infarction. The sensitivity is highest at 48–72 h, when at least 95% of transmural and 80% of sub-endocardial infarcts are detected. Although the intensity of uptake usually fades within 14 days, it persists for several months in up to one-third of cases. Imaging may be performed 2 h after radio-pharmaceutical injection, but contrast may be higher at 4 h.

Technique

The usual activity is 250–500 MBq. SPECT is preferable to planar imaging. Both transverse and coronal reconstructions should be examined. For planar imaging anterior, left anterior oblique and left lateral projections are preferred, collecting not less than 500 000 counts with a high resolution collimator.

Interpretation

Uptake is graded by comparison with adjacent bones. A simple but adequate system is:

- Grade 0, no uptake visible in heart region
- Grade 1, definite but faint uptake (Fig. 4.12).
- Grade 2, definite uptake of similar or greater intensity than adjacent ribs (Fig. 4.13)
- Grade 3, intensity equal to or greater than sternum.

Only grades 2 and 3 are evidence of myocardial infarction. When assessing grade from planar views the highest uptake in any projection determines the final grade.

Clinical applications

In most patients myocardial infarction can be diagnosed earlier by electrocardiographic and biochemical criteria. There remain a small number in whom these are not adequate and scintigraphy is useful, especially those with left

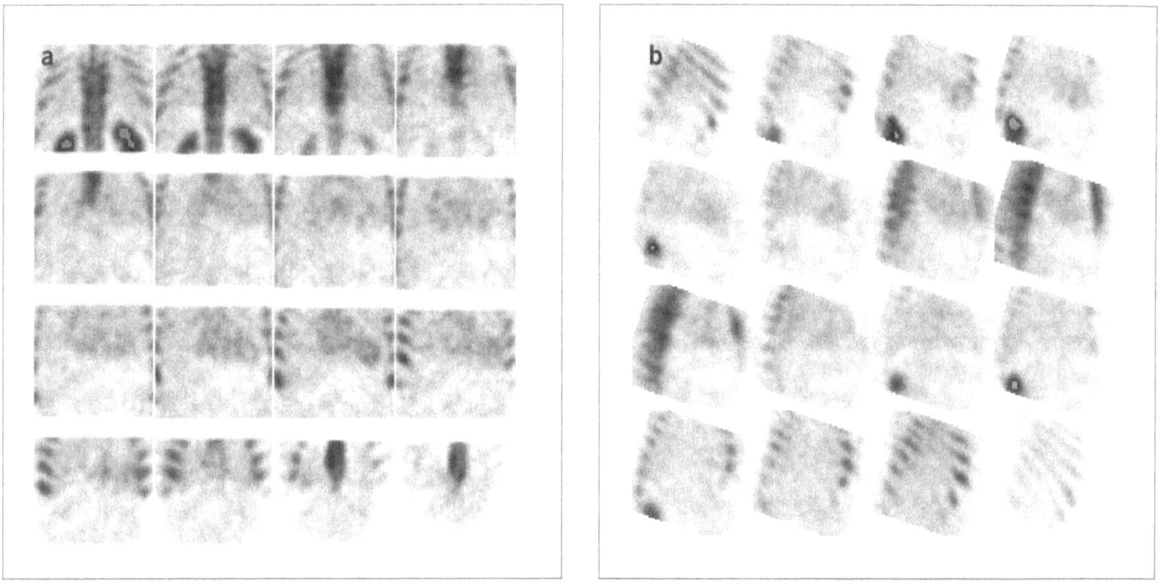

Figure 4.12 (a) Coronal and (b) sagittal SPECT slices following pyrophosphate injection, showing non-specific generalized grade 2 uptake in myocardium, best seen in third frame of third row of coronal slices and second frame of second row of sagittal slices.

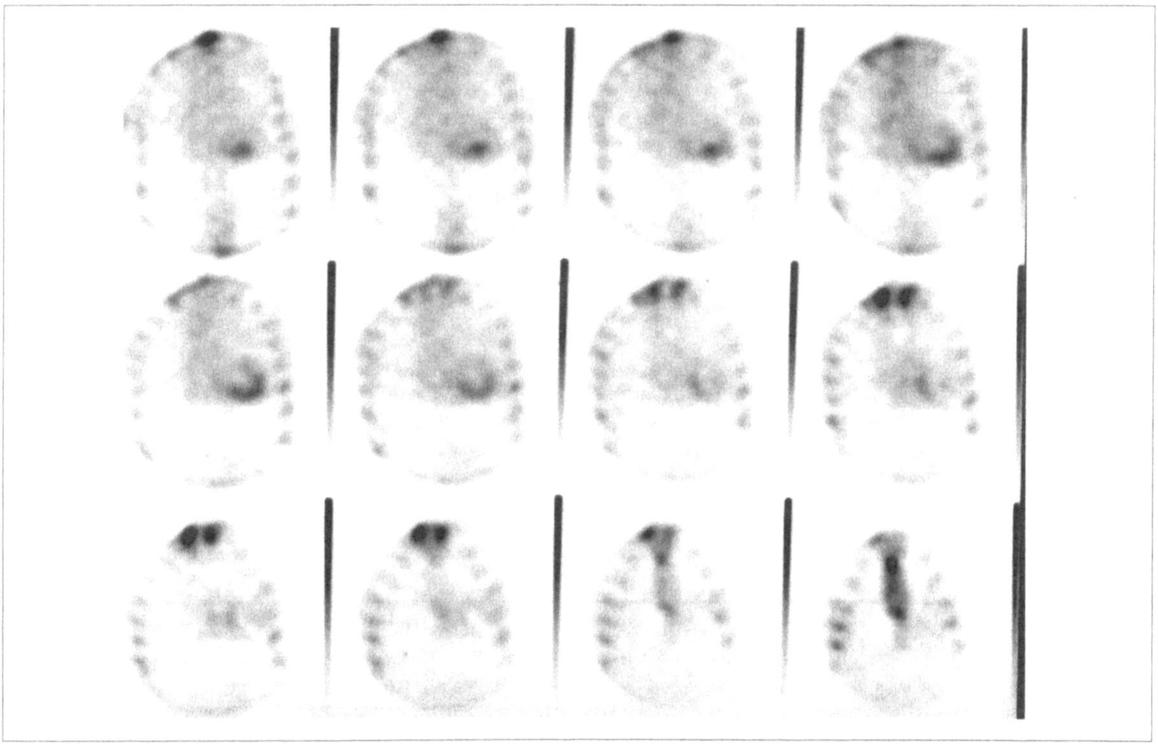

Figure 4.13 Coronal SPECT slices following pyrophosphate injection, showing infero-apical grade 3 uptake in myocardium, indicating recent infarct.

bundle branch block, previous myocardial infarction, immediately following open heart surgery, when presentation to hospital is delayed for several days and in some patients following cardio-pulmonary resuscitation. Uptake has been demonstrated *in vitro* in non-infarcted regions after as little as 30 min ischaemia, although contrast may not be sufficient to be detectable *in vivo*.

Further reading:

*Journal of Nuclear Medicine 1994; **35**: 1366–70*

Ventriculography

Ventriculography is a general term covering techniques of evaluating the left and right ventricles. Anatomy is best demonstrated by magnetic resonance imaging (MRI), x-ray contrast angiography or ultrasound. Ventricular function can be assessed angiographically, ultrasonically or by radioisotope techniques. The angiographic ventriculogram is commonly but mistakenly regarded as the reference technique against which other methods should be compared. However angiographic measures of ventricular volume make assumptions about completeness of mixing of contrast in the chambers, symmetry of ventricular anatomy and the extent to which trabeculation affects definition of the outline which, although not grossly inaccurate in the normal, become progressively less applicable as pathological states intervene. There are also a number of errors and approximations, especially about symmetry, in ultrasonic assessment of ventricular shape and function. Thus there is no single reference technique. All techniques have inherent limitations and there is no absolute standard against which all other techniques can be compared. Radionuclide ventriculography is simple, safe, quick and reproducible. Provided that proper attention is paid to technical aspects of the examination, consistent and useful clinical results can be obtained.

Radiopharmaceuticals

Any radiopharmaceutical which remains intravascular may be employed. For practical purposes only two technetium-labelled agents are common, autologous erythrocytes (RBC) and human serum albumin (HSA). There is little to choose between them and selection may be determined by local availability. Not all technetium HSA preparations are stable *in vivo*. However it is easier to prepare the required activity of Tc-HSA (350–550 MBq) in a volume of 1–2 ml, which is necessary if a satisfactory bolus is to be given for a first-pass examination. If the simple *in vivo* labelling technique is used with labelled red cells (page 134), the activity may be given as a small bolus of pertechnetate, which remains intravascular long enough for a first-pass study to be performed but not for calculation of cardiac output. First-pass ventriculography may also be combined with a bolus of any of the technetium agents used for myocardial imaging (page 116).

Red cells may be labelled in two ways, either *in vitro* or a combined *in vivo* and *in vitro* method. For either technique an intravenous injection of 1 mg of a soluble stannous salt which becomes attached to the surface of red cells, such as stannous pyrophosphate or stannous fluoride, is administered. For the *in vivo* technique the patient is injected after an interval of 15 min with 350–500 MBq sodium pertechnetate. About 85% of the pertechnetate becomes attached to red cells. Unlabelled activity is excreted principally via the urinary tract but some accumulates in iodine trapping tissues and some remains intravascular. For the combined technique, after the same initial interval a 5–7 ml venous blood sample is withdrawn into a syringe containing the radioactivity and 2–5 IU heparin as an anticoagulant; excess heparin reduces labelling efficiency.

This is gently shaken and left to stand at room temperature for 10 min. It may then be reinjected to give an appreciably higher (90–95%) labelling efficiently than the simpler *in vivo* technique. If the combined technique is employed the final volume is commonly 5–10 ml, which is too large for bolus injection into a peripheral vein unless a very large (>16 G) cannula is employed. For most purposes the simpler technique is adequate. However unbound activity diffuses out of the vascular compartment, giving higher background and lower contrast of vascular structures.

Techniques

Two methods are employed, first pass and gated. The former is associated with a lower extra-cardiac background, which is one of the principal sources of error, but the number of counts which can be collected from the heart is limited by the characteristics of the detector. Theoretical considerations indicate that, to obtain the same statistical precision as a first-pass study, it is necessary to collect only a 2 min gated resting equilibrium study or 1 min during stress. Other factors being equal, the gated study has better precision when cardiac output is high and the first pass when it is low. However when cardiac output is low, residual lung activity during the first pass is relatively higher and correction for background correspondingly less accurate. Gated studies are therefore performed more commonly but, as the patient has to receive an injection of tracer for the gated study, two separate estimates of ventricular function can be obtained at no extra cost at the same session by preceding the gated with a first pass. This provides an internal quality control. If they agree, there is clearly a much greater level of confidence in the result. If they do not, both need to be checked in order to determine why the discrepancy has occurred. When one of the technetium agents is used for myocardial imaging (page 116) it is possible to use the injection as a bolus and thus obtain a first-pass measurement of left ventricular function accompanied by imaging of regional myocardial perfusion.

First pass

The considerations when performing a first-pass study are identical to those when measuring cardiac output, namely delivery of a fast bolus of a small volume of tracer into a convenient large peripheral vein. The patient is positioned supine. If a dual-headed camera with heads at right-angles is available, two sets of views can be obtained simultaneously. If a single-headed camera only is available, 30° right anterior oblique is the projection of choice as it gives the best visualization of left ventricle. The right and left ventricles are partially superimposed in this projection, but filling of the two is separated in time unless there is a poor bolus injection or right ventricular failure. Reproducible positioning is best achieved by angling the camera while the patient remains supine. If orthopnoea precludes a supine examination it can be performed erect, although accurate repositioning is more difficult. Unless a very short-lived radionuclide such as ^{195m}Au is available repeat measurements can be obtained only by using incremental doses of ^{99m}Tc, for example 150 MBq for the first injection, 300 MBq for the second and, if permitted, 600 MBq for a third. Increasing background limits accuracy when this technique is employed.

Activity administered must be related to the characteristics of the camera and collimator in use in order to avoid excessive dead time losses, especially in the right ventricle. By the time the bolus has passed through lungs and returned to left ventricle it is diluted into a greater volume, so that peak count rate observed from left ventricle is approximately one-third to one-half of that from right ventricle. A higher activity must be administered to obtain an adequate count rate from left ventricle. With many cameras it is difficult to obtain diagnostic curves from both chambers at a single examination. Unless there is a specific need to measure right ventricular ejection fraction, distortion of the right ventricular time activity curve may be ignored and administered activity adjusted to give an adequate count rate in the left ventricular curve. It should not be necessary to exceed 500 MBq ^{99m}Tc. Dead time limits maximum usable activity. When using a camera with a field view greater than 30 cm diameter, much of the detected count rate originates from lungs and superior vena cava. Dead time losses depend on the total number of counts in the entire field of view. If acquisition is zoomed, as is commonly the case, counts reaching the detector but outside the matrix being collected still affect dead time and must be taken into account when calculating the correction. Covering the excluded part of the collimator face by 2 mm thick lead considerably reduces dead time losses and may thus improve the study.

The optimum frame time is half the duration of end systole, (approximately 0.05 s, depending on heart rate). Thus a 20 s acquisition contains 800 frames with few counts in each. There is nothing to be gained by using a matrix finer than 64 × 64, or a pixel size smaller than 10 × 10 mm. When cardiac output is low and transit from right to left ventricle correspondingly slow, acquisition times in excess of 30 s are occasionally required. However under these circumstances studies are often non-diagnostic as it may not be possible to differentiate adequately between first pass and recirculation.

Data processing

For analysis, regions of interest are defined by creating a summed view from all frames in which activity is identified in the chamber of interest. The selection of frames is not critical, although if too early a portion of the study is selected the left ventricular outline may be partially obscured by right ventricle. When defining a left ventricular region of interest, care must be taken to exclude aortic root, descending aorta and left atrium. If the camera has been positioned correctly these should be readily identifiable. A background region as close as possible to ventricle, but not including any part of the chamber, descending aorta or pulmonary veins must also be identified. This should be crescentic, along the left border, apex and inferior border of the heart (Fig. 4.14a). Time–activity curves for ventricle and adjacent background area are then calculated. Two major peaks can be identified in the ventricular curve, one early when activity is in the right ventricle and the other later as it passes through the left. Superimposed on these are a number of higher frequency peaks of smaller amplitude due to individual heartbeats (Fig. 4.14b). Individual maxima are due to end-diastole and minima to end-systole. Thus the first-pass study allows intrinsic rather than extrinsic gating. Individual beats should not be identifiable in the background curve. If they are seen, the background area has been drawn too close to ventricle and either overlaps it or contains too high a fraction of counts scattered from it. The use of such a background will over-correct, giving a falsely high value of left ventricular ejection fraction (LVEF). Because of the small number of counts within each region of interest in each frame, random statistical noise makes precise identification of systolic and diastolic frames difficult. Smoothing or filtration of curves facilitates identification of frames containing maxima and minima but, because smoothing tends to decrease maxima and increase minima, it is preferable to revert to original unsmoothed data for calculation of results.

Depending on bolus quality and cardiac output it is usually possible to identify between four and ten beats as the bolus makes its first transit through the left ventricle. Often only two to four beats can be identified in the right ventricular portion of the curve. Ejection fraction (EF) is defined as the fraction of the diastolic volume ejected per beat, comparing each systole with its preceding diastole. After subtraction of background, EF is calculated separately for each beat, using as many beats as can be identified during the first pass through the ventricle. Results are expressed as the mean of a stated number of beats; mean beat-to-beat variation is also calculated. Especially when calculating right ventricular ejection fraction, allowance must be made for dead-time errors resulting from the high count rate due to a relatively undiluted bolus. This method does not require ECG gating and can be used whether or not there is sinus rhythm.

In subjects with normal or only slightly reduced ventricular function extra-cardiac activity is low during the first pass. However if circulation time is increased and there is persistence of activity in the lungs, background is higher. Background correction is at best a crude approximation: one of the attractions of the first-pass technique is that when function is good the correction is small and errors are correspondingly reduced.

Causes of error

The commoner causes of error are:

- incorrect positioning of ventricular or background region of interest
- inadequate bolus injection

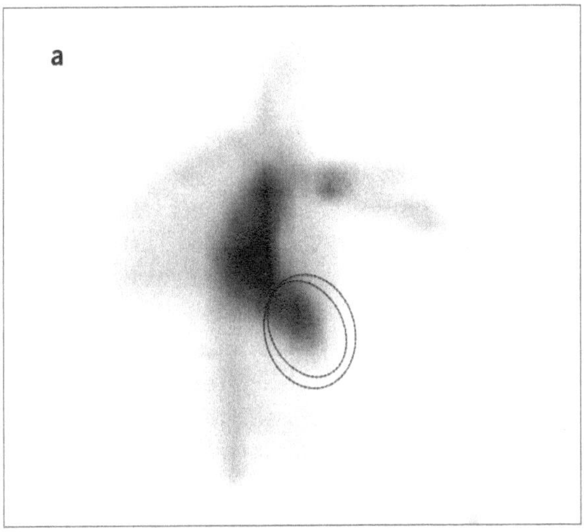

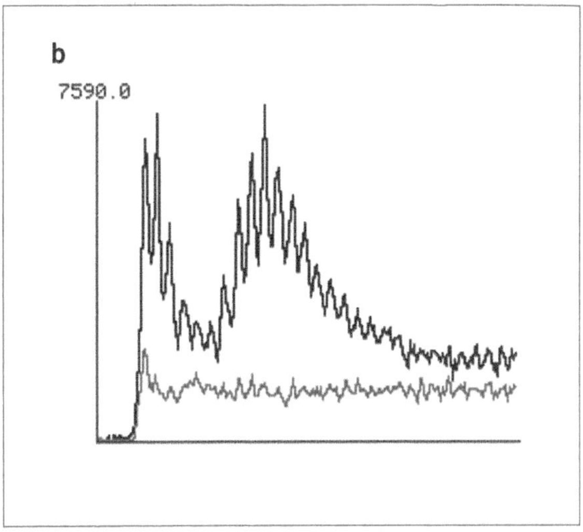

Figure 4.14 (a) Summed view during first pass of bolus through both ventricles, with superimposed ventricular and background regions of interest. (b) Time–activity curves from ventricles and circumferential background region. The count rate during the first pass through the right ventricle is severely attenuated by dead-time losses. Individual systoles and diastoles are clearly visible.

- low right ventricular output
- inclusion of aortic root or descending aorta in the background region
- malfunction of automated methods of region selection, none of which is infallible, especially when function is poor
- incomplete mixing of tracer with the cardiac contents.

First-pass measurements are not applicable if the bolus does not clear from right ventricle before left ventricle is filled, for example in patients in severe cardiac failure. The technique may not be applicable in subjects without remaining peripheral veins of any size, in whom rapid bolus injection is often impractical. This is found most commonly in patients who have received chemotherapy for cancer.

Further reading:

Journal of Nuclear Medicine 1990; ***31****: 1300–2*

Gated acquisition

Non-imaging

Although most commonly performed as an imaging procedure in order to obtain regional as well as global information, non-imaging probes are available for measuring global cardiac function. Their advantages are relatively small size, which permits them to be attached to a belt or jacket to monitor changes over several hours, whereas simple collimation provides much higher sensitivity than a gamma camera, allowing beat by beat time–activity curves to be obtained rather than gated curves averaged over a large number of beats. The principal disadvantages are difficulty of accurate placement and background correction. Most use either ultrasound or a gamma camera to identify the left ventricle and either subtract an arbitrary background or employ a second probe over lung at some distance from the heart. Their applications are principally in research and drug monitoring.

Further reading:

Journal of Nuclear Medicine 1992; ***33****: 448–50*

Planar imaging

When multiple images are obtained throughout the cardiac cycle, frame time is determined by heart rate. Thus 16 frames per cycle at a heart rate of 80 beats per minute limits frame time to less than 25 ms; 25 frames per cycle at an heart rate of 100 gives a frame time of 12 ms. The number of counts collected during each frame in a single beat is correspondingly small. It is therefore necessary to sum counts from a number of beats, to improve statistical quality and obtain images which are analysable both visually and mathematically. There is little point in employing a pixel size smaller then 5 × 5 mm. Any improvement in resolution is offset by increased noise as available counts are distributed between more pixels. If duration of acquisition is restricted, for example in a stress study, a larger pixel size, up to 10 × 10 mm, may be preferable. The modified 45° left anterior oblique (with 10° of caudal tilt) is the preferred projection in the majority of subjects. Additional projections, for example 45° left posterior oblique, are sometimes helpful, particularly to visualize posterior wall. Obliquity must be adjusted for each patient in order to obtain best separation of chambers (Fig. 4.15).

Tomography

Gated blood-pool SPECT is the preferred technique to identify posterior wall motion abnormalities and may improve the assessment at other sites. It should therefore be employed whenever possible. Because of statistical limitations it is usual to acquire only 8–12 frames per cycle. The procedure is otherwise similar to the planar study. Ejection fraction can also be calculated if a gated SPECT acquisition is made during a myocardial study, although this does increase acquisition time appreciably.

Further reading:
European Journal of Nuclear Medicine 1993; ***20:*** *1108–11*

Techniques of gating

There are two methods of gating, intrinsic and extrinsic. Intrinsic gating requires list mode or very short frame-time acquisition, as in the first-pass study (page 135). A region of interest is drawn around the left ventricle and systole and diastole identified on the time–activity curve. This is practicable during the first pass, when the bolus is relatively undiluted and background lower and with non-imaging probes but when imaging at equilibrium the count rate from ventricle is relatively low and background due to activity in lungs and other tissues comparatively high. Intrinsic gating is therefore not usual for equilibrium imaging studies. The alternative, extrinsic gating, makes use of the ECG signal. The R wave, corresponding to onset of contraction, is normally of larger amplitude than other components of the ECG signal and is the most easily recognized electronically. This is therefore used to identify a fixed point in the cardiac cycle, end-diastole. The first step is to measure the average R–R interval and its standard deviation over 100 or so heartbeats. If the heart is irregular a larger number of beats may be required.

To obtain adequate definition of the shape of the time–activity curve in a subject whose heart

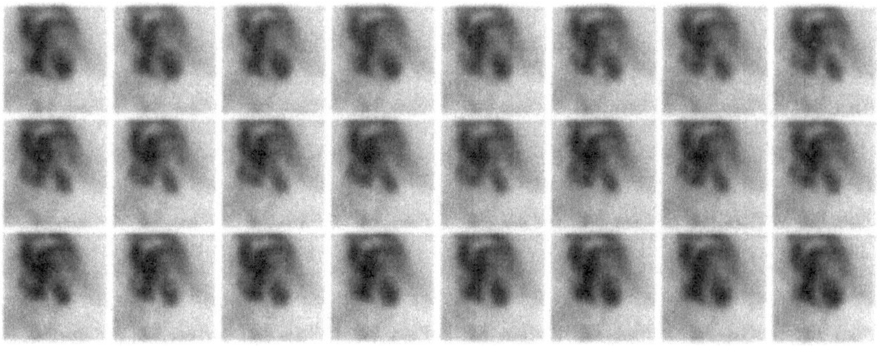

Figure 4.15 24-frame gated study. Systole is in frame 13 and end-diastole frame 2.

rate is less than 80 beats per minute, a minimum of 16 frames/cycle is required. In the presence of a tachycardia up to 32 frames/cycle may be necessary. For most purposes 24 frames/cycle are used.

Four methods, as below, may be used to obtain an ECG-gated image.

Unbuffered, fixed frame time

This is the simplest and although now obsolete is included because it is the easiest to understand and because all currently used techniques are elaborations on the basic principles. The average R–R interval calculated over not less than 100 beats is divided by the number of frames required, to give the duration for each frame. Memory is reserved for the appropriate number of frames.

For the first time-interval from the R wave, all counts are added to the first frame. After the predetermined time has elapsed, all counts are ascribed to the second frame and this is repeated until all frames have been filled or the next R wave occurs, when the cycle is repeated, irrespective of the interval since the preceding R-wave. If the next R wave occurs before counts have been added to all the frames, there will be fewer counts in later frames than earlier, as they will have received a contribution from a smaller number of beats.

If, on the other hand, the interval is longer there may be a delay before starting the next set of acquisitions and late diastolic counts will be discarded. This cycle is repeated for as many heartbeats as appropriate, 250–1000 depending on desired statistical quality of the image and collimator in use. The ventricular time–activity curve is thus an average obtained by summing over the total number of heart beats. This method is adequate in subjects with a regular heart rate but is increasingly prone to error as heart rate becomes more irregular. The first part of the curve, between end-diastole and end-systole, is reasonably accurately delineated and it is therefore possible to calculate ejection fraction. However, the filling phase of the curve is incomplete and it is not possible to calculate parameters derived from the rising part of the curve.

Buffered forward, fixed frame time

In this technique the range of R–R intervals is measured in addition to the mean R–R interval. A window is then defined so that, typically, only beats whose R–R interval is within ± 10% of the mean are accepted. The width of this window is usually adjustable. The method by which bad beat rejection is performed depends on both the computer hardware and software available. In older systems, where computer memory was at a premium, it was not possible to reject bad beats but only the one following. With current systems, counts are held in a buffer until the R–R interval has been calculated. Only counts from those beats which fall within the predetermined limit are used. This reduces variability, but later frames nevertheless contain counts from fewer beats and some must be ignored in analysis. This slightly improves accuracy of ejection fraction measurement, especially if there is some irregularity of heart rate but still does not allow adequate definition of the filling phase of the time–activity curve. In severe dysrythmias the number of discarded beats may extend imaging time considerably.

Buffered forward, fixed number of frames

Having determined the number of frames to be collected, an average R–R interval and range as above, counts from each heartbeat are stored in the buffer initially as a list, not separated into frames. The list is then divided into equal time intervals, depending on the actual R–R interval of each beat and those from beats within the limits set are added to the frames which have been set up. Thus all frames contain counts from all accepted beats, but frame duration varies from beat to beat. Frame time is taken as the mean. This is an acceptable method in patients with a reasonably regular heart rate and gives a somewhat better representation of the filling phase but can be unreliable in the presence of dysrythmias.

Buffered forwards and backwards, fixed frame time

Mean frame time and window are calculated as in the second of the above methods. Counts are initially stored in a list, as in the third. For the first two-thirds of the cycle, frames are timed from the preceding R wave and counts from beats with an R–R interval within the set limits added to corresponding frames. For the last one-third of the cycle frames are timed backwards from the succeeding R wave. This gives a complete representation of the filling curve and discrete units on the time axis, the error falling in a part of the curve, at the junction of the first two-thirds and the final third of the cardiac cycle, where no important measurements are made. This therefore shifts the error from late diastole to a point where it does not affect calculation of parameters of left ventricular diastolic function. This method gives the most accurate delineation of the filling phase and should always be employed when diastolic parameters are to be calculated. Wider limits of acceptable R–R interval may be set, allowing more beats to be accepted if the R–R interval is variable.

Sources of error

Incorrect trigger timing

When an ECG signal rises above a predetermined threshold voltage defined for the R wave, it triggers a pulse generator in the ECG or interface to send a logic pulse to the computer, which then starts acquisition for that heartbeat. It is possible for delays to occur anywhere in this chain of events. If the start of acquisition is not truly at end diastole when ventricular volume is maximum and the start is delayed to some later time, end-diastolic volume will be underestimated. An underestimate may thus occur of LVEF. In most commercial systems the trigger is set up by the manufacturer. It can be difficult to determine whether it is correct. There is no standard or simple way of obtaining this information.

Incorrect R-wave recognition

High amplitude T waves may exceed the threshold and be erroneously interpreted as R waves. This can result in two pulses per beat, giving invalid time–activity curves. If large T waves are observed, alternative lead combinations should be assessed. The triggering mechanism is not sensitive to which ECG lead is in use and the combination chosen should depend on whichever gives a good R wave without confusing high voltage T waves.

Rhythm anomalies

Some dysrythmias, in particular those with coupled beats, may be difficult to gate and indeed the choice of frame length may be arbitrary. Under these circumstances the limitations of any calculations must be recognized.

Identification of end-diastole

The onset of depolarization corresponding to the R wave may not correspond to maximum end-diastolic volume. Calculations should not automatically use the first frame but search the first half of the study to identify the frame with the maximum count rate in the ventricular region of interest.

Projectional errors

Incorrect positioning may lead to superimposition of other chambers over the left ventricle.

Background

A major source of error and probably the major factor contributing to differences between departments is assessment of background. Widely used techniques include a fixed region of interest drawn around the left ventricle with an adjacent crescentic area, avoiding aorta and spleen, taken as background (Fig. 4.16).

Alternatively separate regions of interest are drawn around the left ventricle in each frame using a smaller ventricular region in systole than diastole. The count rate in the systolic frame of pixels included in the diastolic but not in the systolic region of interest (ROI) may be taken as background. This is nearer to the ventricle and may give a better estimate of counts coming from structures in front of and behind the heart.

The modified Goris technique utilizes two lines of pixels, one just clear and to the left of the heart and a second at right angles to the first, just clear of and below the heart. The mean in the cardiac region for each pixel of the two corresponding background rows of pixels is taken as background for that pixel (Fig. 4.17). A separate background count is thus calculated for each pixel.

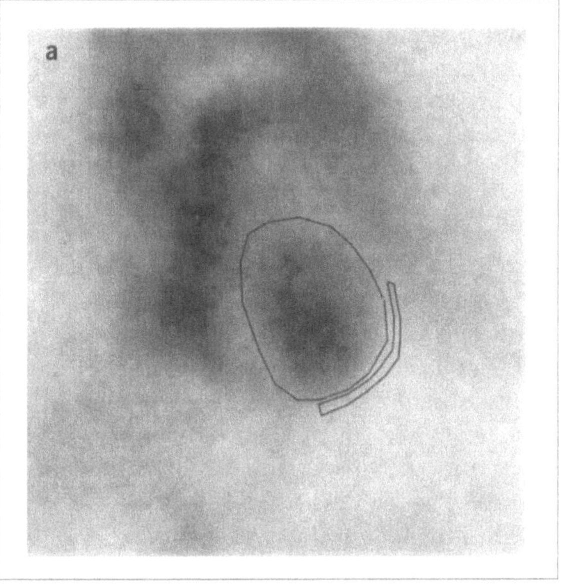

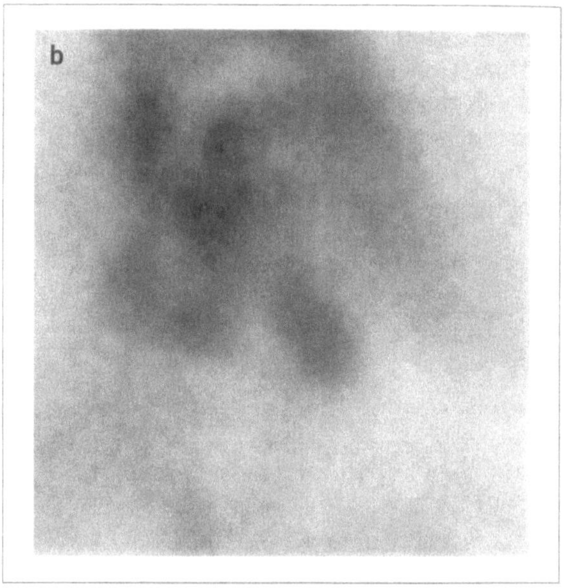

Figure 4.16 (a) Typical left ventricular and background regions superimposed on diastolic frame from gated study. (b) Systolic frame.

There have been many evaluations of techniques of background correction. None is accurate. All are crude but reproducible approximations which overestimate true background. Major sources of error in all estimates of background include scatter in both directions between ventricular and background regions. If a fixed position background is used, the contribution due to scatter may differ between systole and diastole. None of the methods allows for absence of lung in the volume occupied by heart, leading to overestimation of background. In practice, reproducible and clinically useful results can be obtained with most of the techniques provided that they are applied with understanding.

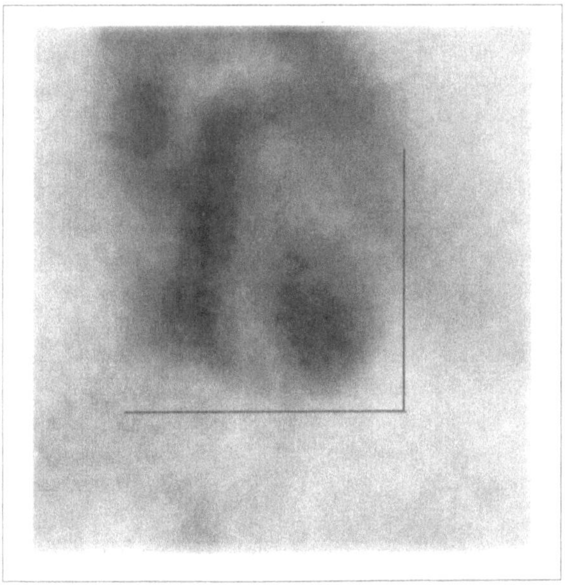

Figure 4.17 Axes for Goris interpolative background subtraction.

The Goris interpolated method appears less prone to inter-observer error than others. The most important consideration is that all members of the department who are concerned with processing cardiac studies should use similar standards and regularly compare results. Automated methods supplied with many commercial systems are prone to error, especially when cardiac function is impaired, and should thus be employed with caution.

Further reading:
Journal of Nuclear Medicine 1989; ***30****: 1870–4*

Calculation of results

Ejection fraction may be calculated using either the difference in volume or difference in background corrected counts in the ventricular ROI between end-systole and diastole. In practice, because of the limited resolution of imaging techniques, ventricular boundaries are poorly defined and ventricular volume measurements are less accurate than ventricular count measurements. Ejection fraction should therefore be calculated using counts, not volume.

$$\text{Ejection fraction} = \frac{(\text{diastolic counts} - B) - (\text{end-systolic counts} - B)}{(\text{end-diastolic counts} - \text{background})}$$

where B = background.

Stress testing

For many clinical purposes measurement of left ventricular ejection fraction while the patient is at rest provides sufficient information. However, patients with mild ischaemic heart disease or cardiomyopathy may have a normal LVEF at rest, but not respond normally to stress. Stressing the normal subject causes LVEF to increase. In contrast, ejection fraction decreases on stress in patients with impaired ventricular function. In normal subjects reproducibility of a single measurement is better than ± 5%. A fall greater than this thus indicates impaired ventricular reserves.

Methods of stress

Sustained exercise

In contrast to myocardial perfusion (page 118) this must be sustained for a sufficient period for measurement to be completed unless the first-pass technique is employed. Thus whereas myocardial perfusion studies employ peak stress, ventriculography normally employs a lower level which the patient is able to sustain, to produce a constant but increased heart rate for 3–5 min. Exercise may be supine bicycle, erect bicycle or treadmill. Each has a different physiological effect, so that results using different techniques may not be comparable.

With all methods it is difficult to keep the patient's thorax sufficiently still for imaging to be performed during exercise. Thus, although physiologically sustained exercise is in principle the preferred stimulus, it is often difficult to obtain a diagnostic study. For this reason several alternatives are widely employed.

Cold pressor

This uses cold as a stimulus. A limb, usually a hand or more conveniently a foot, is plunged into a bucket of iced water. This causes vasoconstriction and therefore increases the resistance against which the heart must work. This differs from exercise, which increases venous return. Thus the cold pressor test is not equivalent to exercise.

The other disadvantage of the cold pressor test is that response is often short-lived. Acquisition cannot start until heart rate has stabilized and by this time much of the response may have worn off. Thus false negatives due to adaptation or failure to respond are more common with cold pressor than with exercise.

Isometric contraction

The patient grasps a balloon, for example the partially inflated bag of a sphygmomanometer, in order to maintain pressure at some fixed value. It is difficult to sustain isometric contraction and the test is often limited by muscle fatigue in the arm. The physiological response to isometric exercise is complex and does not mimic exactly either exercise or cold pressor.

Pacing

Sustaining an increased heart rate by electrical pacing has been employed but is an invasive procedure not applicable in the majority of patients.

Pharmacological stress

This is widely used and allows better statistics as a longer acquisition time is possible. The same four drugs can be used as for myocardial imaging (page 119). Dipyridamole by constant infusion at 0.14 mg/kg/min for 4 min is the best documented, the peak effect occurring within 2 min of the end of infusion. It is difficult to maintain a constant heart rate. Sub-maximal dobutamine infusion is also widely employed.

Measurements

A number of measurements can be made, including:

- ejection fraction, i.e. the percentage of the diastolic volume ejected per beat
- peak filling rate
- peak emptying rate
- the times at which these occur in the cycle
- regional ejection fraction
- regional wall motion.

In most clinical situations and the majority of patients only peak filling rate adds useful information to ejection fraction and regional wall motion parameters. An ischaemic ventricle is less elastic than a normal one and fills more slowly, so that a reduced peak filling rate is additional evidence, occasionally the only evidence, of ischaemia. Images also contain important information on regional function, which can be assessed by viewing consecutive frames as a cine loop, to visualize cardiac contractility. This provides a useful impression of cardiac function and is a simple method of assessing abnormalities of the free border and apex. However, visual inspection is subjective, operator dependent and prone to error, especially when the heart is large. Alternatives which have been proposed include superimposition of systolic and diastolic isocount contours of the left ventricle in systole and diastole, comparison of radial ejection fraction profiles (Fig. 4.18a, Plate I) and phase and amplitude parametric images (Fig. 4.18 b, c, Plate I).

Phase and amplitude images

These are obtained from single-pixel time–activity curves calculated separately for each pixel in the image. Because of the statistical quality of single-pixel curves, these require mathematical analysis (filtration using a Fourier filter) to distinguish statistical noise from the true signal. Only the first harmonic is usually calculated. Two characteristic features are identified from each filtered curve, namely the difference between the highest and lowest counts (amplitude) and timing (phase), i.e. the frame in which the minimum value occurs for every pixel. These are presented as grey-scale or colour-coded images, one showing timing in each pixel and the other magnitude of change, each scaled to the whole image. These are the phase and amplitude images. Because ventricles and atria contract alternately, the phase image is the most accurate for identification of the atrio-ventricular plane. The rest of the ventricular outline is usually best identified on the amplitude image, but much is also well seen on the phase image. There is a larger change in count rate over the cardiac cycle in pixels corresponding to cardiac chamber and great vessels than over lung. Most software has a threshold which sets any pixels with an amplitude below some preset value to zero. Regions of interest drawn on these parametric images can be applied to the original frames to calculate ejection fraction. This is less subjective than other methods of determining left ventricular outline as the plane between left atrium and left ventricle is unambiguously defined on the phase image. It is also the best method of identifying right ventricular outline, although the plane of the pulmonary valve cannot be unambiguously defined by any technique.

Interpretation is usually straightforward. A circular colour scale, one whose maximum and minimum are similar and the greatest difference is between the middle of the scale and its extremities, is preferable for the phase image. One complete cycle is equivalent to one circle, 360°. The mid-point of the cycle, which is the furthest in time from either end, or any opposed points in the cycle are at 180°. In the normal subject the phase image shows homogeneous timing for the whole of both ventricles and, at 180°, both atria. Amplitude shows a smooth gradation and is usually greater in the left than the right ventricle. Not uncommonly there appears to be a phase difference limited to the septum (Fig. 4.19, Plate I). This is an artefact due to rotation of the heart and is not pathological.

Further reading:
Journal of Nuclear Medicine 1987; ***28****: 1536–9*

Data presentation and interpretation

The images required for interpretation are the phase and amplitude images, a cine display of the gated study and background corrected time–

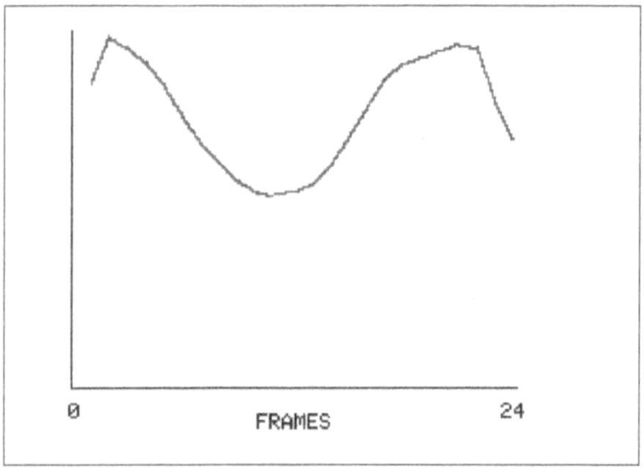

Figure 4.20 Filtered, background-subtracted, gated left ventricular time–activity curve.

activity curves from the left ventricle (Fig. 4.20). A time–activity curve can also be obtained from right ventricle, from which the right ventricular ejection fraction can be calculated. The plane of the pulmonary valve cannot be identified unambiguously. Estimates of right ventricular ejection fraction based on gated studies have an appreciable subjective element.

In the normal subject the lower limit of resting LVEF is 40–50%, depending on the method of background subtraction used. It is therefore essential to establish a normal range locally. Many other parameters can be calculated from the curve. Clinically the most useful is the peak filling rate, which is the steepest portion of the filling phase or up-slope. This can be calculated with accuracy only if the fourth method of gating (page 139) is employed. The normal value is greater than 2.5 end-diastolic volumes/second. In the presence of ischaemia there is loss of compliance by muscle; peak filling rate may be reduced even though LVEF is within normal limits.

Regions of abnormal wall motion, for example following myocardial infarction, can be visualized on the cine display when the apex or free border are involved. Anterior or septal defects can best be identified on phase and amplitude images. Posterior wall abnormalities are seen on gated blood-pool SPECT or if multiple planar projections are obtained. Phase and amplitude images must always be viewed together, when they comprise the most sensitive and least subjective method of analysing wall motion. They may also demonstrate premature ventricular contraction and accessory conduction pathways. Reduction (hypokinesia) (Fig. 4.21, Plate II)or absence (akinesia) of contraction (Fig. 4.22, Plate II) may occur without change in phase or hypokinesia may be associated with a difference in phase between the abnormal area and the rest of the ventricle (dyskinesia). Isolated dyskinesia without hypokinesia is less common. Areas of paradoxical motion show a phase change such that their timing is similar to atria rather than the remainder of the ventricle. A left ventricular aneurysm can be identified as an area with a 180° phase change (dyskinesia) but with normal or only slightly reduced amplitude (Fig. 4.23, Plate III). There is usually an area with no movement (akinesia, no phase and no amplitude) between the paradoxical and normal regions. Akinesia can also occur as an isolated finding. The site and extent of anterior, septal and inferior myocardial infarction can readily be demonstrated. Posteriorly placed abnormal regions are difficult to identify by planar imaging unless at least two projections are obtained.

Further reading:

Journal of Nuclear Medicine 1991; ***32****: 777–82*

Sources of error

These can be:

- Incorrect positioning of ventricular or background regions. This is particularly a problem when automated programmes for defining ventricular outline are used in subjects with poor cardiac function.
- Incorrect timing of the gating signal (page 140).
- Extensive areas of anterior hypokinesia or akinesia, especially associated with dilation, can lead to underestimation of LVEF as counts originating from the functioning posterior part of the heart may be partially absorbed in non-moving blood more anteriorly. This is a recognized limitation of planar studies, to which there is no entirely satisfactory solution. Very low figures for LVEF in patients with anterior hypokinesia should therefore be

regarded with reserve.

- Rotation of the heart during contraction, leading in particular to an artefact on the phase images.
- Statistical limitations in the gated SPECT study, in particular the restricted number of frames per cycle, may lead to over-estimation of the end-systolic volume with consequent underestimate of LVEF.

Clinical applications

Coronary artery disease

Most patients with coronary artery disease have a normal LVEF at rest. The severity of reduction in the resting value correlates with extent of irreversible muscle damage demonstrated by myocardial scintigraphy (page 132). The two methods are complementary as myocardial perfusion imaging provides no information on myocardial contractility unless gated SPECT is employed, when LVEF can be calculated from end-systolic and end-diastolic volumes of the ventricular cavity. This considerably prolongs the duration of SPECT acquisition unless the administered activity is substantially increased. Global or regional abnormalities which appear on stress are evidence of ischaemia, duplicating data obtained from myocardial perfusion studies. Considering in addition peak filling rate improves sensitivity of the resting study to detect myocardial ischaemia, but this has not been adequately evaluated against stress and redistribution (or re-injection) perfusion studies. Perfusion and blood-pool methods have similar sensitivity and specificity for the detection of coronary artery disease. The latter are performed with peak exercise, which is more readily reproducible than sustained stress and may give a clearer demonstration of the extent of disease. Perfusion studies are now in most circumstances the preferred investigation.

The principal indications for ventriculography in the management of coronary artery disease are evaluation of myocardial function and risk assessment, especially post-infarction and in patients being considered for beta blockade, angiotensin converting enzyme inhibition or surgery. A reduced LVEF in patients with chronic stable angina or at day 10 after myocardial infarction is associated with a poor prognosis, as is a normal resting LVEF which falls on stress or the appearance of regional wall motion abnormalities in patients with triple vessel disease but no objective evidence of ischaemia. Ventriculography is sometimes helpful when distinguishing relative importance of left and right ventricular dysfunction.

Further reading:

Journal of Nuclear Medicine 1994; ***35****: 721–5*

Non-coronary heart disease

Valvular regurgitation can be measured from the ratio of stroke counts in the left and right ventricles, which should in principle be equal. In practice a left-to-right ratio up to 1.5:1 can be found in normal subjects, due to overlap from the right atrium and right ventricle. More accurate values can be obtained with gated blood-pool SPECT. A normal resting LVEF in symptomatic patients with aortic regurgitation is associated with a better surgical prognosis than a reduced value, whereas in asymptomatic patients a reduced resting LVEF indicates a poor prognosis. No additional prognostic information is provided by stress studies. A reduced right ventricular ejection fraction in patients with mitral regurgitation may precede other evidence of decompensation and be reversible. In contrast a reduced LVEF whilst often reversible is not of prognostic value in patients with aortic stenosis.

Syndrome X

This term is applied to an ill-defined group whose existence is not universally agreed. It is usually applied to patients with chest pain and angiographically normal coronary arteries in whom other cardiac abnormalities such as cardiomyopathy or coronary vasospasm have been excluded. Many have ischaemia-like electrocardiographic changes. Prognosis is good. In the majority, ejection fraction fails to respond to exercise. About one-third show reversible perfusion defects but none develop any abnormality of wall motion on stress.

Further reading:

European Journal of Nuclear Medicine 1994; ***21****: 95–7*

Hypertension

In hypertension, as in aortic stenosis, systolic function of the hypertrophied ventricle may remain normal. Some patients show a fall in LVEF on exercise which may be reversed by beta blockade. Up to 40% of patients with congestive cardiac failure, especially if due to systemic hypertension, have normal systolic function but impaired diastolic filling; exercise capacity and symptoms may be improved by calcium channel blockade. Although data are incomplete these findings suggest that there may be some clinical importance in distinguishing systolic from diastolic dysfunction.

Cardiomyopathy

The majority of patients with hypertrophic cardiomyopathy have a normal or high resting LVEF but some, with severe symptoms, have a non-obstructed dilated ventricle with a low ejection fraction. This distinction has therapeutic implications. Abnormalities of diastolic filling parameters, consequent upon reduced distensibility or impaired active ventricular relaxation, include a reduced peak filling rate and increased atrial contribution to end-diastolic volume.

Drug toxicity

The maximum tolerable dose of drugs such as doxorubicin and other anthracyclines is determined by their cardiotoxicity. A fall in LVEF precedes other evidence of toxicity and is reversible. ECG changes occur only when there is irreversible damage. It is therefore advisable to measure LVEF before starting treatment and three weeks after the total administered dose has reached 450 mg/m^2 in subjects without cardiac risk factors or 400 mg/m^2 in those with known heart disease, an initial LVEF between 30% and 50%, hypertension or who have previously received cyclophosphamide or irradiation to a field which included the heart. It should be repeated after each subsequent dose and treatment discontinued when the resting LVEF falls by 10% or below the local reference range. These agents should not be used if the baseline LVEF is below 30%.

Further reading:

Journal of Nuclear Medicine 1990; ***31****: 10–22*

Transplantation

Ventriculography has been performed in cardiac transplants. Conventional methods of analysis and interpretation can be employed. Both left and right ventricular ejection fraction and peak filling rates are reduced during rejection episodes.

Further reading:

European Journal of Nuclear Medicine 1991; ***18****: 879–84.*

Shunts

Shunts and congenital heart defects can be evaluated using the first-pass technique, but this has largely been superseded by ultrasound and magnetic resonance imaging (MRI). In the normal subject, despite substantial discrepancies in peak count rates, the areas under left and right ventricular first-pass curves should be equal unless there are errors in dead-time correction. Inequality of these areas is one means of detecting and measuring size of both left to right and right to left shunts. Radionuclide angiography has been employed to follow up small shunts which are expected to close spontaneously. Right-to-left shunts may be measured from the fraction of an intravenous bolus of particulate tracer (preferably microspheres, page 57) not retained in the lungs. A simplified approximation is to measure the ratio of activity in lung to head, assuming a constant fraction of cardiac output to the brain. Meticulous control of particle size and numbers is essential if this procedure is to be undertaken safely.

Further reading:

CRC Critical Reviews in Clinical Radiology 1975; ***6****: 217–51*

Journal of Nuclear Medicine 1994; ***35****: 1328–32*

Cardiac output

Measurement of cardiac output was one of the first applications for radio-isotopic tracers in experimental physiology and subsequently became the first clinical investigation in nuclear cardiology. Cardiac output can be measured following bolus intravenous injection of any substance which mixes within the circulation but remains intravascular, provided that the tracer employed is neither metabolized nor excreted during the few minutes that the study requires. Although now most commonly performed as part of an imaging investigation, for example to measure left or right ventricular ejection fraction during first-pass studies of the heart or brain, imaging is not required to measure cardiac output alone, in which case alternative non-radioactive tracers such as heat or a coloured dye may be employed. Stroke volume and left ventricular volume can also be calculated if cardiac output, heart rate and LVEF have been measured.

Further reading:
Seminars in Nuclear Medicine 1977; ***7****: 85–100*

Radiopharmaceuticals

The usual radioisotopic tracers are ^{99m}Tc-labelled autologous erythrocytes (RBC) or ^{99m}Tc human serum albumin (HSA). Radiopharmaceutical purity needs to be higher than when measuring ejection fraction alone (page 134) because volume of distribution is assumed to be plasma volume. Ideally every preparation should be assayed before use, as individual batches may contain a substantial amount of unbound technetium, either as pertechnetate or a reduced species which does not remain intravascular. It is difficult to prepare an adequate specific activity (>200 MBq/ml) of labelled red cells with a labelling efficiency >95% for the total activity to be contained in a small enough volume (1–2 ml) for an adequate bolus to be achieved. The technique depends critically both on bolus quality and radiopharmaceutical purity of the tracer. Other labels mentioned in the literature but no longer available include ^{131}I-HSA and ^{113m}In-Cl_3, which labels transferrin. A very short-lived radionuclide of gold (^{195m}Au, half-life 30 s) has also been employed and is the preferred tracer if repeated measurements are to be made on the same day.

Unfortunately it is no longer available commercially. Krypton (^{81m}Kr) dissolved in saline can be used for measuring right ventricular output but as it is almost 100% exhaled on a single pass through the lungs it cannot be used by intravenous injection for assessment of left ventricular output.

Technique

The tracer must be injected as a rapid bolus (page 231). Any bolus becomes progressively more spread out as it moves away from the site of injection. The more peripheral the site of measurement, the greater the difficulty in distinguishing first pass from recirculation (see Fig. 8.2). Accurate measurement requires a well defined first-pass time–activity curve. If a non-radioactive tracer is employed the concentration of dye can be estimated by measuring absorption of light of an appropriate wavelength as blood passes through a conveniently translucent region such as the lobe of the ear. Radioactive tracers are measured by positioning a detector over the heart or any convenient more peripheral site. The collimator should be designed to minimize changes in sensitivity when viewing an extended source in scattering media between 5 cm and 15 cm from its face. Alternatively regions of interest can be drawn on images obtained with a gamma camera, but the sensitivity of a gamma camera is much lower than that of a simple probe. Against this must be set the more accurate positioning of the region of interest so that only a single tissue is included. The time–activity curve is calibrated by taking a blood sample at an exactly noted time after injection, not earlier than 3 min and preferably at 5 min. From this blood volume can be calculated by the relationship:

$$\text{Blood volume} = \frac{\text{(activity administered)}}{\text{(activity per unit volume of blood at equilibrium)}}$$

provided that activity is uniformly mixed with blood and there has been no significant diffusion into the extravascular space, excretion or metabolism during the period. By identifying the corresponding point on the time–activity curve the count rate in that curve can be related to the concentration of activity in blood. If efficiency of tracer labelling is inadequate, some will have been lost from the circulation and the calculated volume of distribution will be greater than the blood volume, with corresponding error in calculated cardiac output.

An idealized curve, representing what would have been obtained had there been no recirculation, can be obtained by deconvolution (page 94) or by fitting a gamma variate function to the up-slope and early part of the down-slope of the first-pass curve (Appendix 2). The latter is a mathematical function which has been found empirically to give a good fit to experimental first-pass curves provided that sufficient of the first pass can be identified, before there has been significant recirculation. Cardiac output can be calculated from the relationship:

$$\text{Cardiac output} = \frac{(\text{blood volume})\,(\text{count rate at equilibrium})}{(\text{area under first pass curve})}$$

The unit in which area under the recirculation corrected, first-pass curve is expressed is the total number of counts detected.

Stroke volume can be calculated from the relationship:

$$\text{Stroke volume} = \frac{(\text{cardiac output})}{(\text{heart rate})}$$

and if the ejection fraction is known (see below):

$$\text{Ventricular volume} = \frac{(\text{stroke volume})}{(\text{ejection fraction})}$$

The latter is the least accurate of these parameters as it combines the errors both in cardiac output and ejection fraction.

Right ventricular output can be measured by obtaining the time–activity curve from the right ventricle or lungs and left ventricular output from the left ventricle, aortic arch or more peripherally. A detector viewing the heart usually enables time–activity curves to be obtained from both ventricles, separated in time, but two peaks may be difficult or impossible to distinguish in patients with right ventricular failure or a right to left shunt. The peak count rate during the first pass through the right ventricle is much higher than that during its pass through the left ventricle, as the volume of distribution at that time is smaller and consequently the concentration is higher. The count rate during passage through the right ventricle may be high enough to give serious errors due to dead-time losses in the detector system. First-pass studies should therefore not be performed until the limitations of the detector system have been established.

It is essential to determine both the dead-time correction curve and the maximum useful count rate of the system under conditions similar to those obtaining clinically. All detectors, without exception, suffer from dead-time losses, but the count rate at which these become significant may be as low as 10000 counts per second or in excess of 200000 counts per second. Dead-time corrections are progressively less reliable when they exceed 20% of the true count rate (page 304).

Circulation time

Circulation time is the time required for activity to pass between two defined points, for example arm to head or right ventricle to left ventricle. Peak-to-peak time is usually measured. Leading edge to leading edge is theoretically more accurate but difficult to define. Although one of the earliest parameters used to quantify the circulation, circulation time is crude and insensitive. It is of limited practical value, for example in determining the timing of angiography. The principal application is facilitation of recognition of the left and right ventricular components of the curves obtained in patients with significant shunts.

5 Endocrine

Thyroid – non-malignant diseases

The thyroid is unique in its ability to accumulate a particular element, iodine, which is not stored to a significant extent by any other tissue. The earliest description of clinical use of a radio-isotope of iodine, in a patient with metastatic thyroid carcinoma, was published in 1942 but the availability of radio-isotopes remained restricted until after the cessation of hostilities in 1945. Since that time many *in vivo* tests have been described. Simultaneously, methods for the assay of circulating thyroid hormones and of pituitary hormones which act on the thyroid have been developed. These assays and their interpretation will not be discussed here but the *in vivo* techniques described must be viewed in the light of the considerable role played by *in vitro* measurements in diagnosis and management of thyroid disease.

Radiopharmaceuticals

Iodine has only one stable isotope, ^{127}I, although 23 radioactive isotopes have been identified. Appendix 1 lists the principal characteristics of those which have, or have had, any diagnostic or therapeutic application or which may be found as impurities in radionuclide preparations used clinically.

^{131}I

^{131}I remains the most widely used radio-isotope of iodine for both therapeutic and diagnostic purposes. Methods of manufacture are well established and it can be produced in large quantities, at high purity and relatively low cost. The half-life of approximately eight days is sufficiently long for wastage due to radioactive decay to be acceptably low during distribution from the production facility to the hospital, while the gamma ray energy of 364 keV, although higher than ideal for use with the gamma camera, is usable with a directional probe or when a pinhole or high energy collimator is fitted to a camera. The principle disadvantage of ^{131}I for diagnostic applications is the large number of associated β emissions. These are however the basis of its therapeutic use. Absorbed radiation dose to the thyroid is correspondingly high (Appendix III. The minimum activity required for imaging, 2 MBq, is associated with a higher absorbed dose than is received from most other diagnostic tests involving radionuclides. Comparable absorbed radiation doses to the skin are received as a result of some radio-logical examinations requiring prolonged fluoroscopy, for example complex angiocardio-graphic procedures. The effective dose from the equivalent activity of ^{125}I is only slightly lower than that from ^{131}I. The gamma ray energy of ^{125}I is too low for scintigraphy or the majority of *in vivo* measurements, for which it is no longer used. Its applications are now principally *in vitro*.

^{132}I

^{132}I gives only 1% of the absorbed radiation dose of an equivalent activity of ^{131}I. Its gamma ray energy is however too high for imaging, the images being severely degraded as a result of septal penetration whilst the short half-life permits mainly trapping to be measured. It can be prepared from a $^{132}Te/^{132}I$ generator, but has never been widely used and is no longer available.

^{123}I

^{123}I has the most favourable physical properties of any radio-isotope of iodine, emitting a single gamma ray of 159 keV and no β particles. It can be produced by a number of routes but if it is to be uncontaminated by longer lived and β-emitting radio-isotopes of iodine a high energy accelerator is required. Supplies of ^{123}I are therefore limited. The 13 h half-life results in substantial wastage of the initial radioactivity during purification and distribution. The positron (β^+) emitter ^{124}I is an important impurity in ^{123}I produced by lower-energy cyclotrons. It has a longer half-life than ^{123}I and may increase absorbed dose to such an extent that the advantage over ^{131}I is lost. ^{125}I may also be present as an impurity which increases absorbed dose.

There is thus no single ideal radionuclide of iodine. ^{123}I, when available, may be employed for both uptake measurements and scintigraphy, particularly of ectopic thyroid. ^{131}I should be restricted to uptake measurements and scintigraphy for functioning metastases in patients with proven thyroid carcinoma, when the half-life of ^{123}I is too short for adequate contrast to be achievable. ^{124}I may be employed if positron emission tomography (PET) is available. The *in vivo* use of other radio-isotopes of iodine for investigating the thyroid is now obsolete.

Iodine substitutes

A number of anions of similar charge and ionic radius to iodide are trapped by the thyroid. The most important, sodium pertechnetate, is readily available, cheap, associated with a low absorbed dose and can be imaged with a gamma camera equipped with a pinhole, converging or conventional parallel hole collimator. Trapping, but not binding of iodide and its analogues, also occurs in the salivary glands, gastric mucosa and choroid plexus. Some thyroid carcinomas retain the ability to trap iodide but are unable to organify it. There are therefore a number of circumstances in which imaging with iodide shows a distribution different from that seen with pertechnetate. Many other anions are trapped by the iodide sequestering mechanism in the thyroid and although not useful for imaging may affect iodine or pertechnetate uptake. Those which may be encountered clinically include perchlorate, bromide and thiocyanate. There are in addition a number of less common anions including perrhenate and selenacyanate.

Although pertechnetate is the most widely used agent for imaging the thyroid, it is less suitable for assessment of thyroid function. Uptake measurements do not reliably separate normal from hypo-functioning glands but can distinguish normal from hyperthyroidism. Measurement of uptake is more difficult than that of iodide because absolute uptake is lower and background higher, giving a lower contrast. Quantification is useful principally when there is a specific need to measure thyroid trapping as distinct from binding, for example in patients taking carbimazole (page 151).

Other tracers which accumulate in the thyroid and have been used for imaging include ^{99m}Tc-labelled isonitriles and related compounds, selenomethionine labelled with ^{75}Se and radio-isotopes of thallium, caesium, rubidium and potassium. These make use of less specific mechanisms and therefore give different information: selenomethionine for protein synthesis and alkali metal ions for local potassium pool, a measure of intracellular volume. These cations have been used to image thyroid nodules and suppressed thyroid but are also taken up many other cell types. The technetium-labelled compounds used for imaging regional myocardial perfusion, including sestamibi and tetrofosmin, are also taken up in thyroid, possibly by a mechanism similar to that responsible for their cardiac uptake (page 166). Their principal application is for parathyroid imaging (page 167).

Adverse reactions

Allergic reactions to iodides may occur. These are most commonly encountered in patients receiving non-radioactive iodinated radiographic contrast media, some of the older preparations of which contained substantial amounts of free iodine. Radio-iodine preparations usually have some carrier added to reduce losses due to absorption on the surface of containers. Even so allergic reactions to radio-iodine are very rare.

Adverse reactions are rare following a single oral or intravenous dose of 200 mg sodium or potassium perchlorate used to block iodide or pertechnetate uptake. Chronic administration carries the risk of aplastic anaemia and other blood dyscrasias. The potassium salt is usually used for oral administration. Potassium salts may not be given by bolus intravenous injection and potassium perchlorate is sparingly soluble in water. If intravenous administration is required, 200 mg sodium perchlorate in 10 ml water for injection should be administered.

When long-term blocking of the thyroid is required, as a consequence of industrial leaks of radioactivity or the administration of a radio-labelled pharmaceutical such as MIBG (page 172) or fibrinogen, from which iodine is slowly released *in vivo*, the thyroid should be blocked with an iodine-containing preparation such as 0.3 ml Lugol's solution three times a day (a total of 120 mg iodine per day) or 85 mg sodium iodate twice daily until a combination of excretion and radioactive decay have reduced residual radioactivity to acceptable levels. In the case of a ^{125}I-labelled compound such as iodinated fibrinogen this may be three months or more. For ^{131}I compounds, the combination of excretion and radioactive decay is usually sufficient to permit blocking to be terminated after 7–10 days. Compliance is difficult to achieve with Lugol's solution because it is unpalatable. Potassium iodate is not licensed for clinical use and may be difficult to acquire, even though large quantities are kept in reserved stores for use in a civil emergency.

Iodine uptake and thyroid blocking agents.

Iodide is trapped by the thyroid and oxidized to iodine, which reacts with tyrosine residues in peptides, initially to form monoiodotyrosine and subsequently diiodotyrosine. Pairs of diiodotyrosine residues combine to form thyroxin, from which triiodothyronine is synthesized. These reactions are enzyme-mediated. Anti-thyroid drugs can act by preventing uptake of iodide (perchlorate, thiocyanate) or block any one or more of the stages in production of thyroxine (carbimazole, propylthiouracil). Organification commences immediately. In the normal subject thyroid iodine uptake reaches a peak about 24 h after oral administration and declines slowly thereafter as labelled hormone is released. Some iodine may be discharged if 200 mg potassium perchlorate or 40 mg stable iodine (0.3 ml Lugols solution or 85 mg potassium iodate) is given up to 6 h after radio-isotope administration. The amount organified and therefore not discharged in response to perchlorate increases with time. The size of the fraction discharged decreases correspondingly. The rate of organification is faster in subjects with a small iodine store and in thyrotoxicosis. At any time perchlorate and iodide discharge pertechnetate, which is not able to enter the organification pathway as it is fully oxidized. Late discharge of radio-iodide by perchlorate or iodide is abnormal and indicates a defect in the organification pathway. A normal range for discharge must be obtained locally as the rate of organification depends, amongst other factors, on dietary iodide. Administration of perchlorate has little or no effect on absorbed dose (dose equivalent) from pertechnetate as the critical organ is colon, not thyroid.

Measurement of thyroid function

Thyroid function is best assessed by selecting from the available range of *in vitro* assays of circulating levels of thyroid hormones and the pituitary hormones which control their release. However these require laboratory facilities which are not universally available. Measurement of thyroid uptake of radio-iodine is useful where reliable *in vitro* assays are not available and is indicated in a small number of patients where there is discordance between *in vitro* results and the clinical picture. It is unnecessary to perform thyroid uptake measurements or thyroid scintigraphy in the majority of patients with thyroid disease, especially those who are hypothyroid.

Iodine uptake

Measurement of thyroid radio-iodine uptake is

one of the simplest *in vivo* assays because the thyroid is comparatively small, is situated just below the skin and uptake of iodine is much greater than that in adjacent tissues. The contribution to detected counts from extra-thyroidal activity is thus negligible. Uptake within the first hour is substantially influenced by thyroid trapping. During this early period there is usually appreciable activity in salivary glands and other soft tissues. When radio-iodine is given orally, as is usual, early uptake is influenced by the rate of gastric emptying and whether or not the subject is fasting. Early measurements of thyroid iodine uptake as an index of trapping have now been largely superseded by pertechnetate uptake following intravenous injection, which gives a measure of trapping alone not influenced by organification or extrinsic factors.

A clear distinction between normal and hyperthyroid subjects is usually obtained in measurements made between 2 h and 4 h after oral administration of radio-iodine. False negative values in the normal or hypothyroid range may be due to delayed gastric emptying. The distinction between normal and hypothyroid is clearest at 24 h but this requires a second visit by the patient. A reasonable compromise when thyrotoxicosis is suspected clinically is to perform uptake measurements routinely at a set time, preferably 4 h after oral administration to a fasting patient and reserve 24 h uptake measurements for those whose 4 h uptake is in the equivocal range or unexpectedly normal or low. By 4 h extrathyroidal activity in the neck is low enough to be ignored, eliminating the need for background 'correction' in patients with normal renal function. Uptake measurements are no longer widely used to diagnose hypothyroidism but may be useful if laboratory facilities to provide accurate assays of circulating thyroid hormones are not available.

Technique

Fairly simple equipment is required, namely a sodium iodide or equivalent detector preferably not less than 5 cm diameter × 2.5 cm thick fitted with a simple cylindrical collimator restricting its field of view to a 10 cm diameter circle at a distance of 10 cm from the face of the detector. A phantom similar to that recommended by the American National Standards Institute is useful for initial calibration. This is a cylinder of tissue-equivalent material such as polymethylmethacrylate (Perspex), 12.7 cm in diameter and 12.7 cm high. A hole 3 cm in diameter and 0.5 cm from the surface contains a vial in which the radioactive solution is placed. Diametric to this hole the cylinder is flattened so that it can be placed on a flat surface without movement. The detector is calibrated by correlating the measured count-rate with that from the same sample measured in a calibrated well counter. Using a suitable scintillation counter the ratio of activity in the vial of standard to that about to be administered to the patient is measured. This should be approximately 1:2. The usual administered activity is 200–400 kBq.

The phantom containing the vial of standard is placed in front of the counter used for the thyroid uptake measurement at the same distance as the patient and the count rate measured. Room background must always be measured. The expected count rate per unit activity can thus be calculated. Provided that the position of the counter itself and surrounding large pieces of furniture are standardized, so that the scatter pattern is not changed, this figure should remain constant. It is not necessary to measure the phantom for every patient provided that administered activity is accurately measured. It is however advisable to perform a check phantom measurement periodically, to confirm that no change has taken place, frequency depending on stability of the detector and the environment in which it is situated. It is unnecessary to measure 'background' count rate from non-thyroid tissues. Many of the counts detected originate from the thyroid and are in reality scatter rather than background, especially in hyperthyroid subjects, in whom by 4 h thyroid uptake is high and background low. Under most circumstances background correction is small and the error introduced by ignoring it negligible compared with the error in estimating it.

Pertechnetate uptake

^{99m}Tc-sodium pertechnetate has the attraction of

a lower absorbed radiation dose than any iodine isotope. However it is exclusively a measure of trapping. Normal uptake is therefore lower than that of iodine, with a peak value of between 1% and 3% of the administered activity which reaches a plateau earlier, approximately 15 min after intravenous injection. Pertechnetate uptake is not a measure of the same function as 4 h radio-iodine uptake and there is no constant or simple relationship between them. Unlike iodide, extrathyroid activity contributes substantially to the observed count rate. Measurement of uptake is therefore more complex and optimally requires imaging to identify thyroid and suitable background areas above and below it, although non-imaging techniques have been described. Because of low uptake in the normal subject, pertechnetate cannot reliably distinguish between normal and hypothyroid subjects. It has been used to diagnose hyperthyroidism, but has been largely superseded by *in vitro* and simpler *in vivo* techniques. The only important remaining indication for measurement of pertechnetate uptake is assessment of thyroid trapping in patients receiving anti-thyroid drugs such as carbimazole, which block organification rather than trapping. Reversion of pertechnetate uptake to normal suggests that the patient has passed into remission and medication may be curtailed or terminated.

Precautions

Replacement therapy with thyroxine must be discontinued for 4–6 weeks before radio-iodine uptake measurement or thyroid scintigraphy, to allow residual thyroid to recover fully from any suppression (page 163). Pertechnetate uptake can be measured without stopping carbimazole or propylthiourea but not whilst the patient is receiving thyroid replacement. This is sometimes used to determine if the underlying toxicosis is in remission or when considering modifying treatment. Thyroid blocking drugs should otherwise be discontinued for long enough for their effects to wear off. Radiological examinations involving administration of iodinated contrast media should wherever possible be scheduled after radio-isotope investigations of the thyroid, in case of free iodide in the preparation. This is a less serious issue with modern than with older iodinated contrast media but the problem does persist. Where such scheduling is not possible the nuclear medicine examination should be performed not less than one week after the radiological investigation. Low uptake may however persist for many weeks and is thus difficult to interpret.

Interpretation

The normal range of uptake values varies between populations, depending in particular on the dietary content of iodide. Over 35% uptake at 4 h is usually regarded as hyperthyroid and under 10% hypothyroid, but each laboratory must confirm its own reference range. Misleading high uptake values may be found in subjects whose dietary intake is substantially lower than that of the reference population, as a result of an unrecognized previous administration of radio-iodine or some other radio-isotope with a similar gamma ray energy, for example ^{51}Cr. This can only be excluded if a measurement of background count rate is made for every patient prior to administration of radio-iodine. The detection rate of such abnormalities will not normally justify the time involved. It is however worth enquiring what other tests the patient has undergone within the previous three months and measuring the background and gamma ray spectrum in anyone with a history of previous radio-isotope administration or ambiguous haematological investigation.

There are many causes of falsely low uptake, the commonest being delayed gastric emptying, a high dietary or iatrogenic iodine intake. Some 'cough mixtures', seaweed or mineral supplements contain substantial amounts of iodine. Other possibilities include recent administration of iodinated radiological contrast media, thyroid replacement therapy, treatment with anti-thyroid drugs, a diet high in goitrogenic foods such as kale or turnip, previous thyroid surgery or an ectopic thyroid not within the field of view of the detector. Glucocorticoids, phenylbutazone, sulphonylureas, salicylates, amiodarone, some sedatives, opiates and tranquillizers also reduce thyroidal iodine uptake.

The principal abnormalities causing a low

radio-iodine uptake are hypothyroidism, autoimmune thyroiditis (Hashimoto's disease), acute or sub-acute thyroiditis and Basedow's disease (factitious thyroiditis due to ingestion of large quantities of iodide). Either thyroiditis or Basedow's disease should be suspected if uptake is low but the patient is clinically thyrotoxic. Thyroiditis is usually associated with a history of recent thyroid enlargement and tenderness. If no explanation is found for a low uptake in a patient who is clinically and biochemically euthyroid, scintigraphy with ^{123}I is indicated to locate ectopic thyroid tissue.

Clinical indications

The commonest clinical indication is prior to radio-iodine therapy of thyrotoxicosis, to determine the appropriate therapeutic dose of radio-iodine. Therapeutic dose is determined by gland size as well as uptake (page 162). Uptake measurements are useful for differential diagnosis of acute thyroiditis (in which there is low uptake in a patient with a tender thyroid and raised circulating levels of thyroid hormones) and when Basedow's disease (thyrotoxicosis due to ingestion of excess quantities of iodides) is suspected.

Further reading:
European Journal of Nuclear Medicine 1991; ***18:*** *761–78*

Perchlorate discharge test

This is used to diagnose defective organification. Iodine uptake is measured 2 h after intravenous administration of 200–400 kBq ^{131}I or 4 MBq ^{123}I. 200 mg sodium perchlorate is then administered intravenously and a further measurement of thyroidal radio-iodine is made 1 h later. In normal subjects iodine uptake remains constant or may continue to rise slightly even after perchlorate. However, in subjects with Hashimoto's disease, defective organification, a high iodine intake or who are receiving drugs which inhibit organification, the radio-iodine content of the thyroid falls by 10% or more. The principal indication is to diagnose Pendred's syndrome and other rare inherited conditions associated with defective organification.

Further reading:
European Journal of Nuclear Medicine 1995; ***22:*** *1005–8*

Total thyroid iodine

Two techniques are available, although neither has achieved wide acceptance. X-ray fluorescence analysis involves irradiation of the gland with an external gamma ray source of an energy at which there is maximum photoelectric absorption by iodine, such as ^{241}Am. Absorbed incident gamma rays are re-emitted as iodine k x-rays of characteristic energy. Because of their low energy (approximately 25 keV) a detector with a low base noise such as a silicon semiconductor is preferable to a conventional sodium iodide detector. If the gamma ray source is well collimated only a small volume of gland is irradiated at any instant. It is thus possible to determine regional iodine content by scanning the source over the gland. The iodine content of a malignant nodule is low; that of normal thyroid and benign adenomas is higher. The equipment required is comparatively simple and inexpensive and the absorbed radiation dose low: a few microSieverts. An obsolete rectilinear scanner can readily be adapted. The technique suffers from the disadvantage that only 18 mm of soft tissue is sufficient to absorb half the radiation emitted. Iodine content is therefore underestimated if the gland is enlarged, retrosternal or the neck is fat.

Irradiation of the thyroid with thermal neutrons activates not only iodine but all other elements present, especially calcium, sodium, potassium, phosphorus and chlorine. This method is capable of high accuracy but neutron sources are of limited availability and the technique can measure only total rather than regional iodine content, a parameter of little clinical value.

Scintigraphy

Because of its low cost, ready availability and low absorbed radiation dose, pertechnetate is the most commonly employed radiopharmaceutical. It does however suffer from some disadvantages especially that some thyroid carcinomas continue to trap iodide and its analogues even though organification is impaired. Thus thyroid carcinoma does not always appear photon-deficient with pertechnetate. Uptake of pertechnetate in metastatic thyroid carcinoma is rarely sufficient for secondaries to be imaged. Iodine isotopes are thus essential for imaging spread of thyroid cancer.

Technique

Scintigraphy should be started 15–30 min after intravenous administration of 75–100 MBq ^{99m}Tc sodium pertechnetate or 18–24 h after 10–25 MBq ^{123}I sodium iodide. A gamma camera fitted with a converging or pinhole collimator is employed. A small, high resolution pinhole, approximately 3 mm in diameter, is preferable. Some centres still use rectilinear scanners for imaging the thyroid, but these are obsolete for all other purposes and better resolution can be obtained with pinhole views. Magnifying an image obtained with a conventional parallel hole collimator, even a high resolution collimator, gives inferior resolution and although encouraged by some manufacturers is not recommended. A study may comprise up to four components, best obtained with the patient supine to minimize movement during the extended acquisition times.

- A dynamic study consisting of 120 frames of 2 s, started at the instant of injection to determine if any nodule is vascular or avascular. Avascular photon-deficient nodules may be cysts whilst most thyroid carcinomas are vascular. A gamma camera with a converging collimator is preferable to a pinhole for this part of the examination. A parallel hole general purpose collimator is an acceptable alternative. This part of the examination is usually omitted, especially if there are no palpable nodules.
- An anterior view 15 min after injection with the pinhole as close as possible to the thyroid. The distance of the pinhole from the skin is determined by the size of the gland. It should be as close as possible without excluding any of the gland. This gives maximum magnification and therefore the best possible resolution but does not permit gland dimensions to be determined. It is possible to identify nodules which are impalpable.
- If nodules are detected which are neither clearly photon deficient nor photon rich, further projections may be made with the head of the patient rotated to the left and then to the right, or with the camera head rotated 45° to each side. Many centres now omit these oblique projections as the differentiation between cold and warm nodules is of little practical value.
- A projection with the pinhole to skin distance equal to the pinhole to crystal distance should always be obtained. This gives a magnification factor of unity, enabling the size of the gland to be assessed and demonstrating its relationship to other structures, especially the outline of the neck, the salivary glands and the sternal notch. The latter should be marked either with a radio opaque marker or a well collimated point source. It is impossible to assess retrosternal extension without an accurate indication of this landmark.

A total of 50–100 000 counts should be collected in each of the magnified views. A lower count density may be acceptable if views are to be filtered; 10–25 000 counts are adequate for the localizing view.

Interpretation

Normal

The left and right lobes of the thyroid are usually almost symmetrical. Uptake should be uniformly distributed throughout the gland. There may be slight non-uniformity; some difference in size but not of count density is common (Fig. 5.1). Additional ectopic thyroid tissue may be located anywhere along the line of the foetal thyroglossal duct. The commonest sites are in the root of the tongue and the anterior mediastinum. Most ectopic thyroid tissue can be identified adequately using pertechnetate but in cases of uncertainty ^{123}I is the radio-isotope of choice as by 18–24 h there is little or no uptake in salivary glands (Fig. 5.2). A pyramidal lobe is present in up to 40% of subjects and may originate from the isthmus, but about half arise from one or other of the upper poles (Fig. 5.3). Many show comparatively faint uptake and are visualized only if the patient is hyperthyroid or if the image

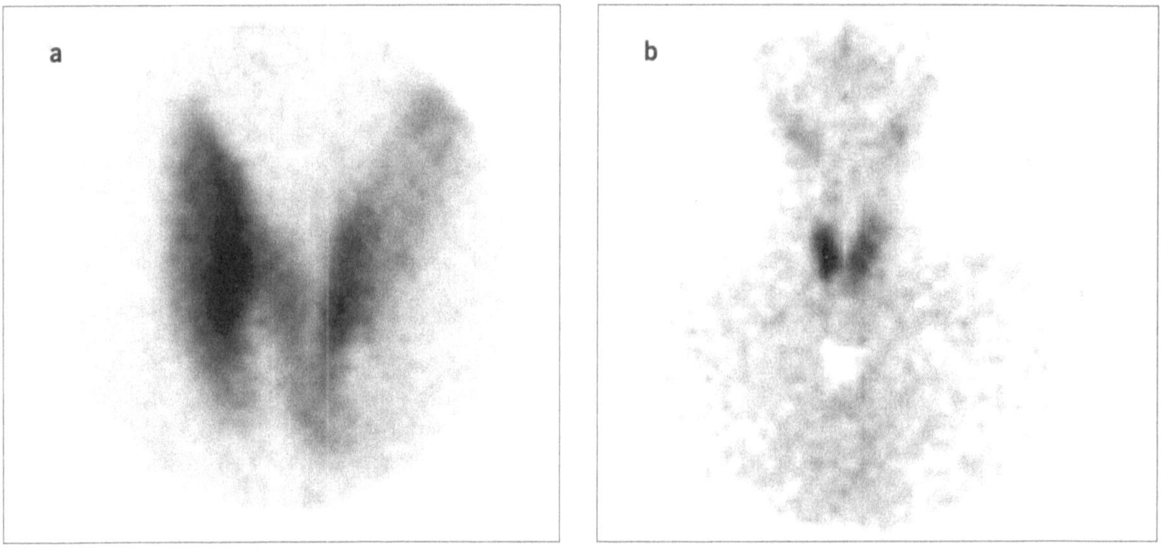

Figure 5.1 Normal thyroid (pertechnetate). (a) Magnified image, (b) localization image. The asymmetry is within normal limits.

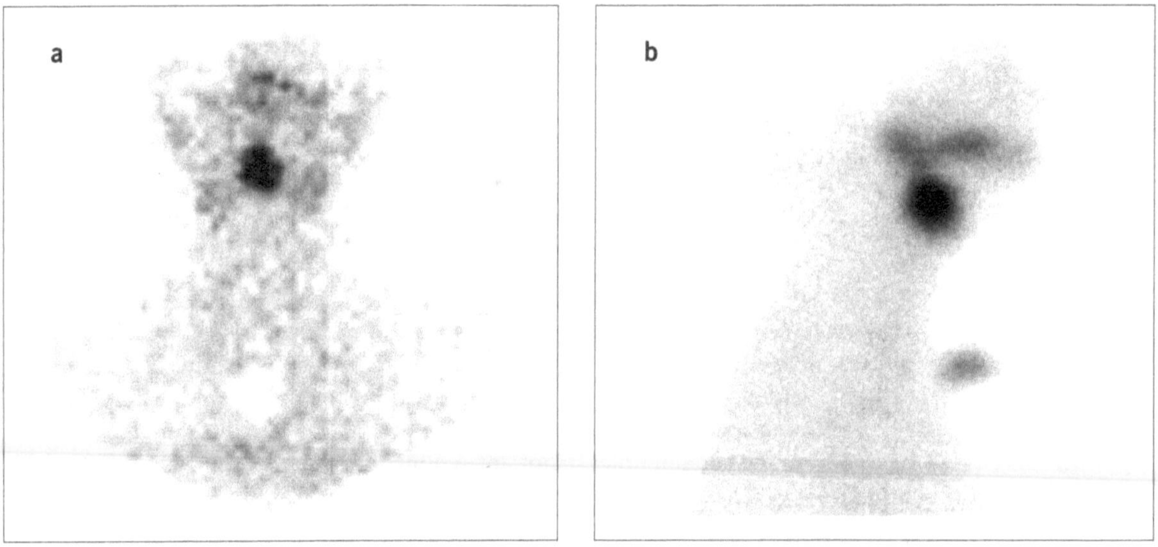

Figure 5.2 Ectopic thyroid in base of tongue imaged 1 h after injection of ^{123}I. Uptake in other tissues is low. The projection is a syringe marking the sternal notch. (a) Anterior projection, (b) lateral.

of the gland is deliberately over-saturated in order to bring out low count areas. There are no clear-cut criteria to distinguish between normal regional variations in uptake and multinodular glands, particularly when there are few nodules and these are small.

The normal thyroid when imaged with pertechnetate shows a higher concentration than the salivary glands, which however should still be visualized. Higher intensity of uptake in the salivary glands than the thyroid is indicative of hypothyroidism (Fig. 5.4).

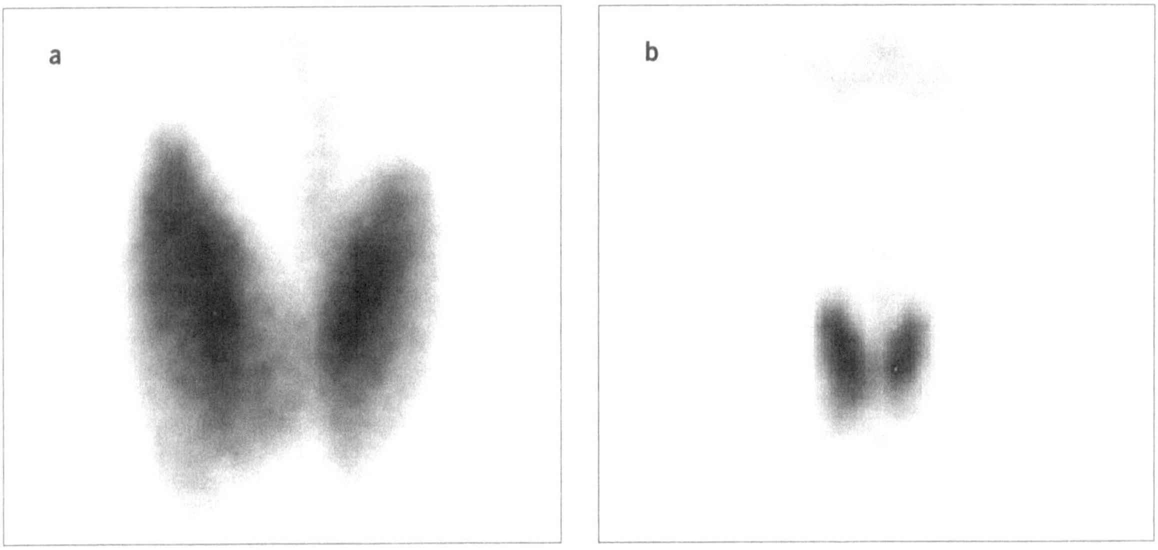

Figure 5.3 Thyroid showing pyramidal lobe. (a) Magnified, (b) localizing.

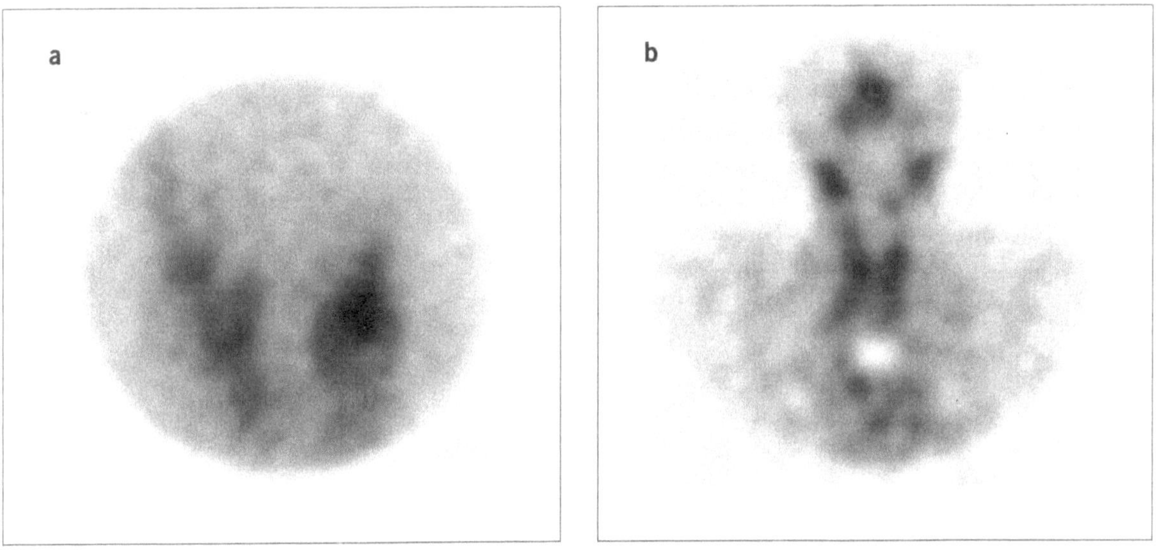

Figure 5.4 (a) Magnified and (b) localization image in hypothyroidism. The thyroid is enlarged but uptake is low, with higher intensity of uptake in salivary glands than thyroid.

Thyrotoxicosis

Thyroid uptake of pertechnetate is usually so high in the thyrotoxic patient that salivary glands may not be visible on the printed image because the range of contrast most media are capable of displaying is insufficient (Fig. 5.5). The appearance may resemble that seen with ^{123}I or ^{131}I. The salivary glands are usually visible on a well-set-up monitor or in colour if a sufficiently long scale (at least 50 levels) is employed. Thyrotoxicosis due to pituitary hyperfunction or rebound following cessation of anti-thyroid therapy gives a generalized increase in uptake, with the gland clearly visualized against a low background. A similar effect is found if the patient has received thyroid stimulating hormone

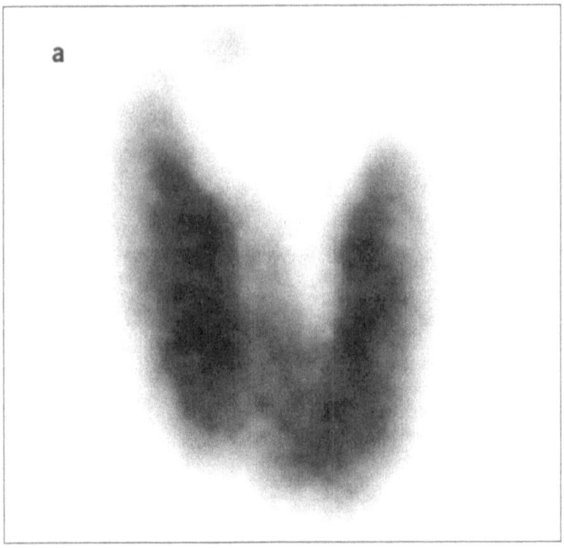

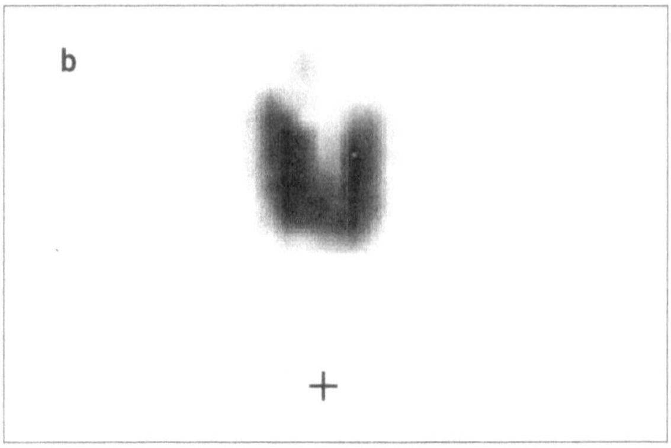

Figure 5.5 Pertechnetate images in patient with Graves' disease. (a) Magnified.Uptake in the thyroid is so high that other structures are barely seen. (b) Localizing. The sternal notch is marked '+' in the localization image.

(TSH) or thyrotrophin releasing factor (TRF). Imaging is necessary to differentiate hyperthyroidism due to an autonomous toxic nodule from the more common Graves disease.

The rigorous criteria required for diagnosis of a toxic nodule are a palpable thyroid nodule with much higher uptake of radioactivity than residual normal thyroid tissue, which may be completely suppressed. There should be biochemical evidence of autonomous thyroid hyperfunction which is not suppressed by an adequate dose of exogenous thyroid hormone (80–120 μg daily of triiodothyronine for 7 days) and increased uptake by the suppressed normal thyroid tissue in response to TSH (10 units intramuscularly daily for 3 days). Stimulation and suppression are not routinely necessary to confirm the diagnosis. Autonomous functioning nodules are usually solitary (Fig. 5.6) but are occasionally multiple (Fig. 5.7). They are almost always benign. If the patient is thyrotoxic the rest of the gland is almost completely suppressed. If the nodule is

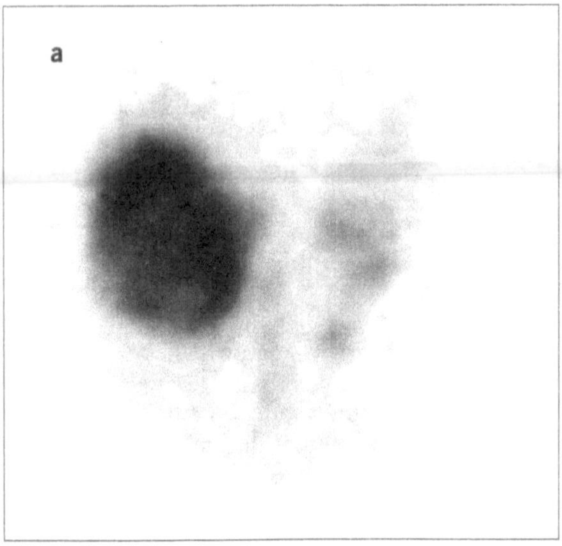

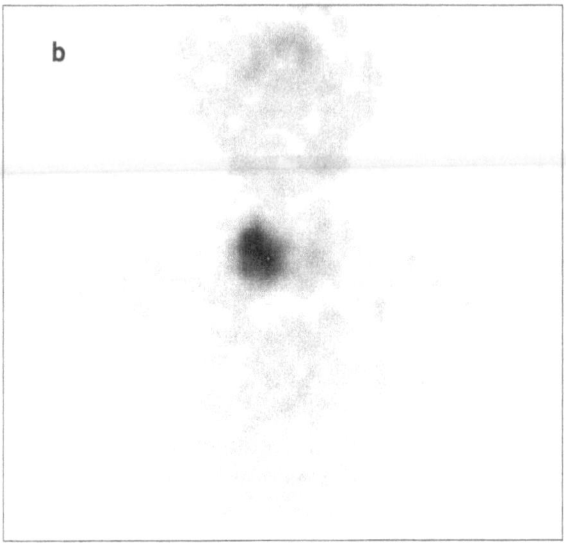

Figure 5.6 Autonomous toxic nodule. The rest of the gland is largely suppressed and salivary glands are only just visible. (a) Magnified, (b) localizing.

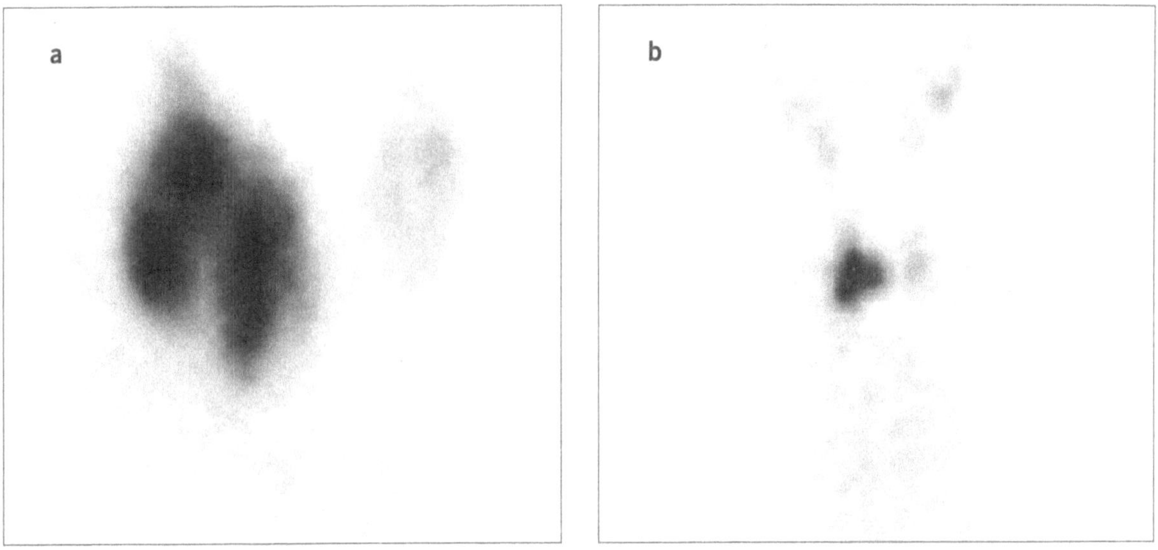

Figure 5.7 Toxic nodular goitre. (a) Magnified, (b) localizing.

not producing sufficient hormone to inhibit pituitary secretion of TSH, the rest of the gland may be faintly visualized. Toxic nodules are associated with a relatively high incidence of toxicosis due to excess production of triiodothyronine rather than thyroxine and may thus be missed by some routine first-line biochemical tests.

Further reading:
European Journal of Nuclear Medicine 1996; ***23****: 587–94*

Ophthalmopathy

Ophthalmopathy sufficient to endanger sight is a potentially serious complication occurring in a minority of patients with Graves' disease. Symptoms may in some cases be controlled by somatostatin. Uptake of pentetreotide has been suggested to identify those patients likely to respond but is probably a measure of the inflammatory response rather than of the underlying disease process.

Further reading:
Journal of Nuclear Medicine 1995; ***36****: 550–4*

Hypothyroidism

Neither imaging nor *in vivo* uptake measurements have any continuing role in the routine diagnosis or management of hypothyroidism.

Hypofunctioning (cold) nodules

These may be solid or cystic, benign or malignant. The commonest causes are colloid cysts, adenoma, non-toxic nodular goitre and carcinoma (Fig. 5.8). Less common causes include parathyroid adenoma, haemorrhage, abscess, artefacts of various sorts and localized thyroiditis. Ultrasound can usually distinguish between solid and cystic but not between benign and malignant. Nodules which take up pertechnetate are not always benign. Some thyroid cancers have an unimpaired or only slightly impaired ability to trap iodide although their capacity for organification is reduced. Thus a palpable nodule which does not appear photon-deficient with pertechnetate may be photon-deficient if imaged 18–24 h after administration of ^{123}I. This is however an inconsistent and unreliable sign and is not a sufficiently frequent or reliable finding to warrant its use as a diagnostic test. Scintigraphy may demonstrate more nodules than are palpable (Fig. 5.9a,b), in which case they are likely to be benign; but the most reliable test is fine needle aspiration biopsy of any nodule clinically suspicious of malignancy. Imaging of clinically solitary nodules often reveals additional non-palpable nodules. Any solitary thyroid nodule should be biopsied to obtain a definitive diagnosis. Definitive diagnosis of most nodules

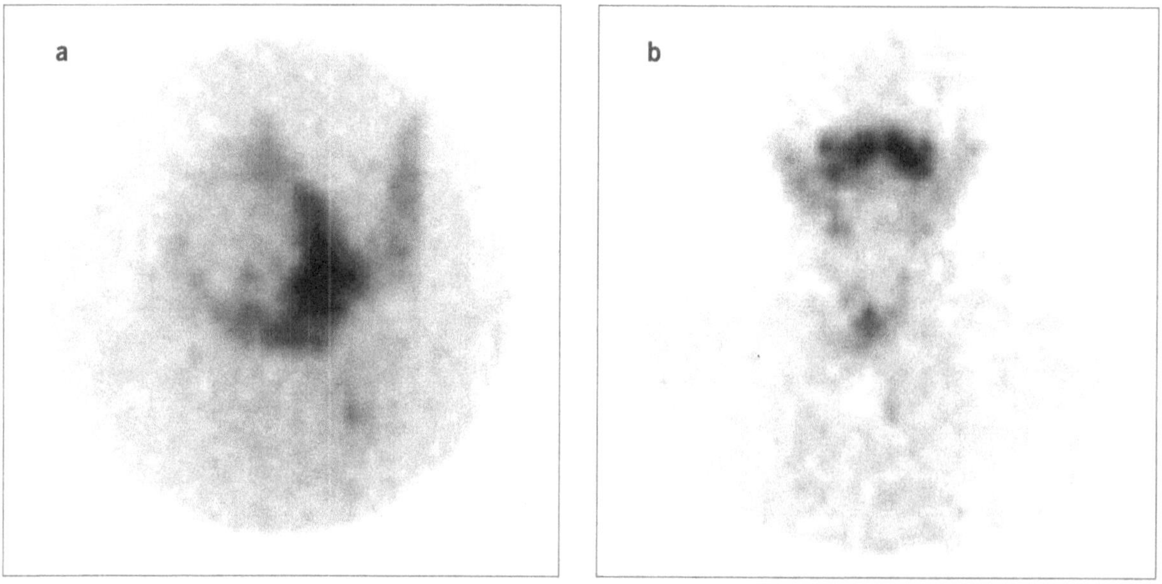

Figure 5.8 Cold nodule visible only on magnified view (a). Note also pyramidal lobe. (b) Localizing view.

can be obtained by fine needle aspiration, without imaging.

Further reading:

Journal of Nuclear Medicine 1991; ***32****: 2181–92*

Multi-nodular goitre

Uptake is often so uneven that it may be difficult to decide which parts are normal, which are regenerating nodules and which are cystic or hypofunctioning nodules (Fig. 5.9). A multinodular

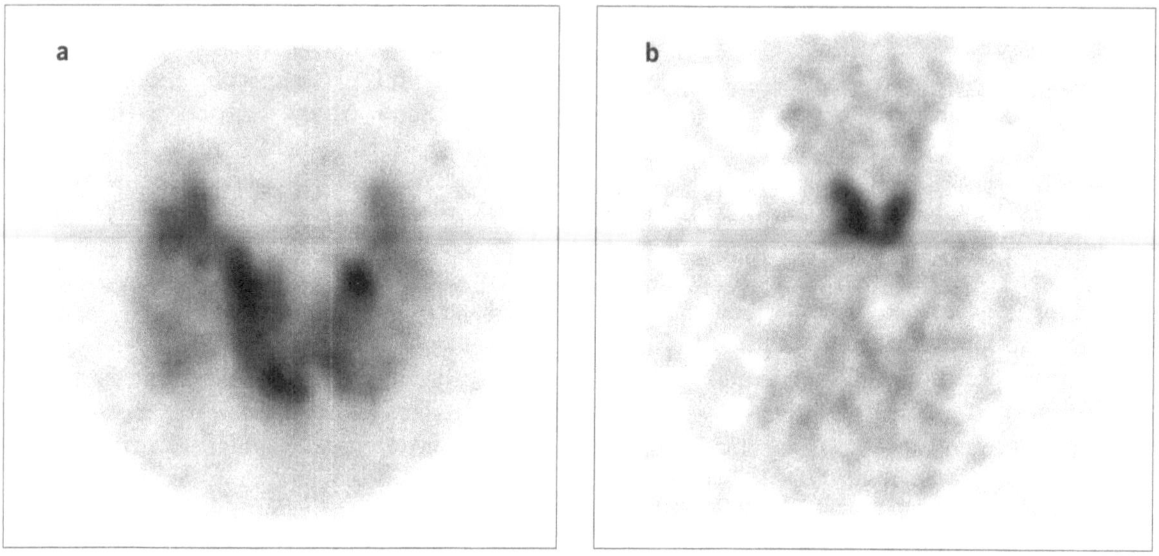

Figure 5.9 Multinodular goitre. (a) Magnified view, (b) localizing view. Few of the nodules are seen without magnification. It should be noted that zooming does not improve resolution and is not a substitute for magnified pinhole views. (c,d) There is retrosternal extension partially concealed by the lead disc used to mark the sternal notch. When uptake is low retrosternal extent of the gland can be more clearly demonstrated with ^{123}I or ^{131}I.

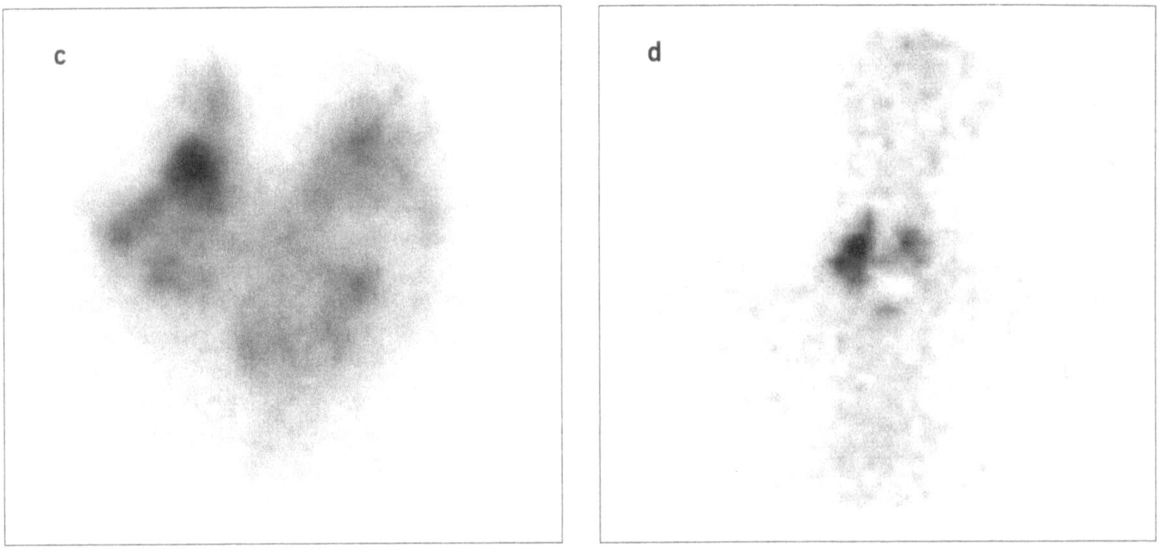

Figure 5.9 *continued*

goitre is rarely associated with malignant change. It is not necessary to image multinodular goitre in the majority of patients. Retrosternal extension cannot be reliably determined unless the sternal notch is accurately identified (Fig. 5.9c, d)

Post-surgery

Regeneration of thyroid tissue following partial or subtotal thyroidectomy gives an uneven and unpredictable pattern. In some the gland may appear almost normal in shape. In others one or more apparent nodules are seen (Fig. 5.10). Because of the disturbed anatomy non-functioning nodules cannot readily be differentiated from lymph nodes and other non-thyroidal structures. Autonomously functioning thyroid nodules can be diagnosed from their failure to suppress following adequate doses of thyroxin.

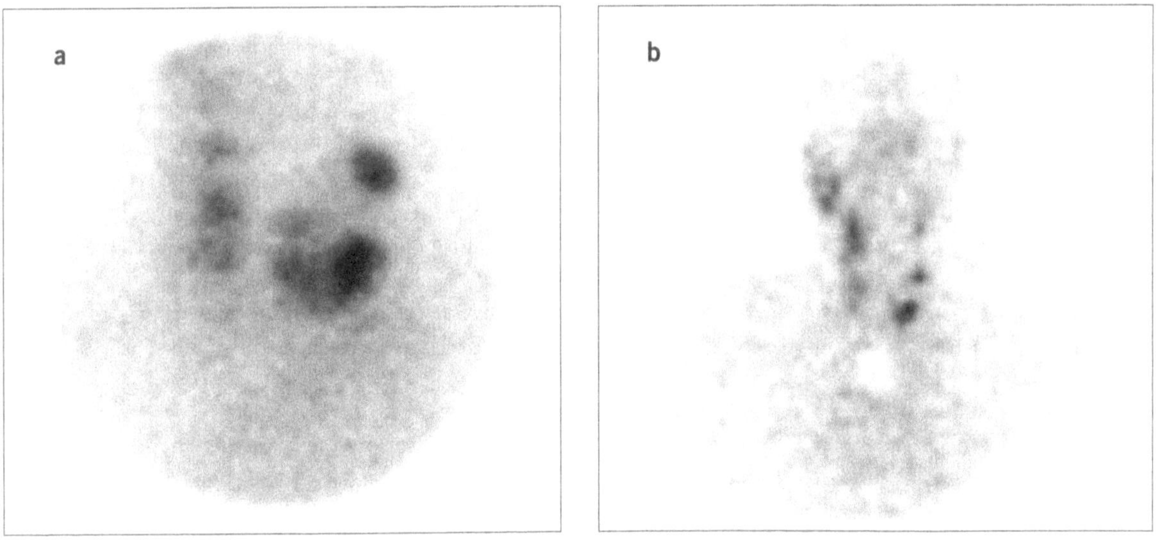

Figure 5.10 Irregular regeneration following partial thyroidectomy some years previously.

Sources of error

False positive diagnosis of toxic nodule may be due to congenital absence of one lobe of the thyroid, previous surgery, asymmetric hypertrophy, residual functioning tissue in patients with auto-immune thyroiditis or lobulation in patients with treated or untreated toxic diffuse goitre or auto-immune thyroiditis. Radio-iodine uptake is occasionally observed in lung carcinoma and lymphoma.

Therapy

The objective of treatment is to re-establish a euthyroid state, preferably with a single administration but avoiding hypothyroidism. Scintigraphy is generally recommended prior to radio-iodine therapy, principally to differentiate Graves' disease from toxic nodule but also to assess the size of the gland. It is however disputed whether scintigraphic estimates of gland size are more accurate than palpation. Uptake measurements are also commonly performed, although there is less unanimity on their role. Radio-iodine, almost always ^{131}I, is employed extensively for treatment of both thyrotoxicosis and of those carcinomas of the thyroid which concentrate iodine. Some feel that, prior to administration of treatment for thyrotoxicosis, dosimetric evaluation is required to determine functional parenchymal volume, for example ultrasonographic measurement of lobe or nodule size, thyroid binding rate and biological half-life of ^{131}I. An estimate of effective half-life from a single measurement after 24 h is of limited clinical value. A more accurate figure can be obtained from three readings over 4 or 6 days. The volume of the thyroid is a major variable; radiation dose is modified according to volume of the gland. A common protocol is to adjust the administered activity to give a calculated absorbed dose of 50 Gy, 75 Gy or 150 Gy according to whether thyroid volume is estimated to be less than 50 g, between 50 g and 80 g, or greater than 80 g. The absorbed radiation dose recommended for treatment of toxic adenoma and toxic multinodular goitre is 300 Gy.

Further reading:
*Journal of Nuclear Medicine 1996; **37**: 228–32*

A more widely held view is that imprecision of estimates of thyroid volume is such that standard activities may be administered depending on the clinical impression of gland size. Although uptake and biological half-life of administered radio-iodine can be measured with precision, all techniques of estimating volume are inaccurate. There is no evidence that estimates based on imaging give a lower incidence of complications such as recurrence or hypothyroidism than clinical estimates of gland size based on palpation by an experienced physician. The typical therapeutic activity of ^{131}I for an adult with Graves' disease and a gland considered clinically to be of 'normal' size is 400 MBq. Double this may be given if the gland is unequivocally clinically enlarged or if there is a toxic nodule with suppression of the rest of the gland. At these dose levels, although some subjects will require further treatment to be rendered euthyroid, the incidence of late hypothyroidism is minimized. Long-term follow-up has not shown any difference in late complication or failure rates between the two approaches.

Treatment with ^{131}I may be employed as the primary modality, after failure of or intolerance to medical treatment, or in combination with medical treatment by antithyroid agents. There is wide agreement that, when treating the majority of patients with thyrotoxicosis, the balance of risks and benefits is in favour of radio-iodine rather than surgery. Medical preparation is justified by concern to minimize the transient increase in hyperthyroidism caused by sudden release of preformed hormone triggered by irradiation. Rarely, this can be dramatic and life-threatening (thyroid storm). Medical treatment should be discontinued 48 h prior to administration of ^{131}I and may if necessary be reinstituted after 10 days for 3 months.

Hazards

Long-term follow-up studies have failed to find any increase in incidence of any of the malignancies looked for, with the possible exception of carcinoma of the bladder or stomach. Even this risk is unproven and is small compared with that associated with thyroidectomy. Transient acute radiation thyroiditis with neck swelling, soreness and exacerbation of thyrotoxic symptoms peaking at 24 h is observed in up to one-third of patients. There may be biochemical evidence of siladenitis, but permanent sequelae are rare except in patients receiving repeated large doses for recurrent carcinoma. A transient increase in serum FSH (follicle-stimulating hormone) levels and reduction in sperm count is observed in some men for 6 to 12 months, subsequently returning to normal. Repeated doses may give rise to irreversible changes. There is no measurable effect on female fertility or the incidence of foetal anomalies. Thyroid storm, due to rapid release of a large amount of thyroid hormone, is a rare but serious complication occurring both in patients who have received thyrostatic medication for 3 months before receiving the radio-iodine and in untreated patients. It is usually observed 1 to 3 weeks after radio-iodine. Unless promptly treated the condition can be fatal. Death may be due to congestive cardiac failure or bronchopneumonia. At highest risk are the elderly and those with severe toxicosis, weight loss and cardiovascular or cerebrovascular disease.

Further reading:

*Journal of Nuclear Medicine 1993; **34**: 1638–41. 1994; **36**: 27–28*

Carcinoma of thyroid

Three principal types of carcinoma arise from the thyroid, papillary, follicular and medullary. The clinical incidence of thyroid cancer is relatively low (0.05% per year), in contrast to that discovered at autopsy (5.6% in the USA and 28.4% in Japan).

Papillary and follicular thyroid cancer

Differentiated papillary and follicular thyroid cancer often progress slowly. The rate of growth of papillary tumours is variable and there is often early involvement of lymph nodes, but in young subjects they tend to be associated with a better prognosis than follicular. Follicular tend to be slow growing but are associated with a poorer long-term survival. Anaplastic tumours are usually fatal within six months. A poorer prognosis is also associated with age of onset above 40 years, multifocal and/or vesicular histological appearance, diameter greater than 1 cm, capsular or vascular invasion and metastasis. After excision of the primary tumour there is a cumulative recurrence rate of 20% within ten years, rising to 40% at 40 years. Primary thyroid carcinoma of any histological type within the thyroid are usually vizualized as photon-deficient areas with radio-iodine or pertechnetate and cannot be distinguished from the commoner benign cold lesions. Metastatic papillary or follicular tumours may trap iodine or secrete thyroid hormones but most primary tumours take up less iodide and pertechnetate than normal thyroid tissue and appear photon-deficient. Very rarely, enough thyroid hormone is produced by functioning thyroid metastases to suppress the normal gland.

Thyroid metastases are rarely visualized with radio-iodine until the thyroid has been ablated. This is recommended in all patients with potentially curable thyroid primaries to facilitate early biochemical and scintigraphic identification and treatment of occult local and distant metastases. Normal thyroid tissue is ablated by total thyroidectomy followed by radio-iodine. It is not technically possible to remove all thyroid tissue at thyroidectomy. It is therefore reasonable to postpone the start of hormone replacement post-operatively and to administer an ablative dose of ^{131}I four to six weeks after surgery. Symptoms of hypothyroidism may be minimized by physiological doses of T3 for the first three or four weeks; this must be discontinued for at least ten days before administering radio-iodine therapy. Complete ablation of thyroid residues following total thyroidectomy is usually considered to require 3000 MBq, but there is evidence that 1000 MBq is adequate in many subjects. Replacement therapy with thyroxine, at a dose sufficient to suppress endogenous TSH but not to produce symptoms of hyperthyroidism should be started 24 h after giving the radio-iodine. Whole-body imaging three days and seven days later reveals any functioning metastases. It is extremely rare for thyroid metastases to be visualized with pertechnetate.

When imaging for functioning metastases in a patient whose thyroid has previously been ablated, replacement therapy with thyroxine should be discontinued six weeks before administration of the tracer, to allow time for the effects of thyroxine, especially on TSH secretion, to wear off completely. An equivalent dose of triiodothyronine (T3), which has a shorter biological half-life, should be substituted for the first four weeks, to minimize clinical symptoms of hypothyroidism. All replacement therapy must be stopped not less than ten days before administering radio-iodine. Replacement therapy should be restarted not earlier than 24 h after administering the tracer dose. The contrast between metastases and background is usually greatest 72 h after oral radio-iodine, but at this time there is usually appreciable activity in the gastro-intestinal and urinary tracts. Deposits below the diaphragm are better seen at seven days, when there is little residual activity in the gut or urinary tract so that any residual uptake is likely to be in metastases. Because of this delay shorter lived isotopes such as ^{123}I cannot be used.

The activity necessary for diagnosis depends on the equipment available. ^{123}I has the most favourable physical characteristics but the 13 h half-life limits imaging to the first 48 h after injection, when appreciable uptake in stomach, large bowel and breast (even in non-lactating women) are commonly observed. Not less than 400 MBq ^{131}I is required for imaging with a gamma camera for one to two weeks, but this may place restrictions on the patient determined by radiation safety legislation. A lower activity may be administered if a one-dimensional profile scan is available. If the profile scanner is a high sensitivity whole body counter, 40 MBq may be adequate because of the better counting statistics obtained with a one-dimensional profile compared with scintigraphy. Interpretation is however more difficult, especially at three days when much of the retained activity is still in the liver, large bowel and sometimes gastric mucosa. A sensitive combination is profile scanning to screen for the presence of an abnormality (Fig. 5.11) combined with scintigraphy to localize it

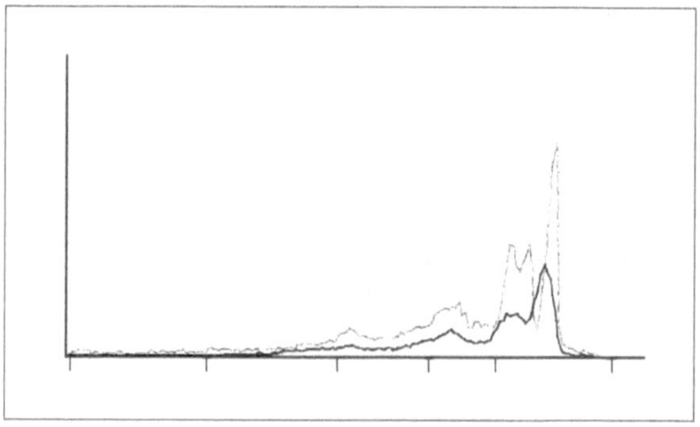

Figure 5.11 Whole-body ^{131}I profile in patient with metastatic carcinoma of thyroid. The paler line is count-rate 3 days after administration and the darker at 7 days (decay corrected). Uptake can be measured in any region by integrating under any part of the curves and comparing with a standard. Markers along *x* axis indicate (from left to right): feet, knees, iliac crest, xiphisternum, sternal notch and top of head.

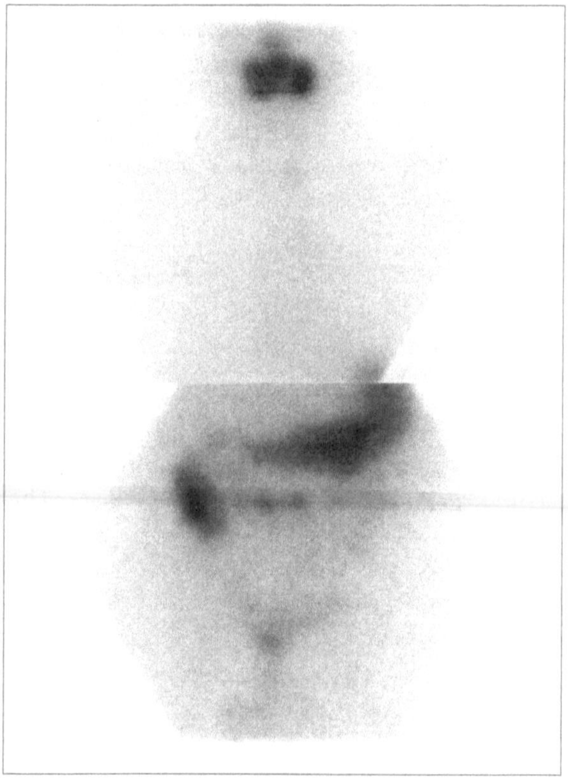

Figure 5.12 Composite anterior projections 3 days after 3000 MBq ^{131}I to ablate residual thyroid after total thyroidectomy, showing uptake principally in residual thyroid tissue in the neck, in gastric mucosa and in colon.

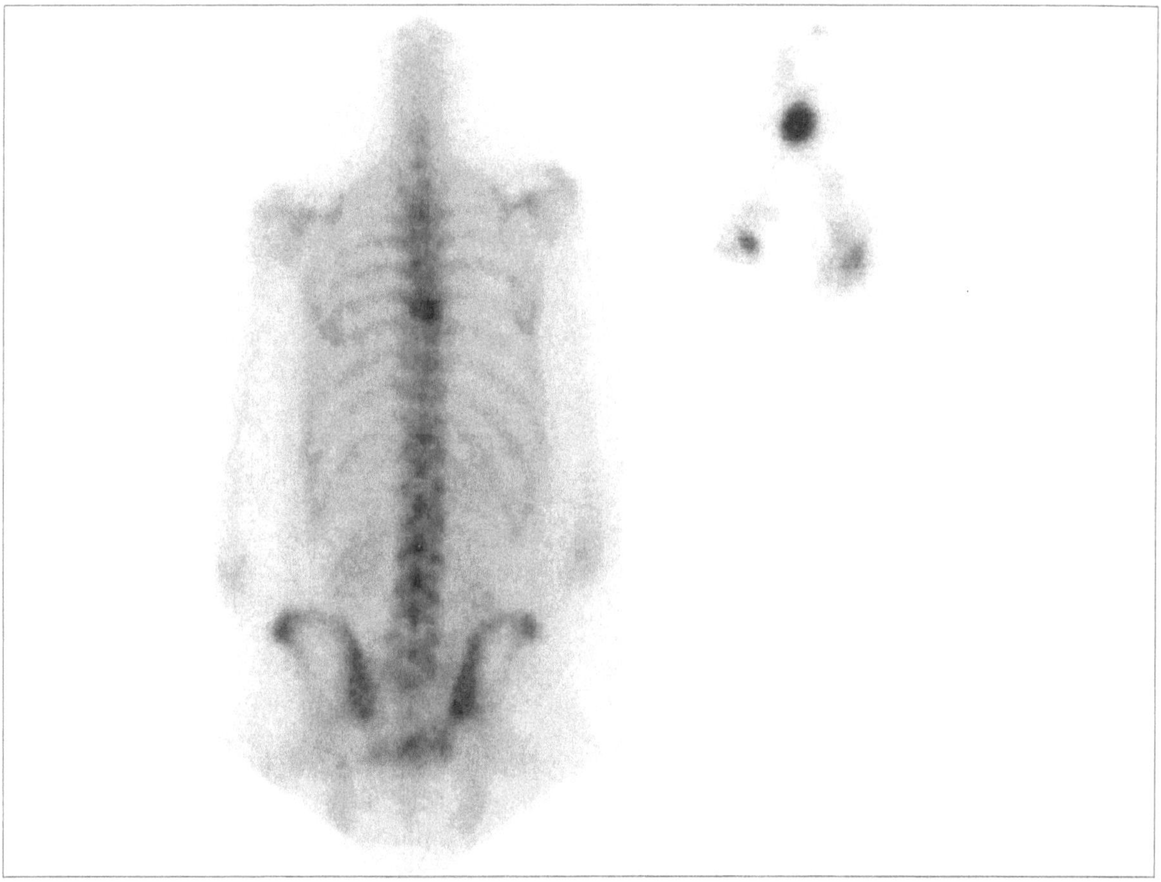

Figure 5.13 Previous total thyroidectomy and residue ablation. Both images are to same scale. Right hand: iodine uptake in neck and lung metastases 14 days after 8000 MBq ^{131}I for recurrent thyroid carcinoma. Left hand: deposit in D7 seen only on the bone scan. It does not take up radio-iodine. It is common for only some metastases to take up radio-iodine.

accurately and in order to distinguish tumour from residual normal structures (Fig. 5.12). In the neck, scintigraphy is necessary to distinguish residual functioning thyroid from laterally placed lymph node metastases. Profile scanning also facilitates measurement of uptake of iodine into metastases. This is of some use in assessing the amount of activity to be administered to obtain a therapeutic effect.

Serum thyroglobulin levels are elevated in many but not all patients with recurrence of metastatic thyroid carcinoma. Some patients with elevated thyroglobulin levels have normal whole-body ^{131}I profiles whereas in others there is scintigraphic evidence of metastases despite a normal thyroglobulin level. The techniques are therefore complementary; higher detection rates are obtained if both are employed. When serum thyroglobulin is elevated but there is no detectable localization of iodine, deposits may in some cases be identified by bone scintigraphy (Fig. 5.13), thallium or sestamibi. Pentetreotide uptake has also been observed in some patients with differentiated thyroid metastases. This was more evident than iodine uptake in some deposits of two patients with the 'insular' form of thyroid carcinoma. Intense persistent uptake of sestamibi has also been reported in Hurthle cell tumour. There is some evidence that a combination of serum thyroglobulin and ultrasound of the neck may be the most effective method of follow-up, reserving follow-up scintigraphy for patients in whom either or both of these is abnormal and patients previously treated for recurrence.

Further reading:
European Journal of Nuclear Medicine 1995; ***22:*** *1330–8*
Journal of Nuclear Medicine 1996 ***37:*** *26–31 & 446–51*

Medullary carcinoma of the thyroid

This is a rare condition, accounting for only 5% of all thyroid malignancies. It may however occur in any one of four distinct settings, namely sporadic, familial medullary thyroid carcinoma or as a component of multiple endocrine neoplasia (MEN) syndrome, of which there are two distinct types, designated 2a and 2b. The sporadic variety is usually unilateral and is not associated with other abnormalities. The familial form is usually bilateral but has no other associated abnormalities and is the least malignant. MEN2a is also usually bilateral, there is commonly a family history and associated phaeochromocytoma and hyperparathyroidism. MEN2b is the most malignant form and is commonly associated also with phaeochromocytoma, mucosal neuromas and ganglion neuromas.

Medullary carcinoma of the thyroid arises from calcitonin-secreting C cells. Measurement of peripheral calcitonin levels is a good marker for the existence of tumour. Approximately half of patients with MEN2a develop phaeochromocytoma and one-third pseudonodular parathyroid hyperplasia. Patients with MEN2b do not have associated hyperparathyroidism but do develop phaeochromocytomas and have a characteristic phenotype associated with a typical neural hypertrophy, usually recognizable within the first months of life. In patients with MEN2b the tumour usually runs an aggressive course, whereas in those with familial medullary thyroid carcinoma the tumour usually develops late in life and grows slowly. The natural history in MEN2a is variable, some patients showing rapid and others slow progression. Before considering surgery in any patient with medullary thyroid carcinoma it is essential to exclude phaeochromocytoma because of the danger of hypertensive crises during anaesthesia.

Untreated, the condition is usually ultimately fatal. Early diagnosis and resection may be curative. The most sensitive diagnostic test is measurement of plasma calcitonin levels after sequential intravenous administration of calcium gluconate (2 mg/kg/min) followed immediately by pentagastrin (0.5 mg/kg/5 s). Peripheral blood samples are obtained at 1, 2, 3 and 5 min after the infusion. Peak values generally occur at 1–2 min. Plasma concentrations above 1000 pg/ml are associated with a high likelihood of metastatic disease whereas under that value there is a greater than 90% probability that the disease is confined to the thyroid, so that total thyroidectomy is likely to be curative.

Radiopharmaceuticals

A range of radiopharmaceuticals have been proposed. Most experience has been obtained with pentetreotide (page 173), MIBG (page 172) and DMSA(V) (page 267). ^{201}Tl and sestamibi (page 116) have been advocated and preliminary results with a monoclonal antibody have been reported. ^{111}In-pentetreotide has the highest sensitivity for detecting both primary and metastatic medullary thyroid carcinoma and any associated phaeochromocytoma. MIBG has a high sensitivity for detecting primary tumour in familial but not sporadic disease, but probably has a slightly lower accuracy than pentetreotide for detection of phaeochromocytoma. Tc-DMSA(V) has a sensitivity of 85–95% for detection of medullary thyroid carcinoma but is not recommended for localization of phaeochromocytoma. In view of the cost differential between Tc-DMSA(V) and pentetreotide the former should be employed first, reserving pentetreotide for those who are negative with DMSA(V). False positives have been reported with pentetreotide due to benign non-secreting adenomas of various types and in sites of chronic inflammation. The false negative rate is 5–15% with Tc-DMSA(V) and pentetreotide. It is somewhat higher with MIBG.

Further reading:
European Journal of Nuclear Medicine 1996; ***23:*** *1367–71*

Technique

Imaging should be performed 2–3 h after 150–250 MBq Tc-DMSA(V) and 4 and 24 h after

20 MBq ^{131}I-MIBG or 100 MBq ^{111}In-pentetreotide. It may be repeated at 48 h with pentetreotide and at intervals up to 7 days with MIBG if abdominal uptake is concealed by excreted activity in the urinary tract or bowel. Contrast improves progressively as the delay increases. SPECT (page 318) may reveal additional abnormalities, especially if contrast is low.

Interpretation

There is a high general background level in all tissues with Tc-DMSA(V). In contrast to Tc-DMSA(III) there is no fixation in the renal cortex although the predominant pathway of excretion is renal, with consequent visualization of the renal collecting systems and bladder. Pelvic lesions may be obscured if the patient has not voided immediately before imaging. The liver is usually identified. Abnormalities appear as focal accumulations of increased count density (Fig. 5.14).

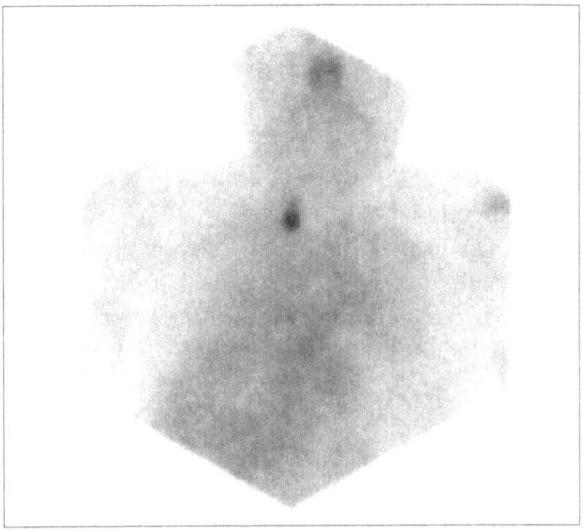

Figure 5.14 Sporadic medullary thyroid carcinoma imaged with DMSA(V).

Further reading:
*Journal of Endocrinology 1995; **144:** 339–45*

The parathyroids

There are two pairs of glands, superior and inferior, the most common location for the superior being on the posterior border of the lateral lobe of the thyroid. The inferior is usually applied to the lower pole. The majority are adjacent to the thyroid but there are many variations in position. They may be found retrosternally in or adjacent to thymus, in anterior mediastinum, within thyroid, within carotid sheath or in posterior mediastinum. An experienced surgeon is able to locate all four glands in 90% or more of patients when exploring the neck.

Radiopharmaceuticals

A number of radiopharmaceuticals have been described for localization of the parathyroids including selenomethionine, iodinated toluidine blue, thallium, sestamibi and tetrofosmin. Only the last three have achieved widespread use. Both thallium and sestamibi may be used in association with ^{99m}Tc sodium pertechnetate. The pertechnetate image, which indicates thyroid, is subtracted from or compared with the image of thallium or sestamibi distribution, which shows both thyroid and parathyroid. In order to obtain a large enough field of view to include all possible sites of the parathyroid a high resolution parallel-hole collimator is preferable to a pinhole. When using thallium, simultaneous dual isotope acquisition is preferable to consecutive acquisition because of the difficulty of eliminating patient movement over the extended duration of the examination. There is however potential for error due to misregistration at the different energies. For this reason thallium subtraction scintigraphy has been largely superseded by the technetium-labelled cardiac agents.

Technique

Doses of 150–700 MBq ^{99m}Tc-sestamibi or tetrofosmin have been employed. Unless SPECT is required 150–250 MBq is adequate. Imaging 10–15 min after injection of sestamibi discloses uptake in both thyroid and parathyroids. Later imaging, at 3–4 h, shows wash-out from the thyroid, revealing the parathyroids as focal areas of higher count rate (Fig. 5.15). Tetrofosmin does not redistribute, and pertechnetate must be used for the thyroid image. The preferred protocol with tetrofosmin is to image for parathyroid tissue 10 min after 75 MBq, administer 75 MBq pertechnetate and image again 10 min later. Visual comparison of the two images is as reliable as subtraction.

When using the dual isotope technique simultaneous dual isotope acquisition is preferable to minimize artefacts due to patient movement. A dose of 40 MBq ^{99m}Tc-pertechnetate is given intravenously. Two windows, centred on the 80 and 140 keV photopeaks, are used. Consecutive 60 s frames are acquired in both windows for 10 min, when 80 MBq ^{201}Tl is injected intravenously. After 5 min a further dual energy acquisition is obtained at one frame per minute for 15 min. Any frames in which there is movement are discarded. The technetium scatter into the thallium window is calculated from images obtained in the thallium window prior to injection of thallium. Choice of normalization factor for subtraction of the technetium images is arbitrary but is conventionally adjusted so that negative values are not obtained.

Clinical indications

The parathyroids can also be imaged by barium swallow, high resolution ultrasound, computed tomography (CT) and magnetic resonance imaging (MRI). High resolution CT and MRI have the highest sensitivities, over 70% compared with approximately 60% for scintigraphic techniques. However no series has separately identified patients with recurrence of hyperparathyroidism following surgery, or in whom the glands had not been identified at surgery, and none of these techniques can reliably differentiate parathyroid from thyroid nodules. In view of the accuracy of identification of parathyroids by an experienced surgeon, routine pre-operative imaging is difficult to justify unless it could be shown to be sufficiently accurate to restrict surgical exploration. It is arguable that, if hyperparathyroidism persists following a neck exploration, the correct investigation should be venous sampling to detect sites of raised parathormone concentration. It is not possible to make out a case for routine pre-operative imaging of patients with biochemical or clinical evidence of parathyroid adenoma. The role of imaging in parathyroid carcinoma is not established. Uptake of both thallium and sestamibi has also been reported in brown tumours.

Further reading:
Lancet 1997; ***349****: 1233-8*

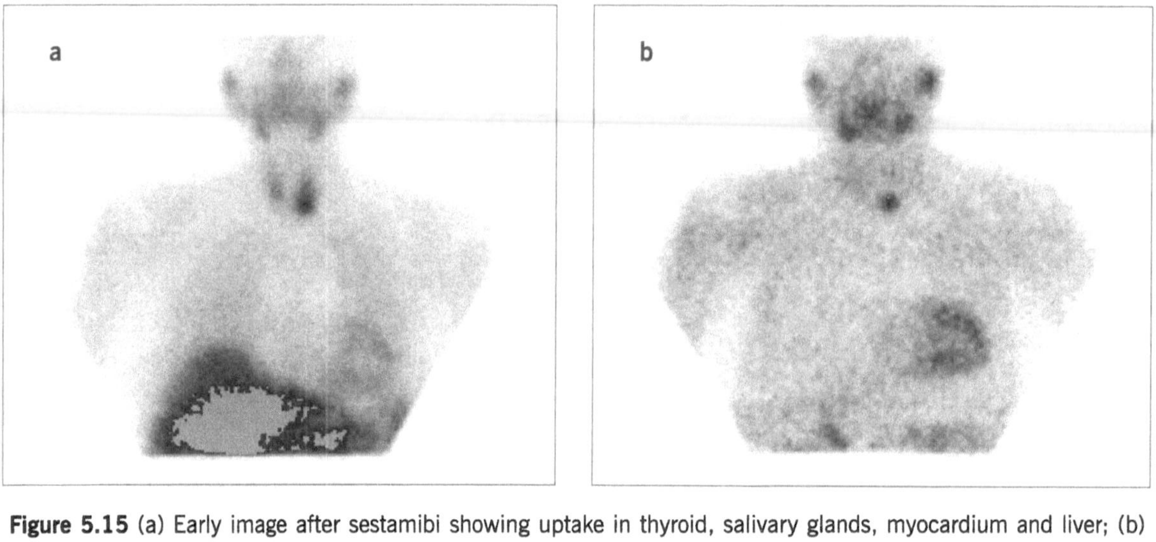

Figure 5.15 (a) Early image after sestamibi showing uptake in thyroid, salivary glands, myocardium and liver; (b) 3 h later uptake has washed out of thyroid, revealing a parathyroid adenoma.

Adrenal glands

The adrenal glands are situated on the medial aspect of the upper pole of each kidney. They consist of two distinct functional and morphological components: an outer cortex and an inner medulla. The former produces a number of steroidal hormones, the latter catecholamines. These have different physiological functions and therefore manifest differently in disease. Adenoma, carcinoma and hyperplasia of either part can occur.

Adrenal cortex

The adrenal cortex has three distinct zones. The outer zona glomerulosa produces mineralocorticoid steroid hormones especially aldosterone, secreted in response to decrease in either total body sodium or plasma volume and also influenced by potassium. Its major effect is on the distal nephron where sodium is exchanged with potassium to maintain salt and water balance. Deep to the zona glomerulosa is the zona fasiculata which produces glucocorticoids, principally cortisol. These have widespread and complex functions and are controlled by a negative feedback loop involving the hypothalamus, anterior pituitary and adrenal cortex. The zona reticularis secretes androgenic steroids. Most functioning adenomas arise from one zone and secrete only one class of steroid.

Radiopharmaceuticals

Three compounds have been used for imaging the adrenal cortex:: 19-iodocholesterol (NM145), 6-iodocholesterol (NP59) and 6-selenonorcholestenol (6-selenomethylcholesterol, 6-selenocholesterol, Scintadren). The 19-iodo compound has now been largely superseded by the 6-compounds which have higher uptake into the adrenal. The iodinated compound is used almost exclusively in the USA. In Europe most experience has been obtained with selenocholesterol; iodocholesterol has been used in some centres but there have been few direct clinical comparisons between the two compounds. All three tracers act as markers of cholesterol and, like cholesterol, may be identified in the low density lipoprotein fraction of plasma either as the free alcohol or esterified with short-chain fatty acids. Within the adrenal cortex there is a high concentration of cholesterol, largely esterified with long chain fatty acids. These esters are storage products which must be hydrolysed before cholesterol can enter the steroid hormone synthesis pathways. Initial uptake is probably dependent on blood flow. The iodine and selenium-labelled cholesterol analogues are able to enter cells of the adrenal cortex where, like cholesterol, they are esterified with long-chain fatty acids. However esters of the labelled analogues are resistant to hydrolysis within the adrenal; radiopharmaceutical that has entered the gland is trapped and not further metabolized. These tracers are thus markers of adrenal perfusion, uptake and storage of cholesterol but not of steroid synthesis.

The principal difference between iodinated and seleno- compounds is in their stability. Selenocholesterol is stable both *in vitro* and *in vivo*. No enzyme systems are known which affect stability of the selenomethyl bond to the cholesterol ring. Circulating selenocholesterol not sequestered in the adrenal is converted to a bile acid analogue. The only pathways of excretion are in bile, either as the unchanged tracer or its bile acid derivative, both of which are reabsorbed and recirculate. In contrast the iodinated compounds are unstable both *in vitro* and *in vivo*. *In vitro* they must stored in solid CO_2 at –20°C and administered immediately after being thawed. The thyroid must be blocked to prevent uptake of released iodine into the thyroid. It has not been established whether breakdown is active or passive and, if the latter, in which tissues deiodination occurs.

This difference in stability has important implications for clinical use. The iodinated compound labelled with ^{131}I is usually imaged 5–7 days after administration, although Cushing's tumours are often visible earlier. By this time much of the extra-adrenal tracer has been deiodinated and the activity excreted. Although there may be residual activity in bowel, hepatic uptake is low and contrast in consequence relatively high. Both selenocholesterol and its bile acid derivative are stable and recirculate around the entero-hepatic circulation, so that the rate of excretion is considerably slower. When using these compounds it is often necessary to wait two to three weeks before the adrenals are clearly visualized above the activity in liver and colon.

Selenium has a half-life of 118 days and emits several gamma rays (Appendix 1) but no β particles. ^{131}I on the other hand has both β- and gamma emissions and a half-life of approximately 8 days. The gammas emitted by selenium, with the exception of the 412 KeV peak, are detected more efficiently by a gamma camera than the 364 KeV gamma ray of iodine. Thus for an equal absorbed radiation dose a higher photon flux can be achieved with selenium-labelled than with iodinated compounds, even allowing for the longer delay before imaging.

Clinical applications

Cushing's disease

The diagnosis of Cushing's disease is established by estimation of plasma and urinary concentration of the steroid hormones and their derivatives. Circulating ACTH (adrenocorticotrophic hormone) levels usually distinguish between pituitary and autonomous adrenal causes. The majority of tumours producing Cushing's disease are large and are readily visualized by CT, ultrasound or MRI. Indeed many can be seen on plain radiographs of the abdomen. It is however not possible to reliably distinguish a functioning adenoma from a lipoma or non-functioning enlargement of the gland by CT or other radiological technique. Scintigraphy with one of the cholesterol analogues may be indicated if there is uncertainty as to the side of an adenoma or if there is doubt that there may be bilateral adenomas. Imaging should be combined with uptake measurements. Visualization of one adrenal only in a patient with biochemical evidence of Cushing's disease is suggestive of an adenoma with suppression of the contralateral gland. Bilateral uptake is usually evidence of hyperplasia, although rarely there may be bilateral adenoma. Bilateral non-visualization in a patient with clear biochemical evidence of Cushing's disease is evidence of adrenal carcinoma, the hormone excess coming from metastases elsewhere. These may take up insufficient cholesterol analogues to be visualized.

Further reading:
Journal of Nuclear Medicine 1990; ***32****: 1627–39*

Conn's syndrome

This is hypertension due to excess secretion of aldosterone. It is usually due to an adenoma (Fig. 5.16) but can occasionally be due to hyperlpasia (Fig. 5.17). Aldosterone-secreting adenomas tend to be smaller than those producing Cushing's syndrome and are less likely to be identified on CT or ultrasound. Two techniques have been employed, with and without dexamethasone suppression. Dexamethasone 1 mg 6-hourly for 7 days can halve uptake of cholesterol analogues. A high salt diet can reduce uptake by a further 10%. The remaining 40% of the uptake cannot be suppressed. It is necessary to suppress the glands for one week before administering the radiopharmaceutical and at least 5 days thereafter. However in practice it is difficult to maintain suppression of the adrenal. In many patients some escape occurs during this period. Uptake measurements are therefore difficult to interpret. The alternative approach is to perform uptake measurements without suppressing the gland and

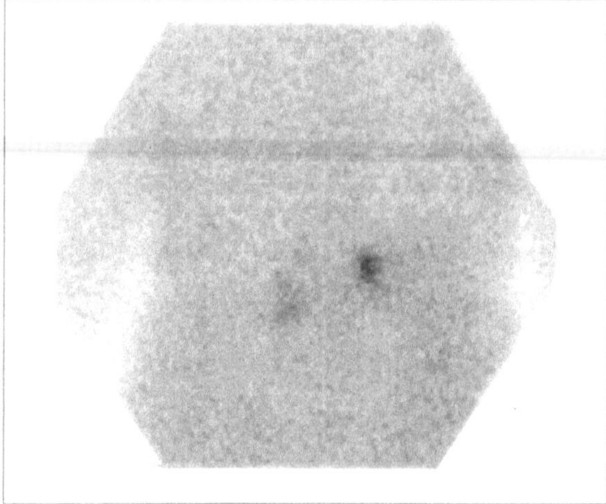

Figure 5.16 Right Conn's tumour three weeks after 20 MBq ^{75}Se-Scintadren. Uptake in the normal left gland is partially suppressed by dexamethasone.

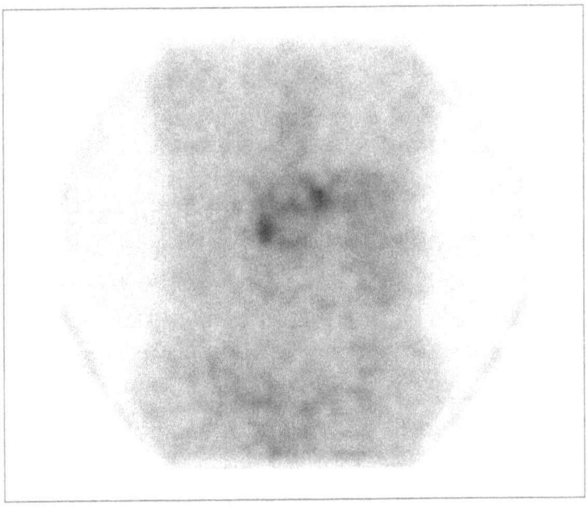

Figure 5.17 High bilateral uptake of Scintadren 14 days after administration, despite dexamethasone suppression. Patient with Conn's syndrome due to adrenal hyperplasia.

to regard asymmetry as evidence of an adenoma, the adenoma being in the gland with the higher uptake. There is insufficient evidence to compare the accuracy of these techniques.

Further reading:
Seminars in Nuclear Medicine 1989; ***19****: 122–43*

Virilizing tumours

Cholesterol analogues have also been used to visualize virilizing tumours arising from the adrenal. Limited information is available; the accuracy and sensitivity of the technique is unknown. The small amount of data which has been published suggests that it may be quite reliable but more information is required.

Further reading:
Journal of Nuclear Medicine 1988; ***29****: 1644–50*

Incidental tumours

Unexpected enlargement of one or both adrenals is an occasional unexpected finding on CT or MRI imaging of the abdomen. Scintigraphy does not assist in differential diagnosis but may assist the decision which side to biopsy, if this is thought necessary.

Further reading:
European Journal of Nuclear Medicine 1995; ***22****: 315–21*

Uptake measurements

A normal adrenal takes up less than 0.3% of the administered activity of iodocholesterol. There is some dispute as to the normal range for selenocholesterol uptake. The same figure, 0.3%, is probably the upper limit of normal but some workers have reported up to 0.45% in normal subjects. However, it has been suggested that these subjects may have been stressed and were thus not normal. In Cushing's disease uptake is usually substantially higher and there is rarely difficulty in making the diagnosis. Conn's syndrome is more difficult to diagnose. Some of the discrepancies between centres no doubt arise from different methods of estimating absolute uptake. There is no agreement as to which method should be employed. The method of measuring renal uptake (page 91) can also be applied to the adrenals. When using iodocholesterol, uptake should be measured at five or seven days. With selenocholesterol it is rarely useful to make a measurement before 14 days because of high and variable uptake in liver and colon. Selenocholesterol uptake measurements are usually best made between two and three weeks after administration, at which time background is usually acceptably low. Selenocholesterol uptake in the adrenals does not change appreciably for at least three weeks and probably somewhat longer.

Adrenal medulla

This arises along with the sympathetic and parasympathetic nervous systems from the primitive neural crest. It is a component of the amine precursor uptake and decarboxylation (APUD) system. Tissues and tumours arising from this have common properties, containing neuroendocrine secretory granules and mechanisms for uptake, synthesis, storage and reuptake of amine precursors and biogenic amines. Tumours arising from this tissue are the neuroblastomas of childhood and phaeochromocytomas and paraganglionomas of adults. There is some confusion in the use of the two latter terms. Some consider paraganglionoma to be a general term for all tumours arising from the neural crest, whereas phaeochromocytomas are specifically those arising from the adrenal. However others regard secretory tumours as phaeochromocytomas and non-secretory as paraganglionomas. About 80% originate in one or other adrenal and three-quarters of the extra-adrenal primary tumours are found between the diaphragm and aortic bifurcation. Reported series variously put the incidence of malignancy at between 2% and 10% for intra-adrenal phaeo-chromocytomas and between 20% and 40% for extra-adrenal. Multicentric tumours are more characteristic of extra-adrenal than intra-adrenal sites. There are rarely both extra- and intra-adrenal tumours in the same patient. When multiple tumours occur they are usually similar in that all are either catecholamine-secreting or non-secreting. Malignancy is most reliably detected by the biological behaviour rather than histological appearance and, in particular, detection of chromaffin tissue at sites at which it should not normally be found, for example, bone, liver, lung and lymph nodes.

Radiopharmaceutical

The only specific radiopharmaceutical is meta-iodobenzyl-guanidine (MIBG) which may be labelled with either ^{131}I or ^{123}I. This is an analogue of noradrenaline and guanethidine. It is taken up predominantly through the neuronal active uptake-1 system and stored in neuro-secretory granules. At high concentrations there is also passive diffusion with eventual storage in the same granules. This does not play any significant part at concentrations used diagnostically.

Uptake-1 is mediated actively by a transport protein in the cell membrane. It is dependent on temperature and sodium concentration. The affinity for MIBG and the physiological neuro-transmitter is high but it has a low capacity and is easily saturated. Na^+/K^+-ATPase inhibitors such as ouabain and metabolic inhibitors such as deoxyglucose inhibit transport, as do selective competitive inhibitors of the transport protein such as desmethylimipramine or cocaine. Uptake can be inhibited by many drugs including tricyclic anti-depressants, labetolol and reserpine. Drugs including amphetamines which cause depletion of storage granules have been shown to block MIBG uptake in animals but interference with imaging has not been reported in man. Uptake is also affected adversely by calcium channel blocking drugs. Any possibly interfering medication should be withdrawn for at least 48 h or more depending on its biological half-life before administration of MIBG. Thallium uptake may be observed in some neuroendocrine tumours which do not take up MIBG. ^{18}F-FDG (fluorodeoxyglucose, page 267) uptake has also been reported in a small number of patients.

Further reading:

European Journal of Nuclear Medicine 1993; ***20****: 1070–7. 1994;* ***21****: 545–59*

Clinical indications

The first investigation should be measurement of urinary excretion of catecholamine metabolites. In patients who have not had previous surgery MRI is the investigation of choice for localization of the tumour, which is usually large enough to be readily visualized. If artefacts due to metal implants from previous surgery are observed CT is recommended. If these fail to identify a source and when there is biochemical evidence of recurrence following surgery, MIBG scintigraphy is indicated. In principle MIBG scintigraphy should be recommended in any patient with biochemical evidence of a tumour secreting adrenaline or noradrenaline. In practice although most published results suggest that the technique has a sensitivity of 80% or better, many centres find a higher false negative. Although most tumours are visible by 24 h, some may not be visualized until seven days. For this reason ^{131}I-MIBG is preferable to ^{123}I, despite the higher radiation dose. Pentetreotide (page 173) may have a higher sensitivity. In children, MIBG scintigraphy is part of the staging protocol for neuroblastoma. There is no evidence that delaying longer than 24 h is of any advantage and ^{123}I-MIBG is therefore the radiopharmaceutical of choice.

Further reading:

Journal of Nuclear Medicine 1993; ***34****: 180–1*

Interpretation

Uptake is normally seen in the salivary glands, spleen, liver and urinary bladder. Liver uptake is probably non-specific because of the excretion

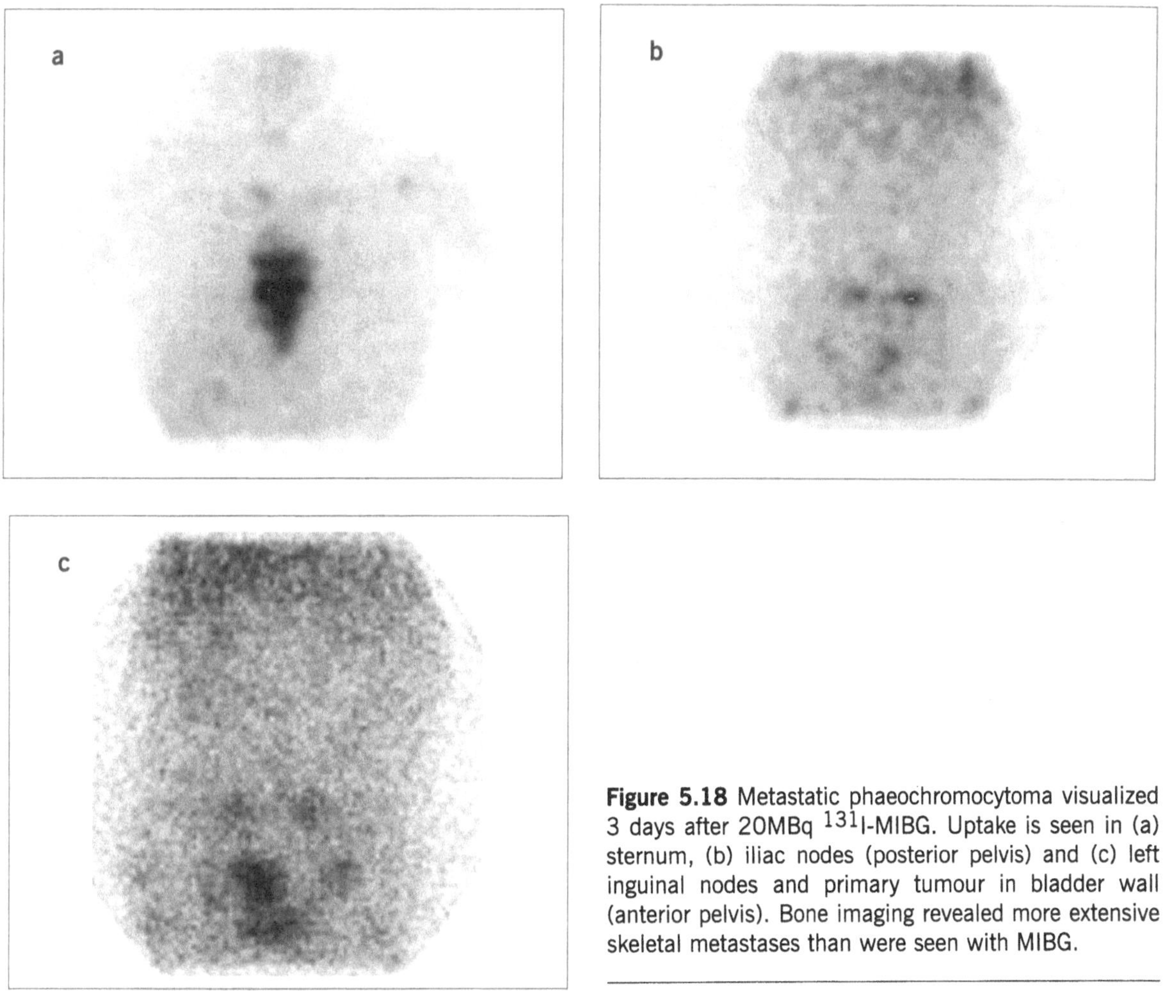

Figure 5.18 Metastatic phaeochromocytoma visualized 3 days after 20MBq ^{131}I-MIBG. Uptake is seen in (a) sternum, (b) iliac nodes (posterior pelvis) and (c) left inguinal nodes and primary tumour in bladder wall (anterior pelvis). Bone imaging revealed more extensive skeletal metastases than were seen with MIBG.

and detoxification which occurs there. Excretion is predominantly unchanged in urine. If the thyroid is not blocked with an adequate dose of Lugol's iodine or sodium iodate, uptake into the thyroid of iodine released during catabolism of MIBG may be high enough to cause thyroid damage. Thyroid blocking (page 151) is therefore essential. The normal adrenals are seen in only approximately 20% of subjects. Uptake is usually visible in the myocardium (page 117). Abnormal accumulations are most commonly seen in the adrenal bed but may be found anywhere along the site of the neural crest and elsewhere if there is metastatic malignant tumour (Fig. 5.18).

Therapeutic uses

In patients with a large but well differentiated tumour burden it is sometimes possible to obtain useful palliation using large (1000–3000 MBq) activities of ^{131}I to achieve a therapeutic effect. Unlike thyroid carcinoma the contrast between tumour and normal tissues is never high enough for cure and the number of patients who can benefit from palliation with this agent is small.

Somatostatin receptor imaging

Somatostatin is a peptide containing 14 amino acid residues. It is present in brain, where it acts as a neurotransmitter, the gastro-intestinal tract and pancreas where it may have an endocrine function. Somatostatin receptors have been identified on many cells of endocrine origin including somatotroph cells of the anterior

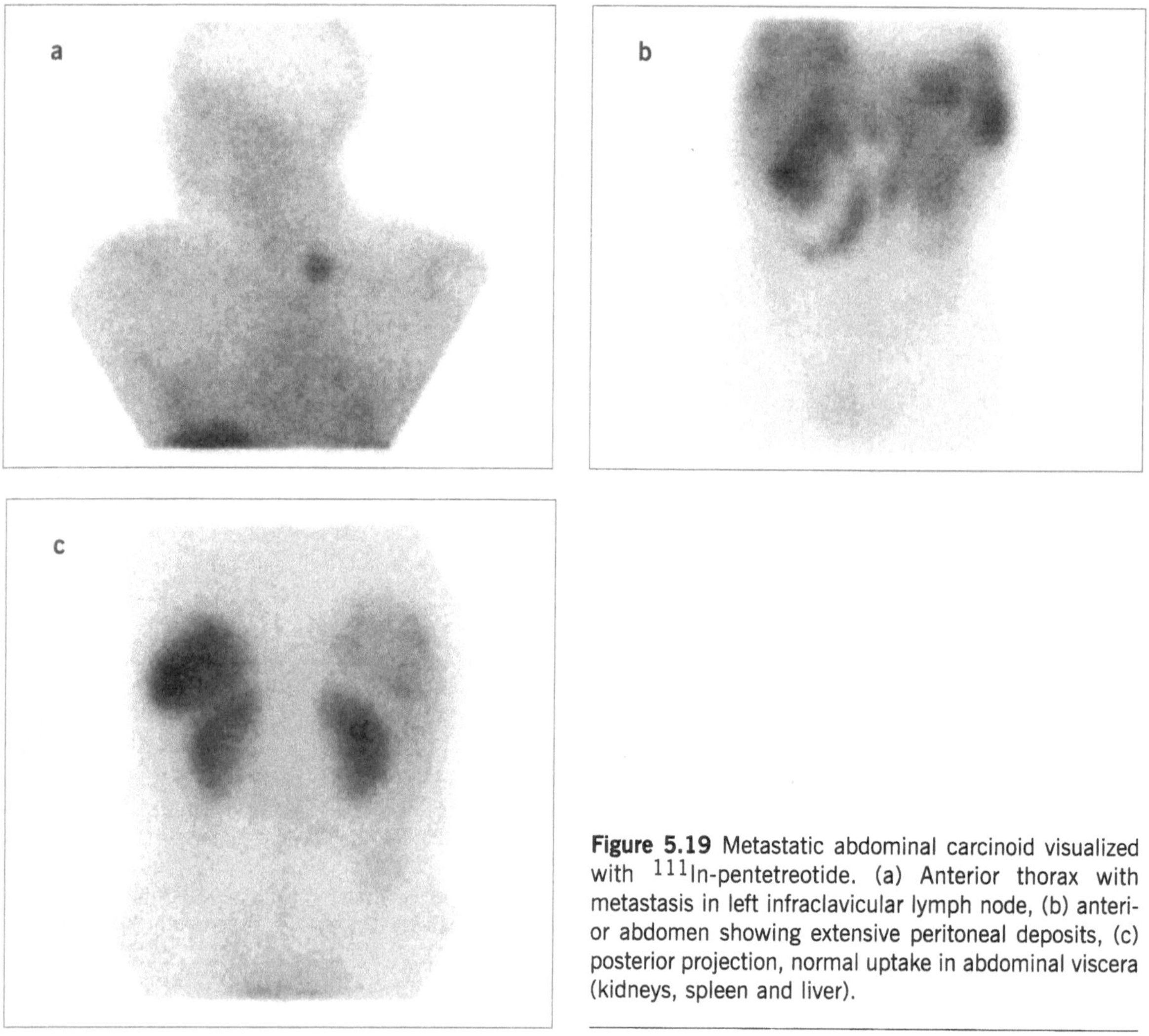

Figure 5.19 Metastatic abdominal carcinoid visualized with ^{111}In-pentetreotide. (a) Anterior thorax with metastasis in left infraclavicular lymph node, (b) anterior abdomen showing extensive peritoneal deposits, (c) posterior projection, normal uptake in abdominal viscera (kidneys, spleen and liver).

pituitary, thyroid C cells and pancreatic islet cells. Somatostatin receptors are also expressed on certain other cells including activated lymphocytes. Four types of receptor have been cloned: type 1 from stomach and jejunum, type 2 from brain, kidney and pancreatic islet cells, type 3 also from pancreatic islet cells and type 4 which has not been cloned in humans although there is circumstantial evidence that it may exist in man. Receptors occur in tumours arising from glial cells in the central nervous system, small cell lung cancer, ovary, cervix, endometrium, breast, kidney, larynx, para-nasal sinus, salivary gland and Merkel cell skin tumours. Octreotide is an eight-residue analogue which acts as an agonist, binding to and blocking type 2 receptors. Pentetreotide is octreotide with ^{111}In attached via DTPA (page 82) to the end-terminal phenylalanine.

Radiopharmaceuticals

Two labelled compounds have been studied, ^{123}I-octreotide and ^{111}In-pentetreotide. The former is relatively unstable *in vivo*, being readily deiodinated. There is also extensive bowel secretion making interpretation in the abdomen difficult. For this reason it has been replaced for diagnostic purposes with indium pentetreotide, excretion of which is principally urinary. Approximately two-thirds of the administered activity can be recovered unchanged from urine within 24 h and over three-quarters by 48 h. Activity in plasma remains unchanged for 24 h but thereafter there is progressive transfer of activity to other plasma components, most probably transferrin which binds strongly to indium.

Interpretation

Pentetreotide is rapidly cleared from blood after intravenous injection. At 24 h, imaging reveals uptake in pituitary, thyroid, sometimes in breast, urinary bladder, liver, spleen, kidneys, occasionally gall-bladder and gastro-intestinal tract. High uptake may be seen in the nasal region and lung in upper respiratory infections and after radiation or chemotherapy. Uptake may also be seen in recent surgical wounds. Imaging at 4 h is sometimes recommended as at this time uptake is not seen in the bowel. However contrast between abnormal accumulations and background is comparatively low. Contrast is higher at 48 h than at 24 h enabling a few additional lesions to be visualized.

Further reading:
European Journal of Nuclear Medicine 1994; ***21:*** *561–81*

Clinical applications

There is sufficient uptake in many pituitary tumours and in most endocrine tumours of the gastro-intestinal tract (with the exception of insulinoma) for visualization. Little more than half of all insulinoma are visualized. Uptake is also seen in most small cell lung cancers, medullary thyroid carcinoma, neuroblastoma, phaeochromocytomas, carcinoid tumours (Fig. 5.19), meningioma, Hodgkin's and non-Hodgkin's lymphoma and many granuloma including sarcoidosis, tuberculosis and Wegener's granulomatosis. Retro-orbital uptake is seen in some patients with Graves' ophthalmopathy. The majority of these tumours can be identified readily by other methods such as CT or MRI. Pentetreotide uptake correlates well with response to therapeutic doses of octreotide. The principle indication may therefore be to identify patients, especially those with secreting gastro-intestinal tumours, who may obtain symptomatic relief from octreotide.

Further reading:
European Journal of Nuclear Medicine 1996; ***23:*** *1448–54*

Sources of error

Tumours are not visualized if there are high endogenous (or exogenous) concentrations of somatostatin, which may block uptake of pentetreotide. Uptake is not seen in tumours which have type 3 rather than type 2 receptors. This latter is probably the most common cause for 'false negative' findings.

6 Gastro-intestinal tract

Salivary glands

These are a group of exocrine glands whose principal function is secretion of saliva into the oral cavity. The parotid and to a lesser extent submandibular glands have the ability to take up iodide and other anions of similar ionic radius (including pertechnetate) in sufficient concentration to be imaged. They also concentrate gallium, almost certainly by a different mechanism. The sublingual and minor salivary glands cannot be visualized in this way, but pertechnetate secreted into the oral cavity adheres to mucosal surfaces too firmly to be eliminated by rinsing the mouth with water. The appearance is easily misinterpreted as visualization of these glands. In contrast to thyroid, but like gastric mucosa, iodide once trapped is not organified in the salivary glands and therefore has no advantage for imaging over pertechnetate. Uptake is blocked by perchlorate and by atropine but salivary glands may still be visualized despite a dose of (stable) iodide sufficient to block pertechnetate uptake by thyroid.

Technique

Perfusion of the salivary glands may be imaged during the first pass of a bolus of any convenient tracer, for example 250–500 MBq sodium pertechnetate. Perfusion is increased non-specifically by any inflammatory process. In order to perform a first-pass acquisition the patient is positioned facing a gamma camera fitted with a general purpose or high sensitivity collimator, with a line from the tip of the nose to the chin parallel to the collimator face and a paired lateral landmark such as the outer canthus symmetrical relative to the collimator face. This places the cantho-meatal line at about 20° from the horizontal if the patient is erect. The thyroid should be included in the field of view as it forms a useful reference organ. Activity is given as a rapid bolus (page 231). Uptake of pertechnetate reaches a plateau 5–10 min after injection.

Imaging to display peak concentration is best started 5 min after injection, as uptake approaches its plateau, to minimize the contribution from activity excreted in saliva. Anterior, lateral and oblique projections (at least 10000 counts in each) should be obtained, obliquity being such as to ensure that the glands are not superimposed. Uptake may be compared with thyroid (provided the patient has a normal thyroid!). Quantitative studies measuring uptake and excretion in consecutive frames of 1 min duration for 15–20 min provide an accurate index of salivary function, although this is rarely clinically helpful. Accurate repositioning is essential for quantitative or semi-quantitative studies. After 10 min the gland should be stimulated to secrete by a sour drink such as lemon juice, which normally discharges most of the parotid activity and rather less of the submandibular uptake within one to 2 min.

Interpretation

Radioactivity is concentrated in parotid and submandibular glands. Sublingual glands are not visualized, although there is usually both adherence of secreted pertechnetate to mucosa and pooling of saliva containing pertechnetate in the floor of the mouth. The concentration in normal thyroid is five to ten times higher than that of salivary glands (Fig. 6.1). Uptake curves obtained from regions of interest including each salivary gland plateau 5 to 10 min after injection and show a rapid fall following acid stimulation (Fig. 6.2). The slope may be compared with a simultaneous uptake curve from thyroid, which does not respond to an acidic stimulus.

Obstructed but still functioning glands have lower uptake than non-obstructed ones and this does not decrease in response to acid stimulation. Stenson's duct may be visualized on the later frames if it is dilated or obstructed but the normal duct is not visualized. Xerostomia is usually associated with generally poor uptake and failure to respond to acid stimulation. In Sjögren's syndrome uptake may be normal or reduced, but response to sialagogues such as lemon juice is usually markedly diminished.

With one important exception, space-occupying lesions are identified as areas of reduced uptake seen most readily in lateral oblique projections. The exception is papillary lymphoid cystadenoma (Warthin's tumour), which has greater uptake than normal salivary tissue but does not discharge this activity in response to oral acid. Most tumours of salivary glands are palpable and are readily identified by ultrasound or other techniques. Scintigraphy thus has little application for tumour localization. Uptake of gallium is increased both in obstructed salivary glands and in uveoparotid fever due to sarcoidosis. This uptake is not discharged by acid stimulation in either normal or abnormal glands. Gallium is therefore of no value in differential diagnosis but uptake is sometimes observed as an incidental finding when the patient being imaged for other reasons.

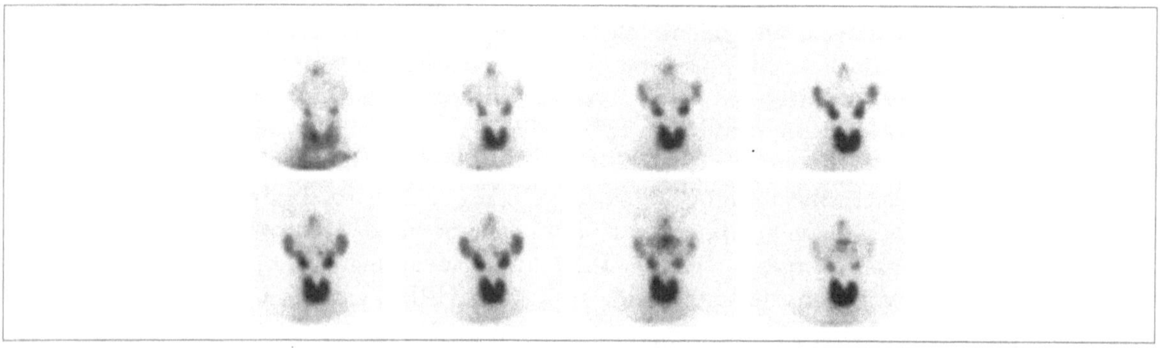

Figure 6.1 Sequence showing uptake of pertechnetate by the salivary glands and its discharge following a drink of lemon juice between frame 6 and frame 7.

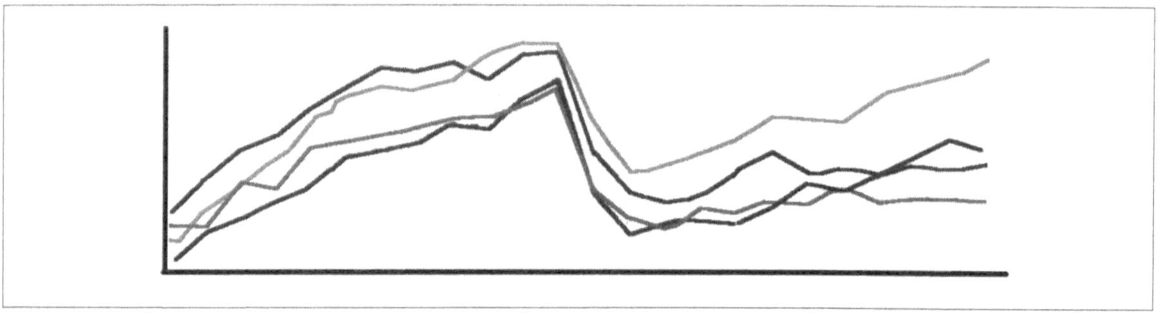

Figure 6.2 Time–activity curves from regions enclosing the salivary glands, showing response to lemon juice. The two upper curves are the left and right submandibular glands and the lower pair the left and right parotid glands.

Clinical applications

The principal roles of salivary imaging are in patients being assessed for asymmetrical salivary function following surgery, radiotherapy, infection or trauma and those in whom adequate clinical examination is impractical, for example in consequence of behavioural disturbance or major deformity of the mouth.

Further reading:
Seminars in Nuclear Medicine 1981; ***11****: 258–65*

Oesophagus

Functional disorders of swallowing fall into three major categories, oro-pharyngeal dysphagia, oesophageal dysphagia and non-cardiac chest pain. Established nuclear medicine methods of assessing oesophageal transit and clearance of a liquid, semi-solid or solid bolus offer a simple and reproducible way of investigating the second and third, namely oesophageal dysphagia and functional disorders of the oesophagus. Their clinical value is not as widely recognized as the techniques deserve, even though there are good correlations with manometry and pH monitoring. Scintigraphic techniques offer an informative and useful assessment of oesophageal motor function in response to a bolus of food or liquid and permit measurement of rate of emptying. An oesophageal cause should be sought in patients with substernal chest pain or a sensation of 'pressure' if a cardiac aetiology has been excluded, especially if pain radiates to the back or left arm. Gastro-oesophageal reflux can also be demonstrated.

Technique

Any gamma camera with a field of view greater than 35 cm diameter fitted with a general purpose or high sensitivity collimator is suitable for studying the oesophagus. The examination may be performed with the patient either supine or erect. Gravity plays no part in the transit of a bolus along the normal oesophagus, but the rate of transit in the hypokinetic or dilated oesophagus may be slower when the subject is supine. It is easier to swallow when upright and physiologically it is probably more realistic to perform the examination erect, but many centres find the examination easier to reproduce when the patient is supine. It is thus desirable for there to be a consistent policy within any department.

The patient should fast for 3–6 h before the test, to allow clearance of any residue. The radioactivity should be in a non-absorbable form, but in practice any soluble or stable solution of a technetium-labelled tracer may be used. It is in general better not to combine oesophageal transit with a gastric emptying study, as the initial requirements of the two are different. Activity may be given in plain water or flavouring may be added. Although some papers recommend a viscous fluid, there is no evidence that this confers any benefit. The volume should be small enough to be swallowed comfortably as a single bolus (5–10 ml); an activity of 15–20 MBq per swallow is adequate. At least three episodes of deglutition should be recorded as there is commonly appreciable inter-event variability. Six per study is preferable. A number of protocols have been used, a typical example acquiring data at 2 frames per second for 60 s and subsequently at 4 frames per minute for up to 15 min. The patient should be rehearsed to swallow on command and not again until 'dry' swallows are requested at pre-determined times, commonly 40 s and 90 s. In normal subjects the oesophagus clears entirely within 60 s. but in patients with achalasia or other motility disorders a prolonged examination is necessary. It is helpful to mark the cricoid cartilage and xiphisternum. Time–activity curves are drawn for the upper, middle and lower thirds of the oesophagus (Fig. 6.3). In addition time-condensed functional images facilitate interpretation and should be calculated in all cases.

Time-condensed images

A relatively long, thin region of interest is drawn enclosing the entire oesophagus. With most computer systems there is no requirement for this to be a rectangle but, as there is no radioactivity outside the oesophagus (unless there is aspiration) the shape of the region is not critical provided that it encompasses the whole of the oesophagus but does not include the fundus of

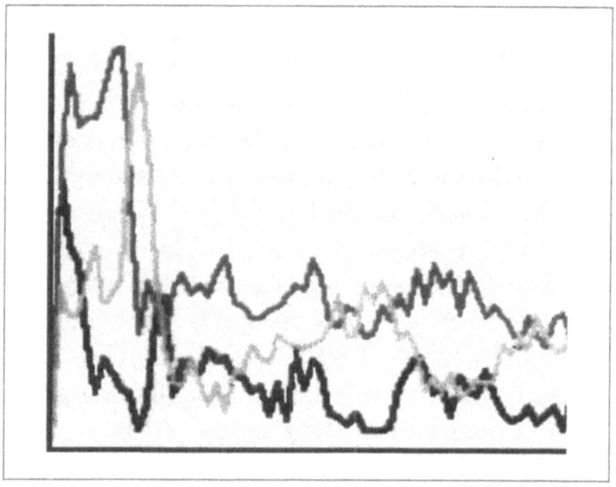

Figure 6.3 Time–activity curves showing normal pattern of oesophageal clearance of a liquid from study in Fig. 6.4. Black curve, upper third; dark grey, mid-third; light grey, lower third.

the stomach. Each horizontal line of pixels within the region of interest is summed separately in every frame, to produce a single vertical line one pixel wide indicating the total number of counts at each level of the oesophagus in that frame. Lines from consecutive frames are then placed adjacent to each other, to produce a two-dimensional image with time from origin on the horizontal axis and distance from cricoid cartilage to xiphisternum or cardia on the vertical axis. The number of counts in each pixel is shown as a colour or shade of grey, as in other isotope images. This is therefore a three-dimensional representation, the x and y dimensions being time and distance, respectively, while the third, amount of radioactivity, is depicted by colour or shade of grey. In the normal subject, activity passes down very rapidly; this is shown as a narrow line sloping steeply downwards from left to right near the left hand edge of the composite (parametric) image – a 'negative' slope (Fig. 6.4). Subsequently residual activity in the stomach should be outside the region of interest and hence not displayed. Reflux is identified as a positive (upward) sloping line and clearance of retained activity as a negative sloping line. Retained activity appears as a horizontal line. The magnitude and regularity of functional oesophageal peristaltic waves can thus be readily appreciated. These correlate with manometric waves greater than 30 mmHg. Smaller amplitude pressure waves may not propel the contents of the oesophagus.

Interpretation

In the normal subject the three regional curves each shows a transient peak, wider in the distal segment than in the more proximal because the bolus spreads as it passes along the oesophagus and there is delay at the cardia (Fig. 6.3). The mean residence time in each segment can be calculated, as can the time for the leading edge or

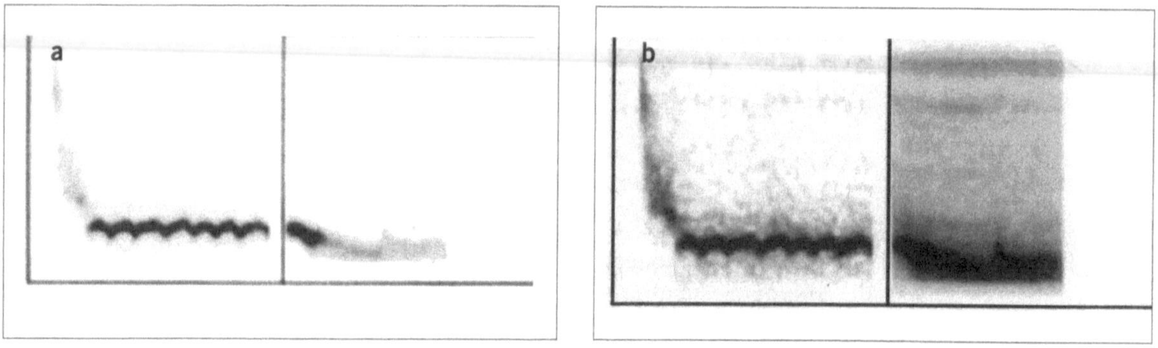

Figure 6.4 Time-condensed image showing normal pattern of a swallowed bolus passing down the oesophagus. (a) Linear grey scale, (b) logarithmic scale which demonstrates low count-rate regions more clearly. The left-hand pair of panels show the initial 30 s, the right-hand pair the subsequent 15 min. The faint horizontal lines near the top of (b) are due to a small amount of tracer retained on the tongue and pyriform fossa. The small peak about 7 min after swallowing indicates clinically insignificant transient gastro-oesophageal reflux.

the centroid of the bolus to pass between any two points. The mean time from arrival of activity in the proximal segment until count rate in the distal falls to 10% of its peak value is 5–15 s. In the absence of peristalsis, for example in achalasia of the cardia or scleroderma, activity in the distal and often in the middle segment rises and then remains constant for an extended period. Diffuse oesophageal spasm shows a characteristic pattern of fluctuating activity as uncoordinated movement causes radioactivity to pass back and forth from one segment to another (Fig. 6.5). In these cases, as count rate falls in one segment there is a corresponding rise in another. All of these changes may be seen more clearly on time-condensed images (Fig. 6.6). Viscous rather than clear fluids or a semi-solid bolus have also been recommended. Insufficient data is available to assess the relative merits of these variations in technique.

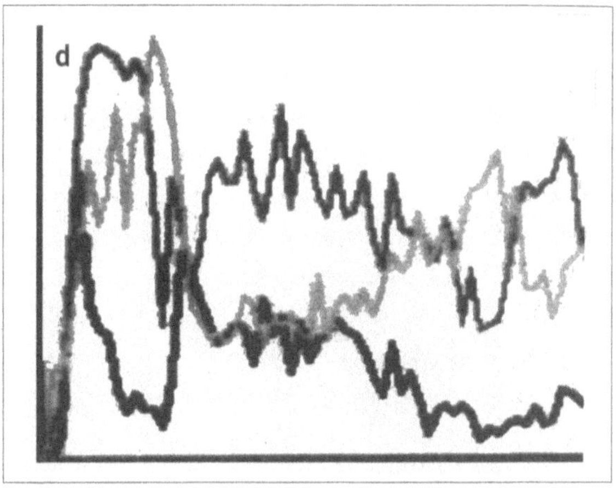

Figure 6.5 Ineffectual to-and-fro movement of a swallowed bolus.

Further reading:
European Journal of Nuclear Medicine 1992; ***19****: 815–23*
Journal of Nuclear Medicine 1996; ***37****: 1799–805*

Clinical applications

Combining oesophageal scintigraphy with manometry allows propulsive activity to be correlated with pressure waves. This is a simple means of determining severity of oesophageal involvement in Parkinson's disease and may have a role in management of motor neurone disease (amyotrophic lateral sclerosis). Different but characteristic abnormal patterns are described in gastro-oesophageal reflux, diffuse oesophageal spasm and nutcracker oesophagus, although these are usually diagnosed by other techniques. Scintigraphic abnormality may precede other evidence of involvement in systemic sclerosis and diabetic neuropathy.

Aspiration can be detected if regions of interest are drawn over the lungs, but curves should be interpreted in conjunction with images as changes in the pattern of scattered radiation can under some circumstances simulate aspiration.

Further reading:
Journal of Nuclear Medicine 1994; ***35****: 1013–6*

Gastro-oesophageal reflux

This can be detected following ingestion of a non-absorbable tracer such as 20–100 MBq of a

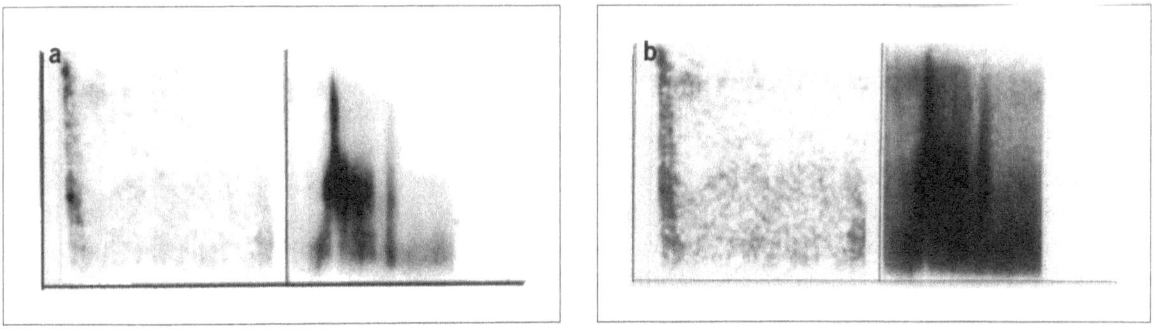

Figure 6.6 Time-condensed image showing ineffectual clearance associated with diffuse spasm.

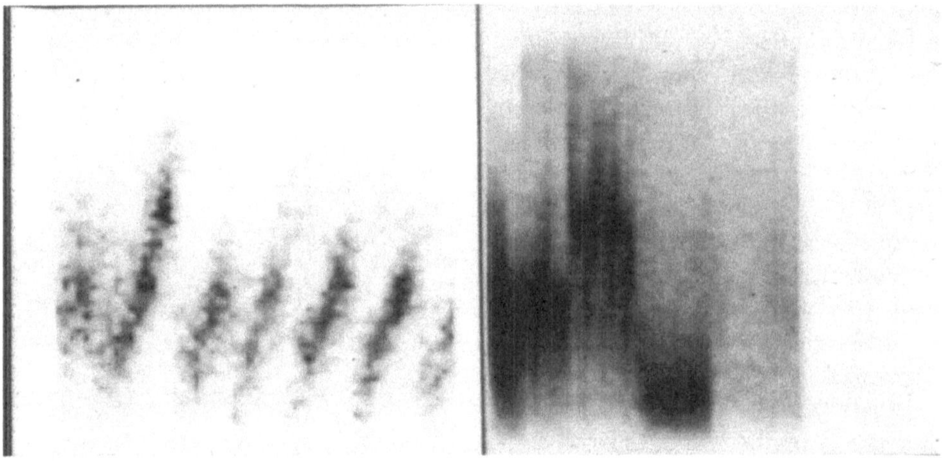

Figure 6.7 Gastro-oesophageal reflux. Left-hand panel, 0–30s; right-hand, 0–15 min.

technetium-labelled colloid. Any of the usual manoeuvres may be employed to precipitate reflux (increasing abdominal pressure, for example with an abdominal binder, inverted posture etc.). The reflux examination may be combined with measurement of oesophageal clearance. Gastro-oesophageal reflux is identified in time-condensed images as positive (upward from left to right) sloping spindles starting at the level of the cardia (Fig. 6.7).

Advantages of nuclear medicine over radiological techniques are:

- a quantitative estimate of severity of reflux
- duration of examination may be extended without increasing radiation dose
- the pattern of abnormal motility can be recorded in an operator-independent manner for evaluation and comparison with follow-up studies, by making use of time-condensed images.

The methodology is particularly suitable for evaluation of functional disorders such as diffuse spasm, 'nutcracker' oesophagus and in follow-up studies. A quantitative measure, for example fraction of gastric contents refluxing, may be calculated but this assumes complete mixing in the stomach before reflux occurs, which may not always be the case. Pulmonary aspiration is identified either by visual inspection of frames or (preferably) by obtaining time–activity curves from regions over the lungs.

Further reading:

*European Journal of Nuclear Medicine 1994; **21**: 1234–42*

Stomach

Pertechnetate is taken up from the circulation, concentrated and secreted into the gastric lumen by gastric mucosa. This is one of the principal pathways of excretion of pertechnetate. Concentration of pertechnetate in gastric mucosa is only slightly reduced in patients with achlorhydria. Uptake is prevented or discharged by perchlorate and other agents which block trapping of iodide. Both rate of uptake and rate of secretion are increased by pentagastrin, whereas H-2 blockers such as cimetidine increase uptake but not secretion, thereby increasing local concentration and contrast between gastric mucosa and background. Pertechnetate behaves like other anions of similar ionic radius, more nearly resembling iodide than chloride but not exactly like either. It cannot be used as an analogue of chloride to assess gastric acid secretion. Uptake probably reflects both mucosal blood flow and the functional state of mucous cells. It is not a marker of acid production, although under many circumstances pertechnetate uptake parallels acid secretion as both are dependent on blood flow. Gastric mucosa extending above the diaphragm, for example in a hiatus hernia or into oesophagus (as in Barrett's oesophagus), can be identified only if the level of the diaphragm is defined. Ectopic gastric mucosa in other sites, for example Meckel's diverticulum, can be detected reliably (page 186).

Gastritis

Infection with *Helicobacter pylori* is a major cause of chronic gastritis, an important factor in development of gastric and duodenal ulceration and implicated in gastric carcinogenesis. Unlike mammalian tissues, which lack any urea-splitting enzyme, this bacterium metabolizes urea, with release of carbon dioxide. Measurement of breath concentration of ^{14}C-carbon dioxide following oral administration of ^{14}C-urea is used as a diagnostic test for infection with this organism.

Technique

Although more elaborate techniques involving multiple samples over several hours have been described, the concentration in a single breath sample 10 min after 175–200 kBq ^{14}C-urea dissolved in water, given after an overnight fast and expressed as the fraction of the administered dose normalized to body weight, is adequate clinically. In the normal subject the ^{14}C activity recovered in this breath sample should not be greater than 0.3% of the administered activity (page 193). Retaining test solution in the mouth is an important cause of false positives. The patient should swallow the tracer dose quickly, followed by plain water.

Further reading:
Journal of Nuclear Medicine 1991; ***32****: 1192–8*

Gastric emptying

The most widely used non-invasive and quantitative methods of measuring the rate at which the stomach empties employ radioactive tracers. Solid and liquid components of a meal leave the stomach at different rates and are influenced to different extents by factors which affect emptying. Solids are initially stored in the fundus and transferred into the antral region for trituration before being expelled by antral contractions, each of which propels an approximately equal volume into the duodenum. In contrast, a constant fraction of the residual volume of liquid is lost per unit time. The time–activity curve of a solid meal is thus initially linear, although there is a delay before excretion starts, whereas the time–activity curve for a liquid marker is exponential. Hence solid and liquid markers are not interchangeable.

Any solid marker must remain reliably on the solid phase, at least while in the stomach and there should be no absorption and re-secretion subsequently. Adding a liquid tracer to a solid meal does not produce a solid phase marker. Before any variation in technique can be accepted it must be shown that the tracer remains firmly adherent to a solid which is, or consistently behaves like, a normal constituent of food. Many preparations which have been described do not satisfy these criteria. Much of the literature is invalid for want of proper controls.

A liquid phase marker must have a high solubility in aqueous solutions, remain stable at low pH, be resistant to digestion, non-absorbable and not exchangeable with or absorbable on any solids present. Chelates of DTPA (page 82) with either ^{111}In, ^{113m}In or ^{99m}Tc are suitable. ^{113m}In is no longer available while ^{111}In is relatively expensive and is associated with a higher absorbed radiation dose than ^{113m}In. Few collimators are designed for use at the energy of either indium isotope. Thus, unless there is an overwhelming reason for performing simultaneous solid and liquid phase studies, it is preferable to perform the solid stage first with one technetium label and if necessary the liquid phase study on another day using Tc-DTPA.

The most reliable solid phase marker is made by addition of technetium-labelled macroaggregated human serum albumin (MAA) (as used for lung perfusion scintigraphy, page 57) to raw egg and cooking the resultant mixture into scrambled eggs or an omelette. This remains insoluble in the stomach for long enough (up to

3 h) for gastric emptying rate to be measured. Other suggestions, such as absorbing pertechnetate onto blotting paper or onto breakfast cereal, are unsatisfactory as they do not reliably label the solid phase, nor has the former been shown to be treated like any normal food constituent. To date no preparation suitable for vegans has been validated, although in principle it may be possible to label tofu.

Technique

The patient attends as early as practicable in the morning after an overnight fast. Diabetic patients should bring their regular medication and administer it before the test meal, at their usual interval and dose level. When considering suitable techniques it is necessary to take account of three potential sources of error.

Scatter and septal penetration are a significant problem only with a high energy gamma ray such as that of ^{113m}In. This is balanced to some extent because attenuation correction becomes less critical as gamma ray energy increases. As attenuation is one of the major sources of error when measuring gastric emptying there is some advantage to be obtained from higher energy isotopes, especially in obese subjects. Scatter occurs with all radionuclides and results in counts which originated from the stomach being ascribed to an extra-gastric origin. As there is no background correction to be made, the best solution is to draw the region of interest as loosely as possible, including any faint scatter surrounding the stomach, although in practice this is limited in many subjects by projection of the duodeno-jejunal flexure over fundus of the stomach. Estimates of gastric volume based on tracing outlines are inaccurate and tend to underestimate volume by 15–25%.

The stomach does not lie parallel to the anterior abdominal wall; its shape is variable and complex. When supine much activity drains into the fundus, which may lie significantly dorsal to the body of the stomach. Transfer of activity from fundus to antrum can, in the worst case, double observed count rate in the anterior projection because there is less soft tissue between radioactivity and detector to attenuate gamma rays, despite no activity having entered the stomach. Alternatively it is possible for emptying to be masked by transfer of residue from fundus to antrum. This error is minimized if the geometrical mean of count-rate in anterior and posterior projections is calculated, as this is almost independent of distance of the source from skin surface in any one projection. This is a more accurate correction than attempting to measure the distance of radioactivity dispersed in stomach contents from skin. When making these pairs of measurements only the camera head, and not the patient, should be moved if imaging is performed with the patient supine. It is of course correct to rotate an erect patient about the vertical axis to obtain anterior and posterior projections. The patient must be allowed to sit up or walk about between observations for normal gastric emptying to proceed. All of the meal may pool in the fundus for an extended time if the patient remains supine throughout.

An alternative technique is to acquire all the images in the left anterior oblique position as this minimizes differences in depth from the surface of the various parts of the stomach. The required obliquity is determined by selecting the projection which gives the greatest transverse dimension of the stomach. Once selected, accurate repositioning is crucial. This technique is adequate in many subjects but may not be accurate in large subjects or those with a transverse stomach. There may be difficulty in projecting the antrum clear of the duodenal bulb. As in the anterior projection, the duodeno-jejunal flexure may be projected over the greater curve.

Particularly in the erect subject it is sometimes impossible to project the stomach clear of the rest of the gastro-intestinal tract, especially duodeno-jejunal flexure. When the stomach is horizontal and transverse it is sometimes difficult to separate activity in antrum from that in duodenal bulb. A more common problem is projection of activity in proximal jejunum over stomach. These errors cannot always be excluded. They can be recognized and discounted when interpreting results only by careful examination of the images prior to selecting regions of interest.

Interpretation

The rate of gastric emptying varies with

population studied, composition of the meal and its calorie content. Liquid meals empty more rapidly than solid but there is no simple or constant relationship between rate of emptying of the liquid phase and that of the solid phase (Fig. 6.8). Solid phase emptying is more often informative than that of the liquid phase. Indeed if solid phase emptying is normal there is little point in performing a liquid phase study unless the patient has symptoms of dumping. When only one marker is available solid phase emptying should be measured first. If this is severely delayed a second measurement, of the rate of liquid phase emptying, is useful. A normal liquid phase with a delayed solid phase indicates impaired antral function, whereas delay of both phases is more suggestive of mechanical obstruction. The rate of emptying is accelerated in patients with dumping, usually of both phases but solid phase emptying is sometimes within normal limits.

The range of normal values for the population under study must be established for a test meal of constant composition and acceptable to local taste. Fat, carbohydrate, protein content and volume of the test meal all affect emptying rate. Small meals which do not distend the stomach give smaller attenuation errors than large meals. High calorie, high fat and hyperosmolar meals tend to empty more slowly. Hyperosmolar liquids may exhibit an emptying pattern resembling that of solids. The rate of gastric emptying is affected by many extraneous factors, for example it is considerably slowed by smoking or if the patient is nauseated. When establishing a local protocol all of these must be defined or controlled. A one-egg omelette or scrambled egg to which labelled MAA has been added before the egg is thoroughly whisked and cooked is a widely accepted and validated formulation.

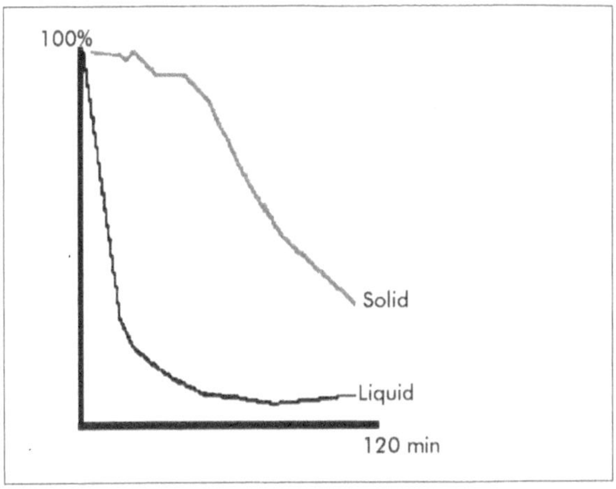

Figure 6.8 Normal liquid and solid gastric emptying curves. There is a lag of up to 25 min before emptying of solids commences. Half of the test meal should leave in the subsequent 50–80 min. There is usually no lag before the start of liquid emptying. Half the activity should leave the stomach between 15–50 min.

Clinical applications

With the reduction in frequency of surgery for peptic ulceration, the principal applications of this technique are no longer assessment of the consequences of gastric surgery but diabetes (Fig. 6.9). Diabetic gastroparesis, one of the manifestations of diabetic autonomic neuropathy, may be symptomatic or may thwart diabetic control. There are less common indications including measuring effects of drugs on gastric motility and in patients with unexplained symptoms possibly related to abnormal gastric motility, but where no abnormality has been found on radiological contrast studies (Fig. 6.10).

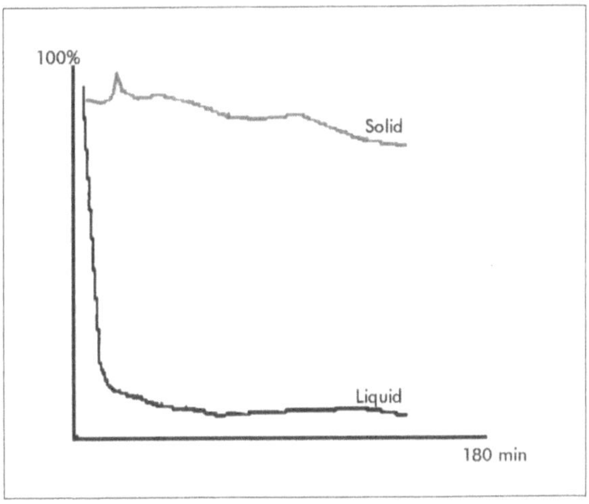

Figure 6.9 Typical pattern of diabetic gastroparesis. There is accelerated emptying of the liquid phase (half-emptying time 5 min, normal 15–50 min) but delayed solid emptying (half-emptying time 675 min, normal 50–80 min).

Further reading:
European Journal of Nuclear Medicine 1994; ***21****: 1263–8*
Journal of Nuclear Medicine 1994; ***35****: 1023–7*

Duodeno-gastric reflux

This can be detected and its severity assessed following intravenous administration of any of the hepato-biliary agents, by looking for activity which has refluxed into the stomach (Fig. 9.9). In the normal subject only duodenum and small intestine are visualized. There should not be detectable reflux of duodenal contents into stomach. When this does occur it is often associated with delayed gastric emptying. Bile reflux into oesophagus is commonly symptomatic. The patient should be imaged continuously for 1 h after intravenous administration of the hepato-biliary agent. If no reflux is observed and the patient is fasting, further views should be obtained after a test meal, which may unmask latent reflux. Interpretation is complicated by activity in the third and fourth parts of the duodenum which commonly overlie stomach. This technique is also of value to demonstrate patency of and direction of flow in an afferent loop following Bilroth 1 or Polya partial gastrectomy and of biliary-enteric and gastro-enteric bypass operations.

Further reading:
Journal of Nuclear Medicine 1991; ***32****: 436–40*

Gastro-intestinal bleeding

Meckel's diverticulum

Meckel's diverticulum is a residue of the embryonic omphalomesenteric duct, present in 1.5–3% of the general population. It is said that 25–40% are symptomatic at some time, most commonly in childhood, but the true prevalence of occult diverticula remains uncertain. Some are lined by acid-secreting gastric mucosa instead of normal small intestinal mucosa, the prevalence being variously reported as 20–60%. A Meckel's diverticulum is usually located in the distal 100 cm of ileum, usually within 30 cm of the ileocaecal valve. Secreted acid ulcerates adjacent small intestinal mucosa, to cause bleeding and more rarely ulceration or perforation. Barium follow-through has a low sensitivity of detection and cannot differentiate gastric mucosa from small intestinal mucosa in the diverticulum. Meckel's diverticula are commonly asymptomatic but may present with overt melaena, as anaemia due to occult blood loss or as an acute abdomen as the consequence of Meckel's diverticulitis. The former occurs only in those Meckel's which contain gastric mucosa and is the commoner presentation in childhood. Meckel's diverticulitis is the more frequent presentation in adults, but either may occur at any age. Scintigraphy is the most sensitive and specific test available.

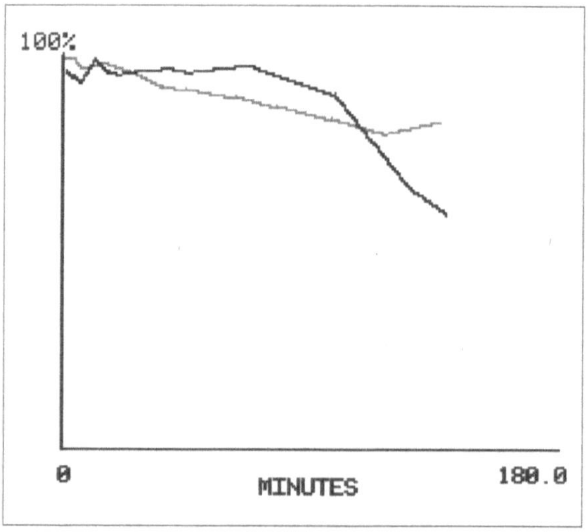

Figure 6.10 A 37-year-old male with recurrent vomiting. Initial barium studies considered normal. There is substantial delay of both solid and liquid emptying. Repeat endoscopy confirmed pyloric canal ulcer.

Radiopharmaceuticals

Sodium pertechnetate is used. Sodium ^{123}I-iodide, although suitable in principle, has no advantages and is more expensive. The usual adult activity of ^{99m}Tc- pertechnetate is 150–200 MBq.

Principles

The mechanism of uptake of pertechnetate by ectopic gastric mucosa is thought to be identical

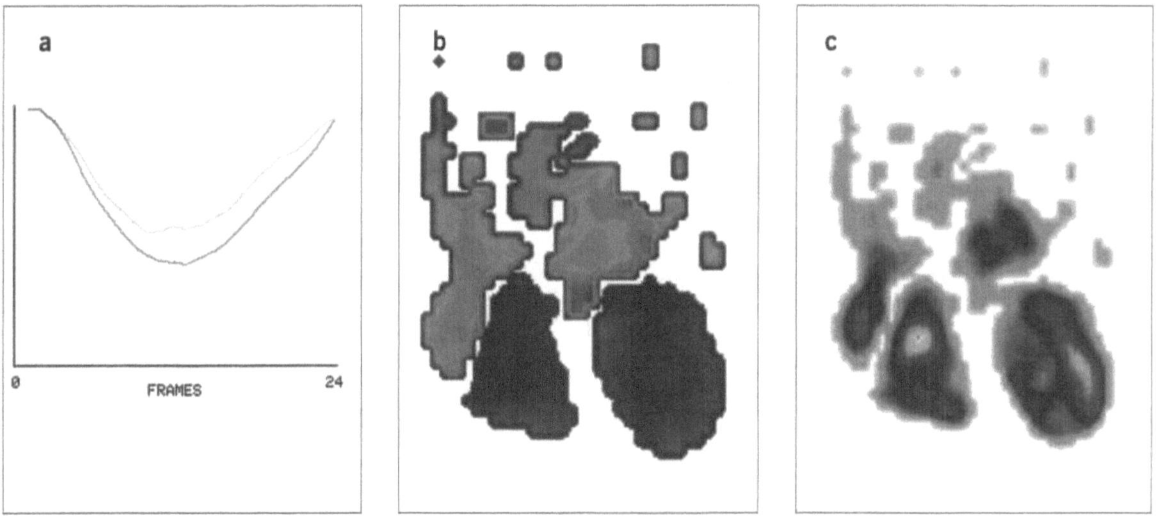

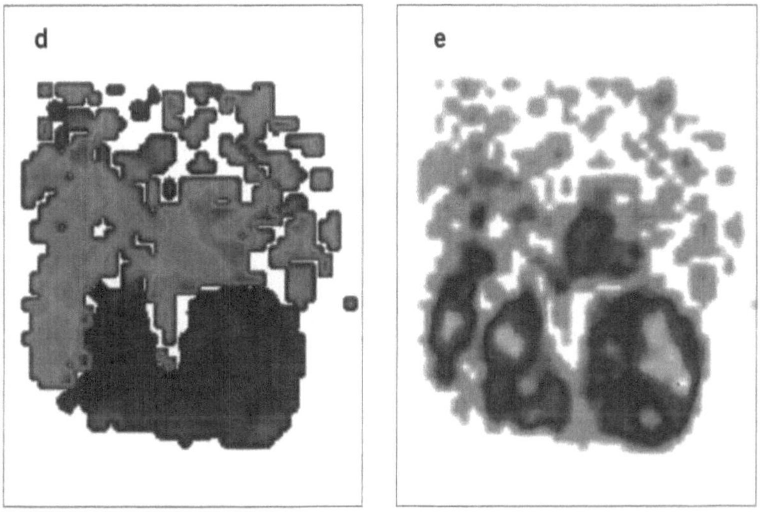

Figure 4.21 Despite ejection fraction of 62% (a, darker curve) and (b) a normal phase image there is (c) a large area of antero-septal hypokinesia. On stress the ejection fraction falls to 48% (a, lighter curve), (d) slight dyskinesia becomes apparent, the left ventricle dilates, and (e) the hypokinetic region becomes larger and more severe.

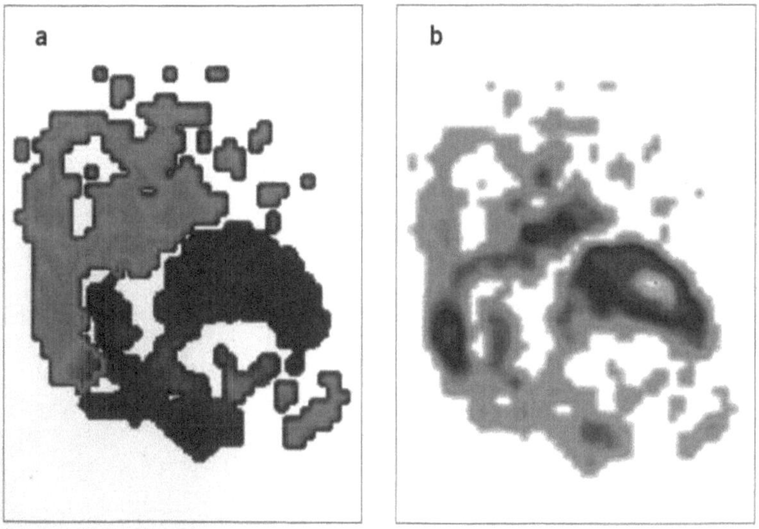

Figure 4.22 Anterior largely akinetic region following myocardial infarction. (a) Phase, (b) amplitude.

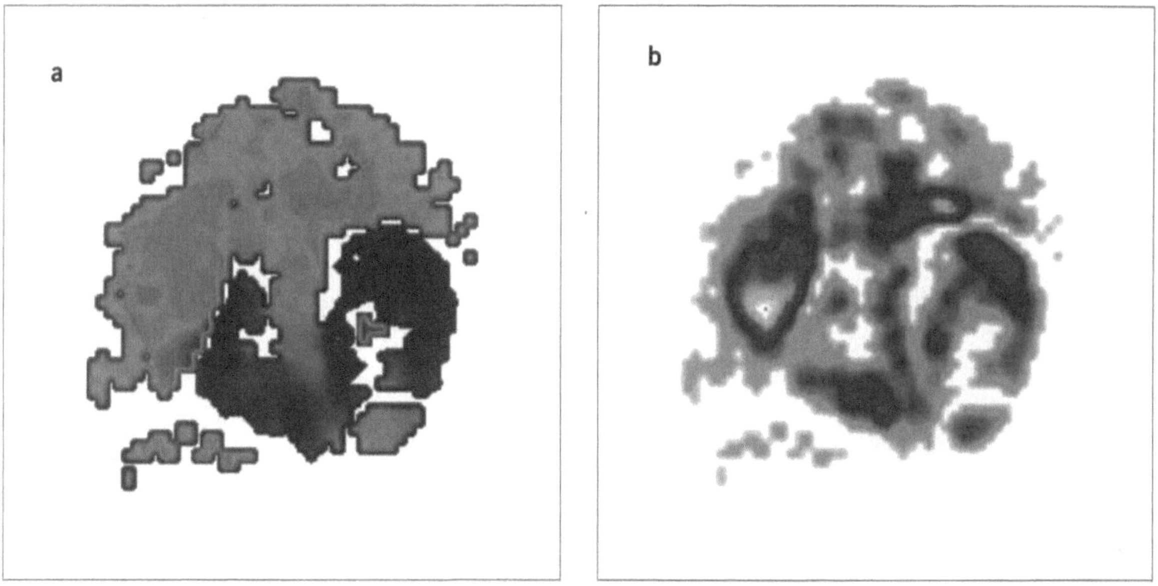

Figure 4.23 Left ventricular aneurysm. (a) Phase (b) amplitude. There is a large anterior region of akinesia. A small region at the apex has similar amplitude to the free border but phase is similar to the atria, i.e. there is paradoxical movement. There is an akinetic region between the normally and paradoxically moving regions.

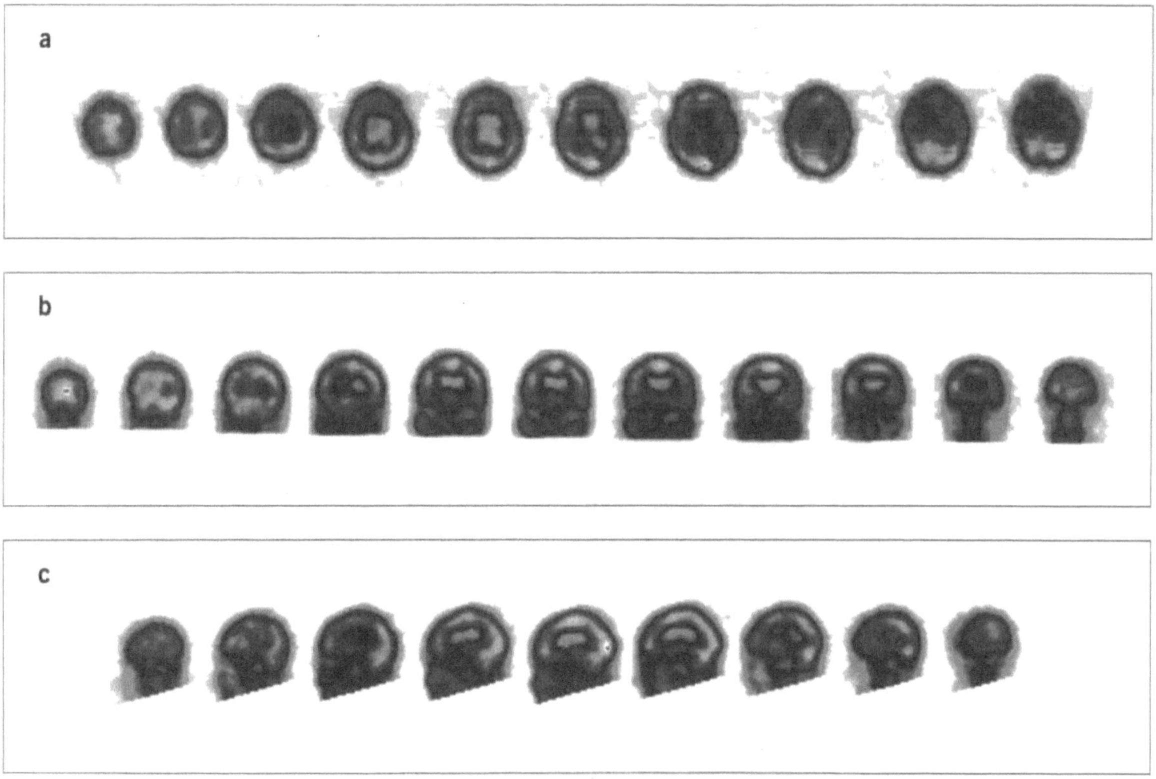

Figure 8.1 Normal exametazime SPECT. (a) Transverse, (b) coronal, (c) sagittal sections.

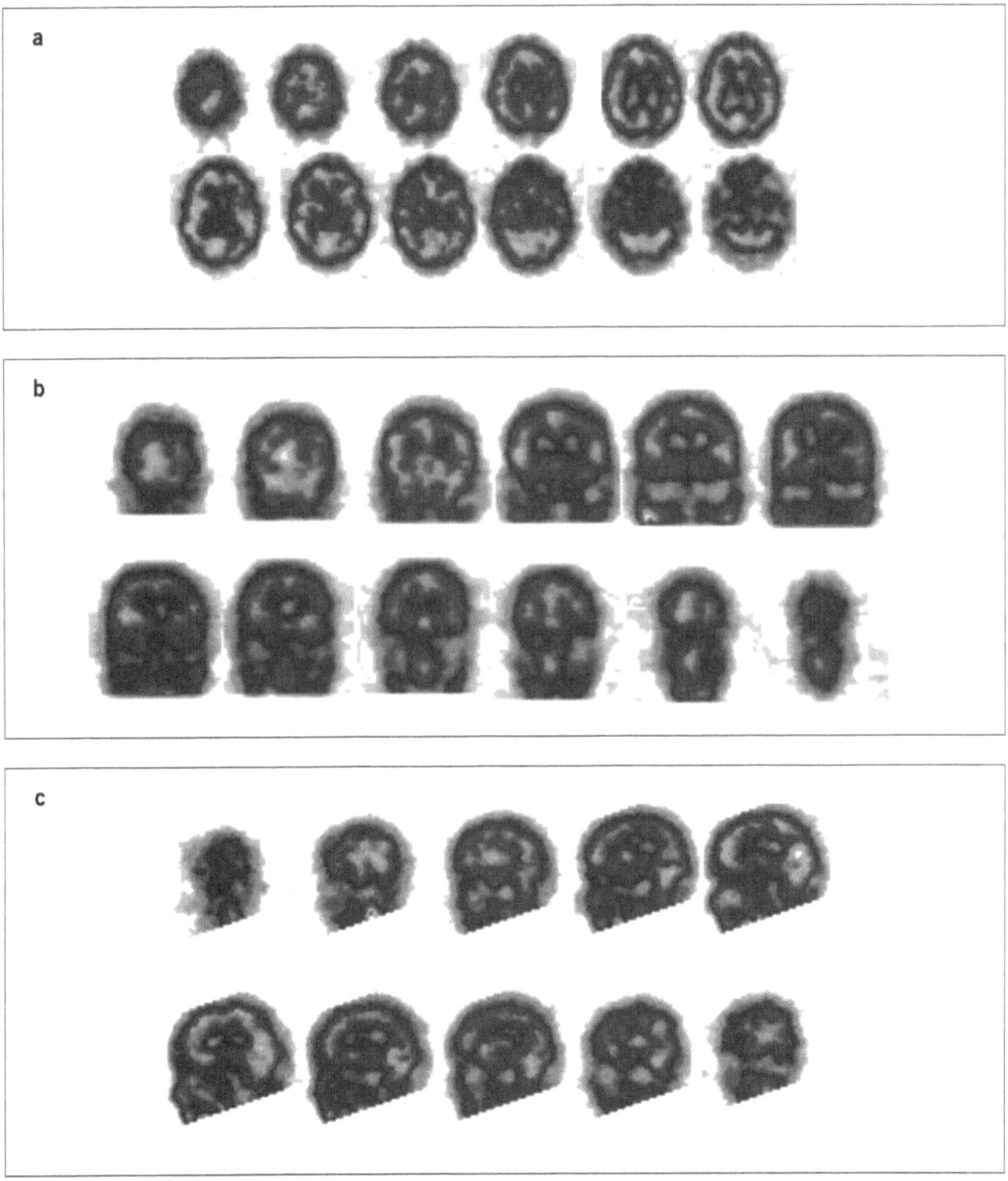

Figure 8.12 SPECT in multi-infarct dementia. (a) Transverse, (b) coronal, (c) sagittal. There are multiple asymmetrical defects bilaterally. Compare with Fig. 8.1.

to that responsible for its uptake in normal gastric mucosa (see above). Pentagastrin accelerates rate of uptake but also increases rate of excretion into duodenum and hence into small intestine. This may conceal a Meckel's diverticulum. Glucagon reduces rate of uptake of pertechnetate by gastric mucosa but accumulation continues nevertheless for at least an hour after injection. Glucagon also reduces gastric motility and therefore minimizes wash-out of secreted pertechnetate from the lumen of the stomach into small intestine. It thus decreases the rate of dispersion of activity from the site of secretion in the Meckel's diverticulum. The highest contrast is obtained when both glucagon and pentagastrin are administered simultaneously, glucagon paralysing the gut and thus allowing secreted radioactivity to pool at the site of secretion. This combination has been used clinically and reduces the incidence of false negative findings compared to patients who have received pertechnetate alone, but has been superseded by H-2 blocking agents, eliminating injections of peptide hormones which are more costly and to which there is a (small) risk of an adverse reaction. H-2 blockers such as cimetidine inhibit intraluminal release of pertechnetate and thus both increase local concentration and minimize small bowel activity. They therefore increase contrast between areas of ectopic gastric mucosa and background, maximizing the probability that such areas can be visualized.

Technique

On the day before the examination an adult patient should take 200 mg cimetidine as early as practical in the morning, before the midday meal and in the evening (600 mg in total). Adults should fast from 10.00 pm on the evening prior to the examination. In the morning, a single dose of 400 mg cimetidine is taken 2 h before the examination with a little water. A dose of 150–200 MBq ^{99m}Tc-pertechnetate is given intravenously and anterior projections of the abdomen, including both stomach and bladder in the field of view, are taken at 5 min intervals for 15 min acquiring at least 500000 counts in each. No useful information is obtained from later films. Meckel's diverticula are most commonly found in the right iliac fossa, but may occur anywhere in the abdomen. If a suspect area is visualized, a lateral – usually a right lateral projection – should be obtained unless the suspect area is to the left of the mid-line, to determine whether the focus is situated anteriorly or posteriorly.

Interpretation

Normal structures visualized are stomach, bladder, kidneys and part or all of the ureters. The uterus may be visualized in adult females. A Meckel's diverticulum, if present should become progressively more clearly identified above background, at the same rate as the stomach itself is visualized (Fig. 6.11). It should be visible in all images but, as gut is commonly freely mobile, it may move between views. Occasionally there is so much movement that it may be concealed in an individual projection by movement blur. Gut movement can be reduced but not eliminated by lying the patient prone during acquisition. If the patient moves, the view is non-diagnostic and must be repeated.

Sources of error

The most common cause of a false negative examination is secreted activity entering small intestine overlying and concealing the Meckel's diverticulum. If activity is visualized in small intestine the examination is non-diagnostic and should be repeated on another day. Other causes of false negatives include:

- patient movement
- a full bladder which may overlie and conceal a Meckel's diverticulum
- too little gastric mucosa to be visualized
- destruction of gastric mucosa by ulceration
- prior administration of drugs which block uptake, such as perchlorate or atropine, which reduces mucosal perfusion.

If the bladder becomes prominent the patient should be instructed to void between films. If movement is apparent it is necessary to repeat the examination. Destruction of ectopic gastric mucosa by extension of the erosion caused by the acid, to form an ulcer, is a rare cause of false negatives. False positives are most commonly due to activity in ureter being misdiagnosed.

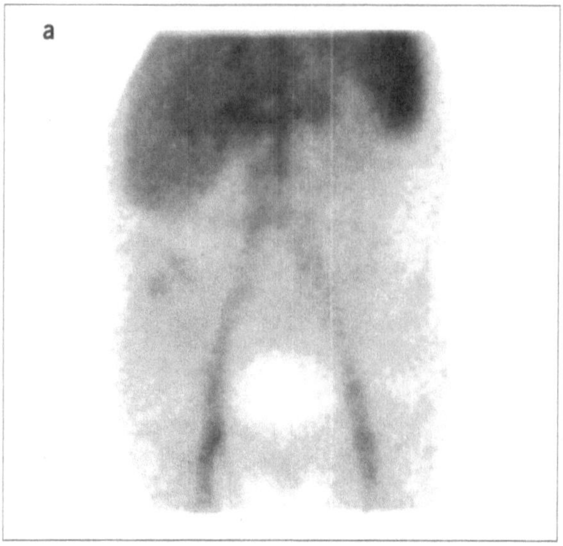

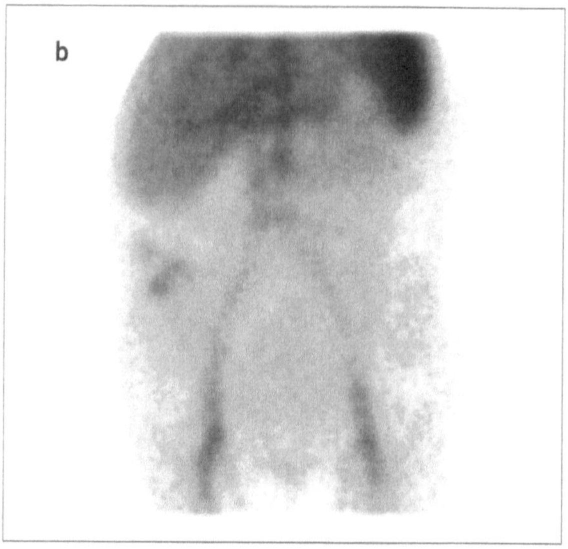

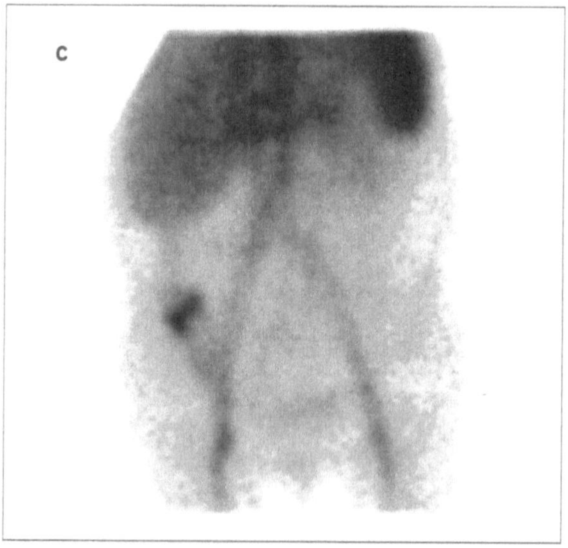

Figure 6.13 Blood loss demonstrated with autologous erythrocytes. (a) At 20 min after injection an abnormal accumulation is visible in the right iliac fossa. (b) This is clearer 10 min later. An additional image 30 min later (c) shows evidence of further bleeding. Angiomatosis of the caecum confirmed surgically.

sometimes occurs (Fig. 6.13). The longer the interval between views, the higher the probability of dispersion occurring and thus the site of bleeding being missed or incorrectly localized. The site of bleeding cannot be identified with the same precision as, for example, angiography but it is usually possible to indicate which quadrant is involved and to estimate whether it is in proximal or distal small intestine or elsewhere in the gut.

False negatives occur if extravasated blood is washed away from the site of bleeding and diluted in the interval between images, if the duration of examination is too short and no bleeding occurs during the period of observation or if images are technically inadequate, for example because of patient movement. In up to 70% of patients bleeding is not identified until more than 90 min after administration of labelled cells. Inadequate stability of labelled cells, with elution of radioactivity or free pertechnetate in the preparation is likely to be associated with secretion of activity into gut lumen, concealing extravasated activity and rendering the study non-diagnostic. If not recognized this may be misinterpreted as a false positive.

Intravenous colloid

If any of the colloidal preparations used for reticulo-endothelial scintigraphy (page 253) is injected intravenously during haemorrhage, some colloid is extravasated at the bleeding site. Because the rest of the colloid is rapidly cleared by the reticulo-endothelial system, any extravasated activity stands out against a comparatively low background. For this technique to work and if the full advantages are to be obtained, the colloidal preparation must contain very low levels of unbound radioactivity (less than 2%). Higher activities than are usual for reticulo-endothelial scintigraphy must be administered (200–400 MBq).

The advantages of colloid over the labelled red cell technique are the potential of the former to detect lower minimum rates of blood loss, about 0.1 ml per minute compared with 5 ml per minute for the latter technique, in consequence of the lower background and rapidity with which a diagnosis can be obtained, often within 5 min of injection. The disadvantages are that the time-window within which bleeding must occur to be detectable is very short (5 min or less following injection while there are still appreciable circulating levels of radioactivity). A further disadvantage is that the upper abdomen is concealed by activity in liver and spleen. The two techniques are therefore complementary. The colloid method may be used first if the patient is thought to be bleeding continuously from a site in the mid or lower abdomen, as it is less time-consuming. If negative it can be followed by the red cell technique not less than 48 h later, to allow activity in the liver and spleen to have decayed sufficiently for imaging of the upper abdomen to be possible. If bleeding is thought to be intermittent, as is more commonly the case, the red cell method only should be employed. Neither technique should be used unless there are reasonable clinical grounds for believing that there is active bleeding at the time of investigation. The pick-up rate is virtually zero in patients in remission.

Clinical indications

This examination is most helpful in patients with prolonged or recurrent bleeding if endoscopy and barium studies have proved inconclusive or negative, especially if high risk candidates for surgery.

Further reading:
Nuclear Medicine Communications 1993; ***14:*** *849–55*

Measurement of gastro-intestinal blood loss

A label with a half-life longer than that of ^{99m}Tc must be employed, to allow for the transit time from caecum to anus (usually one to two days but sometimes much longer) and to permit the study to be extended over several days. ^{51}Cr-labelled autologous erythrocytes (page 203) is the only radiopharmaceutical used widely, despite the unfavourable physical characteristics of ^{51}Cr, in particular the high internal conversion coefficient. Red cells labelled with ^{111}In-oxine or ^{111}In-tropolone have also been used to a limited extent. When ^{51}Cr is employed the absorbed radiation dose precludes administration of sufficient radioactivity for the site of bleeding to be visualized scintigraphically (unless one of the techniques employing technetium described above is employed simultaneously). This has been largely superseded by direct measurement of red cells in effluent obtained by whole bowel lavage.

Technique

A dose of 1 MBq ^{51}Cr-labelled autologous erythrocytes is administered intravenously and all faeces passed for the next three to five days are collected. A venous blood sample is taken not earlier than 10 min after injection, to allow for complete mixing and a second sample at the end of the collection period. The specific activity of red cells is taken as the mean of the two readings. If the patient has delayed bowel transit or is constipated there may be no radioactivity in stool for 24–48 h.

Interpretation

The radioactivity in those samples which contain ^{51}Cr is expressed in terms of the equivalent volume of venous blood. In normal subjects blood loss of up to 1.5 ml per day (or occasionally up to 3 ml per day) may be detected. Loss of greater than 1.5 ml per day should be considered suspicious and greater than 3 ml per day definitely abnormal.

Small intestine

The function of the small intestine is absorption of digested food. The majority of substances are absorbed by passive diffusion, which occurs throughout the duodenum and ileum. A number are absorbed by active transport or facilitated diffusion, bile salts and vitamin B_{12} in particular forming the basis of useful diagnostic tests. The small intestinal lumen is normally sterile.

Any bile salts which enter the colon are subjected to bacterial action, principally deconjugation to release unconjugated bile salts and dehydroxylation with the production of secondary bile salts. After absorption from portal blood by the liver, free salts of both primary and secondary bile acids are reconjugated and recirculate in the bile salt pool. In the normal subject, primary bile salts predominate and only small amounts of secondary salts are present. In some pathological states secondary salts predominate.

Some bile salts have a direct action on colonic mucosa, inhibiting reabsorption of sodium and consequently of water. Small intestinal malabsorption of bile salts thus produces diarrhoea. In the presence of bacterial colonization of small intestine, deconjugation may occur in the small intestine. This forms the basis of the ^{14}C-glycocholate breath test, which has been used to detect both colonization and bile salt malabsorption. The rate of synthesis of bile acids is controlled by a feedback mechanism. It increases if there is a reduction in the amount reabsorbed.

Bile salt malabsorption

Bile salt malabsorption can be detected by five techniques, breath counting, stool counting using a labelled bile salt, chemical assay of stool, measurement of circulating levels of bile acid precursors and measurement of whole body retention of a gamma-labelled bile salt analogue. ^{14}C-Taurocholate labelled in the 23 position of the cholic acid molecule is the standard labelled natural compound against which any other markers must be judged. Absorption of ^{14}C-taurocholate is measured after oral administration of 200 kBq by collecting all stool passed for the next five days. Each day's collection is homogenized, an aliquot dried and oxidized. The ^{14}C content is assayed as CO_2 using standard β-counting techniques. This is impractical as a routine clinical test and has never been widely employed, but must be regarded as a standard reference technique, the alternative being direct chemical assay of the bile acid content of stool. This latter is the only means of determining the spectrum of bile salts and conjugates present but also necessitates collection and manipulation of stool.

Glycocholate breath test

A dose of 200 kBq ^{14}C-glycocholate in 50 mg glycocholate is given orally, the ^{14}C label being on the glycine moiety. In the presence of bacteria glycocholate is hydrolysed. ^{14}C-glycine released is oxidized either by bacteria or more probably in host tissues following absorption. $^{14}CO_2$ is measured in breath over the next 4 h. In the normal subject little $^{14}CO_2$ appears in the breath for 12 h. Each laboratory must establish its own normal range but as an approximate guide less than 2% of the activity should be recovered in breath within the first 4 h.

Interpretation

Breath output of $^{14}CO_2$ is raised when deconjugating anaerobic bacteria are active in the small intestine. This may be idiopathic but is more commonly the result of previous partial gastrectomy, duodenal or small intestinal diverticulosis, blind loops, strictures resulting from Crohn's disease, resection of the ileo-caecal valve, tumours or radiotherapy and conditions such as systemic sclerosis and diabetes mellitus. Bile salt malabsorption from any cause, resulting in increased excretion of bile salts into the large intestine, also increases breath $^{14}CO_2$. ^{14}C-Glycocholate is a good test for the presence of bile acid deconjugation, but does not distinguish between possible causes because it does not take account of the rate of bowel transit. It cannot differentiate between accelerated transit into large bowel, where deconjugation is a normal phenomenon and colonization of small bowel. The sensitivity of the test is thus high but specificity low.

Bile acid precursors

In the presence of bile acid malabsorption the concentration in serum of the bile acid precursor cholestenone is increased above the normal upper limit of 35 ng/ml. Direct comparison with Se-HCAT retention (see below) indicates that the predictive value of a negative test is >0.98, although the predictive value of a positive test is only 0.74. False positives are associated with recent alcohol ingestion, small or large bowel diverticulae and age >70 years if female. Serum cholestenone measurement should therefore precede other tests of bile acid malabsorption, which may be omitted if serum cholestenone concentration is within the normal range.

Se-HCAT retention

The alternative and much simpler technique is to measure whole body retention of ^{75}Se-HCAT, the taurine conjugate of the synthetic trihydroxy bile acid, 23-selena-25-homocholic acid. Absorption and excretion of this compound is virtually identical to that of taurocholate. However because the label is ^{75}Se, an efficient γ-ray emitting isotope, rather than the β-particle emitter ^{14}C, it is comparatively simple to measure the percentage of an administered dose remaining in the patient at any time after ingestion. It is thus unnecessary to collect and process stool. Being the taurine conjugate of a trihydroxy acid, passive diffusion is minimal. Active transport is for practical purposes the only means of absorption and it is thus an ideal marker for the active transport mechanism, which is confined to the terminal one metre of ileum. Each bile salt molecule makes on average five circuits per day of the enterohepatic circulation: liver to bile to intestinal lumen to liver. There is a greater than 95% probability of reabsorption on each transit of the circuit. Recirculation amplifies the effect of small changes in absorption efficiency. Thus after 7 days, assuming no renal excretion, 17.5% of the original activity will still be present if the absorption efficiency of the ileum is 95% but only 8.5% if it is 93%. Bile salt absorption is thus a very sensitive measure of the functional state of terminal ileum.

There is some deconjugation and dehydroxylation of Se-HCAT by anaerobic bacteria in large intestine (or indeed in small if they are present there), but the extent to which this occurs is small compared with native taurocholate. Ultimately if retained long enough Se-HCAT is transformed into a mixture of 23-selena-25-homocholate, its taurine and glycine conjugates, 23-selena-25-homodeoxycholate and its taurine and glycine conjugates. Presumably further degradation occurs, but the products are present at such low concentrations in the gut lumen that their nature has not been characterized. During the normal period of study in man most administered Se-HCAT remains unchanged. There is some renal excretion, but in subjects with normal hepatic function this amounts to less than 15% in 7 days following administration, usually much less. For practical purposes this does not affect sensitivity, specificity or accuracy of Se-HCAT as a test for bile acid malabsorption. In principle this could affect the normal range in patients with severe hepatic failure, but these are not commonly subjects requiring this investigation.

Technique

The activity is administered in a gelatine capsule absorbed onto an inert excipient. Provided that count rate in the patient can be related to activity in a standard or phantom, it is in principle necessary to make only one measurement. In practice it is usually easier to make an initial 100% measurement and a second at 7 days. A standard is always required to correct for any drift of the counting equipment. Long-term measurements have shown that the rate of excretion is a single exponential function for at least two months after administration. A pair of measurements of retention made at 0 and 7 days allows the half-time of excretion to be calculated with good accuracy. Intervening measurements are subject to substantial irregularities because of variations in duration of retention of activity in large bowel and reduce the accuracy of the measurement. In the normal subject retention of more than 10% of the administered activity at 7 days indicates that normal absorptive capacity of the terminal ileum is present, whereas retention of less than 5% is evidence of bile salt malabsorption. Between these values 50% of patients have bile acid malabsorption whereas the others have rapid transit. Relatively few patients fall into this equivocal range. Diarrhoea due to disease affecting jejunum or colon does not influence retention of bile salts. Slightly reduced retention values are found in patients with an ileostomy, possibly because absence of a functioning ileocaecal valve reduces the time intestinal contents spend in terminal ileum.

counted both facing and with their back to the detector and that the geometric mean of the two is used. This minimizes any effects due to changes in distribution of radioactivity on detected count rate.

The initial measurement should be made 30–60 min after ingestion, to allow time for absorption. The count rate is initially measured using a wide window (up to a least 1 MeV) because when whole-body counting it is desirable to include scattered as well as primary counts. This both improves sensitivity and reduces variation of count rate with changing distribution of tracer. The baseline of the analyser is then reset above the high energy γ-ray of selenium, i.e. above 450 keV, and the second count made. Alternatively if a multi-channel analyser is available several windows can be counted simultaneously. This enables the ratio of scattered counts appearing in the selenium window to photopeak counts to be calculated. Se-HCAT is then administered, activity depending on size of the detector. With a large whole-body counter 40 kBq is adequate; with a smaller detector 400 kBq may be required. The patient is recounted 30–60 min later, to allow activity to be dispersed, using both windows. A suitable phantom is then counted to establish standards for both isotopes. One week later both patient and phantom are recounted using the same window settings.

Interpretation

The most common reason for suspecting bile salt malabsorption is persistent diarrhoea for which no explanation has been found following full clinical, stool, endoscopic and barium examinations, especially if symptoms include nocturnal diarrhoea. Many of these patients have idiopathic bile salt malabsorption and respond to specific treatment with cholestyramine. They are otherwise often mislabelled as the diarrhoeal form of irritable bowel syndrome. It is not known how long idiopathic bile acid malabsorption persists. Some patients do remit spontaneously within 2–3 years. However the prevalence of subsequent malignant disease and of overt inflammatory bowel disease is unduly high in these patients and there is some suspicion that idiopathic bile acid malabsorption may be a precursor of gut lymphoma. Bile salt malabsorption is usual after terminal ileal resection and it is therefore unnecessary to measure salt excretion routinely in these patients, unless there is clinical uncertainty whether the extent of resection is sufficient to account for the symptoms. It is often helpful to perform a test for colonization simultaneously.

Bile salt malabsorption is only rarely a significant factor in Crohn's disease unless the terminal ileum has been resected, possibly because of the presence of skip areas, but should nevertheless be considered in patients who appear refractory to conventional treatment. Malabsorption of B_{12} is more commonly due to colonization than primary loss of absorptive surface. Other diseases affecting the terminal ileum for example coeliac disease, sprue, neoplastic infiltration of the ileum or previous ileal resection may be associated with reduced retention of either B_{12} or Se-HCAT or both, depending on the length of gut involved and whether or not there is colonization. Absorption of Se-HCAT may be impaired despite a normal B_{12} retention. This pattern is sometimes seen following radiotherapy and may indicate differential radio-sensitivity of the two transport processes. Reduced absorption of B_{12} in the presence of normal Se-HCAT absorption is evidence of bacterial overgrowth associated with previous surgery, a blind loop or small bowel diverticula.

Small bowel bacterial colonization

The primary test should be measurement of breath hydrogen following an oral glucose load, as this does not require any radioactive tracer. If facilities for this measurement are not available ^{14}C-glucose, ^{14}C-glycocholate or any one of several other ^{14}C-labelled sugars may be administered and breath output of $^{14}CO_2$ measured (see above).

Not all bacteria capable of colonizing the small bowel release hydrogen. Malabsorption of B_{12} is a useful secondary test if clinical suspicion of colonization persists despite a negative hydrogen breath test.

Pancreatic function tests

Those employing selenomethionine are now obsolete. A test has been described involving oral administration of a triglyceride analogue containing 15-(*p*-iodophenyl)-pentadecanoic acid. In the presence of pancreatic lipases this is released and following absorption its principal metabolite (*p*-iodobenzoic acid) recovered in urine. ^{131}I can be employed as the label. Urinary recovery is decreased in the presence of pancreatic dysfunction. Clinical experience with this compound is as yet limited.

Inflammatory bowel disease

Although initial diagnosis is normally made on the basis of radiographic and endoscopic findings, radioisotope tests have an important role in management. Autologous leucocytes labelled either with ^{99m}Tc or ^{111}In accumulate in regions both of active Crohn's disease (Fig. 6.14) and ulcerative colitis (Fig. 6.15) but not where disease is quiescent. The technique is sensitive and being non-invasive may be used in patients in whom endoscopy or barium enema is contra-indicated. It should not be used as a primary diagnostic modality, as histopathology is required to differentiate between Crohn's disease and ulcerative colitis, but is useful for determining the extent and activity of newly diagnosed or residual disease and complications, especially abscess formation. It is regarded by some as the reference standard for assessing disease activity.

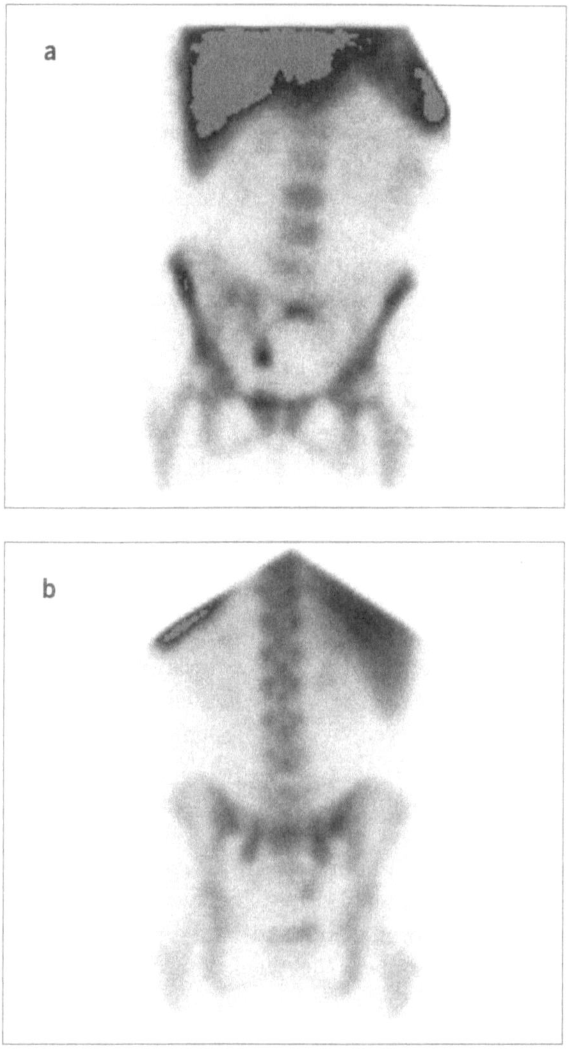

Figure 6.14 Terminal ileal uptake and skip lesions in Crohn's disease. (a) Anterior, (b) posterior 3 h after administration.

Radiopharmaceuticals

There are differences in technique depending on whether ^{111}In oxine or Tc-HMPAO-labelled cells are employed. Other tracers including labelled anti-granulocyte antibodies and labelled immunoglobulin have also been described but have not achieved clinical acceptance. Indium, whether as oxine or as tropolone, is securely attached to cells and does not elute. However labelling with ^{111}In-oxine must be carried out in a plasma-free environment. Cells can be labelled in plasma using ^{111}In-tropolone, but this is not commercially available or product licensed. There are some advantages in using a pure granulocyte preparation which can be obtained using centrifugation in a density gradient, but cell separation adds appreciably to complexity of the labelling procedure. Mixed leucocytes are therefore usually employed. Cells labelled in oxine take a while to recover from the period in a plasma-free environment and spend some time sequestered in the lungs after reinjection.

Technique and interpretation

Imaging is usually performed 3–4 h after

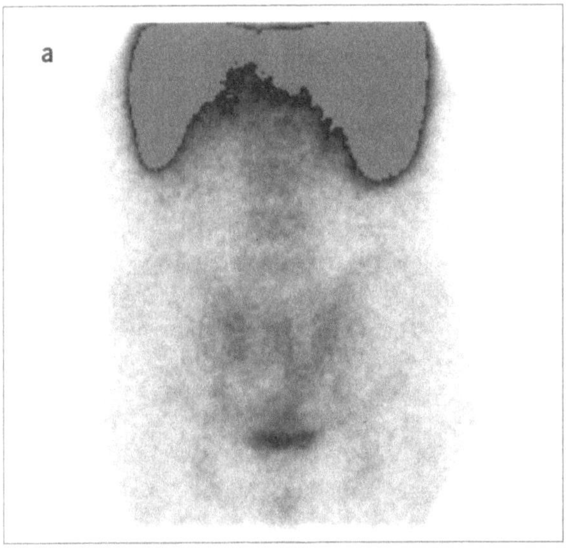

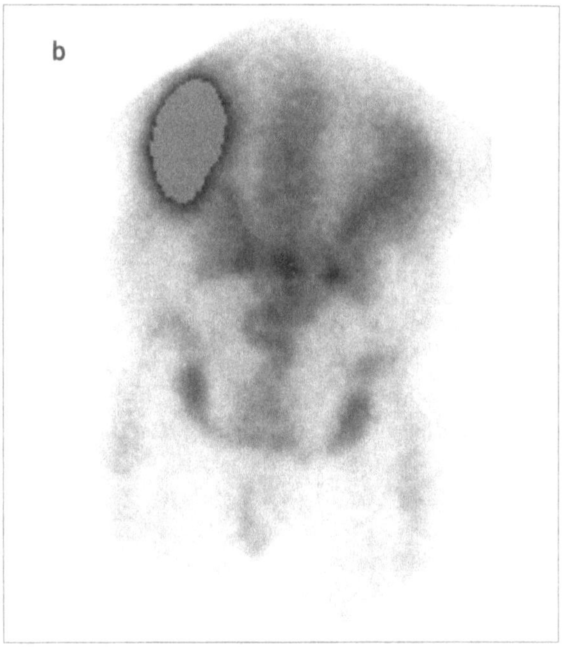

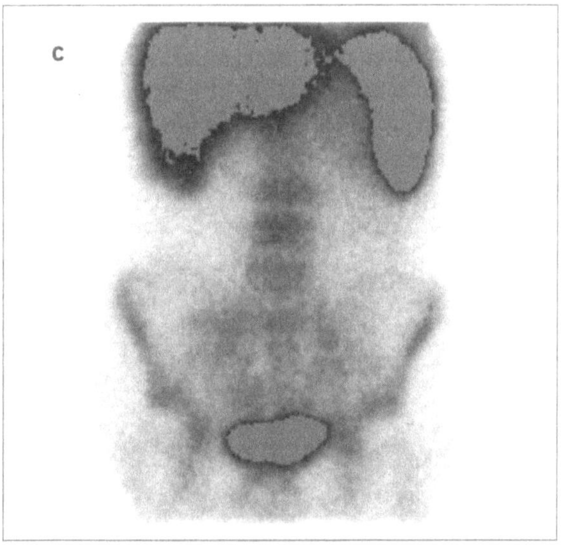

Figure 6.15 Ulcerative colitis with proctitis. (a) At 20 min after injection of Tc-leucocytes, uptake is faintly seen in the mid-line over the sacrum; (b) 3 h pelvic outlet projection; uptake is visible in recto-sigmoid which is displaced towards mid-line. (c) The abnormality is obscured by bladder in the anterior projection. The pelvic outlet projection should always be obtained when investigating inflammatory bowel disease. The bladder must be empty.

reinjection of 10–20 MBq ^{111}In. Later images, beyond 24 h, are possible and are sometimes helpful as uptake in areas of active inflammation increases for many hours. Despite the low count rate from the usual activity, which is limited by considerations of radiation dose and compounded by the lower efficiency of the gamma camera and collimator required for ^{111}In, abnormal uptake in bowel is clearly visualized because contrast between normal and abnormal regions is high. There is no bowel excretion of free indium; faecal counting to measure the quantity of labelled cells shed and whole-body counting to measure retention have been used as an index of disease activity. The normal subject should excrete less than 8(±5)% of the administered activity in 4 days. These refinements add little to the information obtained by imaging.

In contrast, after labelling unenriched buffy coat in plasma, Tc-HMPAO is eluted from lymphocytes fairly rapidly, although it adheres better to granulocytes. Because cells are less traumatized by labelling in plasma, uptake in areas of inflammation is initially more rapid. Contrast between normal and abnormal areas is not as high as with ^{111}In but the better dosimetric characteristics allow much higher activities to be administered, typically 150–250 MBq.

There is both renal and hepatic excretion of activity which is not attached to cells, with temporary storage of the latter in gall-bladder. Excreted activity is almost always observed in bowel lumen in views taken later than 6 h after injection. Consequently when imaging with technetium-labelled cells initial images should be taken 20–30 min after injection. If there is active disease there is usually at least a hint of abnormal bowel uptake by this time (Fig. 6.15a). Later imaging should be performed at 3–4 h, at which time there is higher uptake and contrast in abnormal accumulations. Imaging later than this is rarely helpful in inflammatory bowel disease but may facilitate identification of pus collections. The earlier imaging time is an advantage of technetium-labelled cells. Because of the higher photon flux small accumulations can usually be seen as easily as with ^{111}In despite the lower contrast.

Clinical indications

The principal applications are assessing extent and activity of inflammatory bowel disease and identifying collections of pus not accessible to other imaging techniques. Extent of inflammatory bowel disease is more easily recognized in colon as it is often impossible to unravel the convolutions of small bowel, but the extent and distribution of disease in small bowel is usually evident. Focal collections of pus can also be detected, but many of these can be also identified by ultrasound or computed tomography (CT), which should be the first investigations if a collection is suspected. White cell scintigraphy may detect focal pus collections in about 10% of patients with clinical evidence of occult infection but negative CT and ultrasound. It is most likely to be useful in patients who have previously undergone abdominal or pelvic surgery, when ultrasound and CT may be difficult to interpret. False negative white cell imaging is rare in ulcerative colitis and small bowel Crohn's, but is more likely to be found in large bowel Crohn's. No explanation is available for this observation but there may be a correlation with findings on whole bowel perfusion, when white cells are always found in the effluent in active ulcerative colitis and small bowel Crohn's but are commonly absent in large bowel Crohn's.

Further reading:

European Journal of Gastroenterology and Hepatology 1994; ***6****: 78–84*

Other tracers

Gallium is taken up in chronic inflammatory tissue and in many tumours. It accumulates in high concentration in many colonic carcinomas. Unfortunately gallium is excreted via the liver into colon and it is in practice impossible to distinguish excreted activity in the lumen of the colon from uptake in abnormal areas of bowel. Imaging on consecutive days and after an enema is of some assistance but the difficulty of interpretation is such that gallium scintigraphy has not achieved acceptance for the study of bowel disease.

Small and large bowel transit

The colon is not an inert reservoir for temporary storage of waste-products of digestion, but has essential absorptive functions for salt, water, unconjugated bile salts and other products of digestion. There is no fixed relationship between changes in pressure, movement of the wall of the colon and transit of faecal material. The various parts of the colon differ both morphologically and in their physiological functions,. The physical state of the lumenal contents varies in different parts of the colon and in the same subject from time to time. Study of colonic function is thus complex. Transit through the small intestine is less subject to variability and is less affected by external stimuli, but because of the complexity of its convolutions the only landmarks which can be reliably identified are the pylorus and caecum.

No clinical role has been established for studies of small bowel transit. Consecutive images of non-disintegrating capsules containing a convenient marker enable the pathway to be traced and the time between landmarks measured but this is of limited clinical value. Large bowel transit studies are finding a role in differential diagnosis of chronic constipation, but ingested non-absorbed tracers are mixed into a large volume before they reach the caecum, complicating analysis of the observations. Direct instillation of tracer into the caecum is impractical as a clinical test, but the equivalent effect can be achieved using pH-sensitive

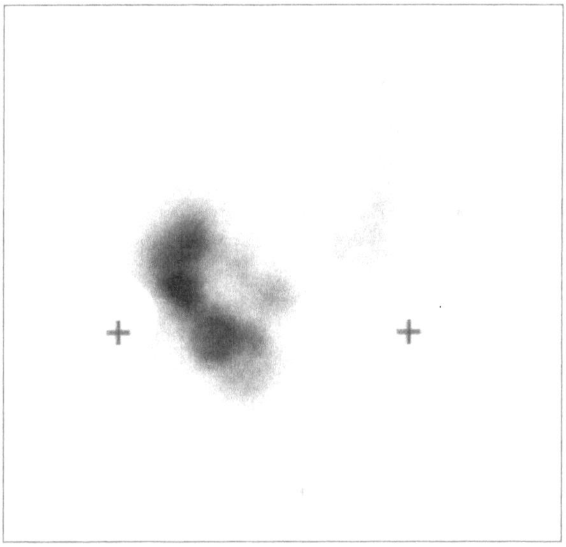

Figure 6.16 Distribution of tracer at 56 h in patient with right-sided delay. No activity had been excreted at this time. Markers indicate anterior superior iliac spines.

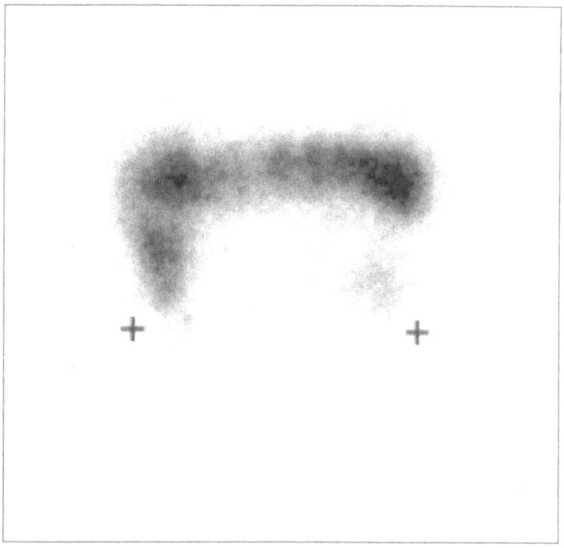

Figure 6.17 Distribution of tracer at 56 h in patient with generalized delay. No activity had been excreted at this time.

polymer-coated capsules filled with an isotopically labelled non-absorbable material, usually a resin on which ^{111}In is absorbed. These disintegrate in the terminal ileum or caecum where there is a change in pH, to give a less diluted, but nevertheless far from perfect bolus. Anterior and posterior pairs of images (to give a depth-independent value) are taken at intervals. The older techniques described were impractical for routine use because of the large number (up to 20) pairs of images required over three days. More recently it has been shown that a maximum of three pairs, at 10 h, on the second morning and on the third day provide all necessary information for proper classification of colonic transit.

Interpretation

Five patterns can be identified, normal, rapid transit, generalized delay, right-sided delay and left-sided delay. The colon is divided into four regions: (1) caecum, ascending colon and hepatic flexure, (2) transverse colon, (3) splenic flexure and (4) descending colon to rectum. In the 10-h images activity is usually seen in the caecum but it extends to region 4 and is present in faeces only in subjects with accelerated transit and a minority of those with normal transit. By day 2 most of the activity has been excreted in rapid transit subjects but is present in all regions in those with a normal transit rate. At this time there is no difference in distribution between the three delayed groups but by day 3 almost all of the activity has been passed by subjects with a normal rate of bowel transit. Over half is found in the in the sigmoid and rectum in patients with left-sided delay. In those with right-sided delay most is in the ascending and transverse colon (Fig. 6.16) whereas in those with generalized delay little if any has been excreted and it is distributed throughout the colon (Fig. 6.17).

Further reading:
Gut 1994; ***35:*** *976–81*

Ano-rectal function

Ano-rectal function can be evaluated by instilling a labelled stool substitute having similar physical properties to stool. Reconstituted dehydrated potato is suitable. Clinical applications are not established.

Further reading:
Nuclear Medicine Communications 1994; ***15:*** *1–3*

7 Blood, infection and inflammation

Red cells

The principal indications for labelling red cells with a radioactive tracer are to:

- measure the volume of erythrocytes in the circulation
- measure their life span
- image their distribution in order to detect sequestration or extravasation (page 189)
- measure cardiac output and other parameters of ventricular function (page 134)
- calculate blood flow (page 230).

Radiopharmaceuticals

Chromium is the most persistent label. Almost all of the initially absorbed activity remains associated with cells for the rest of their life span, but its physical properties are not suitable for imaging. In contrast there is appreciable elution of technetium from cells even within 24 h and it is therefore not suitable for quantitative or long-term studies; it does remain attached long enough for qualitative imaging and many semi-quantitative procedures. Thus either label may be appropriate depending on the application.

Chromium

In order to label cells with chromium, blood is withdrawn into a syringe containing the anti-coagulant 'acid citrate dextrose' (ACD) to give a final composition of one part of ACD to six of whole blood. The required activity of ^{51}Cr is added in the form of sodium chromate, taking care that the concentration of chromate does not exceed 10 μg per ml of blood. At higher concentrations chromate has a toxic effect on red cells. The specific activity of chromium is usually sufficient to maintain the concentration of chromate at about one-tenth of the toxic level. The mixture is incubated at between 37°C and 39°C for 30 min. Next 1.5 mg of ascorbic acid per ml of blood is added to reduce any unbound sodium chromate to chromic ions. The mixture should be centrifuged gently (at 1–2 G) to sediment red cells. The supernatant, which contains white cells, platelets and unbound chromium, is discarded. The cells are re-suspended in fresh plasma prior to re-injection. If the volume is small it is possible to re-inject red cell concentrate, but its viscosity is high and there is a danger of damaging cells if excessive pressure is applied to overcome their resistance to flow. For the same reason the needle employed should not be smaller than 21 gauge.

An activity of 0.2–0.4 MBq is adequate for measurement of red cell volume. Red cell survival requires a longer period of observation and hence a higher initial activity (4 MBq) in order to measure the rate of decrease of circulating activity. Much higher activities would be required for imaging, which is no longer performed with this radioisotope because of the radiation dose. The efficiency with which red cells are labelled with chromium is not affected by age of the cells. Chromium is not recirculated when labelled red cells are destroyed. There is however a small amount of elution from cells *in vivo*. This may be undetectable but is sometimes as high as 5% per day, leading to underestimation of red cell survival.

Technetium

^{99m}Tc is now almost invariably used to label red cells for imaging, although its half-life precludes its use for studies longer than 24 h. The highest efficiency of labelling can be achieved with a combined *in vivo* and *in vitro* method. Alternatively, a lower efficiency of labelling can be achieved more rapidly by performing the entire procedure *in vivo*. The first step in both methods is intravenous injection of 500 μg of stannous tin as an isotonic solution of the pyrophosphate or other soluble salt. For the combined method, 15–30 min later 5 ml of venous blood are withdrawn into a syringe containing the appropriate activity of pertechnetate and 50 IU heparin in a volume of 1–1.5 ml. The blood is mixed with anticoagulant and radioactivity in the syringe and incubated at room temperature for 15 min. During this period about 85% of the pertechnetate is reduced and becomes attached to the surface of red cells. If the entire contents of the syringe are then re-injected with no further manipulation, about 85% of residual unbound activity also becomes attached to circulating erythrocytes. There is thus an overall labelling efficiency greater than 95%. This is adequate both for imaging blood pool and immediate measurement of red cell volume. For the *in vivo* method sodium pertechnetate is injected intravenously 15–30 min after the stannous salt. This gives a labelling efficiency of approximately 85%, which is inadequate for quantitative studies. It is also inadequate for some cardiac function measurements, as extravascular background count is somewhat higher due to diffusion of unbound pertechnetate into the extravascular space.

If labelled cells are incubated at between 49°C and 50°C for 20 min they become pyro-spherocytes, which are sequestered in the spleen (page 210).

Further reading:
Journal of Nuclear Medicine 1991; ***32****: 242–4*

Other labels

^{32}P-Diisopropyl-phosphonate dissolved in propylene glycol labels red cells directly with ^{32}P if injected intravenously at a dose not exceeding 22 mg/kg body weight. Although stable *in vivo* this technique is no longer used because of toxicity of the label and the relative complexity of counting beta emissions from ^{32}P. The indium-labelled compounds used for labelling white cells can also be used to label red cells. Their efficiency for labelling red cells is lower than for white cells. They are rarely used for this purpose as less expensive and more accessible agents are readily available. Carbon monoxide labelled with ^{11}C or ^{15}O is an excellent red cell label for studies that can be completed in the short time permitted by the half-life of these isotopes. Because of their limited availability and high cost this is employed only as a research procedure.

Further reading:
Journal of Nuclear Medicine 1990; ***31****: 2044–5*

Clinical applications

Red cell volume is calculated by measuring the specific activity of a blood sample taken long enough after administration of an accurately known activity of labelled red cells to permit complete mixing. In most patients this is within 10 min, but in patients with congestive cardiac failure, oedema or any condition associated with vascular stasis at least 1 h must be allowed. Red cell volume is calculated from the relationship:

Red cell volume = (administered activity)/(activity per unit volume of blood)

Red cell volume should be considered relative to the ideal weight of the patient rather than actual weight, to avoid misleadingly low values in obese subjects. The normal range is 24–32 ml/kg in males and 22–28 ml/kg in females. Red cell volume is increased in polycythaemia, pregnancy and cardiac failure and is higher in infants than in adults. It is reduced in anaemia and following haemorrhage.

Red cell survival is estimated by measuring specific activity daily for the first week, then twice weekly until specific activity has fallen to less than half of the initial value. The initial labelled sample contains a mixture of cells of all ages. Because destruction depends only on the age of cells and is not a random process, a plot of specific activity against time is linear. A linear extrapolation reaches zero at a time equal to the mean red cell survival time. Loss of label from erythrocytes is however a random process which gives an exponential curve. Thus in the normal subject plasma activity may not fall in a linear manner if there is appreciable loss of label. It is therefore necessary initially to plot residual plasma activity on a logarithmic scale in order to determine extent of elution. The residual component of the activity, which should be much the larger, can then be replotted on a linear scale. The 95% confidence interval for mean red cell survival in normal subjects is 58–161 days.

Further reading:
Journal of Nuclear Medicine 1991; ***32****: 2245–8*

In most haemolytic anaemias destruction is a random process not related to age of the cell. In such cases plotting on a linear scale does not give a straight line and a semilogarithmic plot is necessary. Half-life can be calculated from this, not mean life. In some anaemias more than one population of erythrocytes, each with a different life span, is present. If two populations with different half-lives are present, the semi-logarithmic plot is not linear. It is possible to strip component exponentials, from which half-lives of separate populations can be calculated (page 227). If more than two populations are present this estimate is inaccurate unless a large number of points are obtained and the half-lives of the components differ substantially.

Iron

There are about 5 g of iron in the average adult. Approximately 3 g of this is in haemoglobin and a further 1 g in iron stores, principally in the liver as ferritin or haemosiderin. Iron is absorbed from the gut only when in the ferrous form and in an amount depending on iron requirements at the time. A typical European diet contains about 15 mg of iron per day, of which only about 1 mg is absorbed by a healthy male to balance losses. About two-thirds are minor blood loss and the remainder desquamated cells. Iron loss in the normal menstruating female is about double that in the male. Iron has a number of radioisotopes. ^{59}Fe is usually employed for turnover studies. The half-life of 44.5 days permits long-term counting, whereas the high energy of gamma emissions is well suited to whole body counting. Because of the large number of compartments into which iron enters, it is necessary to measure accumulation and turnover in several compartments simultaneously. A typical ferrokinetic study requires measurement of the rate of plasma iron clearance, its appearance in red cells and surface counting over marrow, liver and spleen. If facilities are available these should be supplemented by whole body counting to measure rate of loss.

For typical ferrokinetic studies 0.2–0.4 MBq of ^{59}Fe-ferrous citrate is injected intravenously. Blood samples are taken at 10 min, 30 min, 1 h, 2 h, 3 h and 4 h after injection and three times weekly for three weeks. The radioactivity over marked areas of liver, spleen and marrow, usually the sacrum, is measured three times a week for three weeks. A number of indices can be calculated, the most useful of which are:

- Plasma iron clearance, expressed as half-time of clearance from blood (normal 80–120 min).
- Plasma iron turnover rate, defined as: [(plasma iron concentration) × (plasma volume) × 1000]/(half-time of plasma iron clearance in days) ×.(body weight in kg) The normal range is 0.42–0.70 mg/kg/day.
- Maximum red cell incorporation of iron, defined as: [(maximum activity per ml whole blood) × (red cell mass) × 100]/ (venous haematocrit × total activity injected). This should be >80% at 5 days and >90% at 7–10 days.
- The ratio of activity in marrow to that in liver and spleen. There should be higher uptake in sacrum than in liver, which in turn should have higher uptake than spleen.

Table 7.1 gives typical values.

Iron radioactivity first appears in erythrocytes within 12 h of administration but does not reach a plateau until one to two weeks later. In the normal subject circulating levels then remain constant for approximately 100 days, followed by a fall which reaches a minimum at 120 days as one generation of erythrocytes is removed. Iron is

Table 7.1 Normal ferrokinetic values.

Half-time of plasma iron clearance	80–120 min
Plasma iron turnover rate	0.42–0.70 mg/kg/day
Maximum red cell incorporation of iron	
5 days	80%
7–10 days	90%
Count rate over sacrum > liver > spleen	
Peak count rate occurs at 12 hours	

however conserved and the plateau is re-established by the 130th day as iron is reissued from marrow. The half-time of plasma clearance is prolonged in aplastic anaemia, leukaemia, erythroid hypoplasia, refractory hypoplastic anaemias and following radiotherapy. It is shortened in iron deficiency and pernicious anaemia, polycythaemia, many haemolytic anaemias and haemoglobinopathies. Plasma iron turnover rate is substantially increased in haemolytic and pernicious anaemia, spherocytosis and polycythaemia rubra vera. It is decreased in erythroid hypoplasia, refractory hypoplasia anaemia and following radiotherapy. The maximum erythrocyte incorporation of iron is low in aplastic anaemia, haemochromatosis, myelofibrosis, leukaemia and following radiotherapy and is increased in iron deficiency anaemia. Thus delayed blood clearance of iron indicates reduced utilization whereas accelerated clearance indicates increased turnover.

Iron uptake by the liver is increased in aplastic anaemia and myelofibrosis, due to extramedullary haematopoiesis and in haemochromatosis, erythroid hypoplasia, refractory hypoplastic anaemia and following radiotherapy. Splenic uptake is increased in hereditary spherocytosis, ineffective erythropoiesis and haemolytic anaemias. Initially increased uptake of iron by liver or spleen followed by subsequent release of radioactivity into circulating red cells is evidence of extramedullary haematopoiesis. Increased uptake by the spleen at later times suggests sequestration of red cells. Prolonged retention following early uptake in the liver is evidence of iron deposition. Blood loss can be measured by whole body counting of iron radioactivity as there is no other major pathway of iron excretion. The minimum loss detectable by this technique is close to the normal loss, approximately 1.5 ml per day.

Vitamin B_{12}

Vitamin B_{12} contains cobalt and can therefore be labelled with radioisotopes of cobalt. ^{57}Co gives the lowest absorbed radiation dose and is suitable both for *in vivo* and *in vitro* counting. ^{58}Co is associated with a slightly higher radiation dose (Appendix 1) and emits gamma rays of higher energy. It is preferable to ^{57}Co for whole body counting as there is less soft tissue absorption. ^{57}Co is preferable for all other applications. The difference in gamma ray energies is sufficient for the two isotopes to be counted simultaneously. ^{60}Co is associated with a higher radiation dose than either ^{57}Co or ^{58}Co and is therefore not used for diagnostic purposes.

B_{12} is absorbed by active transport from the terminal two metres of ileum. Prior to absorption B_{12} must have been bound by intrinsic factor, a mucoprotein with a molecular weight of approximately 50 kDa secreted by parietal cells of the gastric mucosa. Malabsorption may result from failure of production of intrinsic factor, removal or disease of terminal ileal absorption sites, bacterial destruction of vitamin before it can be absorbed, or competition by the fish tapeworm, *Diphyllobothrium latum*. Animal-based foods form the only dietary source of B_{12}, a typical diet containing a few micrograms per day. Body stores, which are principally in the liver, are large by comparison, amounting to several milligrams. Signs of deficiency or malabsorption therefore take a considerable time to become evident.

The active transport mechanism can handle only about 1.5 μg of vitamin B_{12} at a time. If larger quantities are administered orally there is some passive diffusion, but this does not play an important part at normal physiological concentrations. The effect of an active transport mechanism which is easily saturated is that the percentage absorbed falls as administered dose rises, from about 75% when the dose is below 0.25 μg to less than 50% when the dose exceeds 2 μg. There are however wide individual variations. Excess parenterally administered B_{12} is excreted principally via the kidneys, by glomerular filtration. A small amount also appears in bile. This latter can have an important influence on tests of B_{12} absorption if these are carried out within 3 days of a previous large parenteral dose.

Tests of B_{12} absorption

Circulating radioactivity reaches a peak 8–12 h after oral administration. The extent of overlap between normal and abnormal is such that this parameter is not diagnostically useful.

Urinary excretion (Schilling test)

After an overnight fast, 1μg of $^{57}Co\text{-}B_{12}$ containing 20 MBq is given orally. The patient must be fasting at the time of oral administration and should continue to fast for at least 2 h thereafter, to minimize variability of absorption due to dilution of the oral dose with B_{12} present in food. A 1 mg dose of non-radioactive B_{12} is administered intramuscularly 2–6 h after the oral dose. It is often convenient to give the injection at the end of the 2 h fast which follows the oral dose, while the patient is still under observation. All urine passed is collected for the next 24 h. The rate of excretion depends on renal function. In principle, fewer false positive results should be obtained if a 48 h urine collection is used, rather than 24 h which may underestimate absorption in patients with impaired renal function. The longer collection is however more likely to be associated with incomplete collection. A 48 h urine collection is essential if the patient is known to have impaired renal function. In this case a second intramuscular injection of 1 mg B_{12} should be given 24 h after the first flushing dose.

The principle underlying the test is that absorbed radioactive B_{12} is diluted by the large excess of parenteral non-radioactive vitamin and the two are excreted together. The percentage of radioactivity recovered from urine is thus dependent on the percentage of the oral dose absorbed. A recovery of 10% or more of the administered activity in urine is normal, whereas <5% is evidence of malabsorption. A low recovery may also be due to renal disease, failure to administer the intramuscular flushing dose, incomplete urine collection, administration of a large dose of B_{12} within the 3 days preceding the test or a meal with an high B_{12} content shortly before or immediately after administration of the oral dose.

If the test is abnormal it may be repeated using a vitamin B_{12}–intrinsic factor complex, when a normal result indicates Addisonian pernicious anaemia, i.e. malabsorption due to impaired secretion of intrinsic factor. If the result remains abnormal, destruction within gut lumen prior to absorption or a defect of mucosal transport is more probable, although pernicious anaemia cannot be totally excluded from the differential diagnosis as secondary changes sometimes occur in ileal mucosa when this condition is long standing. As an alternative to consecutive tests, $^{57}Co\text{-}B_{12}$ can be administered followed 2 h later by $^{58}Co\text{-}B_{12}$–intrinsic factor complex. Both total excretion and the ratio of the two radionuclides must be measured.

Radiopharmaceuticals administered for other investigations can interfere with measurement of B_{12} uptake. Technetium may be detectable for 2–5 days, ^{131}I for 4–115 days and ^{111}In, ^{67}Ga or ^{201}Tl for 12–44 days.

Further reading:
*Journal of Nuclear Medicine 1996; **37:** 1995–9*

Whole-body counting

This technique is potentially the most accurate, especially if ^{58}Co is employed, as it eliminates the need for urine collection and the flushing dose of B_{12}. However it requires specialized equipment which is not widely available. An uncollimated gamma camera can be used if its energy window can be set to a sufficiently wide value to include most of the scatter as well as the photopeak gamma rays. In normal subjects retention at 3 days is the same as at any subsequent time. False negatives may occur due to constipation. Counting at 7 days is thus preferable to minimize effects due to variations in bowel transit rate. This test requires longer for its completion than the Schilling test, and more complex apparatus. The International Committee for Standardization in Haematology regards whole body counting as the reference technique, although the Schilling test is more commonly employed in practice. Using whole body counting in our laboratories, normal subjects retain more than 23% of the administered dose at 7 days. Some published reports put the lower limit of normal at 35%.

Further reading:
*Journal of Nuclear Medicine 1981; **22**: 1091–3*

Clinical indications

A test for B_{12} absorption, usually the Schilling test, is required to determine whether or not there is malabsorption of B_{12}. This may be primary, as in pernicious anaemia, or the consequence of previous gastric surgery, ileal resection or bacterial colonization. The latter is most commonly seen in Crohn's disease and following limited resection of the terminal ileum. Malabsorption due to resection of the terminal 2 m of the ileum can occur but is uncommon, as resection is not usually so extensive. Malabsorption is occasionally observed in patients whose terminal ileum has been used to create a prosthetic bladder.

White cells

White cells can be labelled with technetium, indium, chromium or a variety of ^{14}C or ^{3}H compounds. The latter are used exclusively for experimental work *in vitro*. The other compounds are for the most part non-specific and will label to a greater or lesser extent most of the cells to which they are exposed. It is therefore necessary initially to concentrate or separate the cell fraction it is desired to label.

Techniques of labelling

The initial stage for both technetium and indium compounds is withdrawal, with aseptic precautions because the cells are to be re-injected, of a venous blood sample through a large bore (19 gauge) needle to minimize risk of damage to the cells. Blood is collected directly into a syringe containing the appropriate quantity of anticoagulant, typically 5 ml of acid citrate dextrose (ACD) to which 45 ml of venous blood is added. It is preferable to use a 'butterfly' or similar device to distance the syringe from the needle, so that the syringe can be rotated to mix blood with anticoagulant while it is being withdrawn. Blood which has clotted is unusable for labelling.

The next step is to concentrate white cells by differential sedimentation, separating them from other formed constituents in blood. A number of methods have been described but all rely on similar principles. If anticoagulated blood is left to stand, red cells, which are more dense, form a bottom layer with a buffy coat of white cells above, whereas platelets remain in suspension in plasma. Buffy coat can be withdrawn and used for labelling with Tc-exametazime. Better separation of white cells can be achieved by centrifugation at low speed (1–2 G) supplemented by substances such as dextran, methyl cellulose or metrizamide which decrease viscosity of plasma. Red cells form a pellet in the bottom of the tube. The supernatant, which contains white cells, platelets and plasma is carefully withdrawn, leaving behind most of the red cells. The disadvantage of using these reagents is an increased risk of reaction to the end-product finally re-injected.

For labelling with one of the indium compounds such as oxine, the supernatant is centrifuged at 150 G for 5 min to form a pellet containing residual red and white cells and a supernatant of platelet-rich plasma. As indium oxine has a greater affinity for white than for red cells some residual red cell contamination is acceptable. The supernatant is withdrawn, the cell pellet re-suspended in saline containing indium oxine dissolved in alcohol and the mixture incubated at room temperature for 4–5 min. It is then spun down again to remove residual unlabelled indium and as much as possible of the oxine and detergents; these are an essential component of the indium oxine preparation (to hold oxine in solution) but are toxic to white cells. The cells are then re-suspended in platelet-poor plasma for re-injection.

There is some evidence that further separation of a 'pure' granulocyte fraction gives higher contrast when imaging infection. The various white cell components can be separated using a density gradient containing methyl cellulose and metrizamide or similar radiological contrast medium. However, these procedures increase the complexity and duration of the labelling procedure and the risk of damage to cells. Most centres use a mixed white cell preparation. Re-suspending the cells in a plasma-free medium itself causes some cell damage. Alternative reagents such as indium tropolone which allow the labelling procedure to be carried out in plasma have been evaluated. They give an appreciably lower labelling yield, wasting up to half of the indium, and are not commercially available.

Cell labelling with technetium can be carried out in plasma using Tc-exametazime (HMPAO, Ceretec). A small amount of red cell contamination does not affect image quality with this preparation. The label itself remains securely associated with granulocytes but leaches fairly rapidly out of lymphocytes. Thus only the first of the stages described above, namely sedimentation to remove the majority of red cells, is required; this makes preparation less labour intensive and less likely to damage white cells. Partly for this reason but also because of the readier availability of technetium and the higher photon flux which can be obtained at acceptable levels of radiation dose, most centres now use Tc-exametazime as the labelling agent of choice. No other technetium compound is currently available for labelling white cells.

Administered activity
150–250 MBq ^{99m}Tc-HMPAO or 10–25 MBq ^{111}In-oxine.

Monoclonal antibodies

A number of monoclonal antibodies and antibody fragments to white-cell associated antigens, labelled with ^{123}I, ^{111}In or ^{99m}Tc, have been evaluated for their potential to label white cells without first separating the cells. All those so far tested label cells in bone marrow to a greater extent than circulating cells, as the antibodies target immature cells in preference to mature granulocytes. Between 60% and 90% of the circulating activity is not cell-bound, giving a high background level of activity. Good results have been reported in small series of patients, but these compounds are not generally available and experience is limited.

Further reading:
Journal of Nuclear Medicine 1996; ***37****: 673–9*

Antibiotics

The majority of antibiotics are not concentrated at sites of infection but the fluoroquinolone ciprofloxin is said to be taken up by, and concentrated in, certain bacteria. It has been labelled with technetium and a small number of patients imaged. Its role is not established.

Further reading:
European Journal of Nuclear Medicine 1996; ***23****: 459–65*

Interpretation

Data processing

Undamaged labelled white cells concentrate principally in the spleen. To visualize abdominal and thoracic structures it is usually necessary to over-range the spleen.

Normal appearance

In normal subjects the highest concentration of white cells is observed in spleen. A similar intensity of activity in liver and spleen indicates that cells have been damaged, in which case the study is non-diagnostic. High uptake in the spleen is therefore an internal quality control. Uptake is also seen in liver and bone marrow. Some intravascular activity persists and it is usually possible to see major vessels. Immediately after injection, cells often sequester in lung. This may be related to the severity of damage sustained during the labelling procedure. Wash-out of technetium-labelled cells is usually rapid but with indium there is commonly substantial uptake in lung if imaging is performed earlier than 4 h. Some lung uptake is usually seen with technetium-labelled cells for the first hour but this is usually of lesser intensity and fades progressively in later images.

Sources of error

Because of the relatively high but variable and non-uniform uptake in normal marrow, osteomyelitis in the spine and other marrow-containing parts of the skeleton is difficult to differentiate from normal marrow uptake. Faulty preparation leading to devitalized cells is associated with relatively lower uptake in the spleen. Technetium radioactivity eluted from cells is excreted both via the gall-bladder into the gut and via the kidneys. This makes interpretation of abdominal images progressively more difficult as time progresses.

Bone marrow

Marrow is visualized when any colloidal preparation is injected intravenously. The relative distribution between marrow, spleen and liver is to some extent determined by particle size; small particles are associated with relatively higher uptake into marrow and larger particles into spleen. Relative blood flow is however by far the major factor (Fig. 7.1). There is no size-specific preparation enabling visualization of one only of these tissues. Colloid shows the distribution of reticulo-endothelial marrow and has been used to identify para-vertebral masses as sites of extramedullary haematopoiesis. Under most circumstances the distribution of reticulo-endothelial and haemopoietic activity is identical, but the two rarely differ. The only true markers of erythropoietic marrow are ^{52}Fe and ^{59}Fe. The former is no longer generally available. Indium (^{111}In) also localizes in marrow. Its distribution differs both from that of iron and that of colloid and may represent distribution of transferrin or lactoferrin.

Marrow imaging is usually with colloid; on some occasions ^{111}In has been used to demonstrate the distribution and extent of functioning marrow in order to obtain an indication of a suitable site for aspiration in patients whose distribution of marrow is thought to be asymmetrical or atypical. Skeletal metastases commonly replace marrow before they cause any new bone formation and thus may be detectable as photon-deficient areas of marrow on scintigraphy before they are visible on bone scintigraphy. When using colloid a large part of the thoracic and lumbar spine is obscured by activity in liver and spleen so that, in most situations, bone scintigraphy provides more clinically useful information. Marrow imaging has also been used to determine viability of the femoral head following fracture, but as the distribution of marrow in the adult is variable and many adults do not have marrow in the proximal femur, this has not proved to be clinically reliable. There is no common clinical indication for marrow imaging.

Further reading:
Journal of Nuclear Medicine 1996; ***37****: 473–5*

Spleen

Roles of the spleen include removal of nuclear fragments from maturing erythrocytes and destruction of effete red cells. It is also one of the major sites of reticulo-endothelial activity. These functions are however distinct. Increased haemolysis, irrespective of origin, may saturate the functional capacity of the spleen producing temporary functional asplenia. Common causes include sickle cell crisis or accelerated neonatal haemolysis. Absence of the spleen may occur as an isolated anomaly or may be associated with congenital abnormalities of the heart, malrotation of the intestine and malposition of the liver. More commonly, asplenia is suspected clinically following observation of Howell–Jolly bodies in circulating erythrocytes. In functional asplenia the ability to take up colloid may be retained despite a temporary or permanent loss of ability to sequester pyrospherocytes. As functional asplenia may be a temporary phenomenon which reverts rapidly to normal, there is little point in imaging healthy subjects because of an isolated observation of Howell–Jolly bodies in peripheral

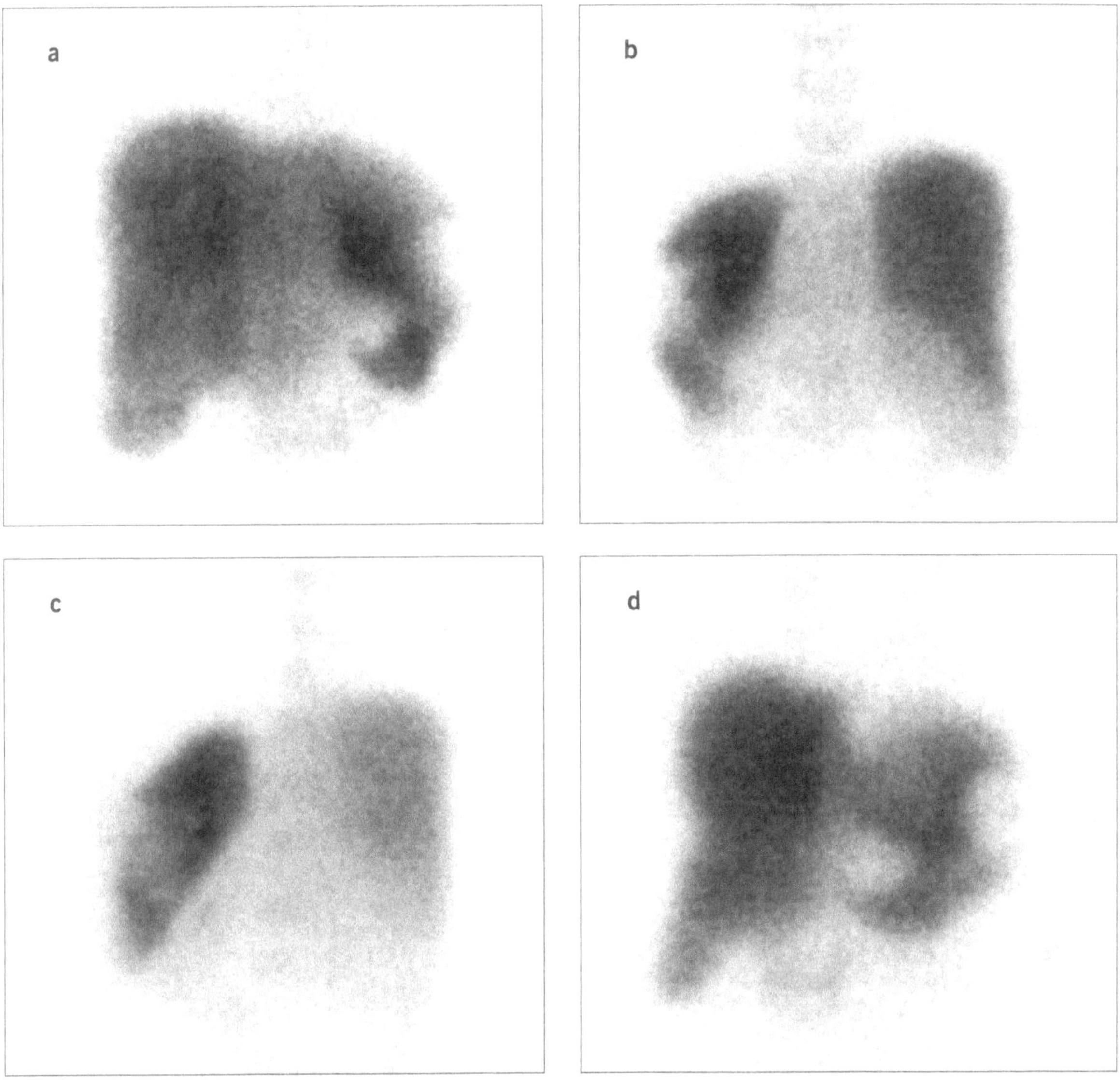

Fig. 7.1 Abdominal distribution of reticulo-endothelial function in 62-year old woman with splenomegaly and splenic infarcts due to chronic lymphatic leukaemia. (a) Anterior, (b) posterior, (c) left posterior oblique, (d) left anterior oblique.

circulating blood. Imaging is indicated only if haematological abnormalities persist after several weeks, or to demonstrate a splenunculus following splenectomy for idiopathic thrombocytopoenic purpura. Although the normal spleen can be visualized with colloid (Fig 9.6) it may be difficult to distinguish from liver. Oblique projections may help. Splenic infarcts can be imaged with colloid (Fig 7.1). Although imaging of spleen for other indications can be undertaken with colloid (Fig 7.2), pyrospherocytes (page 204) (Fig 7.3) are more specific and therefore often preferable. Sites of platelet sequestration or destruction may be identified by labelling autologous platelets with ^{111}In-oxine. In thrombocytopoenia this may confirm whether there is likely to be a response to splenectomy.

Further reading:

Clinical Radiology 1994; ***49****: 115–7*

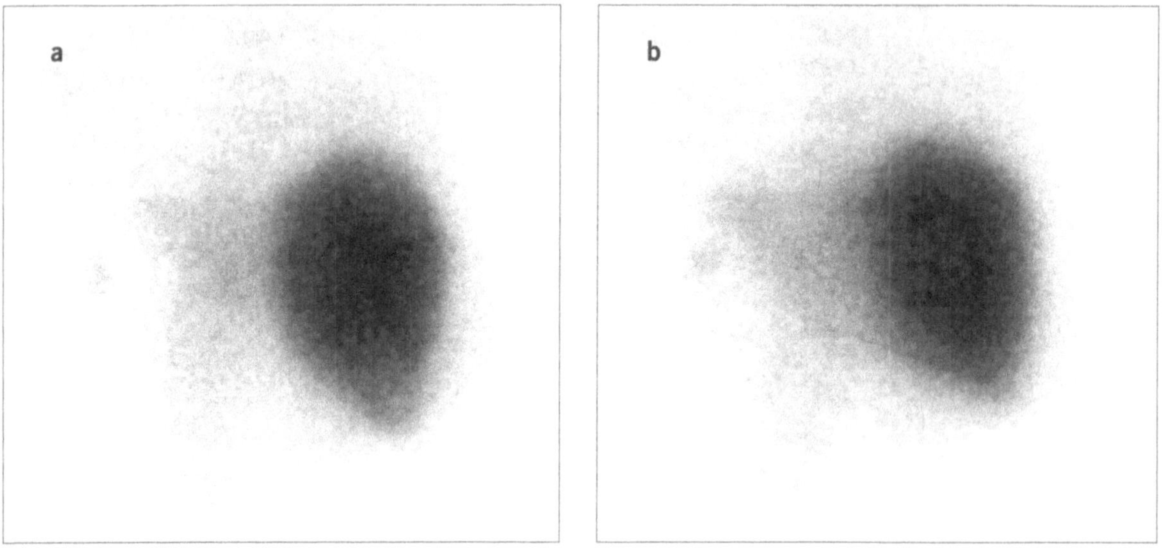

Fig. 7.2 Previous splenectomy for idiopathic thrombocytopenic purpura. A splenunculus is visible in the splenic bed. (a) Posterior, (b) left posterior oblique.

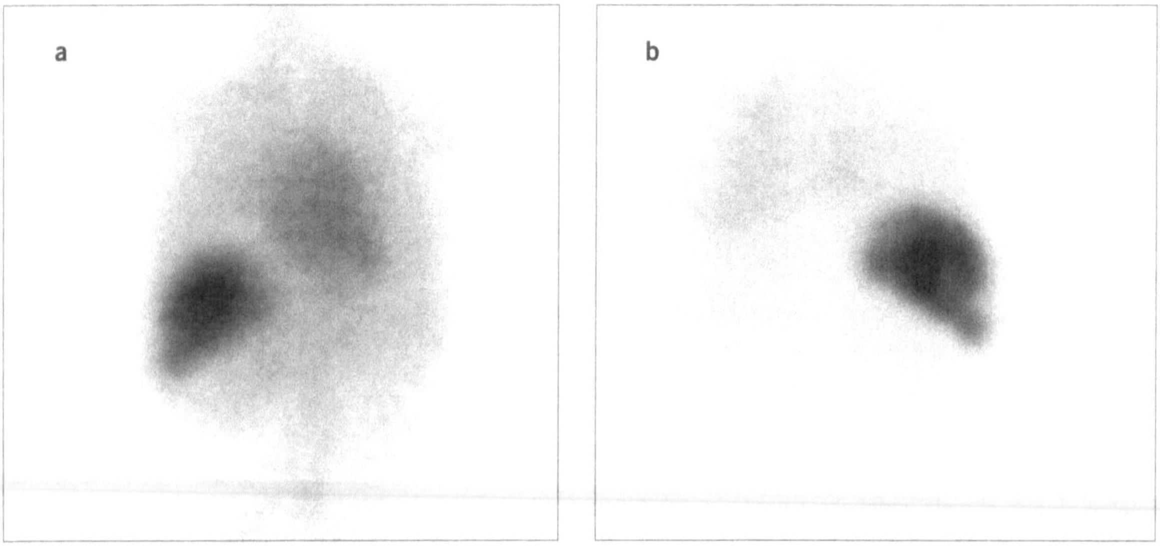

Fig. 7.3 9-year-old boy; pyrospherocyte imaging; 12 MBq; ^{99m}Tc autologous pyrospherocytes. Three persistent masses found on ultrasound in the right upper quadrant of the abdomen after otherwise apparently successful treatment for abdominal Hodgkin's disease. (a) Anterior. Abdominal situs inversus. Spleen is on right side. (b) Posterior. 'Masses' are accessory spleens.

Inflammation labels

Markers of permeability

Apart from white cells, a number of substances have been described which concentrate in areas of acute inflammation. In general they do not distinguish between infectious and non-infectious causes. Diffusible tracers such as pertechnetate and DTPA accumulate in areas of acute inflammation but indicate only a non-specific local increase in extracellular water. Plasma proteins also leak non-specifically into inflamed regions, possibly associated with locally increased capillary permeability. Polyclonal human immunoglobulins labelled with either ^{111}In or ^{99m}Tc have been employed. It has been suggested that the indium compound is more stable and therefore provides better contrast clinically but conflicting results have been reported. Accumulation may well be non-specific and the clinical role of these agents is unclear.

Further reading:

European Journal of Nuclear Medicine 1994; ***21****: 1135–40*

Gallium

The original observation that gallium concentrates in areas of inflammation was serendipitous. Although much effort has been expended, the mechanism remains incompletely elucidated. Gallium is administered as the citrate. However, as with most trivalent metals, all circulating activity is bound to the β_1-globulin transferrin within a few minutes of injection unless circulating transferrin has previously been saturated with iron or some other trivalent cation. Local concentration is thought to be the consequence of a combination of factors, including increased extracellular fluid, greater capillary permeability in inflammatory (and some neoplastic) tissue allowing large molecules such as transferrin to diffuse into the extravascular space, increased binding by another metal-binding protein, lactoferrin, which may be present in high concentration in inflamed tissues and has a higher affinity than transferrin for gallium, and binding by bacterial siderophores. There are probably additional factors.

Gallium accumulates both at sites of inflammation and in some tumours. Uptake both in soft tissues and in bone tends to be higher in chronic than acute inflammation and has been reported in tuberculosis, sarcoidosis, infected joint prostheses, connective tissue diseases such as scleroderma and systemic lupus erythematosis, blastomycosis, *Pneumocystis carinii* pneumonia secondary to AIDS, inflammatory pseudo-tumours, leprosy and pyelonephritis. In soft tissue, especially chronic interstitial lung disease, gallium has been advocated as a means of distinguishing an active but chronic inflammatory process from residual changes associated with inactive disease and to demonstrate the extent and distribution of disseminated disease. In conjunction with conventional (bisphosphonate) bone scintigraphy it is more sensitive than labelled white cells for diagnosis of chronic osteomyelitis (page 44).

Further reading:

Seminars in Nuclear Medicine 1994; ***24****: 128–141*

European Journal of Nuclear Medicine 1996; ***23****: 459–65*

Clinical applications

Pyrexia of unknown origin (PUO)

A focal collection is identified with white cells in about 50% of patients with spontaneous PUO. The detection rate may be as high as 90% in patients with post-operative pyrexia but in the majority the collection is also found by ultrasound and CT. If white cell imaging is limited to patients with negative CT and ultrasound, the pick-up rate is less than 10% in both spontaneous and post-operative pyrexia. The majority of negative examinations are true negatives in that the patient has no focal pus collection. White cell scintigraphy is thus of clinical value in excluding focal collections missed by other imaging modalities (Fig. 7.4).

Further reading:

Seminars in Nuclear Medicine 1994; ***24****: 92–127*

Intra-abdominal inflammation

Accumulation of white cells outside liver, spleen and bone marrow is indicative of an active inflammatory processes (Fig 7.5). Indium-labelled cells should be imaged at 4 h and 24 h. When using technetium-labelled cells an initial abdominal image should be performed within 30 min of injection and repeated at 3–4 h. There is some indication in the early images of the majority of abnormal collections, which become more intense and easier to visualize by 3 h. Collections separate from the gastro-intestinal tract are often more clearly visualized at 18–24 h, but when using technetium-labelled cells the distribution of excreted activity in gastro-intestinal tract must be taken into account. There is almost always some suspicion of a focal

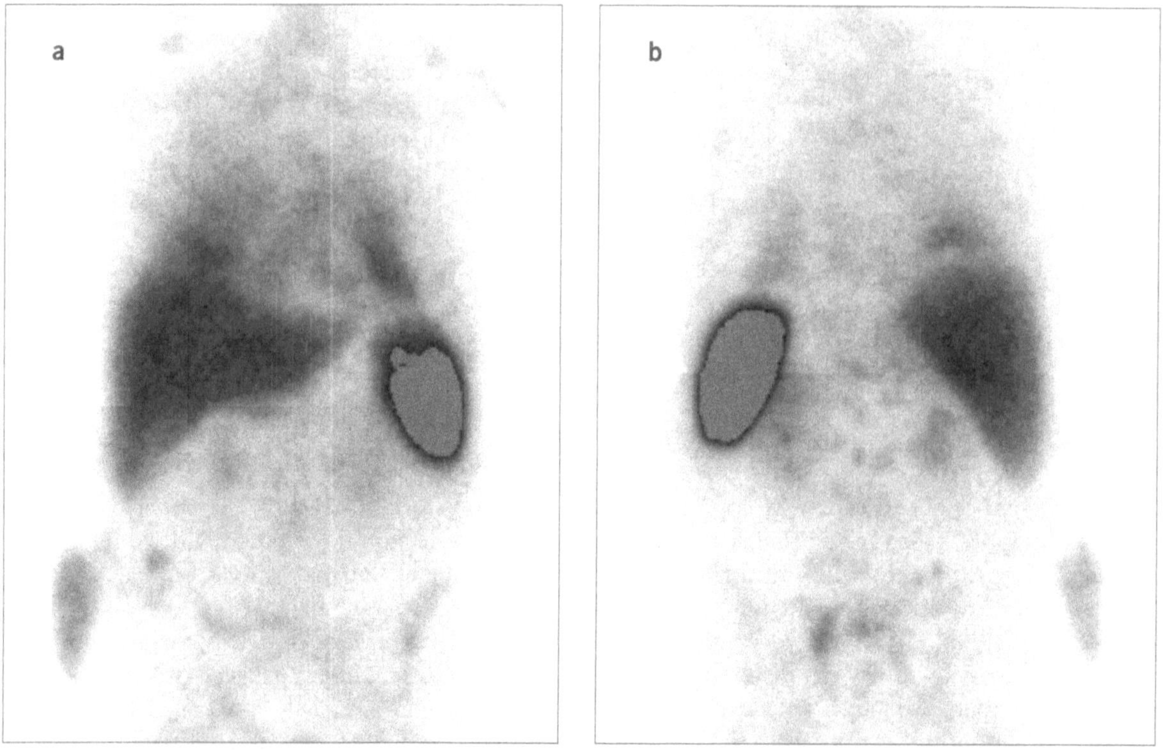

Fig. 7.4 Tc white cells (a) posterior (b) anterior composite images. Thorax showing uptake in the lower zones of both lungs. Consolidation had been present and unchanged at the right base for several weeks. There was doubt whether the intermittent pyrexia was due to continuing lung infection or an occult infection elsewhere. Note patchy distribution of uptake in spine. Catheter in bladder and drainage bag.

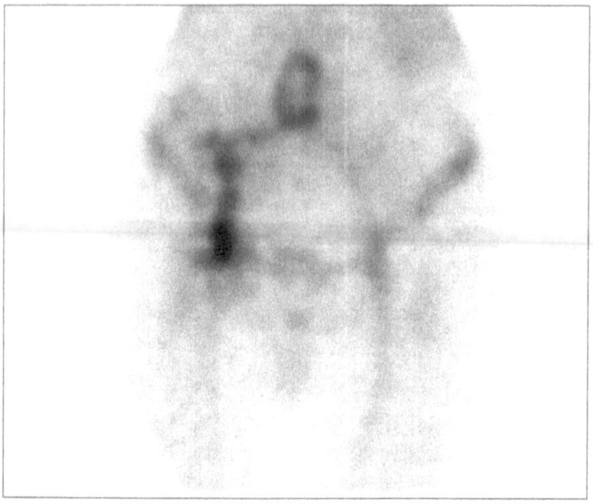

Fig. 7.5 Persistent discharging sinus right groin following aortic bifurcation graft. Infection extends to the aortic portion of the graft.

abnormality in early images, but contrast is often poor. It is necessary to identify abnormal accumulation in both early and late images to confirm intra-abdominal sepsis. Abnormalities seen only on later films must thus be treated with some reserve. If an extraperitoneal focus is suspected an additional image at 24 h is sometimes helpful as contrast continues to increase. At this time there is always substantial activity in bowel lumen, principally colon.

Further reading:

Seminars in Nuclear Medicine 1996; ***26****: 51–64*

Liver

Intrahepatic and perihepatic collections are difficult to differentiate from normal uptake, but in the normal liver uptake is lower at 4 h than 1 h after injection. In contrast, intensity of uptake increases in pathological collections. Activity is excreted via both hepato-biliary and renal pathways. This may be difficult to distinguish from abnormal collections, particularly in the post-operative patient with an assortment of drains and catheters. Uncomplicated pancreatitis is not usually associated with accumulation of white cells but these do concentrate if there is an infected collection.

Further reading:
Journal of Nuclear Medicine 1992; ***33****: 65–7*

Inflammatory bowel disease

White cell imaging with indium or technetium is commonly used to determine extent of disease activity in Crohn's disease of small bowel and in ulcerative colitis. There is a higher incidence of false negatives in Crohn's disease of large bowel than in the other conditions, in which false negatives are rare. Even so, most active large-bowel Crohn's disease is visualized with labelled leucocytes (Figs 6.14, 6.15). This discrepancy is unexplained but may be related to the observation that desquamated white cells are commonly absent in whole bowel perfusion studies of patients with Crohn's of the large bowel although white cells are almost always present in patients with small bowel Crohn's and those with ulcerative colitis. Differential diagnosis is normally made by biopsy but the occurrence of skip lesions and skip areas is indicative of Crohn's disease. White cell imaging does not identify strictures or fibrosis but does demonstrate the extent of currently active disease. It is therefore complementary to radiological contrast studies and endoscopy.

Further reading:
European Journal of Gastroenterology 1994; 6: 78–84

Osteomyelitis

White cell imaging is of little value in the axial skeleton because of variability in the normal uptake pattern in marrow and because high uptake in spleen conceals a variable length of the spine. Sites of infection may present either as photon-deficient or photon-rich areas. Photon-poor areas however also occur in normal marrow where there is no inflammatory focus. Assessment of active infection following insertion of a joint prosthesis is similarly complicated by the variable pattern of marrow regeneration. Gallium is often more useful than white cell imaging in prosthetic joints, particularly those which become painful months after insertion. Interpretation is subjective as the criterion is uptake of gallium 'out of proportion' to that of bone-scanning agent. There is however no clear definition of the limit of normal proportionality.

Cardiovascular infections

Vegetations on heart valves are often small and relatively avascular. The majority are readily observed angiographically or by ultrasound. White cell imaging is not a reliable technique for determining endocarditis although occasional positives have been reported. Uptake has also been observed around infected prosthetic grafts of the abdominal aorta both with white cells (Fig. 7.5) and with gallium. Gallium is however difficult to visualize in the abdomen because of variable and often extensive bowel retention. Indium-labelled cells have a role in diagnosis of mycotic aneurysms, permitting distinction between infection, thrombosis and haemorrhage which cannot be made by MR or CT.

Further reading:
Journal of Nuclear Medicine 1992; ***33****: 1493–5*

Amyloidosis

Amyloid deposits are associated with a number of conditions including long-standing inflammatory disease, plasma cell dyscrasias and long-term haemodialysis. Amyloid fibrils are chemically heterogeneous but associated with a pentraxin protein, serum amyloid P component, the function of which is unknown. This is a pentameric dimer composed of ten identical sub-units each with a molecular weight of approximately 250 kDa. It can be labelled with iodine isotopes and has been employed labelled with ^{123}I to image the distribution of amyloid, but may underestimate the extent of disease. Uptake of DMSA(V) (page 267) has also been described in amyloid deposits.

AIDS

AIDS (acquired immunodeficiency syndrome) is usually the result of infection with one of a small group of retroviruses, the human immunodeficiency viruses (HIV). Carriers may remain asymptomatic but infective for many years and the HIV status of patients being referred for investigation, often of unrelated conditions, is usually unknown. Transmission is parenteral or transmucosal. Infectivity is relatively low compared with hepatitis B, but there is as yet no immunization or cure. Provided that precautions recommended when handling unsealed radioactive sources and when dealing with blood are strictly observed, additional precautions should not be necessary.

Nuclear medicine has a limited role in evaluation of some consequences of AIDS or of its treatment. Pulmonary alveolar permeability (page 77) is increased in patients with *Pneumocystis carnii* pneumonia, and this may precede other evidence of infection. Unfortunately, although there is a clear separation between uninfected and infected non-smokers, there is no distinction between infected and non-infected smokers. The clinical use of this test is therefore restricted to non-smokers. Gallium uptake in lung is increased in individuals with *Pneumocystis* pneumonia. Imaging is usually performed 24–48 h after injection of 75–200 MBq ^{67}Ga. The typical appearance is uniform diffuse uptake in both lungs of greater intensity than in liver, heart and mediastinum appearing relatively photon-poor. There is however considerable individual variation in intensity of uptake. Low uptake in a subject in whom infection is confirmed carries a poor prognosis. Uptake may be more pronounced in the upper lobes, especially in subjects who have been treated with pentamidine, but a similar appearance is also seen in infection with atypical mycobacteria. Lung uptake of ^{111}In-labelled polyclonal antibody has also been reported in subjects with *Pneumocystis* pneumonia. *Cytomegalus* infection of lung is rare but may give a similar appearance to *Pneumocystis*. It is more commonly associated with high uptake of gallium in the eyes due to retinitis. Cutaneous lesions of cutaneous angiomatosis, due to a *Rickettsia*-like organism similar to that causing cat-scratch fever, and zidovudine-induced myositis are both associated with high uptake of gallium. This may be used to differentiate these conditions from Kaposi sarcoma, in which gallium is not usually taken up. The small number of cases reported suggest that thallium uptake is visible in Kaposi sarcoma within minutes of injection and persists for at least 3 h.

Opportunistic infections in other sites may occasionally require white cell imaging to localize a focus. Because of hazards to staff handling infected blood, alternatives to white cell labelling such as gallium or antigranulocyte antibodies should be considered. There is insufficient data on the use of the latter in AIDS for their role to be assessed. Gallium uptake in unsuspected lymphomas has been reported, as have unexplained 'false positive' accumulations. Sclerosing cholangitis due to cryptosporidiosis may be confirmed using hepato-biliary agents (page 248).

Neurological complications of HIV include primary and metastatic tumours, infections, thrombosis, embolism, haemorrhage and dementia. Investigation is no different to other patients in whom these conditions are suspected clinically. The pattern seen with cerebral perfusion agents in patients with AIDS complex dementia is of hypoperfusion in the frontal and parietal regions, often bilateral and associated with relatively increased uptake in the basal ganglia.

Further reading:

Nuclear Medicine Communications 1993; ***14****:* *830–48*

Platelets

Platelets may be labelled either with ^{51}Cr or ^{111}In. The latter has a shorter half-life, which is nevertheless long enough for studies of platelet turnover, a higher yield of gamma rays which permits imaging and a lower absorbed radiation dose per useful photon. Oxine also has a greater affinity for platelets. It has therefore largely replaced chromium, except when indium is not available. The method of labelling is initially similar to that when labelling white cells with indium oxine. The pH of the ACD solution must be adjusted to 6.5 to avoid activation of platelets which are kept throughout at 37°C. After separating red and white cells, platelet-rich plasma is drawn off and centrifuged at higher speed, 800–1300 G for 10–15 min, to separate platelets from plasma. The platelet pellet is washed without re-suspension in approximately 5 ml of phosphate-buffered saline and the pellet re-suspended in a further aliquot of phosphate-buffered saline to which sufficient indium oxine is added to give a 60 μmolar (approximately 5 μg per ml) concentration of oxine. After incubation for 10–15 min platelets are separated from residual unbound radioactivity and oxine by repeat centrifugation at 800 G for 5 min. The supernatant is discarded and the platelet button suspended in 5 ml of platelet-poor plasma for re-injection.

Further reading:
European Journal of Nuclear Medicine 1994; ***21****: 1141–7*

The best control that platelets have not been damaged is their biological behaviour. Normal platelets give an high concentration in spleen and blood pool. Damaged platelets concentrate equally in liver and spleen. There is however often some reversible damage, manifest as a plateau of hepatic activity persisting for about 30 min after injection and associated with a count rate in peripheral blood which initially falls and subsequently rises.

Clinical application

Plasma measurements allow survival time of platelets to be measured. The normal range is 7.3–9.5 days and recovery from plasma of the initial labelled platelets, calculated by extrapolation of the plasma curve back to zero time, ranges from 55–77% of the injected dose. The survival time is considerably reduced in idiopathic thrombocytopoenic purpura (ITP) in which survival time is hours or minutes rather than days and the platelet disappearance curve is exponential. Imaging allows sites of platelet destruction to be identified. This is usually the spleen, but in some patients with ITP destruction occurs in other organs. These patients do not benefit from splenectomy. Platelet labelling has also been investigated for identification of sites of venous thrombosis. Several small series suggest a fairly high accuracy in patients who are not receiving treatment but there is no visible accumulation in patients receiving heparin. This technique has therefore not achieved clinical acceptance.

Clot and thrombus

Fibrinogen

Clot can be labelled while it is forming if labelled fibrinogen is administered before the event. ^{125}I-fibrinogen is most widely used but ^{131}I and ^{123}I have also been employed. Because this does not label established clot, the technique is only applicable in prospective clinical trials. Counting is repeated at fixed points over the legs and compared with count rate over a reference area such as blood pool of the heart. This is a labour intensive investigation, ideally performed twice a day to obtain maximum sensitivity. Although a useful research procedure it is impractical for routine clinical use. Accuracy using ^{125}I for clots in the calf veins is over 90%. This falls to approximately 70% in the thigh because count rate measurable from ^{125}I is limited by the depth of veins below the surface. ^{125}I is not applicable proximal to the inguinal ligament. ^{131}I has been used but experience is limited and although ^{123} I has also been used its short half-life limits its application. Iodinated fibrinogen is no longer commercially available.

Antibodies

Monoclonal antibodies to various of the compounds present in clot and thrombus have been employed and some small series have reported a high success rate. However these compounds are not generally available and their true usefulness remains unknown. Overall there is at present no satisfactory direct radio-isotope method for detecting clot or thrombus.

Further reading:
Journal of Nuclear Medicine 1996; ***37****: 744–8*

Radionuclide venography

This can be performed by injecting ^{99m}Tc-MAA (page 57) into a distal vein in a foot, at the same sites as are used for radiographic venography. The technique is essentially similar and high sensitivity has been reported for femoral and iliac thromboses but not in the calf or popliteal veins. The technique is not widely used. Patency of vessels can also be demonstrated by blood pool scintigraphy. Although less sensitive than Doppler ultrasound or contrast venography the method is useful when these techniques are unavailable or contraindicated.

Further reading:

Journal of Nuclear Medicine 1991; ***32****: 2324–8*

Nuclear Medicine Communications 1993; ***14****: 1014–22*

Clinical Radiology 1994; ***49****: 382–90*

Lymphatics

A diffusible tracer injected directly into a cannulated lymphatic vessel distributes as if it had been injected by any other parenteral route. Non-diffusible soluble tracers remain within lymphatics, permitting lymph nodes to be visualized for an hour or more and lymphatics for a rather shorter period. This technique is little used as resolution is inferior to radiographic lymphangiography which is no more difficult to perform.

Non-diffusible tracers, especially small particle colloids injected subcutaneously, are cleared by the most proximal lymph nodes in their lymphatic pathway. In most cases the amount passing more distally is insufficient for useful diagnostic information to be obtained. Thus injection of a colloid into the web spaces of the foot gives visualization of popliteal, inguinal and external iliac but rarely of more proximal lymph nodes. The pattern of lymphatic pathways can also be visualized for several hours. Clearance from the injection site is accelerated by exercise (Fig. 7.6). Injection into a web space of the hand visualizes antecubital and lower axillary nodes. The internal mammary chain can be visualized by

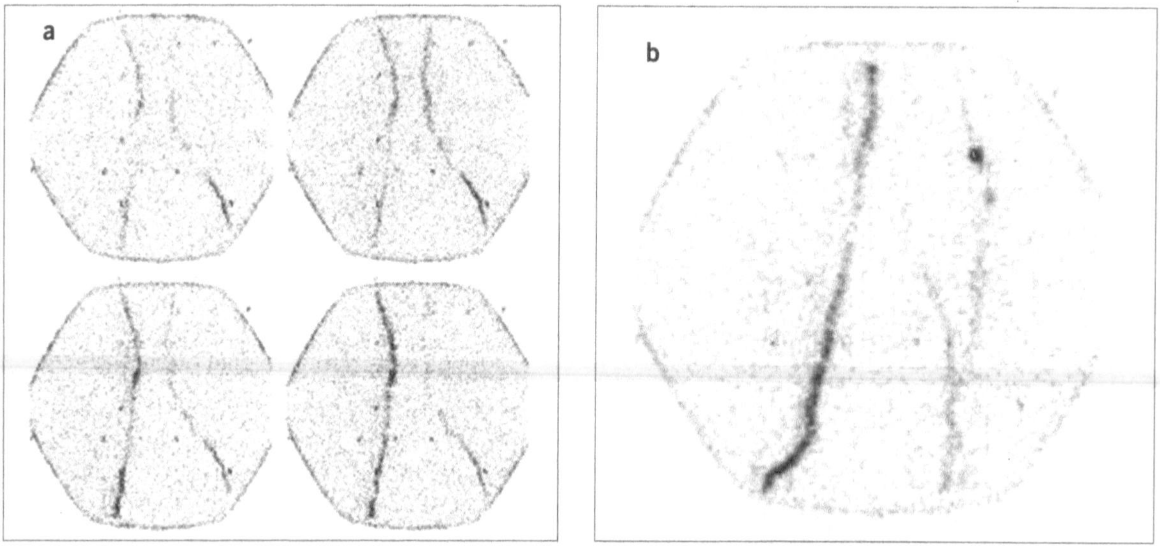

Fig. 7.6 Bilateral leg lymphogram in 18 year-old female; six-month history progressive swelling of left leg. Previous bilateral leg venograms were normal. (a) Anterior projection centred on knees: consecutive 10 min frames following injection of 20 MBq ^{99m}Tc millimicrospheres in 0.1 ml into the 1st and 4th web spaces of both feet. Normal right leg drains mainly via median vessels to popliteal fossa. Only lateral vessels fill on the left. (b) Similar projection. After walking briskly for 10 min there is augmented filling of the same vessels on the right and some small collaterals are seen. On the left there is filling of some additional collaterals and a popliteal lymph node but no augmentation. (c) 2 h later lymphatics are still visible and a number of additional lymph nodes are seen on both sides. (d) Posterior pelvis immediately after (c). The highest activity is in a right inguinal node. Setting this over-range reveals some external iliac node filling. Excreted activity in bladder. (**c,d** *overleaf*)

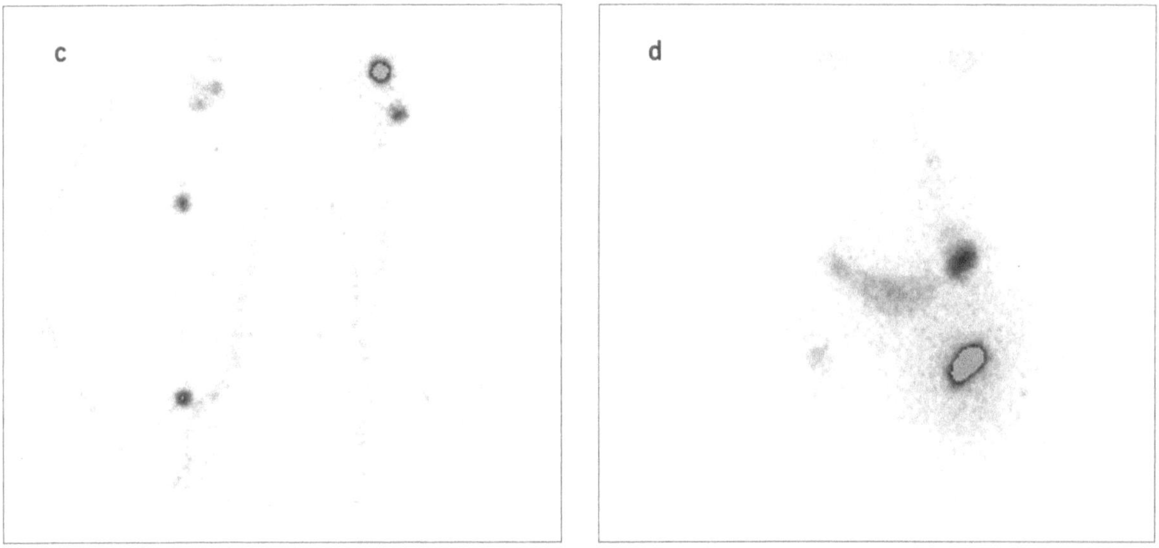

Fig. 7. *continued*

injecting colloid into the upper part of the rectus sheath or peri-areolar region of the breast. There is no filling of nodes which are replaced by infection or tumour. As the normal anatomy is variable the technique has not achieved general acceptance. It may however have an intra-operative role assisting localization of proximal draining (sentinel) nodes in melanoma and carcinoma of breast.

Plasma, extracellular space and volumes of distribution

In principle it should be simple to measure plasma volume by administering a label which remains intravascular and does not leak into red cells. Iodinated human serum albumin (RISA) labelled with either ^{125}I and ^{131}I is regarded as the reference compound and comes nearest to the ideal. Other large molecular weight proteins such as transferrin labelled with ^{111}In can also be used. This is however sequestered in liver and bone marrow and therefore has a volume of distribution larger than the plasma volume. There is no compound which remains wholly intravascular and is not metabolized. Nevertheless, provided that radiopharmaceutical purity of the tracer is carefully controlled and measurements are made under standardized conditions, for example at similar times after administration, reproducible results can be obtained with RISA. The normal plasma volume is 42.4 ± 4.7 ml/kg body weight.

Extracellular volume is less well defined and depends on the tracer used. Theoretically this should mix with extracellular water but not enter cells. In practice some regions, for example skeleton and fatty tissue, may equilibrate more slowly than others. There is no single true value for extracellular fluid volume. Using sodium ^{35}S-sulphate, a value of 14.1% ± 3.2% body weight is obtained in elderly males and 19.1 ± 1% in young healthy males. The initial volume of distribution of other tracers, for example those used for measuring GFR (page 100) is also a measure of extracellular fluid volume but not necessarily the same volume as measured with sodium sulphate. When reporting extracellular volume is therefore necessary to specify not only the tracer but also the duration over which measurements were made and the composition of the population studied.

8 Brain

Introduction

The role of nuclear medicine in investigating diseases of the central nervous system is currently in flux. Older techniques which demonstrated regional abnormalities in capillary permeability have been almost entirely superseded by x-ray computed tomography (CT) and magnetic resonance imaging (MRI), both of which relate pathology more precisely to adjacent anatomical structures. Tracers which demonstrate the distribution of specific neuroreceptors within the central nervous system are becoming available, but their clinical applications are not yet established. Currently interest is focused mainly on imaging regional cerebral metabolism. The effects of ischaemia become apparent rapidly in the brain because, despite its high rate of oxidative metabolism, it has effectively no reserves either of oxygen or of glucose and a very limited capacity for anaerobic metabolism. It is therefore critically dependent on maintenance of an adequate blood supply for its continued function. Oxygen and glucose consumption can be measured directly with positron emission tomography (PET) tracers but, because cerebral metabolism is dependent on immediate availability of both oxygen and glucose, blood flow and perfusion are clinically useful surrogates which can be measured with single-photon tracers. Blood volume and perfusion reserve can also be determined. All of these measurements can be made either globally or regionally and expressed either in absolute units or relative to other parts of the brain.

Normal values

At rest, whole normal brain perfusion in both man and experimental animals is 50–60 ml/min/100 g. Grey matter perfusion is 65–85 ml/min/100 g and white matter 27–33 ml/min/100 g. Limb weakness occurs when cerebral perfusion falls below 23 ml/min/100 g; EEG activity ceases below 17 ml/min/100 g and failure of the Na^+/K^+ pump, leading to cell death, below 10 ml/min/100 g. There is more variation between different neurones in the same individual than between species. Electrically quiescent but viable brain can exist adjacent to infarction. This may account for the sometimes dramatic improvement observed clinically when perfusion is restored. Using PET techniques, regions whose perfusion is in the marginal range between cessation of electrical activity and cell death can be identified by a local increase in both blood volume and oxygen extraction ratio.

Definitions

For the subsequent discussion to be understood, the proper definitions of four terms which are often used incorrectly must be stated.

Blood flow

This term should be used only when volume (of whole blood) per unit time, most commonly expressed as millilitres per second, is measured. It is however often used incorrectly, especially when referring to Doppler ultrasound measurements, which give velocity. Blood flow is not the same as perfusion (below).

Velocity

The dimensions of velocity are length (distance) per unit time and the direction of flow must be specified. This is thus not interchangeable with blood flow.

Perfusion

There is a subtle but vital difference between perfusion and blood flow. Perfusion refers to volume of whole blood per unit time *per unit mass of brain*, typically millilitres per second per 100 g of brain. The distinction between blood flow and perfusion is important because, although some techniques measure blood flow, many of those more commonly performed measure perfusion. It is rarely possible to define precisely the weight of brain being assessed, so that it is not usually possible to make a direct conversion between blood flow and perfusion.

Perfusion reserve

This is a measure of the capacity of the cerebral capillary bed to vasodilate in order to increase local blood flow, for example in response to increased metabolic requirements or a reduction in perfusion pressure. It is expressed in various ways, most commonly fractional change in blood flow per 1% change in arterial pCO_2 or mean transit time of blood through brain (page 230).

Blood flow and perfusion

Radiopharmaceuticals

Soluble tracers

Two types of soluble tracer are used for measuring blood flow; one variety remains intravascular whereas the other is diffusible. Measurements are made during the first transit of intravascular tracers. To calculate absolute flow it is necessary to measure both the volume in which the tracer is mixed, the *volume of distribution* and *mean transit time*. Freely diffusible tracers mix in a 'space' which is much larger than the blood volume. They are usually employed to measure perfusion and are assayed either during wash-in or wash-out. Very short-lived diffusible tracers measured at equilibrium during a constant rate infusion are a special case of diffusible tracers. Perfusion or flow can be derived from rate of wash-in or of wash-out, total quantity of tracer delivered or variation in quantity of tracer present with time.

Particulate tracers

In principle one of the simplest methods of measuring blood flow in any tissue should be to administer a tracer which is completely extracted on a single pass through the circulation and which subsequently neither redistributes nor washes out. Distribution is then proportional to the fraction of cardiac output reaching the tissue in question: the Sapirstein principle. Absolute flow can be calculated if cardiac output is measured simultaneously. Microspheres are the reference tracer for single-pass techniques. These are small insoluble particles which are trapped in the first capillary bed they meet and are therefore totally extracted, provided there is no arterio-venous shunting which bypasses all capillary beds. The rate of breakdown of particles or of elution of tracer from them must be slow compared with the time scale of measurement. This method is applicable in most tissues except those with a portal circulation, for example liver, posterior pituitary or adult skeleton (in some long bones periosteal vessels draining attached muscles make a significant contribution). In other sites this is one of the best reference methods.

An important limitation is the requirement for intra-arterial injection at a site which is sufficiently proximal to ensure complete mixing of particles with the output from the left ventricle. Ideally this should be left atrium, although in practice aortic root is often adequate. Intravenous administration is used for lung perfusion imaging, which demonstrates the distribution of right ventricular output in the lungs (page 64). The range of particle size must be maintained within narrow limits because this affects their distribution across the diameter of larger vessels, leading to preferential loss of larger particles into proximal branches. As the quantity of radioactive tracer carried by each particle depends on its size, a small number of very large particles will spuriously alter apparent distribution. Particles which pass through capillaries are subsequently taken up by the reticulo-endothelial system, further altering measured distribution. This can be determined *in vivo* by quantitative positron or single-photon emission tomography or whole body counting. Intra-arterial microspheres have been used in man and are safe provided that stringent controls are taken to limit the number of particles administered to less than one million. They should be 50–100 nm in diameter. This technique is technically demanding and requires meticulous attention to detail to obtain accurate and reliable results. It is principally a reference standard against which other and simpler techniques are validated. A number of clinically applicable methods are based on approximations to this method.

Lipophilic tracers

A number of soluble tracers have a high first-pass extraction by brain. These are lipophilic (fat-soluble) compounds of sufficiently low molecular weight to pass rapidly across normal cerebral capillaries, which are largely impermeable to passive diffusion by water-soluble molecules. The main disadvantage of such compounds is that, unlike microspheres, they are able to diffuse in both directions. During the first pass, diffusion is almost exclusively outwards from blood, but subsequently the concentration gradient is reduced and diffusion occurs in both directions. Unless back-diffusion can be prevented distribution represents blood flow only for a relatively short time following injection. As well as blood flow, uptake depends on the quantity and characteristics of lipid present as this affects partition coefficient: the ratio of solubility in lipid to that in water. The most useful measure of lipophilicity is the *octanol/water partition coefficient*, often expressed as a decimal logarithm of the ratio of solubility of the substance in octanol (a medium chain length alcohol) to its solubility in water. Any substance insufficiently soluble in octanol has a low extraction by brain on a single pass unless there exists a mechanism for active transport. If the octanol solubility is too high, much of the tracer dissolves in red cells, also reducing first-pass extraction efficiency. The optimal range in decimal logarithm units is 0.9–2.5. Brain has both substantial lipid content and high blood flow and there is therefore high single-pass extraction. Most other (extracranial) tissues have a lower lipid content or blood flow and therefore a lower uptake. In consequence recirculation complicates measurement.

The distribution of tracer within brain can be displayed by planar or, more commonly, tomographic imaging. Blood flow is calculated using the Fick principle, which states that the change in amount of a tracer in any volume of tissue over a measured time is equal to the product of the difference between arterial and venous concentrations and blood flow. To relate this distribution to blood flow the perceived distribution of count-rate must be corrected for attenuation, back-diffusion, scatter and cross-talk and related to administered activity and blood concentration. If only arterial and venous concentrations are measured (indirect Fick method), global flow is calculated. To obtain regional information it is necessary to employ the direct Fick method and image regional distribution quantitatively.

In vivo autoradiography

Autoradiography is one application of the Fick principle. If a freely diffusible non-metabolized tracer is added to the arterial supply of any tissue or organ, the concentration after any time interval in any small volume depends on the total arterial concentration during that time, local perfusion (flow per unit volume to that region) and effectiveness of the diffusion equilibrium between capillaries and tissue. This technique was originally developed *in vitro* but has been applied *in vivo* using quantitative PET or single photon emission computed tomography (SPECT) to determine the sectional distribution of tracer. Data must be collected over a short enough period for brain concentration to be effectively constant. This is more easily achieved with dedicated multi-headed systems than with a single rotating gamma camera. The accuracy of attenuation correction is an additional limitation of quantitative SPECT.

Compounds employed *in vivo* include a number of PET tracers such as ^{11}C-butanol and single-photon tracers including ^{131}I-iodo-antipyrene, ^{99m}Tc-propylene amine oxime (PAO), ^{123}I-isopropyl-iodo-amphetamine (IMP) and ^{123}I-hydroxy-iodobenzyl-propane-diamine (HIPDM). The two latter undergo complex metabolism *in vivo* including deiodination, with release of free iodine, necessitating thyroid blocking but brain uptake is due to the parent compound. Although initial uptake of IMP is blood-flow dependent, there is both specific and non-specific binding. Later images may represent distribution of amphetamine receptors so that it may be both a blood flow and a receptor labelling radiopharmaceutical. Both IMP and HIPDM (and possibly other compounds) are extracted by lung following intravenous injection. It is not clear to what extent the relatively constant uptake in brain which permits imaging is a fortuitous consequence of brain wash-out balanced by uptake of activity washing out of lung rather than fixation in brain, but the measured retention of IMP is higher than of exametazime or bicisate (see below).

To estimate arterial input, blood is drawn at frequent intervals from an in-dwelling arterial catheter or passed through a continuous-reading detector to obtain an arterial time/concentration curve. Allowance must be made for distortion of the curve due to dead-space in tubing between the artery and the measuring device. 'Arterialized' venous blood obtained by warming an arm to cause vasodilatation is sometimes employed, but it cannot always be assumed that the form of the arterial input curve to brain is the same as that at a peripheral site. The partition coefficient may be measured separately by maintaining a constant arterial concentration and making sequential measurements of regional brain uptake up to 1 h, by which time there is complete equilibrium between blood and brain. This is more accurate than assuming a uniform partition coefficient for the whole brain, which is an acceptable approximation only when low spatial resolution detectors are employed.

Further reading:

Journal of Nuclear Medicine 1995; ***36****: 531–6*

Equilibrium methods

A modification of this approach is to use a tracer which is fixed in brain, thereby reducing or eliminating back-diffusion and allowing uptake to be measured following bolus injection but when the concentration is relatively stable. If this can be combined with rapid external inactivation or removal, thus minimizing recirculation, the distribution following intravenous bolus injection correlates closely with blood flow. By analogy with microspheres these are sometimes called 'chemical microspheres'. A number of compounds have been developed for this purpose. Exametazime (HMPAO, Ceretec) and bicisate (ECD, Neurolite) are in clinical use, both labelled with technetium. IMP is less widely available and more expensive but has an higher extraction efficiency and better retention in brain.

Exametazime

The formulation used clinically comprises the d,l diastereomer of hexamethylpropylene amine oxime (HMPAO) labelled with ^{99m}Tc. This compound is sufficiently fat-soluble (log partition coefficient 1.90) to be taken up readily into brain, with an extraction efficiency of 77%. It is however comparatively unstable, converting *in vitro* at room temperature to more polar secondary complexes which are not lipophilic and do not cross the 'blood–brain barrier'. The speed of this change is such that the radio-pharmaceutical must be injected within 15 min of preparation unless stabilized. A number of techniques of stabilization have been described including the addition of 1 mmol gentisic acid or 0.4 mg sodium iodide to the eluant immediately after elution and before labelling of exametazime, or addition of 200 mg cobalt chloride hexahydrate in 2 ml water immediately after reconstitution of the kit. The mechanism of action of these stabilizers is uncertain but may be due to their ability to remove free radicals and thus inhibit radiolysis. Although widely used, none of these modifications is yet licensed for clinical use. Conversion of the lipophilic to an hydrophilic compound, the nature of which has not been established, occurs also *in vivo*. The product appears to behave *in vivo* like other water-soluble technetium compounds, diffusing into regions where the blood–brain barrier is disrupted. The rate of conversion is influenced by the pH of blood, occurring more rapidly at low pH.

This instability has two advantageous effects. When decomposition occurs within brain, activity which has been taken up is unable to diffuse out. Uptake is thus more closely representative of first-pass distribution than when using a tracer able to diffuse readily in both directions. However conversion is not instantaneous and appreciable wash-out can occur, particularly at high flow rates. The relationship between blood flow and uptake is thus non-linear and under-estimates high flow unless a correction is applied. The flow rates at which this becomes important are above those found in normal brain; the error becomes significant when imaging high flow lesions such as arterio-venous malformations (AVMs). Peripheral metabolism has the effect of reducing the amount available for recirculation, so that the distribution more closely resembles a true single-pass tracer. Although distribution is not identical to that observed with microspheres, the difference is small and for practical purposes is not usually significant. The disadvantage of this instability is that absolute, as opposed to comparative, measurements are difficult and in practice are not usually performed. *In vivo* autoradiography with exametazime requires measurement of changes in lipid-soluble, not total, arterial blood concentration in the period between injection and measurement of distribution in brain. Blood samples must be immediately extracted with ether and only the ether-extractable fraction measured. Although this technique has been used successfully, it is impractical for routine clinical use. For most clinical purposes imaging relative distribution of uptake which approximates to relative regional perfusion, rather than absolute measurement, is adequate. Within 48 h after injection approximately 40% of administered activity is excreted in urine and about 15% via the gall-bladder into gut, indicating the extent of conversion to the hydrophilic species peripherally.

Bicisate

Bicisate is ethylene dicysteine dimer. Its log partition coefficient (1.64) and extraction efficiency (60%) are lower than exametazime but the retention fraction is higher (72% compared with 50%). Brain retention is thought to be due to metabolism by an esterase to neutral mono-acid and di-acid products which are ionized at physiological pH and are therefore unable to cross back across the blood–brain barrier. The enzymes responsible for this reaction have not been identified. Unlike most tracers there are important differences between the metabolism of this compound in man and higher primates and its metabolism in other species, in which it is not retained in brain. Bicisate differs from exametazime in being stable *in vitro* for some hours after preparation. *In vivo*, elimination is principally urinary. There is appreciable wash-out from brain, resulting in a decrease of approximately 25% in brain activity 4 h after injection compared to peak uptake within the first few minutes. Nevertheless retention within brain is sufficient for imaging to be possible with conventional single-photon tomographic equipment. Sequential comparisons of bicisate and exametazime in human subjects have shown some differences in distribution. The explanation for these is not fully established but is possibly in part associated with water-soluble metabolites of exametazime which may diffuse into areas where the blood–brain barrier is abnormal. The clinical significance of these differences is not established.

Further reading:
Nuclear Medicine in Clinical Diagnosis and Treatment (ed. IPC Murray, PJ Ell) Churchill Livingstone, Edinburgh 1995 pp 457–518

Technique

No special preparation is required. Prescription medications should be continued. The distribution of cerebral blood flow varies rapidly depending on both physical and mental activity. Examinations should therefore be performed under controlled conditions. As neither bicisate nor exametazime redistributes it is necessary with these compounds to maintain control only until just after injection. With other tracers these conditions must be maintained throughout the period of measurement. For most purposes the subject should rest in a quiet room with a low level of lighting, low ambient sound level and minimal distractions. A small in-dwelling needle is inserted into a peripheral vein and the subject left for 5–10 min to relax and recover from any stress associated with venepuncture. Activity, typically 400–800 MBq of the ^{99m}Tc complex, is administered through the prepared access, which should not be withdrawn for 2–3 min, to allow tracer to be cleared from the circulation. The needle may then be removed and the patient positioned for imaging. For ictal examinations (see below) injection should be made whilst the patient is fitting, or at worst within 30 s of the end of a fit. For post-ictal examinations the injection should be administered 5–10 min after a fit. Inter-ictal examinations should be performed with the patient under basal conditions and at rest. SPECT is essential for all clinical purposes. The highest available resolution should be used (high resolution collimators, 128 × 128 matrix) and acquisition time kept as short as practicable to minimize patient movement. Attenuation correction should be employed if available.

Further reading:
Journal of Nuclear Medicine 1994; ***35****: 359–63, 2003–10*

Interpretation

Reconstructions are most commonly displayed in transverse (orbito-meatal) and coronal planes but additional projections such as sagittal or through the temporal lobes are sometimes helpful. The normal pattern of relative regional perfusion in a young adult is demonstrated in Fig. 8.1 (Plate III). There is symmetry about the mid-line in most brain regions, but the pattern is dependent on age. Cortical metabolism and perfusion decrease with age, especially in the frontal lobes. There is more inter-subject variability in the temporal, parietal and occipital lobes and in all age groups there is relative hypometabolism in the left temporal lobe relative to the right. No systematic gender differences have been reported. It is therefore necessary to compare with age-matched controls. The distribution depends on three independent variables: distribution of blood flow, partition coefficient of tracer between blood and brain at each point, and efficiency with which tracer is retained. Mental stimulation, whether visual, auditory or intellectual is associated with measurable local increase in metabolism and perfusion. A further complicating factor with exametazime is the possibility of uptake of water-soluble breakdown products in areas where there is a disturbance of the blood–brain barrier. Caution must therefore be exercised not to over-interpret the images.

Further reading:
Journal of Nuclear Medicine 1992; ***33****: 696–703. 1995;* ***36****: 1141–9*

Diaschisis, reduction in local blood flow remote from the site of focal brain injury, can exhibit a number of patterns, for example cortex ipselateral to a thalamic lesion, thalamus ipselateral to a cortical infarct, the hemisphere contralateral to a supratentorial infarct, visual cortex distal to a lesion anywhere in the visual pathways or contralateral cerebellar hemisphere following supratentorial infarcts in the carotid distribution. Crossed cerebellar diaschisis is also observed during the Wada test (internal carotid artery injection of 3 mg/kg sodium amytal in order to lateralize speech dominance). The phenomenon may be associated with functional interruption of cerebral motor fibres.

Further reading:
Journal of Nuclear Medicine 1995; ***36****: 399–402*

Wash-out methods

A number of substances approximate sufficiently to the requirements for a diffusible tracer to be clinically useful, although none is ideal. They including isotopes of the noble gases xenon and krypton and many organo-halogen compounds. Only xenon is commonly used clinically. The average log partition coefficient for xenon between blood and whole brain is 1.15 in subjects who are not anaemic; the coefficient for normal grey matter is 0.8 and white matter 1.5. These values are altered in anaemia depending on its severity, in the presence of hypothermia and in hyperlipidaemia. Highest accuracy is achieved by techniques which measure the partition coefficient in each volume element (voxel) and do not assume uniform global values. Two isotopes of xenon have been employed: ^{133}Xe and ^{127}Xe. The former emits gamma-rays principally of 81 keV. At this energy 2 cm of soft tissue approximately halve the detected count rate. Measurements are therefore disproportionately influenced by radioactivity in the scalp and more superficial parts of the cerebral cortex. ^{127}Xe emits more and higher energy gamma emissions per disintegration than ^{133}Xe (Appendix 1) and gives a lower absorbed dose per detected photon, but is not widely available. Because of its higher energy, thicker detectors surrounded by heavier shielding are required, hampering positioning of multiple detectors close to the head. Equipment designed for use with ^{133}Xe is in general not suitable for use with ^{127}Xe. Xenon may be administered dissolved in saline, either by intravenous or intra-arterial injection or by inhalation. Intravenous administration is now rarely used because the solubility in water is relatively poor and up to 70% of administered activity is exhaled on the first pass through the lungs.

Theoretical principles

Perfusion, calculated as the ratio of partition coefficient to half-time of wash-out, can be determined by fitting one or more exponentials to the experimental wash-out curves. The mathematical derivation for this relationship is given in the reference cited at the end of this section. The slope of a single exponential fitted to the first 2 min, usually called the initial slope index (ISI) is the wash-out rate from regions having higher blood flow. This is clinically useful because errors inherent in fitting two exponentials to clinical data often outweigh the theoretical improvement in precision from fitting two exponentials to curves obtained over 15 min. This procedure, known as curve stripping, depends on the assumption that, because tracer is eliminated from different tissues at different rates, by the end of the study the faster compartment has washed out. Counts in the later part of the curve are considered to originate from extracranial structures and white matter which, having lower perfusion than grey matter, lose tracer more slowly. After subtracting this component, a second exponential function can be fitted to the residue, from which the half-time of the faster compartment is obtained. If measurements are extended to 40 min three exponentials can be fitted, ascribed to grey matter, white matter and extracranial structures but which correlate poorly with measurements in dogs using microspheres.

Many curves can be fitted by the sum of two or more exponential functions. Fitting a mathematical function to an experimental curve is not evidence that the curve parameters correspond to any physiological process. Moreover separation of an experimental curve into two or more components does not yield a single unambiguous solution, but depends upon the sections of the curve taken. The distinction is fairly straightforward in normal subjects, in whom flow through grey matter is much higher than through other compartments. However as grey matter flow falls and the difference between compartments decreases 'curve stripping' becomes progressively more subjective. Other errors include the assumption that partition coefficients are homogeneous. This is approximately correct in normal brain and when using detectors with poor spatial resolution but is invalid with high resolution systems and in pathological areas. Compton scattered radiation, which results from incomplete absorption of some emitted gamma radiation originating within the brain, tends to affect white matter curves to a greater extent than those from grey matter, leading to an overestimate of white matter perfusion. Despite the many potential sources of error there is a good correlation between planar and tomographic methods.

An alternative approach is to use the Kety–Schmidt equation, dividing the change in count-rate by the corresponding change in area under the wash-out curve. Unlike curve stripping, which predicates a compartmental model of the circulation comprising as many 'compartments' as there are exponentials, the latter does not assume any particular model of perfusion. The disadvantage is that a longer period of measurement, at least 10 min, is required.

Further reading:

Problems of Intracranial Pressure in Childhood (ed. R.A.Minns) MacKeith Press, London 1991 pp 77–122

Technique

The inhalation method is usually employed. No preparation is required. The subject rests comfortably in a quiet room with subdued lighting. Xenon is supplied using a mouthpiece or, if this is not tolerated, a well-fitting face mask, as for lung ventilation (page 60). The subject inhales for 4–5 min from a closed circuit containing sufficient oxygen (usually as air), a CO_2 absorber and xenon. Perfusion can be calculated from the slope of either uptake or wash-out curves. During the wash-in phase, scatter from inhaled activity in the naso-pharynx is an important source of error. Measurements are therefore usually made during wash-out while the patient inhales room air. Exhaled air should be exhausted to minimize contamination. It is necessary to correct for recirculation of isotope and for radiation scattered from the respiratory tract into the field of view of the brain detectors. Correction for the former may be made from the curve obtained with a detector sampling expired air, to measure end-tidal radioactivity. This reflects the concentration in arterial blood recirculating to brain. The latter is minimized by discarding points on the head curves before end-tidal radioactivity has fallen to 20% of its peak level. This usually takes less than 1 min but, as the distinction between fast and slow components is greatest in the early part of the curve, this reduces accuracy especially when perfusion is low. The highest count rates originate from grey matter, which has higher perfusion and greater solvency for xenon than other structures. Xenon-enhanced computed x-ray tomography utilizes the same principles but as scatter from the oropharynx is not a source of error measurements are usually made during wash-in. An inhaled concentration of 20–35% of stable xenon is required, at which levels its anaesthetic properties may alter regional blood flow.

Perfusion can also be measured from wash-out following intra-arterial injection of ^{133}Xe into a carotid or vertebral artery. The greater part of activity in venous outflow is exhaled as blood passes through the lungs. Very little of the bolus recirculates or reaches the rest of the body. This eliminates the major errors inherent in inhalation or intravenous administration but is no longer widely used, partly because of the need for arterial puncture but principally because only the region perfused by a single vessel can be measured at each injection. Although this might appear an advantage, there is considerable individual variation and overlap in vascular territories. The intravenous technique is now obsolete as most of the bolus is wasted because it is exhaled before reaching the brain.

Administered activity depends on the number, size and sensitivity of the detectors and the intended duration of measurement. It may range from 100–1000 MBq ^{133}Xe. Larger initial activities are required to maintain adequate statistics to the end of a 15 min than a 2 min wash-out because only a small percentage of the original activity remains. The highest activities are required for SPECT. The simplest equipment consists of a pair of detectors fitted with cylindrical collimators, one positioned to view each hemisphere. Regional information can be obtained by using multiple detectors, up to ten per hemisphere, either positioned in parallel planar jigs or, more commonly, in an helmet-shaped holder. Even larger and more complex arrays have been employed. An additional detector continuously samples air drawn from the mouthpiece in order to construct the recirculation correction curve. End-expiratory concentration can be assumed to be in equilibrium with arterial. End-tidal pCO_2 can be measured using the same sample.

As more detectors are employed better information is obtained about regional differences in perfusion at the cost of increased absorbed radiation dose as each detector is smaller and receives counts from a smaller volume of brain, thus necessitating an higher administered activity. The most detailed anatomical information can be obtained using multi-detector systems encircling the head or a gamma camera. The former may oscillate and rotate, making repeated measurements up to six times per minute in multiple projections from which tomographic sections can be calculated by fitting one or two exponentials as when using stationary detectors, although there are fewer points to each curve. The same corrections are made. Tomographic cross-sections showing the distribution of perfusion are reconstructed using standard filtered back-projection algorithms. An important advantage unique to this technique is that, because only slope of the curve is required and not absolute quantity, attenuation correction is unnecessary. The relatively small number of counts originating from deeper structures limits useful spatial resolution to regions greater than 3 cm in diameter.

Further reading:
Seminars in Nuclear Medicine 1985; ***15****: 347–56*
Journal of Nuclear Medicine 1988; ***29****: 348–55*

Steady-state measurements

The continuous inhalation steady-state method devised for use with older PET equipment, which was not capable of making images in rapid sequence, is a variant on this. ^{15}O-labelled water, which is assumed to be freely diffusible between blood and brain, is employed as the tracer. The patient inhales air containing constant tracer quantities of carbon dioxide (at a level which does not affect cerebral blood flow) labelled with ^{15}O. In the presence of carbonic anhydrase, conversion of carbon dioxide to carbonic acid is virtually instantaneous in lung and oxygen in CO_2 is effectively in equilibrium with that in water. Because of the short (123 s) half-life of ^{15}O, after a period of three to four half-lives (about 8 min) an equilibrium is reached such that the rate of arrival of labelled water at the brain is equal to the combined rate of loss by wash-out and radioactive decay. The concentration and distribution of radioactivity thereafter remain constant and a single image of distribution at equilibrium may therefore be obtained, taking as long as necessary to obtain adequate statistics. The other measurement necessary is of arterial blood concentration of ^{15}O.

The relationship between perfusion and tissue tracer concentration is linear only for a tracer with an infinitely short half-life and falls off as the half-life of the radioisotope increases. Moreover the assumption that water is freely diffusible is not correct. For oxygen, uptake is virtually independent of flow at perfusion rates greater than 100 ml/min/100 g. There is a good correlation between this method and intra-arterial microspheres for flows between 10–100 ml/min/100 g. It is therefore most useful for measuring low or normal flow but is unreliable when perfusion is increased. This technique can also be employed using SPECT with ^{81m}Kr. However because of the low solubility of krypton in aqueous solution the infusion must be given into the aortic root, limiting applicability.

Further reading:
Journal of Nuclear Medicine 1994; ***35****: 1878–9*

Autoregulation and reactivity

Under normal conditions blood flow to brain responds rapidly, both globally and regionally, to changes in local metabolic requirements but is unaltered over a wide range of arterial pressure. As input pressure falls, resistance vessels increase in calibre, maintaining flow by reducing vascular resistance, whereas a rise in arterial pressure is accompanied by an equivalent increase in resistance. This has the incidental effect of altering local blood volume, which can double or halve relative to the mean resting value of 4 g per 100 g brain. No further adaptation can occur once vessels are fully dilated or constricted. Thereafter a passive relationship is maintained between pressure and flow. Reflex regulation is impaired for a time following head injury, stroke and at all times in brain tumours which have morphologically abnormal blood vessels unable to respond to the usual stimuli. Increasing inspired pCO_2 to between 5–7% also increases blood flow, by on average 75%, whereas a fall is associated with a corresponding decrease. Similar changes are also effected by alterations in vascular resistance of brain. The responses to hypercapnoea and hypoxia are better preserved following head injury and stroke than are responses to changes in arterial pressure. The immediate stimulus is probably altered pH of extracellular fluid, a rise in pCO_2 producing vasodilatation and a fall vasoconstriction. In hypercapnic states, as in hypotension, once there is maximal vasodilatation no further compensation is possible and a passive relationship between pressure and flow ensues. Cerebral blood flow also responds to changes in pO_2. Inhaling 100% oxygen at normal atmospheric pressure reduces global perfusion by 13% in normal adults, whereas decreasing inspired oxygen concentration to 10% increases blood flow by 35%. Acute hypoxia over-rides the vasoconstrictive effect of hypocapnia. Perfusion increases in chronic hypoxia, at least partly due to an increase in the number and diameter of capillaries and arterioles, whereas basal levels are reset in hypertension.

This is the physiological rationalization for the limited clinical value of cerebral blood flow measurement in cerebrovascular disease. Failure is neither gradual nor progressive. The circulation compensates until conditions are grossly abnormal, when there is a sudden catastrophic failure. Correlations between arterial pCO_2, vascular resistance and perfusion form the basis for tests of cerebral vascular reactivity, which indicate residual capacity of the cerebral circulation to respond to further insults. This is expressed as fractional change in perfusion relative to alterations in $paCO_2$ induced by oral administration of 250–500 mg acetazolamide, a drug which blocks carbonic anhydrase, or inhalation of air enriched to a CO_2 content of 5%. Reactivity is also influenced by blood pressure, age and sex . In contrast to normal brain, regions where flow is maintained by reflex vasodilatation exhibit a reduction in perfusion on hypercapnia due to 'steal'. This may be sufficient to induce symptomatic ischaemia. No change in a region of low perfusion implies that irreversible changes have occurred, such as infarction.

Further reading:

Cerebral Blood Flow (ed. JH Wood) McGraw Hill, New York 1987 pp 402–12

Perfusion reserve and mean cerebral transit time

Patients with cerebro-vascular disease are usually asymptomatic as long as they are able to autoregulate. Any symptoms are as likely to be due to emboli from an atheromatous plaque in the carotid as to ischaemia. In consequence ultrasound imaging of the carotid bifurcation supplemented by a measure of residual capacity of the cerebral circulation to respond further is more useful than blood flow or perfusion measurement to stratify risk and demonstrate the extent to which flow is sustained by vasodilatation. Studies employing PET to measure simultaneously regional oxygen consumption, blood flow and blood volume have shown that the most sensitive index of the extent to which the cerebral circulation is maintained by vasodilatation is the ratio of blood flow to blood volume (cerebral vascular reserve, CVR). This may be measured by determining the response to an increase in $paCO_2$ by separate consecutive assays of blood flow or perfusion (above), or by measuring the ratio of blood flow to volume.

Theoretical principles

The reciprocal of the ratio of blood flow (but not perfusion) to volume, that is the ratio of volume to flow, is the mean transit time (MTT) of blood through any tissue. This can be measured directly provided the first pass of a bolus is distinguishable from recirculation. It eliminates the need to measure either blood volume or blood flow. MTT is the mean time any small volume of blood spends in that region during each circulation. If flow is unchanged then a longer transit time indicates a larger blood volume, that is vasodilatation. However, unless bolus dispersion can be prevented before it enters the region under investigation, only MTT of the bolus from the start of injection, no matter how remote from the site of measurement, to the time it leaves the field of view of the detector is measured, not that through the region visualized by the detector alone. In practice any intravenous bolus always spreads as it passes through the heart and lungs and is diluted into a progressively larger volume, different parts of which traverse pathways of different lengths. The mean from arrival of the leading edge of the bolus in the field of view to its final departure is mean residence time, a parameter in itself of no physiological importance. MTT from start of injection and including the measurement site is obtained by adding the time between start of injection and arrival at the detector to the calculated mean residence time. Correction for dispersion of the bolus, in order to calculate mean cerebral transit time (MCTT), can be achieved in two ways: either by convolution of the cerebral time–activity curve with the incoming arterial curve (the input function) or, more simply, by subtraction. MTTs are additive. MCTT can therefore be calculated if MTT of the bolus as it enters the cerebral circulation is subtracted from MTT including the cerebral circulation.

When the blood–brain barrier is intact, any soluble tracer, for example sodium pertechnetate, can be employed. If this barrier is disrupted a larger molecular weight tracer such as albumin or red cells labelled with technetium may be preferable, provided that sufficient activity can be obtained in a sufficiently small volume (less than 1 ml) to achieve an adequate bolus. When tracer remains intravascular the transit time of blood is calculated. If the volume of distribution is greater than the plasma volume, MTT of the larger volume is obtained. These 'volumes of distribution' are mathematical abstractions with no anatomical equivalent. The relationship between flow and transit time is valid for all tracers, but different values of transit time are obtained if tracers with different volumes of distribution are employed. This technique is impractical in atrial fibrillation and other conditions which slow the central circulation, for example right-to-left or left-to-right shunts, aortic or mitral stenosis or incompetence or low output failure, in which the first transit may be so prolonged that it cannot be clearly separated from recirculation.

Further reading:

European Journal of Nuclear Medicine 1991; ***18****: 171–7, 259–264*

Technique

The head is positioned on the face of a gamma camera fitted with an high-sensitivity collimator. An 18-gauge cannula is inserted into a large antecubital vein to facilitate transfer of a tight bolus into the superior vena cava. It is not possible to influence the bolus beyond this point by injection technique. One method of achieving a good bolus is to attach a syringe containing the radioactivity to one inlet of a three-way tap and a 10 ml syringe of saline to the second. A tube large enough to contain the pertechnetate bolus is inserted between the saline-filled syringe and cannula. The three-way tap is used to fill the tubing gently with radioactivity, which is then flushed rapidly into the vein with the saline. Lying the patient supine maximizes venous return and accelerates transport of activity from superior vena cava into right atrium. A poor bolus injection results in the separate phases merging into a single indistinct and indeterminate blur.

The aortic detector, a 1–2 cm diameter probe fitted with a simple cylindrical collimator, is positioned on the sternum 2–3 cm below the sterno-manubrial joint, to view aortic arch and record a time–activity curve as the bolus enters the extracranial portion of the cerebral circulation. It would be preferable to measure MTT of arterial input at the base of the brain, but because of the physical proximity of carotid, vertebral and jugular vessels and overlap in time of arterial and venous curves following intravenous injection, arterial and venous curves are rarely sufficiently distinct distal to the arch. The error introduced by use of aortic arch is small because distance between the two is comparatively short and the vessels are of large diameter. Two clearly separated peaks are usually distinguishable, the first due to activity passing through superior vena cava and the second aorta (Fig. 8.2). An intermediate pulmonary artery peak is occasionally identified. Arrival time and first transit through the head are identified by plotting count rate in the whole head region against time. Mean residence time of the arriving bolus is calculated, for example by fitting a gamma variate function. The interval between start of injection and start point of fit must be added to obtain mean transit time. However as the difference between MTT at the arch and that at the head is the required parameter, it is not necessary to identify start time accurately, provided that the same start time is used both for head and aortic curves.

Because of statistical limitations data are acquired at a pixel size of 8–10 mm (64 × 64 matrix not zoomed), at three frames per second for 60 s from the start of injection. The vertex projection gives the best overall view of cortex. A

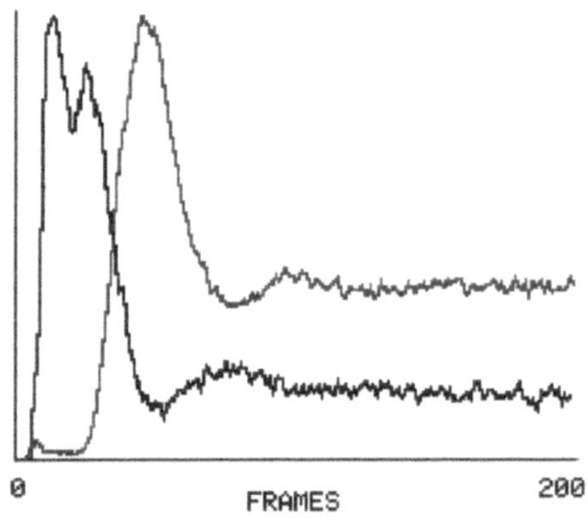

Figure 8.2 First-pass curve at the aortic arch (black) and brain (grey) following rapid bolus intravenous injection of an intravascular tracer. The first peak of the black curve is due to tracer passing through superior vena cava, the second aorta. Note progressive widening of the curve as it moves further from the site of injection. This is a general finding.

lead collar placed around the neck, which must be extended and over the shoulders minimizes scattered radiation from the trunk and considerably improves image quality. The rest of the body must be out of the field of view of the camera. Posterior or anterior projections can be employed if neck extension is not possible, but under these circumstances the principal vascular territories of cortex are partially superimposed. The lateral view provides a good distinction between arterial territories but does not permit simultaneous evaluation of both sides unless a dual-head camera is available. There is always a certain amount of 'break-through' of activity visualized from the contralateral side. Depending on age and condition of the patient, frame duration varies between 0.3–3 s. Sets of frames are added so that the first pass is condensed into about 20 frames. Summation is necessary to improve statistical quality; even so the maximum in any single pixel derived from summed frames is usually less than 150 counts. Dead-time correction is usually necessary for both aortic and head curves. Losses as low as 10%, by selectively reducing peak values, substantially increase calculated MTT and thus systematically overestimate MTT in regions with shorter transit times. A time–activity curve is produced for each pixel, correction made for recirculation and mean residence time of the first pass calculated. The interval from start of injection to arrival is added to give MTT of each pixel; the calculated value of MTT from start of injection to aortic arch is subtracted to obtain a value of MCTT for each pixel. When calculating mean transit time start time of fit must be calculated separately for every single-pixel curve. The resultant values are stored as a new matrix which can be displayed as a parametric image after interpolation to 256 × 256 (Fig. 8.3, Plate IV) using a colour scale with 30 discrete levels, each colour change calculated to represent a 0.5 s difference in mean transit time in the range 0–15 s. The speed of the first pass precludes SPECT but as the greater part of perfusion is to cerebral cortex and there is appreciable attenuation of counts originating from deeper structures, planar imaging in the vertex projection allows good representation of the distribution of MCTT in cortex, but not of brain stem or other deep structures. If absolute values are to be compared it is desirable to measure end-tidal pCO_2 at the same time, as there is a linear relationship between it and MCTT, which increases by 1.6 s for each 1% rise in end-tidal pCO_2.

Regional permeability (blood–brain barrier)

Regional variations can be demonstrated using any diffusible water-soluble tracer. The principle is now more widely used for contrast enhanced CT and MR imaging. Sodium pertechnetate (400–800 MBq) is used most commonly, preceded 15 min earlier by 200 mg potassium perchlorate orally or followed by 200 mg sodium perchlorate in 10 ml water for injection intravenously, to block uptake by choroid plexus. Technetium glucoheptonate may give higher contrast between tumour and background but not between infarct and background. However the use of these tracers has now been largely superseded by CT and MR with contrast, which give similar information but with more detailed anatomical correlations. Abnormalities of regional permeability (disruption of the blood–brain barrier) are usually imaged 1–4 h after intravenous injection of the radiopharmaceutical.

Further reading:
*Imaging 1992; **4**: 185–191*

Technique

Injection should always be associated with a dynamic study to demonstrate regional MCTT (see above). Equilibrium imaging should be started not less than 1 h and at any time up to 4 h after injection. With all agents pick-up rate increases as the interval lengthens, but beyond 1 h the difference is small. A minimum of four projections must be obtained: anterior, posterior and both laterals. A vertex view is useful to confirm any suggestion of a lesion near the mid-line, whereas obliques are occasionally helpful to confirm that an apparent abnormality is superficial. Sensitivity is improved by SPECT. In the anterior projection the patient is positioned with the orbito-meatal line at a right angle to the face of the collimator. The posterior projection requires the head to be flexed as far as possible consistent with contact with the collimator in order to display the posterior fossa. When positioning for lateral projections care must be taken to include all of the posterior fossa. Images are collected for a pre-set time, typically 200 s. The vertex projection is not obtained as a routine, but is useful in selected cases. The neck of the patient must be extended as far as possible and a lead collar positioned carefully to shield the camera from activity in the trunk, as in the dynamic study. If the projection is acquired looking directly down onto the unextended head, scattered radiation from the trunk greatly degrades the quality of the image, reducing contrast and thus concealing lesions.

Normal appearance

A normal four-projection equilibrium study is shown in Fig. 8.4. This was obtained with pertechnetate and perchlorate, but the appearance with any other agent is similar. Orbits are identified in the anterior projection, whilst in the posterior sigmoid sinus is often asymmetrical. Unilateral hypoplasia or absence of one sigmoid sinus is a common variant of no pathological significance. If the head is well flexed the posterior fossa is seen below the sigmoid sinuses but if adequate flexion is not possible the posterior fossa may be visualized only in lateral projections. Abnormalities show as regions of increased count rate (Fig. 8.5)

Clinical applications

Stroke

Infarction is associated with reduced flow. Dynamic imaging with any intravenous tracer reveals asymmetry of flow in almost two-thirds of patients with recent stroke, but also in up to one-quarter of asymptomatic subjects. Inflammation and hyperaemia around an infarct may initially produce greater rather than reduced flow (luxury perfusion) on the side of the lesion. On follow-up this resolves, leaving permanently reduced flow on the affected side. CT is required to distinguish between haemorrhagic and ischaemic stroke, in particular between subarachnoid haemorrhage and forms of stroke not requiring surgical intervention.

Parametric imaging of regional MCTT demonstrates adaptive changes in perfusion pattern (Fig. 8.6, Plate IV), improves the accuracy of clinical classifications of severity and prognosis after ischaemic stroke and indicates both the extent of compromized but potentially recoverable brain and of regions which although asymptomatic, are approaching the limits of their vascular reserves (Fig. 8.7, Plate IV). MCTT longer than 11 s is associated with an increased oxygen extraction ratio, indicating that the limit of autoregulation has been reached and decompensation is imminent.

The size and severity of perfusion defects following stroke can also be demonstrated with xenon or relative regional perfusion SPECT (Fig. 8.8, Plate IV). Duplicate studies before and after inhalation of 5% CO_2 or administration of a carbonic anhydrase inhibitor such as aceta-

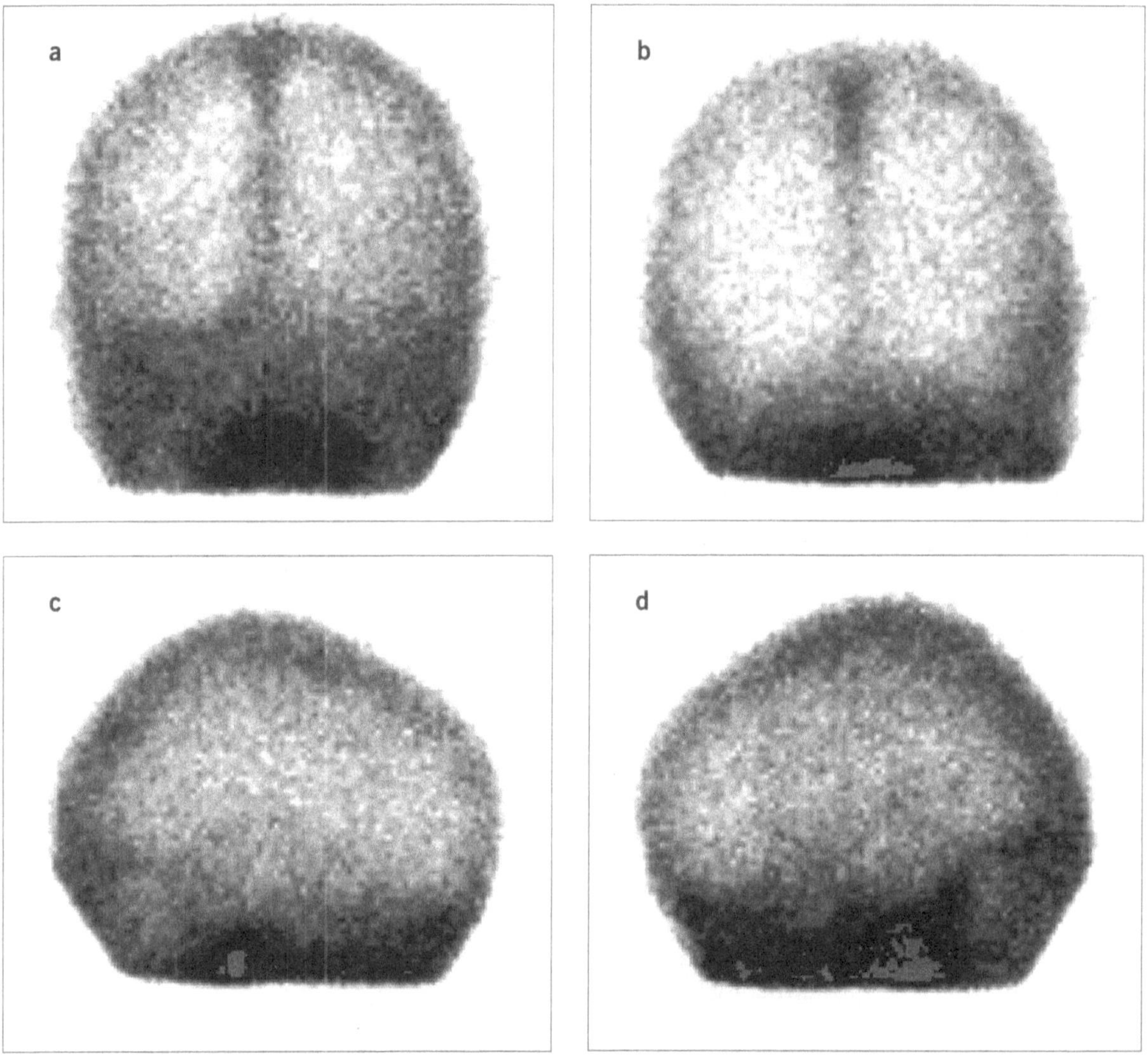

Figure 8.4 Normal pattern of permeability images. (a) Posterior, (b) anterior, (c) right lateral, (d) left lateral.

zolamide have been employed experimentally to demonstrate perfusion reserves. SPECT may also reveal appropriate defects in patients with cortical blindness and in Moya Moya disease despite absence of CT or MRI abnormalities.

Further reading:

European Journal of Nuclear Medicine 1994; ***21****: 455–65*

Blood–brain barrier tracers show increased uptake in an infarcted area within 24 h. This reaches maximum intensity at about 10 days then fades, usually becoming imperceptible by 8–10 weeks (Fig. 8.9). Uptake of permeability tracers, visualized on planar scintigraphy, SPECT, MRI or CT, occurs after irreversible infarction of relatively large volumes of tissue. Many small cortical, posterior fossa, lacuna and internal capsule infarcts are not seen. Not all recent infarcts exhibit luxury perfusion but its presence indicates that an infarct is less than three weeks old. Uptake on the equilibrium views implies that an infarct is not more than ten weeks old. If clearly seen it is likely to be more recent. In the absence of established active therapy, imaging techniques have a limited clinical role.

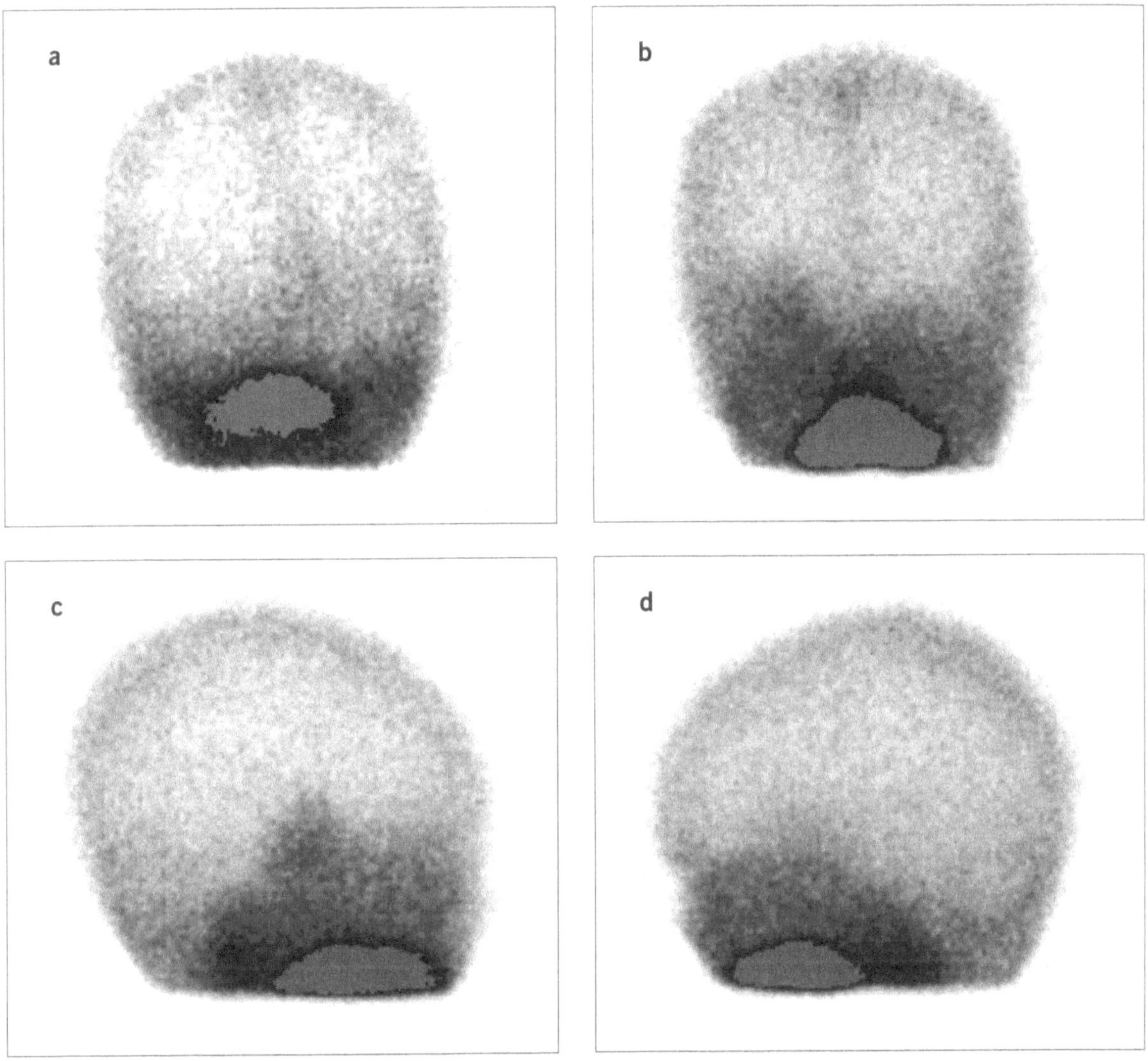

Figure 8.5 Comparable projections showing right temporal tumour.

Further reading:
*British Journal of Radiology 1980; **53**: 1174–6*
*European Journal of Nuclear Medicine 1994; **21**: 189–90*
*Journal of Nuclear Medicine 1996; **37**: 419–20*

Vascular malformations

Arterio-venous malformations (AVMs) are readily demonstrated by parametric imaging of MCTT, which may demonstrate a more extensive disturbance in perfusion pattern than the angiographic appearance of the angioma suggests (Fig. 8.10, Plate V). Follow-up MCTT studies after stereotactic radiotherapy provide a good marker of response to treatment. Similar findings are observed with SPECT of relative regional perfusion. SPECT after temporary inflation of an occlusive balloon has been suggested as a means of assessing haemodynamic effects of therapeutic embolization before irreversible changes have occurred.

Further reading:
*Journal of Nuclear Medicine 1993; **34**: 1243–5*
*Nuclear Medicine Communications 1994; **15**: 461–8*

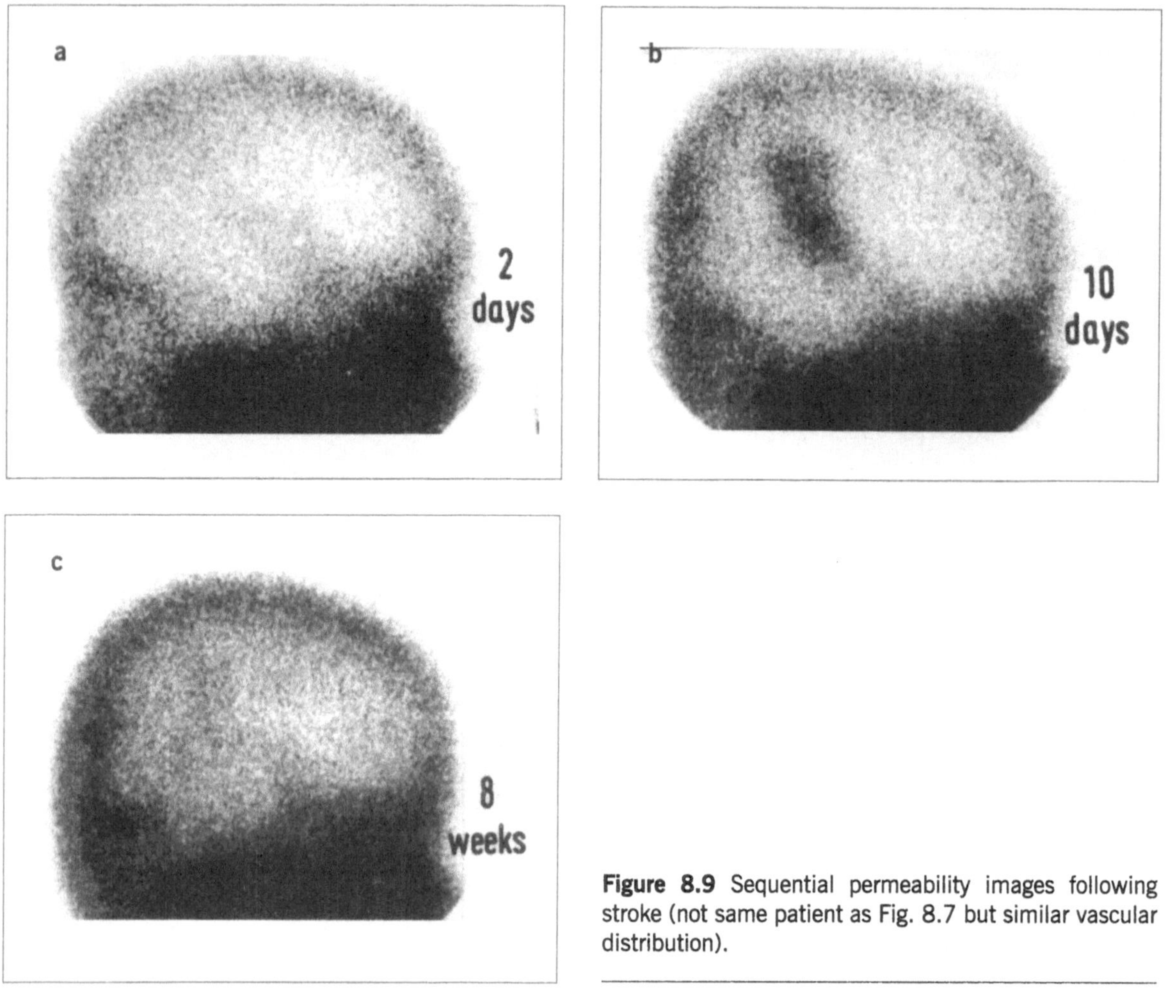

Figure 8.9 Sequential permeability images following stroke (not same patient as Fig. 8.7 but similar vascular distribution).

Brain death

Absence of cerebral perfusion may be demonstrated by dynamic imaging following bolus injection of water-soluble tracers such as pertechnetate or by imaging lipophilic tracers such as exametazime or bicisate. Although absence of perfusion is accepted as contributory evidence of brain death in some countries, this is not a recognized criterion in the United Kingdom.

Further reading:

European Journal of Nuclear Medicine 1991; ***18:*** *75–7*

Epilepsy

Epilepsy is characterized by recurrent seizures consequent upon abnormal electrical discharges arising within the brain. These affect neighbouring neurons to produce further discharges, so that abnormal electrical activity is propagated. Epilepsy may be generalized, with multiple origins of abnormal electrical discharges or partial, in which the abnormality starts from a single focus, commonly within the temporal lobe. Resection of a correctly identified focus may abolish seizures or reduce their frequency in approximately 80% of subjects. Between 35% and 40% of patients with partial epilepsy are resistant to medical treatment and may thus be candidates for surgery.

The origin of abnormal activity is in many cases localized by mapping EEG or magnetic potentials. Imaging is unnecessary in most patients. Measurement of regional perfusion or metabolism is helpful in selected cases, principally those with complex partial seizures, either because the diagnosis is difficult to establish by other methods or because fits have proved refractory to medical management and surgery is contemplated. Interictal foci of reduced glucose consumption identified by ^{18}F-fluoro-deoxy-glucose (FDG) PET may correspond in position to sites of slowed or abnormal EEG rhythms. Perfusion also is reduced at affected sites interictally but increased ictally (Fig. 8.11, Plate V), a finding which can be demonstrated using SPECT as well as PET. Crossed cerebellar hyperperfusion is a common finding in ictal scans and is sometimes useful in assessing lateralization. The majority of foci identified have no corresponding CT or MRI abnormality although pathological examination of the excised specimen typically reveals lesions which are smaller than the metabolic defect. Not all areas of reduced metabolism are epileptogenic and there is no correlation between the magnitude of metabolic and electrical abnormalities. The significance of individual lesions must be confirmed, for example by electroencephalography. The accuracy of interictal FDG PET is about 85%, compared with a sensitivity and specificity for interictal SPECT in temporal lobe epilepsy of approximately 65%. Ictal relative regional perfusion SPECT, injected during or within 30 s of the end of a fit, has been claimed to give results as good as or better than interictal PET, but in practice it is not possible to obtain a diagnostic study in up to one-third of patients. Meticulous technique and considerable experience are required for reliable results. If injected more than 30 s after the fit, but within 5 min, sensitivity falls to 72%. In view of the low prevalence of intractable epilepsy this technique is best confined to specialist centres.

Further reading:

Journal of Nuclear Medicine 1994; ***35****: 1094–6. 1996;* ***37****: 426–9*

Head injury

Following head injury, relative regional perfusion SPECT demonstrates cortical abnormalities in some patients with normal CT or MRI and in others may show more and larger defects. Relative regional perfusion SPECT in patients with symptoms persisting 1–10 years after whiplash injury, typically from a road traffic accident, may show hypoperfusion localized to the watershed zones between territories of large cerebral arteries. Patients in whom defects were found remained symptomatic for longer than those with a normal SPECT study. About one-quarter of defects were no longer seen on follow-up at three months and most patients became asymptomatic, whereas symptoms persisted in the majority of patients in whom relative regional perfusion abnormalities persisted. The distinction was less clear in patients who had received a moderate head injury than following minor trauma. Most available reports describe non-randomized studies of patient groups, often without controls, performed weeks or months after injury. Generalizations from such material must be interpreted with caution. A finding of defects on SPECT may be a criterion to predict if further rehabilitation is possible and to distinguish patients with organic post head injury symptoms. A normal SPECT study reliably excludes clinical sequelae following mild head injury.

Chronic subdural haematoma has a characteristic pattern when viewed in profile in a dynamic study, often present even when equilibrium views are normal. It comprises a crescentic area of absent perfusion peripherally which persists throughout the study into the venous phase and is bounded medially by a rim of increased perfusion due to compressed brain. This is usually seen most clearly in anterior or posterior projections. The region should remain avascular throughout the study, unlike infarcts which commonly fill in to some extent in the late venous phase, possibly due to slow retrograde or collateral flow. A subdural can be identified in equilibrium views as a peripheral biconvex photon-rich area, usually seen in one projection only, most commonly the anterior or posterior. In other projections the edge of the haematoma tapers gently and is not clearly defined. It is thus difficult to visualize on planar projections.

An acute subdural haematomata presents a similar scintigraphic appearance to the chronic form. The 'rim sign' may be more evident in the dynamic study and if present definitive treatment should not be delayed to obtain confirmation from late equilibrium views, which are often inconclusive. CT has replaced scintigraphy for investigation of acute head injury as it enables assessment of the extent of contusion and any penetrating injury, as well as the presence or absence of a subdural collection.

Further reading:
Journal of Nuclear Medicine 1996; ***37****: 1605–9*

Glioma

The probability of detection with permeability tracers depends upon histological grade, size and site of the tumour. Small, low-grade, well differentiated tumours are least likely to be detected by planar imaging, especially if situated near the floor of the middle fossa or in the posterior fossa. Poorly differentiated tumours are rarely missed. Vascular tumours (and arterio-venous malformations) are identified by early filling of veins and superior sagittal sinus on the dynamic study. It is not possible to distinguish between the various types of hypervascular pathology on the basis of radionuclide angiography alone. Non-visualization of the superior sagittal sinus is abnormal. It is a rare finding, but when seen is diagnostic of superior sagittal sinus thrombosis. The most characteristic feature of a glioma is an area of increased uptake more or less circular in the lateral projection but wedge-shaped in the anterior or posterior, apex close to the midline with the base of the wedge towards the surface of the skull. Occasionally a rim of tumour shows greater uptake than the centre: the doughnut lesion. This is evidence of a tumour with an avascular or necrotic core but is not diagnostic of glioma, as an identical appearance is seen with some infarcts. Gliomas large enough to be distinguished scintigraphically are almost invariably solitary. Multiple abnormalities are likely to be emboli, abscesses or metastases rather than glioma. The only glioma to have a characteristic scintigraphic appearance is that arising from the corpus callosum: the butterfly tumour. This enlarges asymmetrically to both sides of the mid-line. Increased uptake is therefore visible on both lateral projections, but may easily be missed on planar anterior and posterior views if the tumour does not project beyond the superior sagittal sinus. A vertex view is often helpful to confirm extent. Sensitivity is substantially improved by SPECT, but the superior resolution of CT and MR have largely rendered scintigraphy obsolete.

Uptake of ^{201}Tl has been employed as a non-invasive method of grading gliomas, the least differentiated tumours having the highest uptake. Uptake is partly a function of cell mass but is also substantially influenced by blood flow and condition of the blood–brain barrier. Thallium has also been advocated to differentiate recurrence from post-radiotherapy changes. Thallium uptake has however also been reported in infarcts. Its specificity as a marker of tumour viability in brain must therefore be regarded as suspect. Uptake of sestamibi (page 116), penetreotide (page 173) and various labelled amino acids has also been described but experience with these agents is too limited for a clinical role to have been defined.

Further reading:
Journal of Nuclear Medicine 1993; ***34****: 2089–90*

Acoustic neuroma

These benign tumours concentrate pertechnetate, but nevertheless are usually too small to be visible on planar imaging, lying as they do close to the floor of the middle fossa. When large enough to be detected by scintigraphy, they usually require craniotomy for removal. Isotope imaging has been superseded by CT and MR.

Meningioma

These arise from the meninges but their shape is variable. Over the vault they are most commonly found close to the superior sagittal sinus and appear round in some projections. Meningioma en plaque is not uncommon and, although clearly visible as an area of increased uptake in at least one projection, cannot be resolved from bone. Other common sites include the greater wing of the sphenoid and olfactory ridge. The majority concentrate pertechnetate and appears as a well defined area of markedly increased uptake. Pentetreotide uptake has also been observed in all investigated to date and there is high uptake of ^{201}Tl in the majority, although the rate of washout is variable. Meningioma should be suspected if there is a solitary, well defined abnormality in any of the common sites. Definitive diagnosis requires biopsy.

Further reading:
Journal of Nuclear Medicine 1995; ***36****: 403–10*

Metastases

Almost any malignant tumour may metastasize to brain. The commonest primary sites to behave in this way are bronchus, breast and kidney. Metastases may be solitary, in which case there are no scintigraphic features which permit their differential diagnosis, or multiple in which case differential diagnosis is from abscess or emboli. Commonly more metastases are present than can be identified scintigraphically. There is little correlation between scintigraphic size or activity of a tumour and its actual size. Scintigraphic activity is affected by vascularity, extracellular water content and capillary permeability and is most commonly an indicator of oedema surrounding a deposit. The false negative rate is high, many tumours not being detected. However, cerebral scintigraphy remains a useful clinical test because those lesions which are not detected are unlikely to respond to simple palliative measures such as steroids, whilst there is a reasonable prospect of a favourable but temporary response to treatment of those tumours which are detectable. The sensitivity of scintigraphy is increased by SPECT.

Further reading:
British Journal of Radiology 1980; ***53****: 1174–6*
Seminars in Nuclear Medicine 1987; ***17****: 214–29*

Dementia

Dementia is a clinical syndrome characterized by persistent impairment of multiple cognitive capacities. Memory is always defective; disturbances of language function and visuospatial skills are common. Mutations in the APP gene cause some cases of Alzheimer's disease, which accounts for 50–60% of cases of late-onset cognitive deterioration. The differential diagnosis includes degenerative, vascular, traumatic, demyelinating, infectious, inflammatory, hydrocephalic, neoplastic, metabolic and toxic conditions. Characteristic neuropsychological deficits include impaired ability to learn new information, decline of language function and deterioration in visuospatial skills. There are frequently changes in spontaneous behaviour with disengagement, indifference and diminished affection occurring early. Paranoia with persecutory delusions occurs in up to 50% whereas agitation is common in the later stages. Depressive symptoms are common but although severe depression is unusual, in one series 8% of British patients who were initially diagnosed with dementia were subsequently determined to have major depression. Two major neuropathological lesions are associated with Alzheimer's disease, extraneuronal amyloid plaques resulting from proteolytic processing of an amyloid precursor and intraneuronal neuro-fibrillary tangles originating from paired helical filaments of microtubule protein. Histological changes are most striking in the medial temporal lobe and neocortex of the temporo-parieto-occipital junction. The most characteristic neurochemical change is a severe deficiency of choline acetyltransferase, which catalyses synthesis of acetyl choline. This follows neuropathological alterations in the nucleus basalis of Meynert, a basal forebrain nucleus.

Vascular dementia

This is a clinical syndrome of intellectual decline produced by ischaemic, hypoxic, or haemorrhagic brain lesions. Diagnosis requires dementia to be associated with cerebrovascular disease manifested by neurological signs and with neuroimaging evidence of stroke or ischaemic brain injury. There must be a temporal relationship between the dementia and vascular disease. Occlusion of a vessel, which may produce a variety of cognitive deficits depending on the site of the ischaemic damage, is the most common cause of vascular dementia (Fig. 8.12, Plate VI).

Alzheimer's disease

Bilateral cortical metabolic deficits in the posterior temporo-parietal regions have been shown in Alzheimer's disease by PET scanning with ^{18}F-FDG. Cortical perfusion defects similar to the PET findings are disclosed on relative regional perfusion SPECT. Frontal and temporal lobe defects are also commonly seen in relative regional perfusion scans, but these are less specific for Alzheimer's. The areas usually affected are the association cortex, parts of the frontal, temporal and parietal lobes involved in higher-order associative thought and memory functions. The primary sensorimotor cortex along the central sulcus, visual cortex in the occipital lobes and cerebellum are usually relatively spared by Alzheimer's. The distribution of relative regional perfusion defects in Alzheimer's disease is similar to that of the degenerative changes shown histopathologically, in areas where plaques and neurofibrillary tangles are found in high density. The posterior temporo-parietal perfusion defects are the most consistent sign of Alzheimer's on relative regional perfusion SPECT, particularly when symmetrical. In many patients, the temporo-parietal defects extend without a break from high over the hemispheres posteriorly into the temporal lobes medially to give a pattern which has been described as resembling an hockey stick. Over 80% of patients with dementia in whom this symmetrical pattern is seen have Alzheimer's but only about two-thirds of patients with Alzheimer's have this relative regional perfusion pattern. When présent in patients with memory loss or dementia, provided that Parkinson's disease with dementia has been excluded the probability of Alzheimer's exceeds 90%. Temporo-parietal defects are not unique to Alzheimer's. The differential diagnosis includes Alzheimer's, Parkinson's disease with dementia, multi-infarct dementia, carbon monoxide intoxication and bilateral parietal infarcts.

Bilateral temporal and posterior parietal abnormalities on relative regional perfusion SPECT (Fig. 8.13) are specifically associated with Alzheimer-type cognitive deficit but common perfusion patterns occur in Alzheimer's and Parkinson's diseases when the pattern of cognitive impairment is similar, localizing the site of cognitive dysfunction generally to the parieto–temporal cortex. Bilateral temporo-parietal perfusion defects are strongly associated with a clinical diagnosis of Alzheimer's disease. Patients with amnestic states such as Korsakoff psychosis, rather than cognitive defects, have different patterns with frontal perfusion deficits but preserved temporo-parietal function. Central cholinergic stimulation with physostigmine produces a focal increase of uptake in the posterior parieto-temporal region in patients but not controls. Most of the differences between Alzheimer and vascular dementia and controls can be detected by both PET and SPECT but only the former correlates with severity of dementia. Patients with depression may have a pattern comprising selective frontal, central, superior temporal and anterior parietal relative regional perfusion abnormalities. In contrast, patients with Alzheimer's demonstrate abnormal relative regional perfusion in four regional networks, cognitive impairment correlating most closely with disrupted parieto-temporal topography. Only two-thirds of individual patients are correctly identified using these parameters.

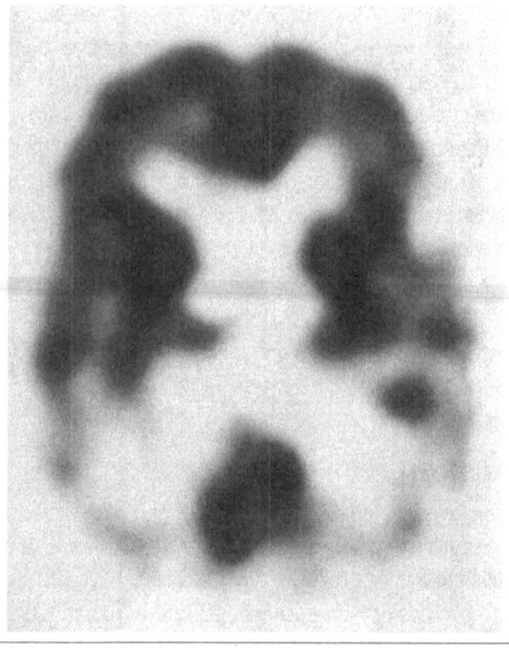

Figure 8.13 Biparietal defects in Alzheimer's. [Reproduced with kind permission of Dr D.M.Hadley]

Perfusion defects involving the frontal lobes or medial and antero-lateral parts of the temporal lobes occur frequently but in a more variable pattern which is less specific for Alzheimer's than bilateral temporo-parietal defects. Substantial asymmetries in the pattern of perfusion defects are also common and tend to be slightly worse on the left side. Unilateral temporo-parietal and isolated frontal defects are found in approximately 20% of Alzheimer's patients. Other aetiologies, most commonly stroke, can cause asymmetric or unilateral perfusion defects; defects which are not temporo-parietal are thus non-specific and neither confirm nor exclude Alzheimer's disease. Other large cortical defects are even less specific, with only an 18% probability of being associated with Alzheimer's, whereas multiple small cortical defects are generally not due to Alzheimer's. MRI may be helpful to determine whether perfusion defects are due to infarction but is often relatively normal in areas involved with Alzheimer's. Computed tomography is less sensitive for small infarcts than MRI and is thus less useful for establishing a diagnosis.

Pre-senile dementia

Posterior temporo-parietal defects are generally more severe in pre-senile dementia (occurring before the age of 60 years) than in senile dementia, occurring after the age of 60 years. Cerebral perfusion is often globally but non-specifically depressed in very severely demented patients. Relative regional perfusion SPECT may be used to confirm the diagnosis of Alzheimer's and may be helpful for excluding other potentially arrestable, reversible or treatable conditions such as multi-infarct dementia. The distinction between multi-infarct dementia and Alzheimer's can be difficult. In some patients both multi-infarct dementia and Alzheimer's may coexist.

Frontal dementias

These are typically associated with frontal signs such as behavioural changes and with bilateral frontal reduction in relative regional perfusion. The usual SPECT pattern consists of unilateral or bilateral frontal lobe perfusion defects, in contrast to the posterior temporo-parietal defects of Alzheimer's. Pick's disease is a rare frontal dementia in which there are generally diffuse perfusion defects in both frontal lobes; similar changes are seen in progressive supra-nuclear palsy.

Human immunodeficiency virus (HIV) dementia

The acquired immunodeficiency syndrome is associated with infection of the central nervous system by HIV and progressive dementia. It is also known as AIDS (acquired immunodeficiency syndrome) dementia complex or AIDS-related dementia. It may eventually develop in the majority of patients with symptomatic AIDS. The cortical pattern of relative regional perfusion is heterogeneous with multiple scattered perfusion defects. In some patients the disease may produce a large focal defect referable to specific clinical symptoms.

Parkinson's disease

Parkinson's disease is a progressive, disabling neuro-degenerative disorder characterized clinically by tremor at rest, bradykinesia, rigidity and postural instability. Of patients with Parkinson's disease, 40% have overt dementia and up to 70% have some degree of cognitive impairment. Pathologically the idiopathic form is characterized by degeneration of dopaminergic neurons in the substantia nigra, with an 80–90% reduction in striatal dopamine concentration; D1 and D2 dopamine receptors are normal. The neurotoxin 1-methyl-4-phenyl-1,2,3,6-tetrahydropyridine (MPTP) also destroys pre-synaptic neurons. In striato-nigral degeneration there is loss of post-synaptic dopaminergic neurons. The earliest abnormality in idiopathic Parkinson's disease is reduced tracer binding to presynaptic dopamine transporters in the anterior and particularly the posterior putamen. Replacement with levodopa or receptor agonists effectively reverses motor deficits. There is an inverse correlation between levodopa dosage and D2-receptor density, most prominent in the least dopamine-depleted region, the ventral caudate; reduced receptor density in anterior putamen may be associated with dyskinesia. Subsequent disabilities may result from drug-induced adverse effects, progressive dysfunction due to continued degeneration of nerve terminals or various non-motor, non-dopamine-responsive symptoms. [^{123}I]β-CIT [2β-carbomethoxy-3β-(4-iodophenyl) tropane] (RTI-55) labels (presynaptic) dopamine transporters which are located on the terminals of dopamine neuronal projections from the substantia nigra to the striatum and is thus a marker for neurons which degenerate in Parkinson's disease. Binding in the caudate nucleus and putamen is reduced also as a function of normal ageing. ^{123}I-iodobenzamide (IBZM) and a number of other tracers label post-synaptic D2 receptors. This allows functional resolution of clinically similar symptomatology due to abnormalities in structures that are only microns apart. The clinical role of these and related tracers and the optimal methodology for their clinical use is not yet established.

Cortical and regional uptake ratios of relative regional perfusion in non-demented patients with Parkinson's disease are not significantly different from controls, but when associated with dementia there is generalized cortical hypoperfusion. Uptake is reduced in the frontal and basal ganglia regions. The only asymmetrical finding in hemi-parkinsonian patients is relative hypoperfusion in the contralateral parietal region, possibly due to deafferentation of the thalamo-parietal pathways.

Further reading:
European Journal of Nuclear Medicine 1994; ***21****:1–5*
Journal of Nuclear Medicine 1995; ***36****:384–393, 1196–1200*

Substance abuse

Although multiple metabolic effects of alcohol on the brain have been demonstrated *in vitro*, clinically relevant changes have not been observed using either PET or SPECT. PET studies in chronic cocaine abusers have demonstrated reduced striatal post-synaptic dopamine receptor binding and reduced presynaptic striatal dopamine precursor uptake. Alterations have also been recorded in relative regional perfusion, glucose metabolism and dopamine receptor function. Both PET and SPECT studies have shown global hypometabolism and relative regional perfusion SPECT small focal abnormalities in men but not women. The frontal and temporal lobes are most severely involved. This may be related to clinical affective changes. Small defects have also been reported in relative regional perfusion SPECT of amphetamine abusers. These were not related to the severity of psychiatric disturbance or admitted dose.

Further reading:
Journal of Nuclear Medicine 1995; ***36****: 1298–1300*

Migraine

Blood flow is reduced in the affected hemisphere during attacks of classical (with aura) but not of common (without aura) migraine, in the periventricular region and centrally in the frontal region in children with the so-called 'attention deficit disorder'.

Further reading:
Seminars in Nuclear Medicine 1985; ***15****: 347–56*

Cerebrospinal fluid

Bromine partition test

This is a simple and sensitive investigation to distinguish tuberculous from other forms of meningitis which are usually diagnosed on the basis of cerebrospinal fluid (CSF) examination. The bromide partition test is indicated in patients with a clinical suspicion of meningitis but normal CSF findings as tuberculous meningitis can occur despite a normal CSF sugar concentration. 2 MBq sodium or ammonium bromide labelled with ^{82}Br is administered orally. Two days later a lumbar puncture is performed and 5 ml CSF, uncontaminated with blood, is collected. At the same time a 5 ml venous blood sample is taken and the radioactivity in both counted. The ratio of activity in serum to that in an equal volume of cerebrospinal fluid should be greater than 1.9:1. In the presence of tuberculous meningitis increased permeability results in a lower ratio. The ratio is not elevated in viral meningitis, although some cases come in the equivocal range between 1.6:1 and 1.9:1.

Further reading:
British Medical Journal 1972; ***4****: 413–5*

Cisternography

The intrathecal route is approximately one hundred times more sensitive to the presence of pyrogens than the intravenous. Substances passed as safe for intravenous injection are not necessarily safe intrathecally. The radiopharmacist preparing the injection must be informed if intrathecal use is contemplated.

Any soluble inert tracer injected intrathecally into the lumbar theca in a normal subject will pass upwards into the basal cisterns within a few hours. The tracer may briefly enter lateral ventricles before passing over the surface of the hemispheres, where it is absorbed through arachnoid granulations. Within 24 h of injection all residual tracer is located over the surface of the hemispheres. Two preparations have been widely used, ^{111}In-DTPA and ^{169}Yb-DTPA, although neither is currently licensed for this purpose. A dose of 7–15 MBq of either is injected intrathecally at any convenient level. To avoid the risk of inadvertent injection into an extrathecal collection of CSF the examination should be postponed if there have been any attempts at lumbar puncture within the previous two weeks. The syringe containing the tracer should be flushed with CSF. As with all examinations involving lumbar puncture the patient must remain horizontal for 8–12 h to minimize CSF leakage from the puncture site. Unlike earlier preparations such as iodinated human serum albumin, there is no radiation danger to the cord if the injection is extrathecal as both of the preparations specified are rapidly absorbed from an extrathecal site and excreted by glomerular filtration.

Imaging should commence 2–4 h after injection into the lumbar theca or within 1 h of cisternal puncture. By this time activity should have reached the basal cisterns. Earlier imaging may be used to demonstrate CSF leakage (Fig. 8.14). Lateral ventricles are sometimes visualized, particularly if they are dilated. Two projections of the head, posterior and lateral, are adequate to determine the location of radioactivity in the normal subject. At 24 h after injection, residual activity appears over the surface of the hemispheres. The count rate may be low as much of the radioactivity will have been absorbed. A single lateral projection is sufficient to confirm normality as regional irregularities of distribution are of no clinical significance. Continued absence of activity over the hemispheres up to 48 h is indicative of a block to free passage of CSF from the site of production in the choroid plexus to the site of absorption, the arachnoid granulations. This is the characteristic abnormality of intermittent or normotensive hydrocephalus, the result of previous infection, sterile arachnoiditis following for example subarachnoid haemorrhage, or head injury.

CSF rhinorrhoea

Gross rhinorrhoea can be diagnosed by demonstrating glucose in the nasal discharge. Less severe cases are best demonstrated by intrathecal injection of 7–15 MBq ^{111}In-DTPA or ^{169}Yb-

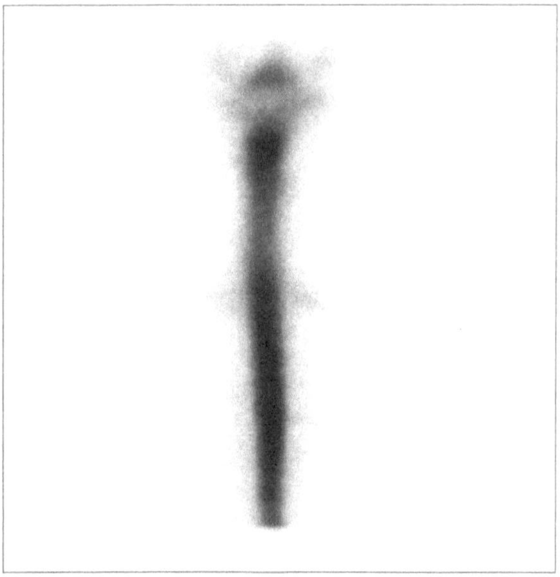

Figure 8.14 A 29-year-old woman with disabling and intractable headaches relieved only by lying flat. Cisternal injection of ^{99m}Tc-DTPA. This is one of a series of images 1–4 h after injection showing leakage of CSF at T2.

DTPA. The nose is then packed, preferably by an ENT surgeon able to define the position of each pack. The packs are identified by coloured threads. Depending on the severity of rhinorrhoea the packs are changed at intervals of 4–24 h and are counted separately for the presence of radioactivity. A venous blood sample should be taken every time the packs are changed and activity in the swabs related to plasma activity. Because of the freely diffusible nature of the tracers used, it is usual for activity in each pack to be equivalent to that in 1–2 ml plasma. In the presence of rhinorrhoea much higher levels are found.

If packs are not changed sufficiently often activity may wash in and then be leached out. The site of the leak can be ascertained from differential activity in the packs, provided that these are changed sufficiently often. If not, activity is spread over all, concealing the site and often even the side of leakage. Imaging has no role to play in this investigation.

Further reading:

European Journal of Nuclear Medicine 1985; ***11:*** *76–9*

Tests of shunt patency

A variety of diversionary cerebrospinal fluid shunts have been developed for treatment of obstructive hydrocephalus. With all there is a risk of obstruction. A number of tests of shunt patency have been developed. These may be considered under three general headings.

Measurement of rate of removal

Measurement of the rate of removal following direct intraventricular injection of the tracer. Within 24 h there should be a clear distinction between normal, arrested hydrocephalus and active hydrocephalus.

Shunt clearance measurements.

Injection directly into the shunt eliminates the potential dangers of ventricular puncture. Because the clearance time for normal shunts is minutes rather than hours, results may be obtained much more quickly. Pertechnetate may therefore be employed. The rate of clearance depends on a number of factors, including the type of shunt and reservoir employed.

Intrathecal injection

In communicating hydrocephalus when the shunt is functioning, radioactivity injected into the lumbar theca enters the ventricles and leaves via the shunt. When the shunt is not functioning, activity ascends over the surface of the ventricles. The shunt reservoir is often visualized. In non-communicating hydrocephalus the ventricles are not visualized and activity ascends over the surface of the hemispheres.

The choice of technique and of radiopharmaceutical depends on the nature of the shunt under investigation. There are many varieties and it is essential to obtain a clear picture of the information required by the neurosurgeon and the type of procedure which the patient has undergone before planning the investigation.

9 Liver and gall-bladder

Hepato-biliary function

Radiopharmaceuticals

The liver is one of the major excretory organs of the body. It differs from kidney in its ability to excrete certain molecules which are not water-soluble and because it possesses the additional capacity of detoxifying or altering many substances, for example by conjugation before they are eliminated. Detoxification is an important aspect of excretion. The biochemical mechanisms involved are relatively non-specific and there is thus a wide range of potential agents for investigating excretory function. In practice only a limited number of technetium-labelled compounds are still used clinically. These are often referred to collectively as HIDA. This acronym, which stands for hepatic imino-diacetic acid derivatives, was originally introduced as an abbreviation for the first useful agent of this type but is now often employed as a generic term to describe a family of related compounds.

All substances which are excreted by liver can to some extent also be excreted by kidney and vice versa. The principal factors which determine whether the predominant route of excretion is biliary or urinary are amount and strength of protein binding in plasma, molecular weight, molecular size, polarity and structure. There are in addition important inter-species differences. Biliary excretion plays a major role in elimination of anions, cations and non-ionized molecules which contain polar and lipophilic groups and which have a molecular weight in the range 300–1000 Da. The polar groups can be carboxyl, sulphonic, quaternary ammonium or anionic metallic complexes which allow a molecule to exist as a water-soluble anion or cation at physiological pH. Suitable molecules usually also possess a lipophilic group. Natural substrates include bile salts, which consist of a large non-polar steroid moiety and polar side chain conjugated to a glycine or taurine residue.

The HIDA group of compounds have the general structure $RN(CH_2COOH)_2$ (Fig. 9.1). The simplest compound of this group, methyl-IDA where R is a methyl group, has principally renal excretion. Substituting a benzene ring gives a compound with a close chemical resemblance to lignocaine and predominantly hepatic excretion. This was the original HIDA compound. Its chemical description is *N*-(2,6-dimethyl-phenyl-carbamoyl-methyl)imino-diacetic acid. Substituting isopropyl for methyl groups on the ring gives diisopropyl IDA (DISIDA, Diisofenin). This is more lipophilic than HIDA and there is substantially less renal excretion. This is of little practical importance in subjects with normal hepatic function but does enable the agent to be used in mild hepatic failure with a serum bilirubin up to 5 mg %. Another analogue, trimethyl-bromo-IDA (Mebrofenin) has even less renal excretion and can be used with serum bilirubin levels up to 10 mg %.

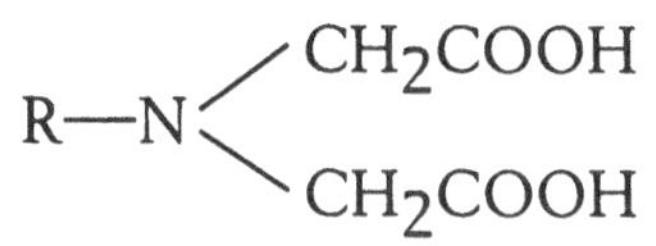

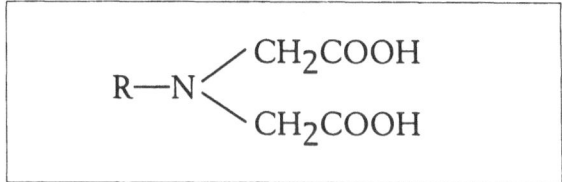

Figure 9.1 General structure of HIDA.

The mechanism of excretion is not fully understood. It is commonly stated that there is competition between bilirubin and technetium agents for a transport mechanism across hepatocyte membranes. However most investigators have failed to distinguish between hyperbilirubinaemia with and without hepato-canalicular disease. There are a number of rare congenital hyperbilirubinaemic states which are not associated with hepatocellular or hepato-canalicular disease. Excretion of these compounds is not impaired in patients with Crigler–Najjar syndrome, a hereditary disorder of bilirubin conjugation. Blood clearance is reduced in patients with the Rotor syndrome, a disorder of uptake and storage of organic anions. The first process involved in excretion of these compounds is transfer of tracer from circulating protein to which it is bound to a membrane-bound carrier. Separate carriers may exist for anions, cations, neutral substances and bile salts. Once within the hepatocyte the substrate can undergo binding and/or metabolism prior to excretion into bile. A number of intermediate binding and storage sites have been identified. It is not clear to which of these technetium compounds bind, or indeed if they can be bound by more than one of these sites. Canalicular secretion involves at least two processes, active transport of excretory products (including bile acids) into the canalicular lumen and production of a bile-salt-independent canalicular fraction which may involve active transport of sodium. As canalicular bile moves along bile ducts composition of the secretion may be further modified by addition or absorption of water and electrolytes. It is unclear if other substances can also be added at this stage.

Technique

The patient should attend as early as practical in the morning having had clear fluids only by mouth from 10 pm the previous evening. Any regular medications likely to affect the hepatobiliary system should be omitted and in the case of diabetics appropriate arrangements made. Excretion is enhanced in jaundiced patients by prior treatment with enzyme-inducing agents such as phenobarbitone (5 mg/kg/day for 5 days). The patient lies supine and the gamma camera positioned as close as possible above (anterior to) the patient, preferably just touching the thorax or abdomen. The field of view should include heart, all of the liver and upper abdomen. An administered activity of 50 MBq is adequate when using a general purpose low-energy collimator and acquiring data at one frame per minute for 60 min. On many cameras higher activities lead to distortion of data due to dead-time losses, especially once administered activity is concentrated in the gall-bladder. Data acquisition should be started immediately before injection of the radiopharmaceutical, which is given as a bolus into any convenient peripheral vein. If the examination is normal it may usually be terminated at 30 min. After the initial hour it is rarely helpful to image more often than once per hour up to 4 h. It is occasionally necessary to take later views. Depending on the clinical problem imaging may be required up to 24 h. If time–activity curves are required regions of interest may be placed over an area of liver distinct from lung, gall-bladder, common bile duct and small bowel. If an input function is required a region of interest can be drawn over the heart. It is sometimes difficult to determine when activity first appears in the duodenum. Using the facilities of the computer system, such as alternative colour scales and manipulating the upper and lower thresholds of the display, is often helpful.

Cholecystokinin augmented cholescintigraphy

Cholecystokinin (CCK) is the peptide hormone principally responsible for gall-bladder contractions. A synthetic 8-peptide analogue (CCK8) is usually employed. Some workers have used ceruletidediethylamine (caerulein). Interpretation of the literature is complicated by the varying dose levels and dose rates which have been employed. The optimum dose appears to be 10 μg/kg infused over 30 min (0.03 μg/kg/min). Higher dose rates may be counter-productive, despite the short plasma half-life (2.5 min) of CCK because, although smooth muscle of the cystic duct is in continuity with that of the gall-bladder, the former has a higher threshold for

CCK than the latter but responds more rapidly. In consequence high dose levels may provoke spasm of the cystic duct and prevent gall-bladder emptying. The duration of action is considerably longer than the plasma half-life. Despite the short plasma half-life, contraction of the gall-bladder continues for 10 min or so after the infusion has been completed. Thus imaging should be continued for at least 15 min after completion of the infusion. Gall-bladder ejection fraction is taken as the difference between maximum and minimum counts over the gall-bladder after correction for background. This method is more accurate than methods based on measuring dimensions as the gall-bladder may change its axis on contraction, making estimates of its volume inaccurate. Using the scintigraphic technique gall-bladder ejection fraction in normal subjects should be between 55% and 85%.

In patients with partial obstruction of the common bile duct the 'reflux sign' may be seen. When this is present radioactivity in intra-hepatic bile ducts more peripheral than the main left and right hepatic ducts either becomes apparent or obviously increases in intensity after CCK infusion. Although an uncommon sign, when present it is strongly suggestive of incomplete common bile duct obstruction. The gall-bladder ejection fraction is commonly reduced in patients with acalculous gall-bladder disease but is not a good predictor of benefit from surgery.

Further reading:

Journal of Nuclear Medicine 1996; ***37****: 261–6*

Filling of the gall-bladder

The normal resting pressure is 15 cm of water at the sphincter of Oddi, 12 cm of water in the common bile duct and 10 cm of water within the gall-bladder. Thus the normal pathway for bile is into the gall-bladder. Following sphincterotomy this gradient is abolished so that both contrast media and drugs may fail to enter the gall-bladder.

Normal appearance

In a subject with normal hepatic function, a normal gall-bladder and patent bile ducts there is uniform high uptake in liver within 5 min of injection, with low extra-hepatic activity. Intra-hepatic bile ducts can be seen faintly as areas of increased uptake by 10 min and are usually clearly seen by 15 min (Fig. 9.2). Common bile duct and duodenum are often visualized within

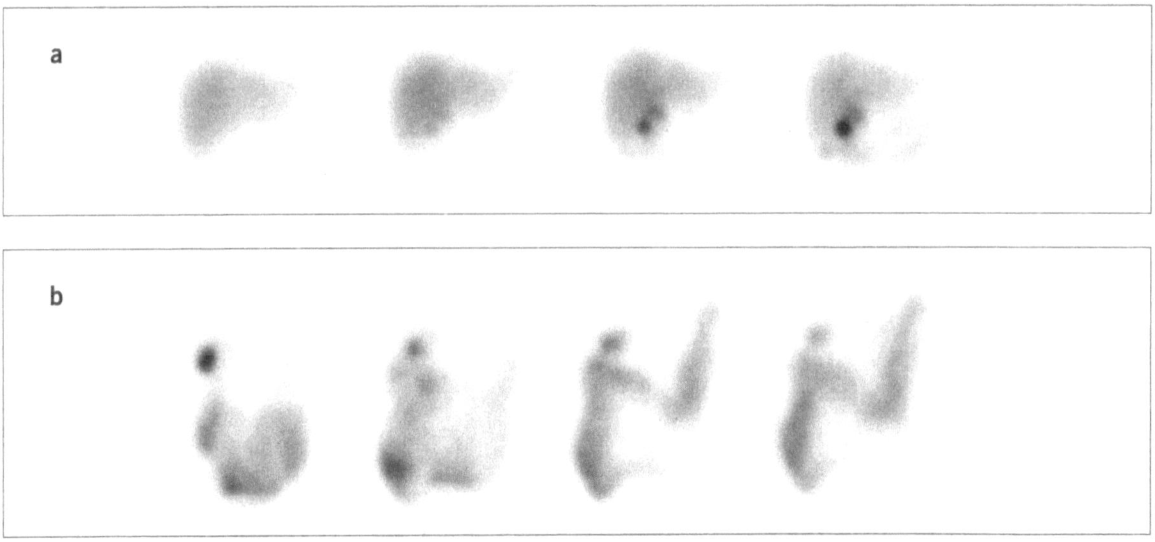

Figure 9.2 (a) Consecutive 5 min frames starting at time of injection and scaled globally, showing rapid transit of tracer (diisofenin) through liver into gall-bladder. (b) Emptying over 30 min in response to slow infusion of cholecystokinin.

15 min and should always be visualized within 30 min. Small bowel activity can often be identified at 30 min, even in fasting subjects. By 1 h most of the residual activity is usually in small intestine although bile ducts, gall-bladder and liver can still be visualized. Retention within gall-bladder with delayed visualization of small intestine is not necessarily abnormal.

Clinical applications

Acute cholecystitis

The role of cholescintigraphy in diagnosis of cholecystitis varies between centres, determined partly by available expertise in ultrasound and partly on surgical practice, depending whether early or late operation is preferred. Most centres with access to good ultrasound facilities now regard the latter as the initial investigation of choice, enabling a definitive diagnosis to be made in the majority of patients. Alternatively cholescintigraphy can be performed. Visualization of gall-bladder within 30 min excludes the diagnosis of acute cholecystitis as this is associated with obstruction of the cystic duct in more than 95% of patients. If the gall-bladder is not clearly visualized within 30–40 min, a low dose (2 mg) of morphine sulphate given intravenously may provoke closure of the sphincter of Oddi, enabling visualization of the gall-bladder within the following 10–15 min. In patients with a previously normal gall-bladder the sensitivity and specificity of cholescintigraphy are greater than 90%.

False positives, that is non-visualization of gall-bladder in a patient who does not have acute cholecystitis (Fig. 9.3), are most commonly associated with chronic gall-bladder disease or previous surgery. The value of cholescintigraphy in cholecystitis therefore depends on the prevalence of chronic gall-bladder disease in the population under study.

Indium-labelled white cells accumulate in the gall-bladder and its bed in acute cholecystitis. Technetium-labelled cells are more difficult to interpret because there is biliary excretion of

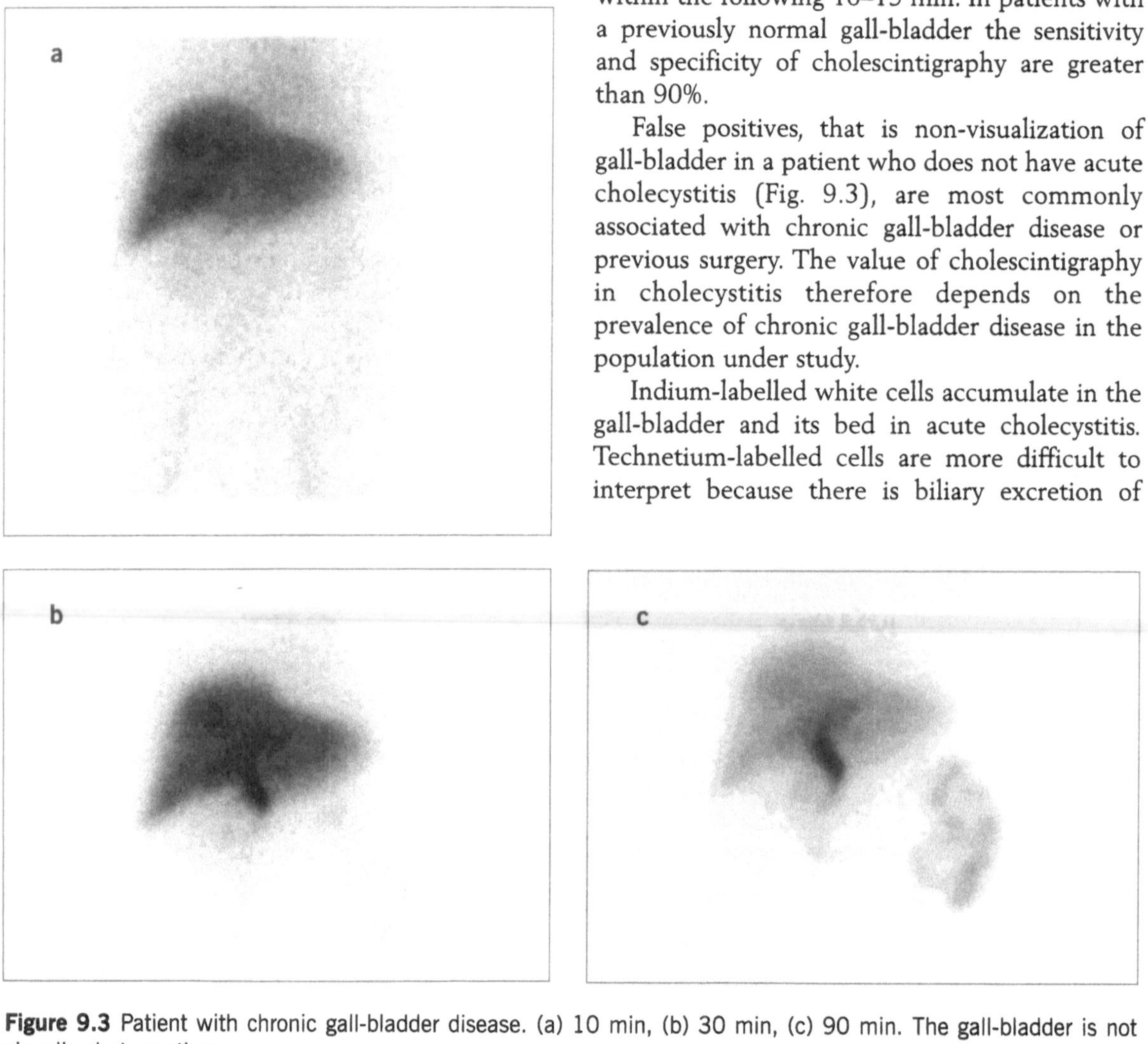

Figure 9.3 Patient with chronic gall-bladder disease. (a) 10 min, (b) 30 min, (c) 90 min. The gall-bladder is not visualized at any time.

tracer eluted from leucocytes. Accumulation of labelled cells in an inflamed gall-bladder occurs much earlier than visualization of gall-bladder due to excretion of eluted technetium and is commonly more intense. Diagnosis of focal information is usually made clinically; technetium-labelled leucocytes rarely add useful information in the acute phase of the illness. The complexity and cost of white cell labelling can thus rarely be justified, although it is useful in occasional patients post-operatively or where diagnosis is in doubt.

Hazards

Morphine should not be administered unless excreted activity has been identified in small bowel, thereby confirming patency of the common duct. Failure to observe this precaution is associated with the risk of perforation of an inflamed and obstructed common bile duct or gall-bladder.

Measurement of hepato-biliary function

The rates of blood clearance, hepatic uptake and biliary excretion of the HIDA group of compounds can be measured both by blood sampling and by time–activity curves over appropriate regions. As with renal time–activity curves, various approaches have been used to analyse these curves, which are in mathematical terms quite complex. So far none of the methods has been universally accepted. Parameters calculated include hepatic extraction fraction and mean transit times through liver and bile ducts. There exist many tests of hepatic function which do not involve the use of radioisotopes, and isotopic methods have not been shown to have any practical advantage over non-isotopic. They are therefore not widely employed. Percentage emptying of the gall-bladder after a stimulus such as a fatty meal or cholecystokinin can also be measured (page 251).

Sclerosing cholangitis

This condition affects both intra- and extrahepatic bile ducts and is often distributed unevenly throughout the liver. Serum alkaline phosphatase is usually raised at the time of presentation but bilirubin may not rise until late in the course of the disease. There are usually multiple sites of obstruction and typically these are more proximal than in patients with isolated common bile duct obstruction. A feature of sclerosing cholangitis which is uncommon in other liver diseases is that the rate of excretion of IDA derivatives differs in different parts of the liver, in keeping with regional differences in disease severity. This is a feature of particular diagnostic importance. Scintigraphy may show multiple levels of ductal obstruction with a typical 'beading' pattern due to accumulation of activity between strictures. Single photon emission computed tomography (SPECT) of the liver 60–90 min after injection often supplements planar images and detects lesions which are not evident in planar images or on contrast cholangiography. There is commonly associated obstruction of the cystic duct and response to cholecystokinin (CCK) is poor. Although usually idiopathic this condition is one of the complications which may be observed in patients with AIDS.

Chronic cholecystitis

The scintigraphic findings are varied. In the majority of patients with minimal symptoms cholescintigraphy may be normal or there may be some delay in visualization of the gall-bladder, which may not be seen until 4 h after injection. If there have been multiple episodes of acute inflammation there may be fibrosis of the wall with resulting contracture of the gall-bladder. Under these circumstances delayed or absent visualization is more probable (Fig. 9.3). In some patients it is possible to visualize the gall-bladder after CCK administration. The principal role of cholescintigraphy in this context is in the patient with an acute exacerbation of recurrent upper abdominal pain; visualization of the gall-bladder suggests that pain is unlikely to be due to acute or chronic cholecystitis. Unfortunately the converse is not true.

Cystic fibrosis

Cholescintigraphy is widely employed in cystic fibrosis to assess hepatic function, cystic duct patency, abnormalities of both intra- and extrahepatic biliary drainage and in monitoring response to stone dissolution therapy. Non-visualization of the gall-bladder is most commonly due to inspissated bile, mucous plugs or gall-stones. Other causes of non-visualization include inadequate fasting, pancreatitis, severe hepatocellular disease and more rarely an ectopic, congenitally absent, micro- or contracted gall-bladder.

Further reading:

Journal of Nuclear Medicine 1994; ***35****: 432–5*

Post-operative patients

Biliary leak is common following cholecystectomy (Fig. 9.4), especially endoscopic cholecystectomy. However the vast majority are small, of no clinical significance and close spontaneously. Cholescintigraphy is the simplest method of identifying a leak. It may be necessary to obtain multiple oblique projections to distinguish between leakage and bile ducts and to identify the likely source. Obstruction, for

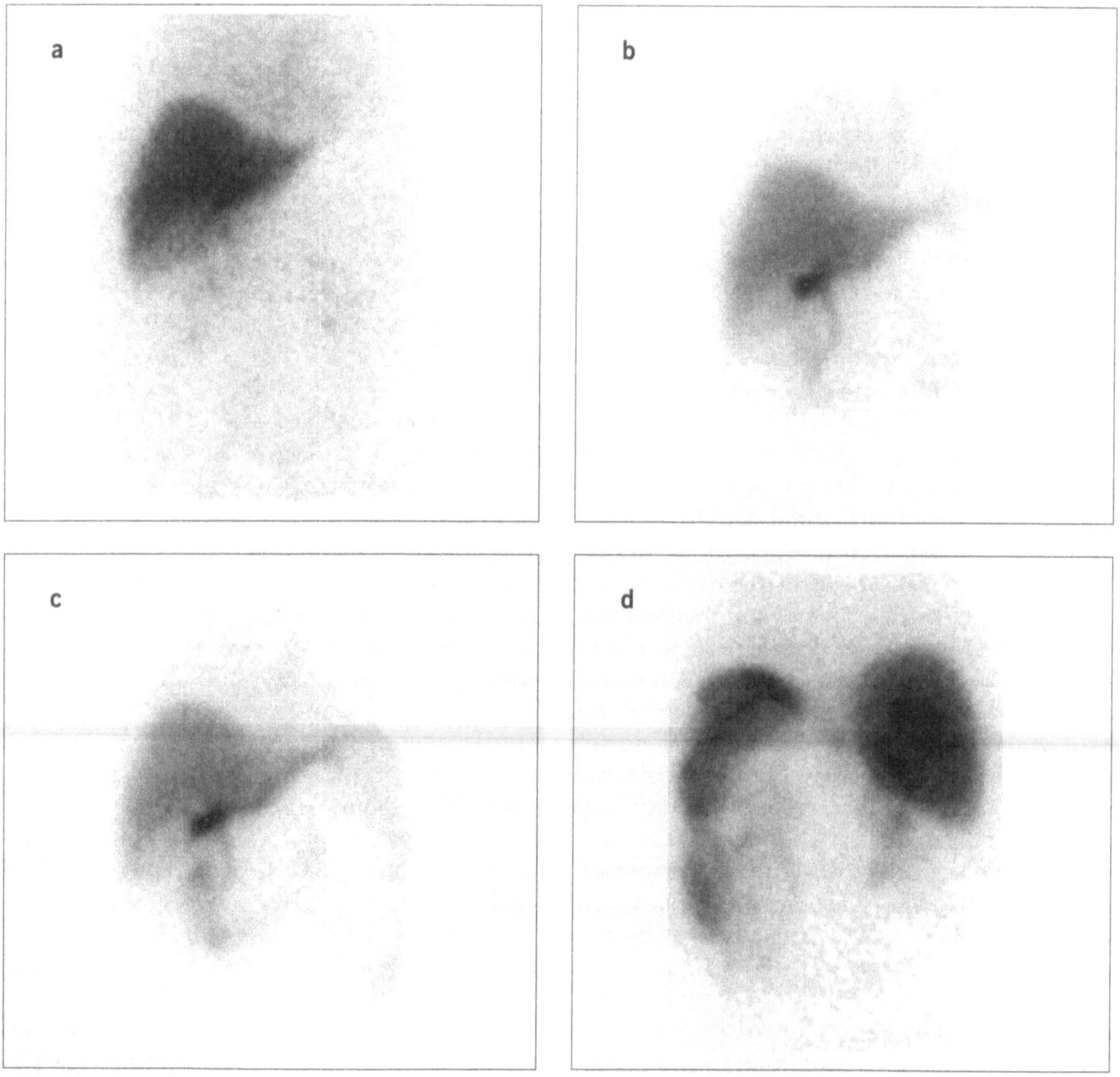

Figure 9.4 Biliary leak following cholecystectomy. There is rapid filling of the common bile duct and duodenum (a, b). (c) Within 15 min activity tracks to the left, outlining liver and large bowel. (d) Posterior projection shows extent of leak into left paracolic gutter.

example due to a retained stone, can also be identified but endoscopic visualization is usually necessary to determine the anatomy and for definitive treatment.

Cholescintigraphy is the best simple technique for determining patency of biliary-enteric bypass procedures. The anatomy is often difficult to distinguish and multiple projections are necessary. Tomography is not often helpful because of the rapidly changing distribution of excreted tracer. Cholescintigraphy may have some role in evaluation of the patient with persistent symptoms following cholecystectomy. This includes identification of hold-up or obstruction, cystic duct remnants and persistent biliary leakage.

However in most centres endoscopy is now regarded as the initial investigation of choice in these patients. Anatomy can be defined with greater precision using oral or intravenous cholecystography in association with computed tomography.

Further reading:
Journal of Nuclear Medicine 1991; ***32:*** *1910–1*

Choledochal cyst

This is an uncommon condition in Europe, somewhat less rare in populations of Chinese origin, usually thought to be congenital and most often presenting in childhood, although it may present at any age. The clinical presentation is variable and it is not always diagnosed preoperatively. The presence of a cystic abnormality in the liver can usually be determined by ultrasound or computed tomography (CT), but cholescintigraphy is required to confirm whether or not there is communication with the biliary tree.
Scintigraphically, rather over half show accumulation of hepato-biliary agents within 1 h of injection. In about half of the remainder the cyst fills if imaging is prolonged up to 24 h. However in a substantial minority of patients the cyst does not fill. Non-filling thus cannot be used as a criterion to exclude the possibility of choledochal cyst. The gall-bladder is usually visualized separately but, especially in adult patients, concomitant chronic cholecystitis is not uncommon.

Acute pancreatitis

Non-filling of the gall-bladder following administration of a hepato-biliary agent occurs in rather over half of patients with acute pancreatitis. This is usually associated with stones in the common bile duct. Visualization of gall-bladder is good evidence that the cystic duct is patent.

Hepatocellular carcinoma

Uptake of HIDA in primary hepatocellular tumours is equal to or greater than that in surrounding normal liver in between one-third and one-half of patients with this tumour. In some this is evident within 30 min, but in others uptake is evident only on delayed scans taken at 3 h, by which time much activity will have washed out of normal liver cells. Uptake is not sufficient for visualization of poorly differentiated tumours, which appear as non-specific photon-deficient areas, but is adequate in about 70% of well differentiated and 30% of moderately differentiated tumours.

Sphincter of Oddi dysfunction

This may be a cause of recurrent biliary pain following cholecystectomy. There have been a number of attempts to establish criteria for differential diagnosis using quantitative hepatobiliary scintigraphy, principally by measuring time to peak uptake or half-time of wash-out under basal conditions and following CCK8 (Fig. 9.5). Most published series are small and the characteristics of the populations studied poorly defined. There appears to be appreciable overlap between normal and abnormal groups and it has not proved possible to define generally accepted and useful normal limits. The problem is exacerbated by lack of agreement on the optimum dose of CCK8 to be administered. It has been suggested that vasodilators such amyl nitrite may help to distinguish between dyskinesia and a functional stenosis, amyl nitrite increasing the rate of drainage in patients with dyskinesia but having no effect on those with an organic stenosis. This is an interesting development but needs to be confirmed by other centres.

Further reading:
European Journal of Nuclear Medicine 1993; ***21:*** *203–8*

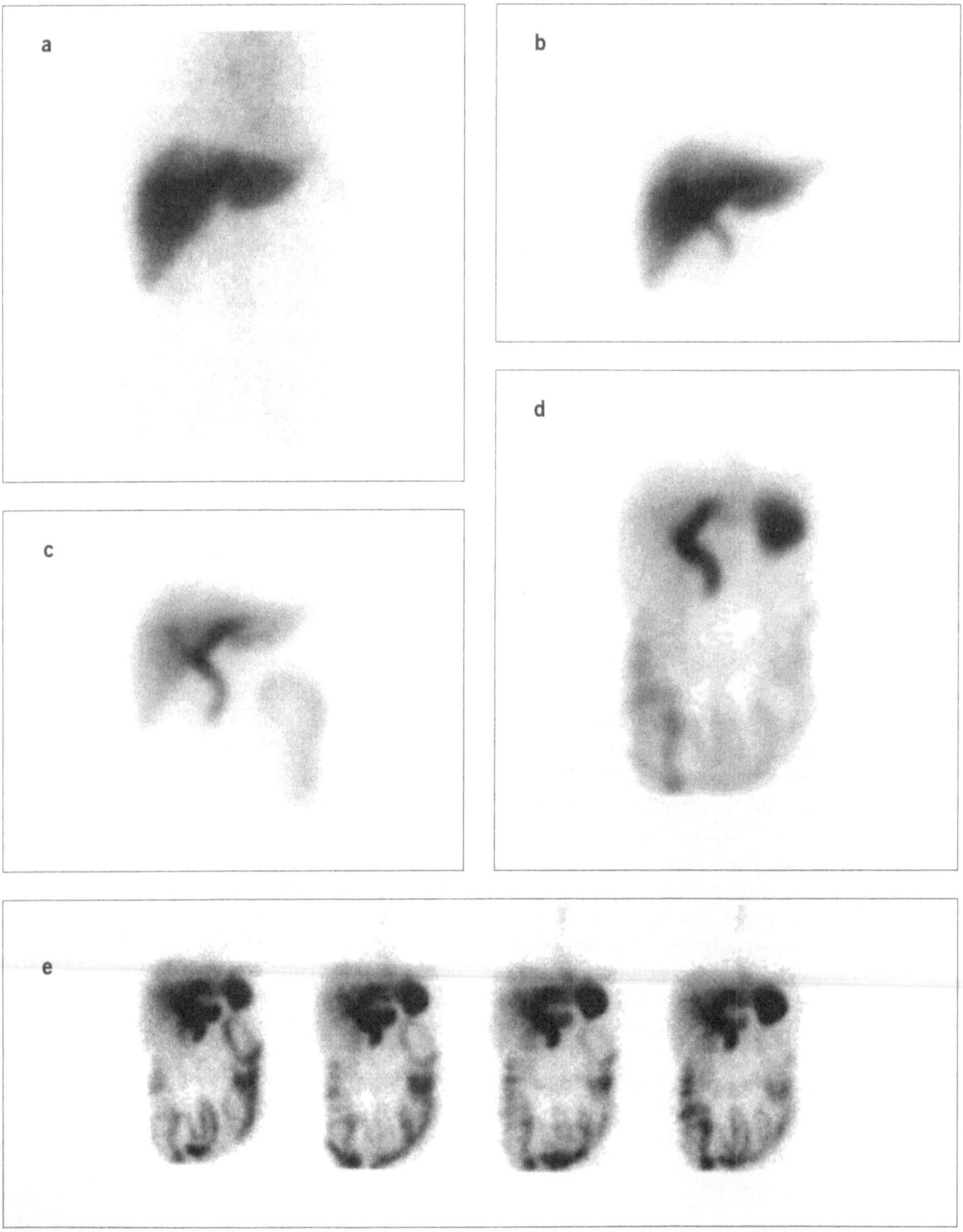

Figure 9.5 Persistent right upper quadrant pain despite cholecystectomy. (a–c) Normal uptake and excretion over 15 min. (d) After 30 min infusion of cholecystokinin there is duodeno-gastric reflux of bile and persisting high concentration of tracer in the common bile duct, compatible with sphincter of Oddi dysfunction. (e) Sequential images during infusion showing enhanced small bowel motility.

Liver transplants

A number of techniques have been described for quantitative evaluation of liver time–activity curves following transplantation, in order to differentiate abnormalities of hepatocellular function from drainage problems. None has become generally accepted and their clinical value is not established.

Further reading:
Journal of Nuclear Medicine 1996; ***37****: 847–51*

Reticulo-endothelial function

The Kupfer cells of the liver form the largest single mass of reticulo-endothelial cells. Any intravenously administered colloidal substance, whether or not it is radioactive, is rapidly cleared from the circulation by cells of the reticulo-endothelial system unless the quantity administered is so large as to saturate the phagocytic capacity of the system. Blood passing through any organ containing a substantial reticulo-endothelial content is almost totally cleared of any colloid which it contains. Thus the first-pass extraction efficiency is almost 100% and uptake is consequently proportional to blood flow. In the case of the liver, clinical application of this principle is complicated by the dual inflow, both arterial and portal, the latter normally predominating.

The final distribution of any particular colloid depends not only on the distribution of cardiac output but also on particle size of the colloid. Smaller particles are to a slightly greater extent cleared by marrow whereas larger particles are more efficiently cleared by spleen. This is because there are slight differences in extraction efficiency, which in practice is always less than 100%. It is not possible to determine site specificity purely by altering particle size and shifts in the pattern of distribution are comparatively small. In the normal subject 80–90% of an administered tracer dose of colloid is taken up in liver and spleen, approximately in proportion to their relative weights; most of the rest accumulates in bone marrow. A small amount is taken up by phagocytic cells in lung and a negligible amount by other sites such as lymph nodes, which have a tiny blood flow. Lung uptake must not be confused with that of the larger particles used for lung perfusion scintigraphy which are filtered out by capillaries or arterioles, not by phagocytes.

The nature of the colloid is of relatively little importance. The choice of radiopharmaceutical is determined principally by convenience of preparation and availability. Formerly, when colloid scintigraphy of the liver was widely employed for detecting space-occupying lesions, a large number of preparations were available as this was a common examination. As this has now been largely superseded by ultrasound and CT the number of available preparations has fallen. For most purposes there is little to choose between available preparations. Inorganic colloids such as tin colloid are not catabolized, whereas millimicrospheres of denatured albumin are eventually digested.

Other tracers

Other tracers, including colloidal gold, antimony sulphide and indium hydroxide colloid are no longer generally available and are of historic interest only.

Instrumentation and technique

Dynamic uptake measurements should be performed using a large-field camera equipped with a general purpose collimator, imaging at 30 frames per minute for 2 min following bolus injection of 75 MBq of any technetium-labelled colloid. The posterior projection is usually preferable as it allows time–activity curves from spleen and kidneys to be compared with liver. Because of statistical limitations imposed by the usual administered activity and the short frame times required, there is no benefit in using a matrix finer than 64 × 64 pixels. If the equilibrium distribution is to be imaged it is conventional to collect 400000 counts in at least three projections, anterior, right lateral and posterior, using a 128 × 128 matrix. Additional obliques and, if the spleen is being visualized, a left lateral, are occasionally also helpful. Resolution is limited by respiratory movement.

SPECT, by improving contrast at depth, increases detectability of larger deep-seated lesions but respiratory movement is again a limiting factor. A photon-poor lesion 2 cm in diameter can be detected at the liver edge under optimal conditions. However lesions of 5 cm diameter in the centre of the liver cannot always be detected on planar imaging. It is these resolution limitations which have determined replacement of this test by ultrasound or CT, both of which have better anatomical resolution.

Normal appearance

There is considerable variation between individuals in shape of the liver and shape and size of the spleen. In the anterior projection, liver appears approximately triangular and the right lobe is substantially larger than the left. A larger left lobe suggests diffuse disease such as cirrhosis, although this is occasionally a normal variation of shape. An area of reduced count-rate may be seen between the left and right lobes due to structures present at the porta hepatis (confluence of the left and right hepatic ducts with the common bile duct, portal vein and its branches and portal lymph nodes). This may be difficult to distinguish from a space occupying lesion at this site or from an extrinsic space occupying lesion displacing the liver. The female breast or breast prosthesis may give rise to an ill-defined, or less often a fairly clearly defined, crescentic defect on the diaphragmatic aspect in the anterior projection. This usually alters if the view is repeated with the patient in a different posture, for example supine rather than erect. A curvilinear area of increased uptake is sometimes seen in the mid or lower third of the liver and may be due to scatter from a roll of subcutaneous fat at this level. Costal impressions on the right border of the liver may be difficult to distinguish from focal replacement

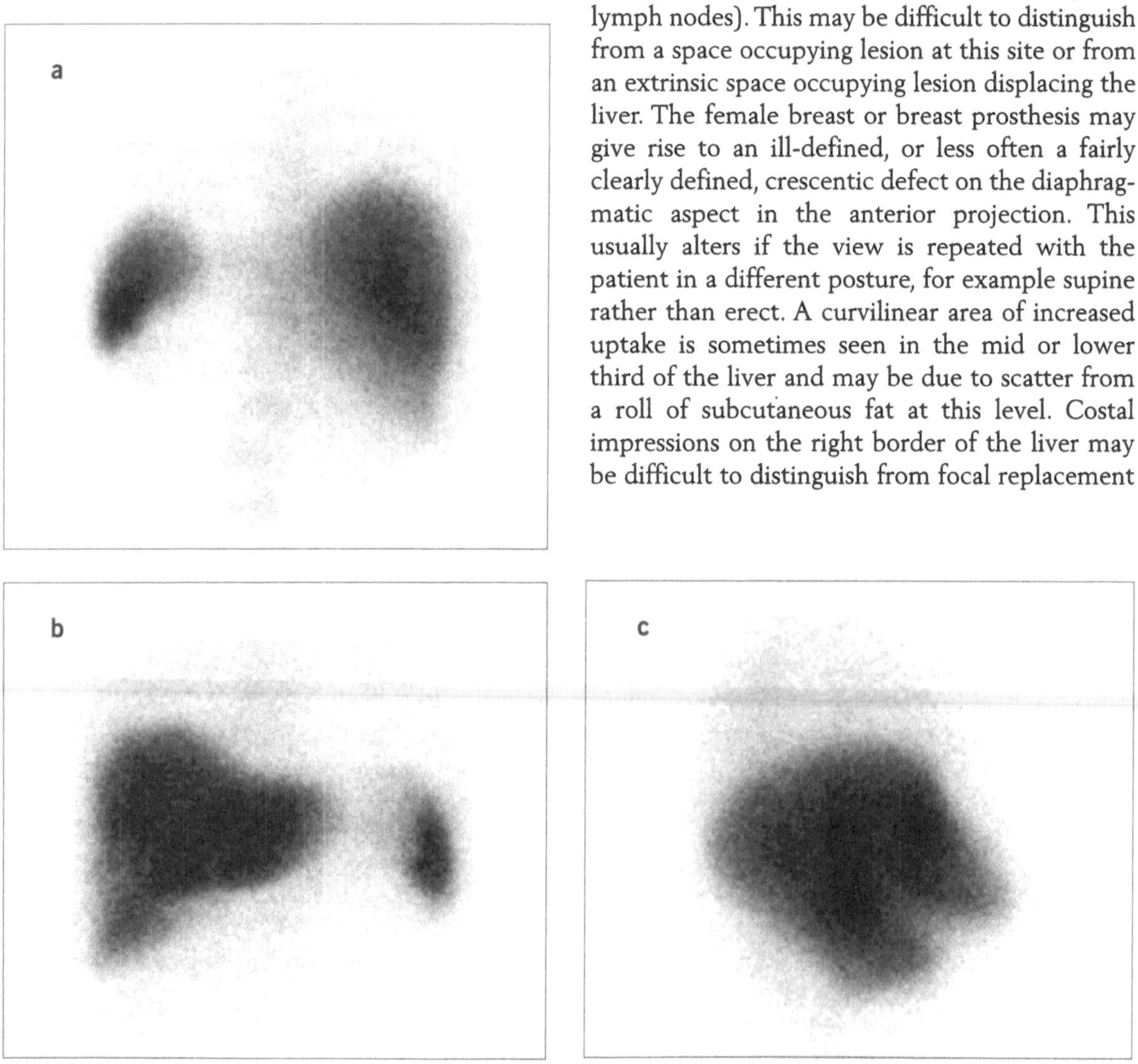

Figure 9.6 Normal distribution of reticulo-endothelial function following any radioactive colloid. (a) Posterior, (b) anterior, (c) right lateral.

whereas the inferior border is often notched to a variable amount by the gall-bladder. In the right lateral projection the right and left lobes cannot be distinguished, but a notch on the anterior surface is due to the lower border of the left lobe projected in front of the right. The diaphragmatic surface is usually a smooth curve but is sometimes unevenly indented by the diaphragm.

In the posterior projection peak count-rates from liver and spleen should be approximately equal. There is considerable variation in spleen size between individuals. An impression can usually be identified on the medial aspect of the right lobe of liver caused by the kidney. The spleen can often be identified separate from the liver, but frequently the two are projected over each other and no border can be distinguished (Fig. 9.6). The spleen should always appear less active than the liver in the anterior projection. It is always abnormal for concentration of radioactivity in spleen to exceed that in liver in the anterior projection. When present this indicates displacement or enlargement of the spleen or infiltration of the liver. Non-visualization of spleen is always abnormal and may be due to previous splenectomy, infarction, replacement (for example by tumour), displacement by a space-occupying lesion or suppression of splenic function, for example following a haemolytic crisis. The amount of detail visible in liver depends on the quality both of the gamma camera and of the display being used. Marrow activity can always be seen with a good display which provides a wide contrast range. However marrow may not be visualized in normal subjects when using more traditional analogue and in poorer quality digital displays.

Clinical applications

Space occupying lesions

Primary and metastatic tumours, abscesses and benign tumours of the liver do not contain reticulo-endothelial tissue and thus appear as photon-deficient areas (Fig. 9.7). However because of the limitations of resolution of isotope imaging systems this phenomenon is no longer widely used diagnostically. A photon-deficient area is non-specific whereas the resolution of

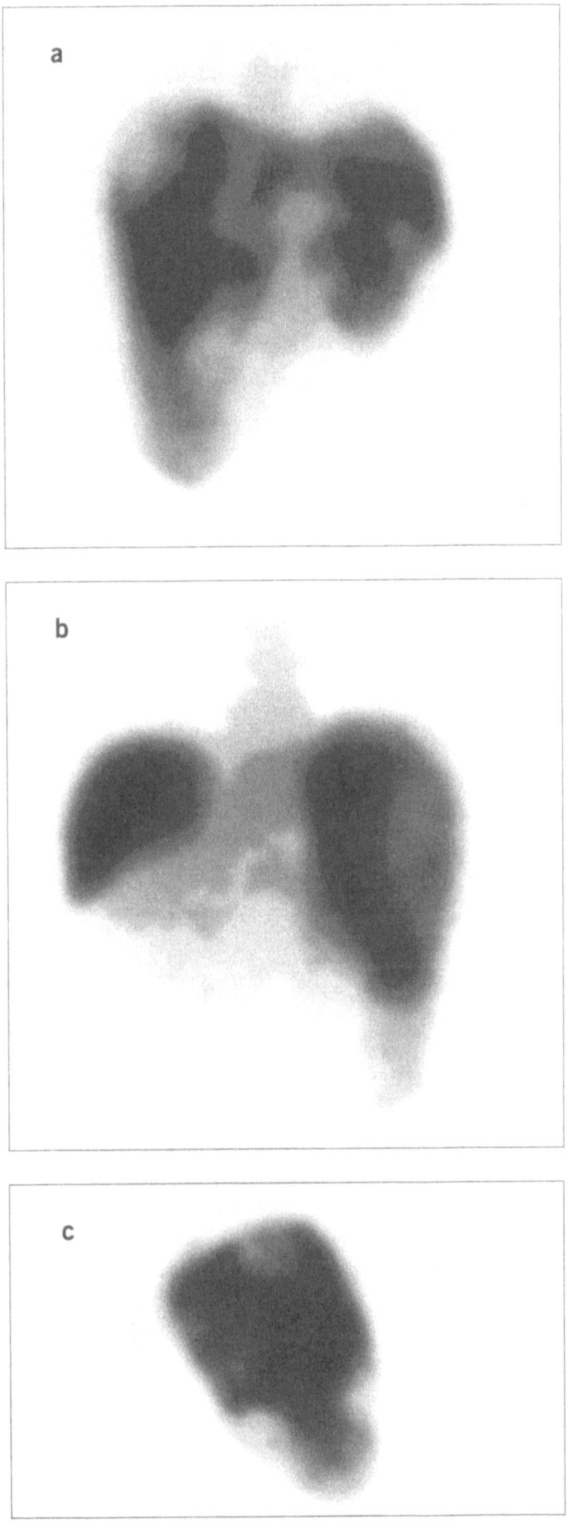

Figure 9.7 Space occupying lesions (metastatic colon carcinoma). (a) Anterior, (b) posterior, (c) lateral projections.

both CT and ultrasound is considerably better than that of scintigraphic imaging.

Diffuse liver disease

Cirrhosis, irrespective of cause or histological type, is recognized scintigraphically by a number of signs which may present singly or in any of the possible combinations. They are in general consequent upon fibrosis, which increases the vascular resistance of the liver. Blood is therefore diverted via collaterals (varices) to the systemic circulation and spleen. The earliest sign may be enlargement and relative increase of uptake of activity in spleen compared to liver. Subsequently as the liver shrinks the discrepancy between liver and spleen increases; as more blood is diverted via collaterals into the systemic circulation the percentage cleared by marrow increases. Marrow visualization thus becomes progressively more marked (Fig. 9.8). Tumours are recognized as photon-deficient areas as in the liver. Initially the increased blood flow causes a relative increase in splenic uptake and size, which in consequence appears of higher activity than the liver. Subsequently as fibrosis progresses and collaterals are established both size and relative uptake diminish whilst relative uptake in marrow increases.

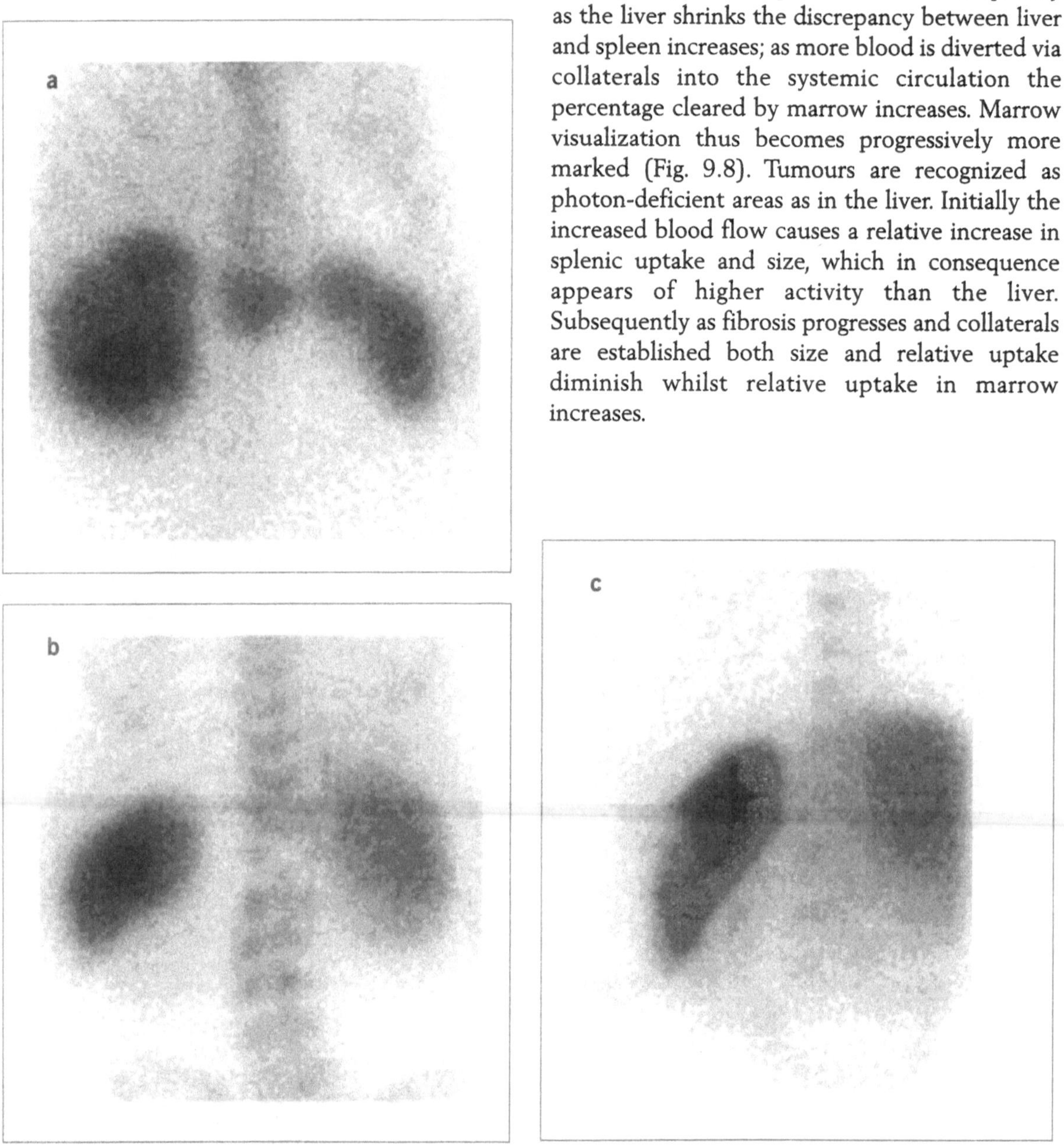

Figure 9.8 Advanced cirrhosis with shrunken liver, enlarged spleen (with filling defect) and high marrow uptake. (a) Posterior, (b) anterior, (c) right lateral.

Hepatic perfusion

Perfusion index

From the first-pass study, time–activity curves can be drawn over liver and kidney. An index can be calculated by comparing slope of the first part of the time–activity curve, before the peak of the first pass of activity through the kidney, to that in the subsequent part of the liver curve. It has been suggested, although without good evidence, that the first part is representative of hepatic arterial and the second of portal venous flow. The ratio is taken as an indicator of the ratio of hepatic arterial to portal venous perfusion. This ratio is higher in subjects with hepatic metastases than those without because deposits are thought to take their blood supply principally from hepatic artery branches and thus increase hepatic arterial flow. Some centres have reported high sensitivities and specificities with these techniques but others have not been able to reproduce their results. Experimental studies have not confirmed the suggestion that the two parts of the curve correspond to hepatic arterial and portal venous blood flow.

Other methods

A number of other techniques have been described, including deconvolution of the hepatic curve with a lung curve, after some rather dubious manipulations to take account of the arterial component of the hepatic curve. A further method compared hepatic with splenic time–activity curves. These latter both give reasonable correlations with measurements using flow probes at medium flow rates but none of the methods were satisfactory at high or low flows. Clinical applications of these techniques are not established. In subjects undergoing arteriography, hepatic blood flow can be measured directly following injection of ^{133}Xe into the hepatic artery. The calculation is identical to that used for measuring cerebral blood flow (page 226).

Portal-systemic shunting

The extent of portal-systemic shunting can be estimated from measurements following absorption of tracers administered into the upper rectum. Tracers which have been used include ^{133}Xe dissolved in saline, ^{123}I -iodoamphetamine and ^{201}Tl in the form of the chloride. Absorption of iodoamphetamine administered directly into the duodenum has been measured. Pertechnetate can also be used. Shunt size can be estimated from time–activity curves over heart and liver.

Investigations employing other tracers

Haemangioma of the liver

The increased blood pool in haemangioma of the liver can be demonstrated in the majority of patients using labelled red cells. They may be visualized immediately after injection if a dynamic study is performed, but the majority are equally well seen at equilibrium, once mixing of the labelled cells is complete (Fig. 9.9). The pick-up rate may be improved by SPECT. Haemangiomas are identified as photon-deficient areas with both colloid and HIDA imaging of the liver but red cells are both more sensitive and more specific. Additional imaging with other agents rarely adds additional information.

Further reading:

Nuclear Medicine Annual 1994. Raven Press, New York pp 55–90

Ascites

A number of surgical procedures have been described for palliation of intractable ascites. Some of these involve drainage via a valved shunt into the venous system. These shunts may become blocked. Patency can be demonstrated by intraperitoneal injection of any soluble technetium compound, commonly DTPA (page 82), and by observing filling of the shunt (Fig. 9.10).

Wilson's disease

This rare disorder of copper metabolism can usually be diagnosed fairly readily, but in some cases additional tests may be required. Blood clearance of a tracer dose of radioactive copper (^{64}Cu) has been used. In the normal subject

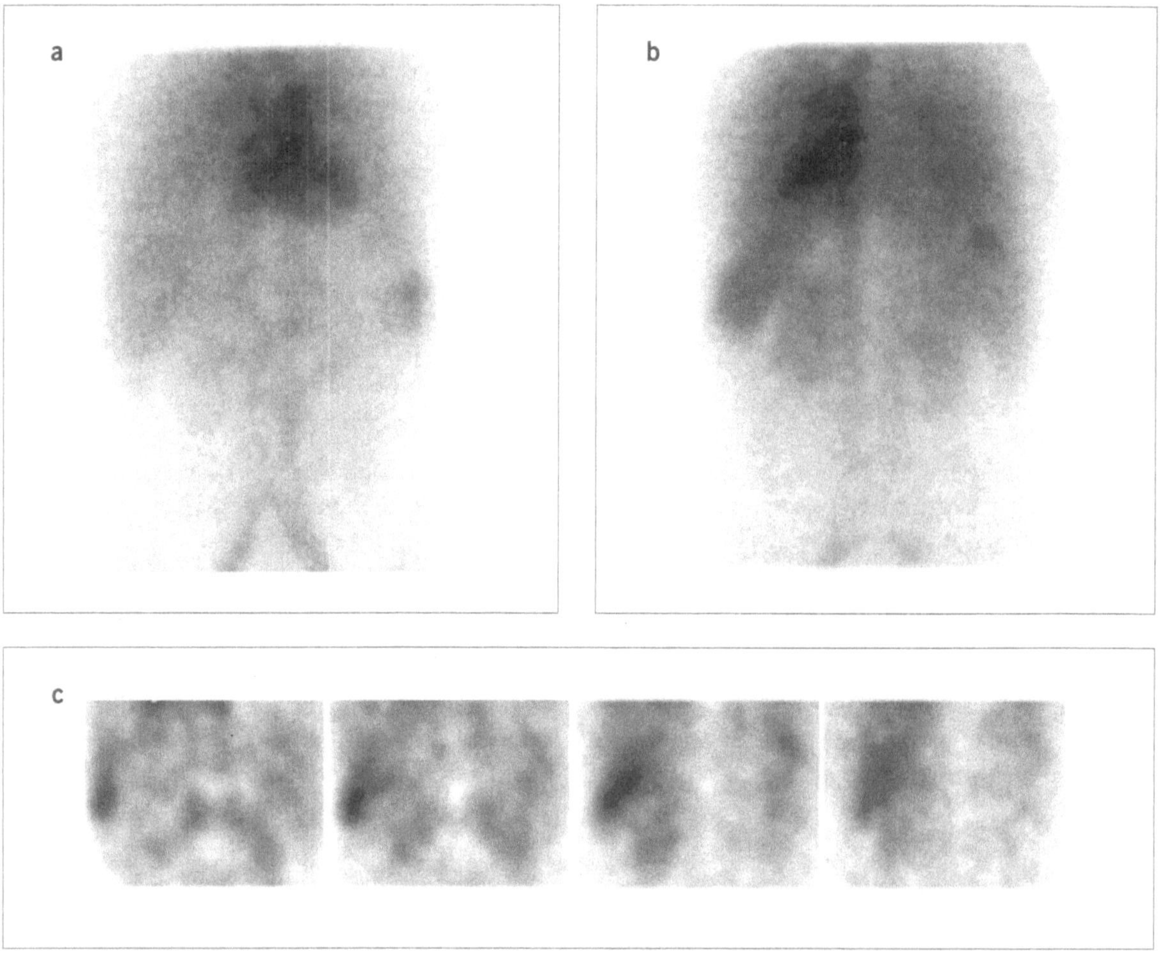

Figure 9.9 SPECT slices 1 h after 150 MBq autologous erythrocytes, showing increased blood pool in hepatic haemangioma. (a) Anterior, (b) posterior, (c) coronal.

following an intravenous tracer dose the plasma level at 48 h is higher than that at 2 h, as copper is initially cleared and subsequently returned to the plasma as caeruloplasmin. In Wilson's disease this does not happen and the blood concentration at 48 h is substantially lower than that at 2 h.

Further reading:

British Medical Journal 1990; ***301****: 331–2*

Galctosyl-neoglycoalbumin

This protein may be labelled with technetium or gallium and is taken up by a receptor, the asialoglycoprotein receptor, on the sinusoidal membrane of hepatocytes. The concentration decreases in patients with chronic liver diseases. The clearance rate of these tracers approximates to hepatic perfusion and can be calculated from hepatic and cardiac time–activity curves.

Labelled low density lipoproteins

Hepatic uptake of this tracer has been used to measure LDL receptor activity. No clinical role has yet been established.

Tumour-specific uptake

A number of tracers are concentrated in tumours to a higher extent than in liver. A large number of

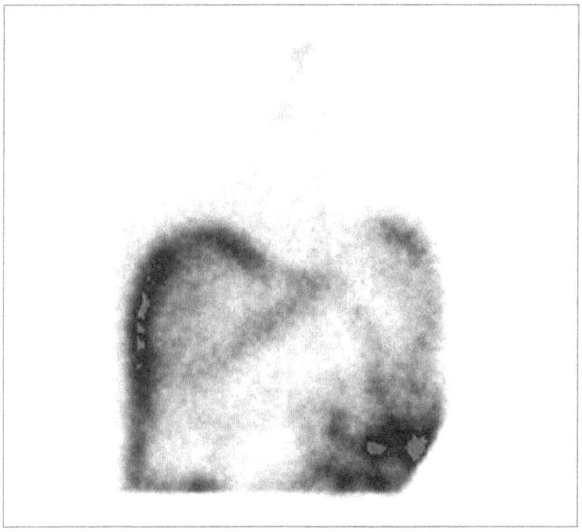

Figure 9.10 30 min after intraperitoneal injection of 50 MBq TcDTPA in patient with suspected blockage of peritoneo-atrial shunt for intractable ascites. The shunt is visualized and is therefore patent.

labelled monoclonal antibodies have been investigated and these do occasionally concentrate in some tumours sufficiently to be visualized above normal liver uptake. None is yet generally available which gives sufficient contrast between tumour and liver to be clinically useful. High uptake of ^{18}F-fluorodeoxyglucose (FDG) is also seen in many deposits, both in the liver and at other sites. Because of the short half-life of ^{18}F this is currently only available close to a source of production. It is currently the best means of assessing metabolic activity of a tumour.

10 Tumours and soft tissues

The concept of using radioactive tracers to locate sites and evaluate extent of primary or metastatic malignant tumours originated from one of the earliest clinical uses of radio-isotopes, namely to identify metastatic thyroid cancer. At the time of the original observations, over 50 years ago, there were few other methods of localizing tumours. Chemotherapy of cancer was in its infancy and surgical excision and radiotherapy were the only treatments generally available. Radio-iodine was a major innovation. Subsequent developments in surgery, chemotherapy and radiotherapy for many other tumours have substantially altered the nature of information required clinically. Ultrasound, computed tomography (CT) and magnetic resonance imaging (MRI) have revolutionized non-invasive demonstration of tumour boundaries. Staging is increasingly important in deciding the most effective plan of therapy and for differentiating between residual or recurrent tumour, fibrosis and other late changes. Radioisotope imaging does not provide the anatomical detail of other imaging modalities and with few exceptions is no longer used to determine the extent of primary or metastatic malignancies. Nuclear medicine retains a role where perceptibility is limited by contrast rather than resolution, when functional information is therapeutically relevant and for therapy of an increasing number of tumours. Radioactive tracers continue to be used when they can provide higher contrast, enabling deposits not visible by other modalities to be seen, or to provide specific data about metabolic or receptor activity.

Radiopharmaceuticals

Mechanisms

Tumour-localizing agents may be categorized in various ways depending on their properties.

Specific uptake

The most important examples are iodine by differentiated thyroid carcinoma (page 163), pentetreotide by tumours which retain somatostatin receptors (page 173) and MIBG (meta-iodobenzyl-guanidine) by tumours which retain nor-adrenaline re-uptake receptors (page 171). Somatostatin receptors are found on many tumours but also on activated lymphocytes in areas of active inflammation. Iodine and MIBG uptake is largely restricted to tumours arising from particular tissues, although iodine uptake has rarely been observed in metastatic bronchogenic carcinoma. Receptors for some other specific peptides have been identified on a number of tumours. The majority of malignant tumours, like the tissues from which they arise, do not have receptors for such substances. A subset of this category is partial retention of a normal function, for example continued ability of some tumours of hepatocellular origin to concentrate substances such as HIDA (page 245), even though they are unable to excrete them.

Metabolic disturbances

These may be incidental consequences of a tumour rather than specific to it. Common examples include the osteoblastic response which follows bone destruction (page 6) and FDG (flurodeoxyglucose) uptake observed in many tumours. Focal replacement of tissue performing a specific function may be identified as a photon-poor defect, the commonest example being 'cold' areas seen on radio-colloid imaging of the liver due to large tumour masses replacing normal reticulo-endothelial tissue.

Non-specific

There are numerous examples of non-specific localization, including uptake of complexes of gallium and a number of other metals into a wide variety of tumours and some inflammatory tissues. Diffusible water-soluble tracers such as pertechnetate are able to pass through abnormal neovascular capillaries in many tumours and around infarcts but are, at least partially, excluded from normal tissues. Thallium is probably a marker of intracellular fluid or sodium/potassium exchange and may indicate viable cells in the presence of fibrosis or other relatively hypo-cellular tissue. Many of these non-specific findings are mimicked in areas of inflammation.

Antibody uptake

Antibodies have been produced to various antigens on the surface of, or in some circumstances within, a variety of tumours. The antigens are never unique to tumour but are normal antigens expressed either in abnormal quantity, at abnormal sites or in abnormal situations.

Gallium

Gallium (^{67}Ga) is usually administered as the citrate, as many simple salts form a colloidal precipitate when administered intravenously at neutral or alkaline pH. Uptake is not specific to malignancy and occurs in many benign conditions, especially those associated with a chronic inflammatory response such as tuberculosis and rheumatoid disease. The mechanism of gallium uptake by tumours and inflammatory tissue remains unclear. A number of factors probably contribute but differ in relative importance in different situations. These include permeability of tumour capillaries to the gallium transferrin complex and the volume, distribution, protein content and composition of extracellular fluid. The concentration in extracellular fluid of proteins such as lactoferrin able to sequester metal ions transported by transferrin, the amount of inflammatory reaction and the permeability of cell membranes of tumour or inflammatory cells to gallium-carrying species are probably also significant factors. There are important differences between tissue distribution of high specific activity preparations used for diagnosis, when most gallium radioactivity is bound to transferrin and its behaviour is that of a moderately stable tracer for this high molecular weight β_1-globulin, and that of low specific activity and 'cold' preparations which have been used therapeutically in milligram or gram quantities, when most of the administered gallium is either excreted in urine or bound to the skeleton. There is substantial hepatic excretion of carrier-free preparations, whereas excretion of larger doses is almost entirely urinary.

Technique

The usual administered activity in the UK is 75 MBq. Higher activities, up to 550 MBq, are usual elsewhere and may be associated with improved accuracy. Imaging of gallium can commence within 4 h of injection. Contrast between abnormal foci and background activity increases with time but by the day after injection excreted activity in large bowel conceals much of the abdominal lymphatic tissue, where tumour deposition is likely. Bowel cleansing with laxatives or a high fibre diet reduces but rarely eliminates this problem. More radical techniques of bowel cleansing such as whole bowel lavage with isotonic solutions containing polyethylene glycol (PEG) are more effective. By increasing the rate of elimination of excreted activity from large bowel this also reduces absorbed radiation dose. The optimal interval before imaging is in most cases 24–48 h. Repeat imaging on subsequent days, if necessary up to one week, often clarifies whether an accumulation is fixed or is in bowel lumen.

Because, unlike technetium, the gamma ray spectrum of ^{67}Ga is not monochromatic but comprises four principal peaks of different energies (Appendix 1), multiple analyser windows and a gamma camera in which the position of every detected event is independent of energy are required. The lesion detection rate is often improved by SPECT (single photon emission computed tomography) because there is relatively low contrast between most tumours and background. However the low permitted activity (because of half-life and relatively high radiation dose) and the comparatively poor sensitivity of high energy collimators limits statistical quality of both SPECT and planar images. Contrast can be improved to a greater extent by employing sophisticated scatter rejection techniques when data processing than can be achieved with multi-channel analysers alone. Although it is reasonable to expect this to result in a higher clinical accuracy and sensitivity, this has yet to be proven. There remains some disagreement as to optimum choice of energy window. Some of the controversy may be due to failure to recognize the importance of scatter from the small (4%) 393 keV photopeak which, although it contributes little to the primary image, makes a substantial contribution to scatter in the lower energy windows, especially the 300 keV peak, and necessitates thicker collimator septa than would be required for this latter energy alone.

Normal appearance

The normal distribution following intravenous injection of a high specific activity preparation of gallium citrate to an adult shows the highest concentrations in tissues which sequester iron, especially bone marrow and liver (Fig. 10.1). Bilateral symmetrical uptake in both lung hila is seen in about 50% of patients. There may be transiently increased uptake (flare) in the thymus of children and other affected regions in both

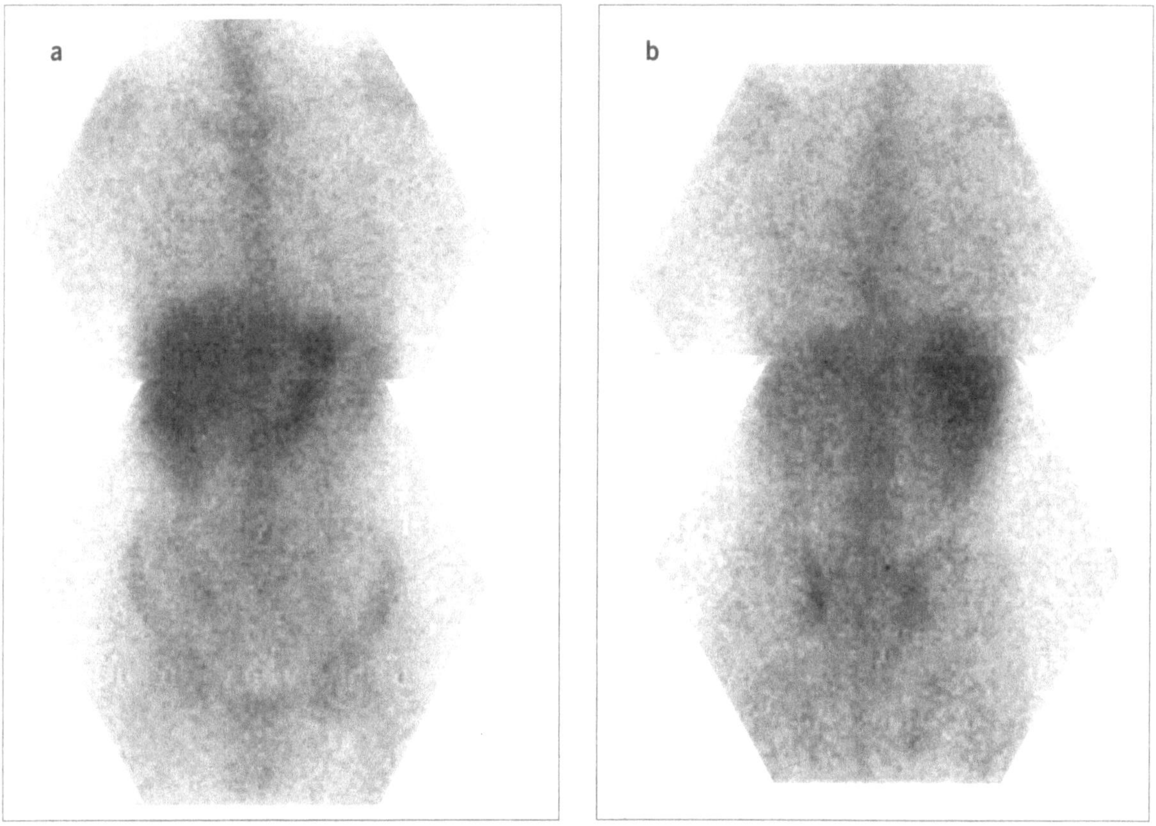

Figure 10.1 Normal distribution of gallium in the trunk 7 days after injection. (a) Anterior, (b) posterior. Retained activity is principally in liver and marrow. At earlier times activity is seen in large bowel.

adults and children following radiotherapy or chemotherapy. Uptake may also be observed in uncomplicated surgical wounds for two weeks or so and in infected wounds for rather longer. There is high uptake in breast tissue during pregnancy and lactation. Accumulation in breast is also seen in some patients taking oestrogen preparations such as oral contraceptives or for treatment of prostate cancer. This may be associated with the high concentration of lactoferrin in active breast tissue, which binds gallium and other trivalent metals more strongly than does transferrin. Excretion is largely hepatic and activity is retained in colon for several days, rendering recognition of abnormal accumulations difficult in the abdomen.

Clinical applications

Abnormal uptake may be observed in lymph nodes affected by tuberculosis, sarcoidosis, other chronic inflammatory conditions, sometimes in acute infections, in lymphoma, lung cancer and many other tumours (Fig. 10.2). Soft-tissue uptake is observed in chronic inflammatory conditions including synovium around rheumatoid joints, lung interstitium involved by sarcoidosis, infections such as tuberculosis, *Pneumocystis* pneumonia or other granulomatous conditions, in patients with AIDS and sometimes following chemotherapy (Fig. 10.3). Uptake in lung parenchyma has been observed in areas of radiation pneumonitis and following chemotherapy with bleomycin, nitrosourea and

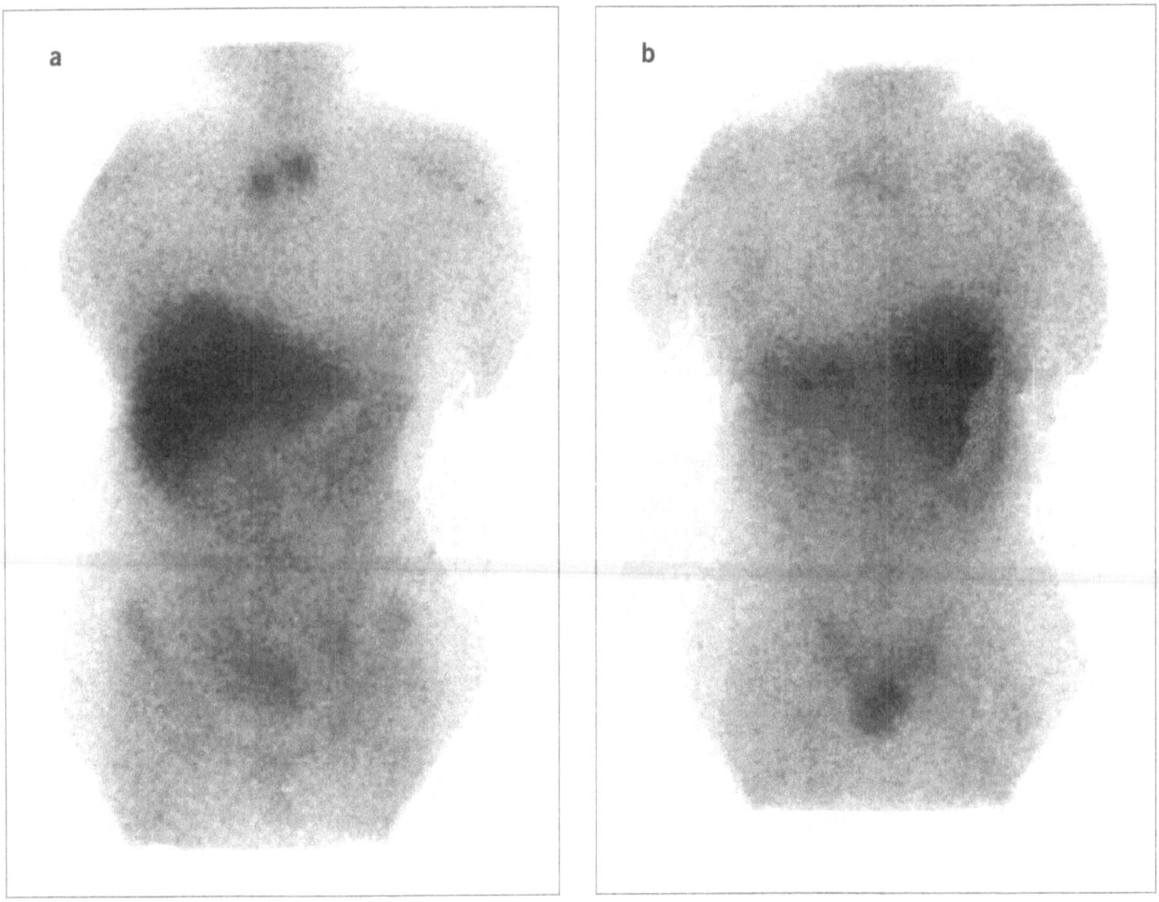

Figure 10.2 Hodgkin's disease treated with radiotherapy and chemotherapy. (a) Anterior, (b) posterior at 7 days. Anterior projection shows uptake of gallium in residual disease in superior mediastinum and left axilla. Note absence of marrow uptake in irradiated volume of thoracic spine in the posterior projection and in sternum in the anterior.

cyclophosphamide in the absence of radiographic changes. Increased uptake in bone, as distinct from marrow, with low soft-tissue uptake is seen in patients with iron overload and following administration of trivalent metals such as gallium, indium, scandium or gadolinium, sometimes following chemotherapy and also in AIDS. Abnormal bone uptake is virtually impossible to distinguish visually from abnormal marrow uptake due to tumour infiltration. Renal parenchymal uptake is increased globally in patients with iron overload, intestinal nephritis, glomerulonephritis and segmentally in pyelonephritis.

A wide variety of other tumours also show uptake of gallium, but in most it identifies only 25–50% of cases. In none does gallium reliably detect tumour extent not previously confirmed by other techniques. The principal clinical application for gallium scintigraphy in patients with a proven malignancy is to distinguish residual viable tumour from surgical changes or fibrosis following chemotherapy, surgery or radiotherapy.

Further reading:

Nuclear Medicine in Clinical Diagnosis and Treatment. Churchill Livingstone, Edinburgh 1994; pp. 711–25

Indium

Indium has close chemical similarities with gallium and ^{111}In has been evaluated as a tumour-localizing agent. Uptake into tumours is usually less than that of gallium and soft-tissue background tends to be higher, giving lower contrast. Simple salts are no longer used for tumour imaging but it is increasingly being employed chelated to specific ligands such as monoclonal antibodies (see below).

Thallium

This is used as its monovalent cation (Tl^+) because of the similarity in biological behaviour to potassium, which has a similar ionic radius. It is however in other respects more closely related chemically to gallium and indium and thus has also a trivalent form. Although used mainly for imaging myocardium and parathyroid (pages 115, 167), it has also been found to accumulate in some tumours.

Technique

^{201}Tl is administered as a simple salt, usually the chloride, at a dose of 1–2 MBq/kg. The lower dose is more commonly used for planar imaging, the higher for SPECT. Its principal disadvantages are

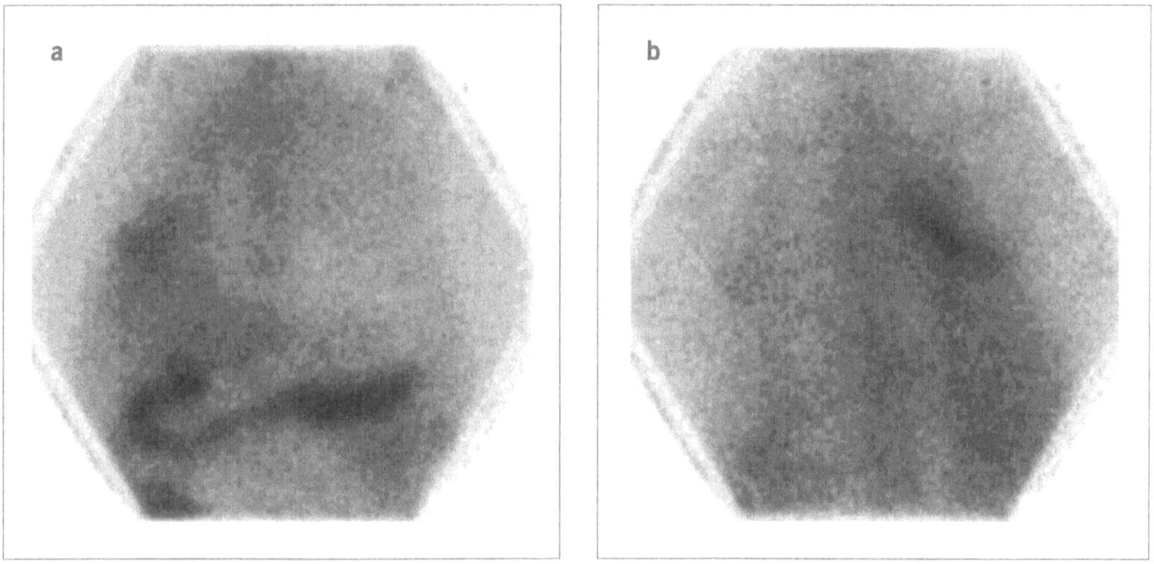

Figure 10.3 (a) Anterior, (b) posterior. There is uptake of gallium in a right pleural empyema. Note residual excreted activity in transverse colon.

the relatively high absorbed radiation dose and low energy of its gamma ray emissions (Appendix 1) giving a relatively poor detected count rate from deeper structures. It is generally considered that uptake is an indicator of viable cell volume and therefore cell mass, although this interpretation is an extrapolation from indirect evidence. There is no significant bowel excretion, facilitating visualization of the abdomen. Imaging can be started within 5 min of injection. The highest contrast is in some cases achieved 10–20 min after injection, but other tumours show up to 20% wash-out within the first 2 h whereas others continue to accumulate thallium for a longer period. Blood-pool activity progressively decreases with time and 3 h has also been recommended as the optimum imaging time. In general, uptake washes out of benign lesions more rapidly than from malignant, so that later imaging is clinically preferable when investigating possible malignant disease.

Normal appearance

The normal distribution of thallium observed on scintigraphy shows uptake principally into actively contracting muscle, liver, kidney parenchyma and myocardium, but with a high general soft-tissue background. Malignant tumours in general show an increasing concentration with time from injection, whereas there is rapid wash-in to and wash-out from benign tumours.

Clinical applications

Uptake has been reported in almost all primary lung and breast carcinomas studied, benign and malignant thyroid and parathyroid tumours (page 271) and most lymphomas (Fig. 10.4); it has been proposed for staging these tumours. The sensitivity for detecting mediastinal or lymph node involvement is however too low to justify routine clinical use, while specificity is relatively poor as benign lesions also are commonly visualized. Wash-out from the latter tends to be more rapid than from malignant tumours. There is, on average, higher thallium uptake by malignant than benign primary bone tumours, but the extent of overlap is such that the distinction is not of practical value in individual patients. Change in thallium uptake has been used as an index of response to chemotherapy or radiotherapy. There is one report of uptake in 29 of 32 patients with pancreatic carcinoma imaged with SPECT starting 5 min after injection. The abnormality was visible on planar images in only four. Contrast is enhanced by fasting, which reduces small bowel uptake. Many soft-tissue sarcomas also show thallium uptake. There is an association between the intensity of uptake into gliomas and the grade of malignancy, higher uptake being seen in the most malignant tumours. Possible clinical applications include distinguishing residual viable tumour from gliosis following treatment of malignant brain tumours and early identification of therapeutic response to chemotherapy of osteosarcoma, but even these are contentious. The specificity of the former must be questioned as thallium uptake has also been observed in infarcts.

Further reading:

European Journal of Nuclear Medicine 1995; ***22****:* *553–5*

Journal of Nuclear Medicine 1995; ***36****: 762–70*

Sestamibi and tetrofosmin

Most experience has been obtained with sestamibi but preliminary reports suggest that tetrofosmin may behave similarly. The mechanism of uptake is uncertain. A glycoprotein, the P-glycoprotein, which has also been implicated in multi-drug resistance of certain tumours, may be involved. The pattern of uptake tends to parallel that of thallium, lower contrast being to some extent at least compensated by better statistical quality of the images. Most published data is from patients with lymphoma or breast carcinoma, but uptake has also been reported in both benign and malignant bone disease including Ewing's sarcoma, osteogenic sarcoma, fibrous dysplasia, Paget's disease and osteomyelitis. The evidence that it may assist in prediction of response to treatment, for example in breast carcinoma, is inconclusive.

Further reading:

European Journal of Nuclear Medicine 1994; ***21****:* *582–6*

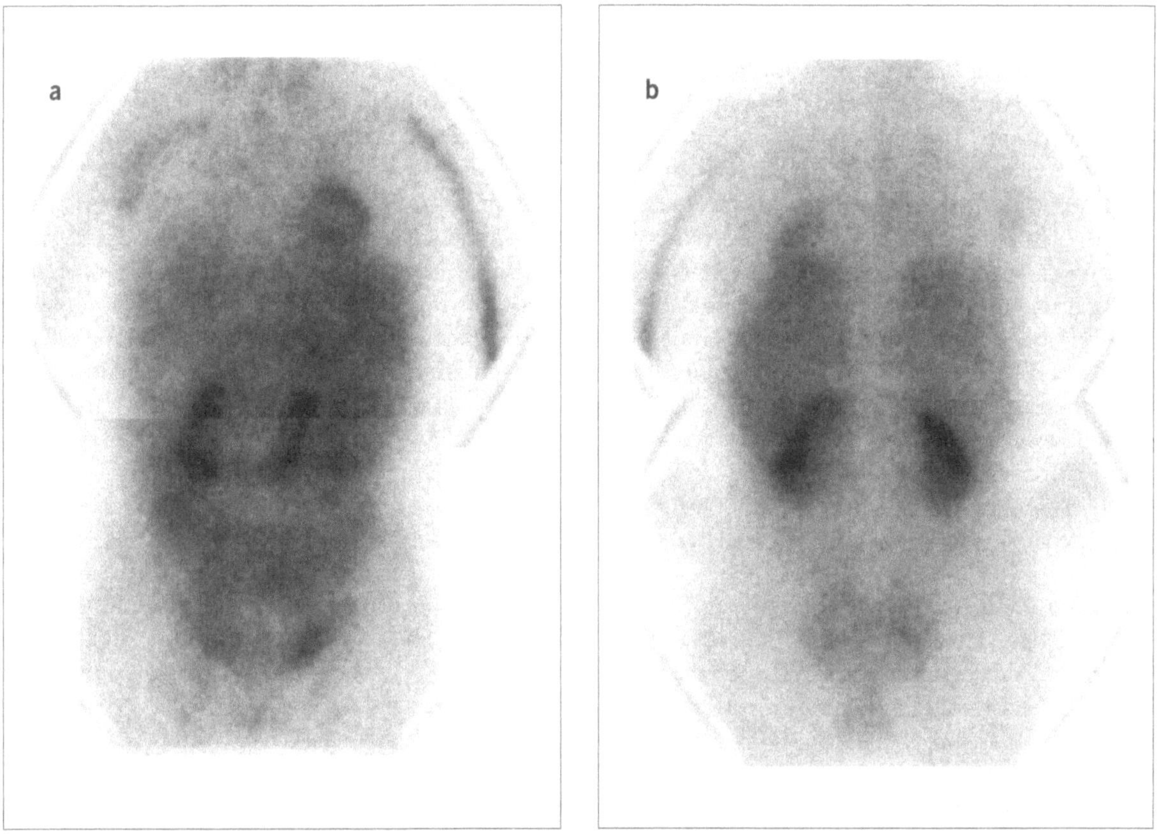

Figure 10.4 (a) Anterior, (b) posterior. In addition to uptake of thallium in thyroid, heart, liver, small bowel and kidneys there is absorption onto the vascular endothelium on the side of injection. There is also uptake in residual non-Hodgkin's lymphoma in the right axilla.

Pentavalent DMSA

Pentavalent technetium dimercaptosuccinate, DMSA(V), localizes in many malignant tumours, both primary and metastatic, including squamous carcinoma arising in the head and neck and most types of lung cancer. Uptake has also been described in amyloidosis, some benign tumours including haemangioma and in organizing pneumonia. In practice, apart from its use in medullary thyroid carcinoma (page 166), the sensitivity for detecting known lesions is appreciably poorer than other modalities such as CT, while specificity to differentiate benign from malignant is inadequate. The number of previously unknown sites detected is too few to justify routine clinical use.

Further reading:

Journal of Nuclear Medicine 1995; ***36****: 207–10*

Flurodeoxyglucose

Until recently, imaging with ^{18}F-labelled compounds was restricted to PET (positron emission tomography) centres. Some newer and especially dual-head gamma cameras are able to image at 511 keV. Coincidence counting is likely to become more generally available, thus improving sensitivity by more than one order of magnitude compared with FDG SPECT and facilitating more general clinical use of these tracers. Accumulation of ^{18}F-FDG in glial tumours is related to the phosphorylation ratio or magnitude of the 'lump constant'. An increased lump constant reflects increased activity of hexokinase enzymes 2 and 3, which are associated with anaerobic glycolysis. It is thus a function of the difference in phosphorylation rates between deoxyglucose and glucose rather than a simple indicator of glucose utilization. An FDG parametric image is principally influenced by the product of the lump constant and the metabolic rate for glucose. FDG uptake in tumours consequently reflects increased FDG accumulation rather than glucose utilization. Uptake is increased in many tumours and contrast may be sufficient to visualize metastases, for example from colon cancer, not detectable by other imaging modalities. Uptake is however also influenced by perfusion and occurs into inflammatory foci as well as tumour. Typical administered activities of ^{18}F when using a multi-ring PET scanner are 100–450 MBq.

Further reading:
Journal of Nuclear Medicine 1996; ***37****: 441–6*

Peptides

Receptors for a number of peptides have been identified on tumour cells. Some peptides, especially those containing tyrosine residues, have been labelled directly with ^{123}I. Analogues capable of being labelled with ^{111}In, ^{99m}Tc or other radio-isotopes have also been synthesized by attaching a suitable chelating moiety. Attractions of using a labelled peptide rather than an antibody include low or absent antigenicity and the possibility of chemical rather than biosynthesis, which facilitates the unambiguous characterization and purification required for licensing. Octreotide is the best characterized. Many types of cancer express somatostatin receptors (page 173). In some patients both somatostatin and vasoactive intestinal peptide (VIP) receptors are found in the same tumour, whereas others express one only. VIP receptors are in addition often found in ovarian, prostate, bladder and pancreatic cancers where there are usually no somatostatin receptors. Substance P is another peptide, receptors for which are found in most astrocytomas, glioblastomas, ganglioneuroblastomas, medullary thyroid and breast carcinomas. Receptors are found not only in tumour but also in surrounding blood vessels. Clinical applications of labels for the majority of these receptors are not yet established.

Further reading:
Journal of Nuclear Medicine 1995;
supplement: 1s–30s

Monoclonal antibodies

A number of surface and sometimes internal antigens are expressed to a greater extent in tumour than in the normal tissues from which they arise and may be shed into the circulation. Attempts to image tumour masses using polyclonal antibodies have in general found these preparations to be no more specific than labelled non-specific serum proteins. Such accumulation as does occur probably reflects increased capillary permeability to large molecules associated with newly formed and immature capillaries compared to those in normal tissue, rather than specific localized binding.

A number of monoclonal antibodies and antibody derivatives have been used experimentally. Antibodies are proteins which are classified, on the basis of molecular weight and electrical properties, as globulins. They are collectively referred to as immunoglobulins (Ig). Five classes have been identified. The basic structure of all consists of four polypeptide chains, two of which are small or light chains and the others large or heavy chains. The classes differ in the nature of the heavy chains, named respectively gamma, mu, alpha, epsilon and delta. There are only two types of light chain, kappa and lambda. In all immunoglobulins the basic structure of the molecule resembles the letter Y. The paired long-chain molecules are positioned adjacent to each other. Each bends at a similar point but angled away from each other. The greater part of the chain is constant in composition. Only the non-contiguous end varies, depending on antigen. This is known as the 'fragment of antigen binding' or Fab. The Fab site is unique to the particular antigen-binding properties of the immunoglobulin. The rest of the molecule is known as the crystalizable fragment and contains a segment specific for biological function, for example, opsonization or activation of the complement cascade. Light chains also contain a Fab site specific to the particular antigen, whereas the rest of the molecule is constant in composition. Light chains are attached to heavy chains so that the Fab sites on the long and short chains are adjacent and the light chain is positioned beyond the point of angulation of the heavy chain.

IgG is concerned with neutralization of toxins, agglutination, opsonization and bacteriolysis. IgM usually occurs as a ring of five Y structures connected in a circular or pentameric configuration by a joining J-protein, giving the molecule ten possible sites for antigen binding, although in most cases only five are used. It has the same functions as IgG, with the exception of opsonization and is the major antigen receptor found on B lymphocytes. IgM antibodies are the first to respond to foreign antigens in a primary response. The major function of IgA is the protection of mucus membranes. Secretory IgA interferes with the adhesion of bacteria to mucosal surfaces. It also neutralizes some viruses and toxins. IgG combines simultaneously both to antigens via the Fab site and to mast cells or basophilic sites via the C receptors, with the release of histamine, serotonin and other mediators. IgD is found on the plasma membrane of lymphocytes, playing an important role in recognition of antigens and initiation of antibody formation.

In addition to intact monoclonal antibodies, several derivatives have been studied. Digestion with the proteolytic enzyme pepsin removes part of the heavy chain, leaving a lower molecular weight fragment known as F(ab)2′. Digestion with the enzyme papain removes a larger part of the constant region of heavy chains, leaving light chains attached to greatly shortened residues of the long chains. Depending on the conditions these may be Fab fragments which are less than half of the molecular weight of the F(ab)2′ fragments or Fab′ fragments. One of the problems with labelled monoclonal antibodies is that if these are derived from non-human cells, as is commonly the case, they are themselves antigenic and may cause a reaction, especially on second or subsequent administration. The fragments are less antigenic and less likely to provoke a severe antigenic response. The risk of reaction can also be reduced by 'humanizing' the antibodies using chimeric hybrid cells to synthesize them. Smaller fragments are cleared mainly by the kidney and thus have a shorter biological half-life than intact antibody, which is metabolized principally in the liver. Label bound to fragments is therefore cleared more rapidly and imaging can be performed earlier with fragments than with whole antibodies. However it has been suggested that under some circumstances the longer biological half-life of intact antibody, by increasing the duration over which it is delivered into tumour, may give higher contrast.

Further reading:

European Journal of Nuclear Medicine 1995; ***22:*** *571–80*

Labels

Antibodies have been labelled with iodine, technetium, indium, copper and a number of other metals possessing radio-isotopes suitable for imaging or therapy. Tyrosine residues are in general required for iodination and all iodinated antibodies are ultimately deiodinated *in vivo*, with release of free iodine. Thyroid blocking is thus always required. Labelling with metallic radio-isotopes usually requires a suitable chelating group such as DTPA (page 82) to be attached to the molecule at a site where it has the least effect on antigen-binding properties. Techniques are well established and the resultant compounds are generally more stable *in vivo* than iodinated antibodies, resulting in higher contrast. Smaller molecular weight fragments, which clear from the blood more rapidly, can be labelled using similar techniques and are preferable when employing technetium, but many tested to date clear too slowly or provide insufficient contrast within the first few hours after injection for technetium to be used. Technetium has been used successfully in some situations, but most experience to date is with iodinated or indium-labelled antibodies.

Blood pool, inflammation and permeability markers

Labelled autologous erythrocytes (page 204) have been used to demonstrate local abnormalities of blood volume, especially in haemangioma of the liver and at other sites. The capillaries in tumours and in inflammatory lesions are more permeable than normal mature capillaries to most soluble molecules. The distinction is greatest in the brain but this phenomenon occurs in all tissues. As a consequence there is non-specific accumulation of many soluble injected tracers at sites where such blood vessels are present. Other factors which affect biodistribution include the extent of protein binding and lipophilicity. Pentetreotide (page 173) is taken up by activated lymphocytes in inflamed tissues (Fig. 10.5).

Specific diseases

Lung cancer

Gallium is taken up into a high percentage of primary lung carcinomas of all histological types. It also accumulates in areas of pneumonic consolidation, including those distal to an obstruction, and in many benign tumours. It is therefore not useful to differentiate tumour from infection or benign from malignant masses. Sensitivity for detecting hilar involvement from peripheral bronchogenic carcinoma is disputed. A few centres have claimed sensitivities approaching 100%, but most have not been able to replicate these results. The sensitivity for detecting extra-thoracic secondaries is probably lower than that for detecting hilar involvement. Reduction in gallium uptake after therapy may indicate response, whereas renewed appearance of uptake is evidence of recurrence.

Although there is detectable uptake of thallium into the majority of primary lung tumours, most comparisons have found thallium uptake to be less intense than that of gallium, giving lower contrast between tumour and normal tissues. The specificity is low, principally because of uptake into sarcoidosis and tuberculosis. Preliminary reports of the use of sestamibi suggest that it gives similar results to thallium. No clinical role is accepted for either in the management of patients with lung cancer.

Most published experience with FDG refers to PET, but similar results may be achievable with the new generation of gamma cameras now becoming available. Various techniques of quantification have been described, many complex. Their clinical utility is unproven and the sensitivity and specificity of the parameters obtained for differentiating benign from malignant, although high, is not good enough to eliminate the need for biopsy. The principal application for FDG may be to monitor response to treatment, as metabolic response may be a more accurate index than measures of volume.

Phase 1 trials of various monoclonal antibodies have reported promising results, with in some cases better accuracy than CT for staging patients being evaluated for surgery. Phase 2 studies are awaited. Uptake of pentetreotide has

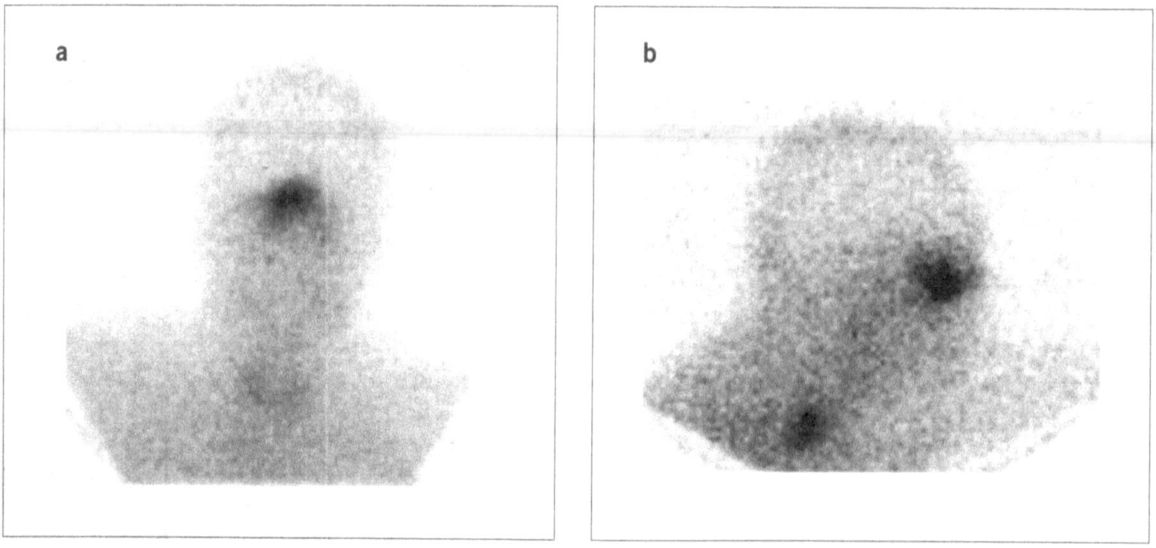

Figure 10.5 (a) Anterior, (b) right lateral skull. High uptake of pentetreotide in ethmoid sinusitis.

been observed into all small-cell lung cancers reported to date. Non-small-cell tumours do not contain somatostatin receptors but uptake is sometimes observed, possibly due to activation of surrounding or infiltrating lymphocytes. No clinical role has been established for any of these agents.

Further reading:

European Journal of Nuclear Medicine 1994; ***21****: 57–81*

Breast cancer

Accumulation of FDG is related to the S-phase of the cell cycle, the period of DNA replication. FDG imaging of lymph node involvement in untreated cancer of the breast is accurate when positive but the false-negative rate is too high for a negative FDG study to replace surgery to detect lymph node involvement. It has been suggested that combining ^{99m}Tc-sestamibi with mammography may be helpful in women with dense breasts. The optimal technique is a matter of dispute. The prone lateral projection provides a good separation between breast and thoracic structures. A positive predictive value of 98% and a negative predictive value of 85% in detecting axillary node involvement have been reported, but others have not achieved such good results in patients with impalpable axillary nodes. Efflux of ^{99m}Tc-sestamibi is related to the degree of P-glycoprotein expression in breast cancers before treatment. Those with a low concentration of P-glycoprotein in the tumours have a slow efflux rate, whereas faster efflux rates are found in lesions with high P-glycoprotein. This may predict invasiveness or, alternatively, which patients are likely to develop multi-drug resistance, but experience to date is limited.

^{99m}Tc-MDP (page 3) uptake is detectable in the primary tumour 20 min after injection in over 90% of patients, but in almost two-thirds this has faded by the usual time for obtaining skeletal images (2 h). Less than 5% of benign lesions show detectable uptake. This may complement mammography in women with dense breasts whose mammograms are difficult to evaluate. Uptake of thallium, labelled oestrogens, pentetreotide and monoclonal antibodies into breast cancers have also been reported. Their clinical role is not established.

Further reading:

European Journal of Nuclear Medicine 1994; ***21****: 357–62*

Journal of Nuclear Medicine 1996; ***37****: 615–30*

The primary pathway of lymphatic drainage of a tumour may be identified by injection of a colloidal preparation around the tumour. Biopsy of the most proximal node containing radioactivity, identified with a collimated probe at surgery, may be as accurate as axillary sampling for detection of axillary spread but carries a substantially reduced morbidity.

Lymphoma

Planar imaging after 75 MBq ^{67}Ga, employing a single photopeak (page 262), reveals uptake in about 95% of patients with untreated Hodgkin's disease in at least one involved site, but commonly uptake is observed in some sites but not others, giving an overall lesion detection rate of 80–85%. Slightly lower figures have been reported in non-Hodgkin's lymphoma. Because of this variable pattern, gallium scintigraphy has been largely abandoned for initial staging of Hodgkin's disease and other lymphomas. CT has a higher sensitivity for detection of enlarged lymph nodes in both conditions but cannot distinguish involved from non-involved normal size nodes.

More recently, using a combination of three photopeaks and higher administered activities to increase the detected count rate, supplemented by SPECT to improve contrast, lesion detection rates approaching 100% have been reported in both conditions, similar to those of MRI. Neither modality can reliably distinguish inflammatory from tumour masses. In Burkitt's lymphoma (small non-cleaved cell lymphoma) a sensitivity of 89% has been reported for site of involvement, with a specificity of 91%. Normal uptake at the lung hila may be difficult to distinguish from residual or recurrent tumour but is usually of lower intensity. False positives may be due to activity in colon, salivary glands or sites of focal inflammation. Gallium no longer forms part of initial staging protocols. It may have some role following chemotherapy or radiotherapy, when continuing gallium uptake indicates residual viable tumour, but the relative accuracy of gallium and thallium in this context is uncertain.

Preliminary results indicate that high uptake of sestamibi in lymphoma is associated with stable disease and a good prognosis, but more data are required before this can be recommended as a clinical procedure. FDG uptake is also a marker of active disease. Response to treatment is difficult to monitor as this depends on both tissue perfusion and metab-olism. False positives have been reported due to inflammatory lesions.

Further reading:

*Journal of Nuclear Medicine 1996; **37**: 46–50, 530–2*

*European Journal of Nuclear Medicine 1995; **22**: 434–42*

Primary liver tumours

Many intravenously administered tracers are taken up into hepatoma, including labelled amino acid analogues such as ^{75}Se-selenomethionine, ^{67}Ga-gallium citrate, ^{18}F-FDG, ^{123}I-insulin, pentetreotide and various monoclonal antibodies or their derivatives. Not all hepatoma have higher uptake than normal liver and high uptake is sometimes seen in non-malignant nodular regeneration. The tumours themselves are always detectable by other imaging modalities and none of the agents tested to date is able reliably to differentiate between hepatoma and metastatic tumours originating from other tissues. Scintigraphy has no accepted clinical role.

Prostate cancer

The monoclonal antibodies tested so far do not appear promising. In one typical study patients at high risk of prostate cancer were imaged with an ^{111}In-capromab pentetide which gave a positive predictive value of 72% and a negative predictive value of 76%. Other agents which have been evaluated but have not established a clinical role include radio-isotopes of zinc and labelled oestrogens.

Further reading:

*Journal of Nuclear Medicine 1996; **37**: 11–15N*

Colon cancer

A number of antibodies and antibody derivatives have been evaluated. A few centres claim to be able to improve on the accuracy of staging achieved by other techniques but the agents they use are not yet generally available. The contrast currently achievable is insufficient for therapeutic purposes but a great deal of effort is currently being concentrated in this field.

Further reading:

*Journal of Nuclear Medicine 1995; **36**: 430-41*

Ovarian malignancies

Ovarian carcinoma is the commonest fatal gynaecological cancer in more affluent countries, in part at least because of the lack of any practicable method of detecting most early disease. The principal incentive for imaging with radioactive tracers has been the limitations of both CT and MRI in the pelvis, especially following surgery, which in some centres has led to routine 'second look' surgery for re-staging. A number of studies employing various monoclonal antibodies have been reported, some labelled with technetium and giving a fairly good correlation with surgery. This must however be regarded principally as a research procedure. Until the therapeutic consequences of salvage treatment have been better defined none of the agents proposed can be considered clinically efficacious.

Further reading:

*European Journal of Nuclear Medicine 1995; **22**: 645–51*

Melanoma

Lymphoscintigraphy (page 218) with a ^{99m}Tc-labelled colloid has been suggested to optimize the surgical approach in patients with clinical stage I tumours. Normal variations in the number and site of draining lymph node make interpretation difficult and this technique has not achieved general acceptance. Initial studies of FDG uptake in small numbers of patients with cutaneous melanoma suggest this may be more accurate and cost-effective than CT to detect lymph node involvement, FDG detecting 96% of involved sites compared with 55% by CT. More data are required. Imaging with an iodinated benzamide derivative related to those employed experimentally to study cerebral dopamine (D2) receptors, ^{123}I-(S)-IBZM, has also been proposed because of the ectodermal origin of melanocytes and presence of melanin in the substantia nigra. Initial trials have been encouraging. Both primary and metastatic lesions have been detected with a maximum tumour-to-background ratio in planar images of 2.6, but hepatobiliary excretion of the tracer may limit detection of intra-abdominal lesions. It is unclear if radiopharmaceutical uptake is due to binding to membrane receptors or interaction with intracellular structures. Radiation dosimetry is similar to that in non-oncological patients. Uptake of gallium is seen in the majority of both primary and metastatic melanoma. It has been proposed as a technique to distinguish residual tumour from scarring but there is insufficient data at the present time to establish its clinical role. Other agents which have been employed experimentally but have not achieved routine status include several monoclonal antibodies, iodinated quinoline analogues and pentetreotide. Intraoperative detection of involved lymph nodes using β-emitting isotopes and specially designed hand-held imaging or non-imaging probes is another area whose potential has not yet been fully explored.

Further reading:
Journal of Nuclear Medicine 1994; ***35****: 1741–7*

Therapy

Antibodies

Although tumour uptake of most antibodies reaches a maximum within one day, slow clearance of residual unbound activity imposes a delay of several days before optimum imaging contrast is achieved. This is one of the most important limiting factors preventing therapy with radio-labelled antibodies. One approach to this problem is pre-targeting with a slowly cleared monoclonal antibody which also has a high-affinity binding site for a small rapidly cleared effector molecule. Contrast may be further enhanced if a second antibody, which accelerates clearance from the circulation of the first and enhances its intracellular uptake, is administered one to two days later, rendering it unavailable to the effector molecule. Pre-targeting may be achieved, for example, by labelling the initial antibody with a marker such as streptavidin. Circulating antibody can then be targeted by biotin. The conditions required to make this a practical therapeutic procedure are not yet adequately characterized.

Further reading:
Journal of Nuclear Medicine 1995; ***36****: 876–879*

Lipiodol

Lipiodol ultrafluide is a radiological contrast agent which has been used for many years. Selective retention by hepatic tumours following intra-arterial injection into the coeliac trunk or hepatic artery improves diagnostic sensitivity of CT. When radio-iodinated and injected non-selectively into a hepatic artery during conventional arteriography via the femoral route, retention by hepatic tumours is generally intense and fairly homogeneous. Between 80% and 85% of the administered activity is retained in liver, most of the residual 15–20% localizing in lung. Treatment usually comprises four to eight injections at intervals of three months. Circulating activity and thyroid binding of released free iodine are not significant. No preparation of the patient is required beyond that required for angiography. Tumour retention is on average four to five times higher than that of normal hepatic tissue. The mean administered activity is in the region of 2000 MBq, ranging from 1000 to 4000 MBq ^{131}I, diluted with 'cold' lipiodol into a volume of 10 ml and injected directly into the femoral catheter. With proper attention to the usual precautions when dealing with therapeutic quantities of radioactivity, radiation to the operator should be negligible and less than that received by the radiologist during arteriography.

A leaded screen is required to protect the operator during compression of the arterial injection point for approximately 5–10 min after removal of the catheter and the patient must be isolated in a licensed facility until radiation dose rate from the patient falls below statutory limits. The biological half-life is approximately 4–5 days and radioactivity is mainly eliminated via the urine. In small trials partial or complete relief of pain has been reported in up to three-quarters of patients, with some improvement in duration of survival.

Further reading:
Journal of Nuclear Medicine 1994; ***35****: 1318–20*

Endocrine tumours

See pages 163–7,.171–5.

Neuroendocrine tumours

These arise from the neuroendocrine cell system, characterized by the expression of certain marker proteins and cell type specific hormone products. They tend to form small clusters within other tissues including lung, thymus, thyroid and gut. The commonest group are those described as carcinoid tumours, but even within this classification there are considerable variations in symptomatology and malignancy. Symptoms are most commonly due to secretion of serotonin, but a number of peptides with hormonal activity may be present in addition or instead. Some produce no hormonal activity. In the majority of patients pentetreotide (page 173) is the radiopharmaceutical of choice to detect metastatic spread and small symptomatic primary tumours not identified by other means, but the majority of primary tumours are readily visualized by other modalities. Imaging is usually performed at 4 h and 24 h. In most cases contrast is high and SPECT is not necessary. Hepatic deposits are difficult to visualize because of uptake in normal liver but pre-treatment with 600 μg octreotide daily may improve their visibility, although it decreases uptake in extra-hepatic sites. The optimum duration of pre-treatment is not established. Other agents taken up in carcinoid tumours include MIBG (page 171) and thallium, but their clinical role is not established.

Tumours of the central nervous system

See pages 238–9.

Male reproductive system

Methods have been described to study the blood supply of the penis, scrotum or testes. Penile blood flow is principally a research procedure. Both inert gas wash-out (page 226) and first-pass techniques with autologous red cells (page 230) have been employed. A clinical role in management of impotence has not been established. Scrotal imaging is used in some centres for differential diagnosis of acute scrotal pain. Imaging of varicocoeles is proving useful when considering embolization.

Testicular disease

To be useful in clinical management of acute scrotal pain an imaging technique must be able to provide reliable information without delaying definitive surgery. Scrotal scintigraphy has established a role only in centres able to perform the procedure at very short notice.

Technique

The patient lies supine with the scrotum supported on a sling taped between the thighs and as nearly as possible in the middle of the field of view. If necessary adhesive tape is used to restrain the median raphe in the mid-line. A bolus of pertechnetate (300–500 MBq) is administered into a convenient peripheral vein and images are acquired at 12 frames per minute until 1 min after activity is identified in the testes. A single high count frame (>500 000 counts) is then obtained followed by a further frame for the same time but with a lead sheet separating scrotum from thighs.

Interpretation

In testicular torsion of less than 7 h duration the only abnormality is a halo of normal or increased activity around a photon-poor centre which is the hypovascular testis. When torsion is of longer duration, increased uptake is also seen in the dartos muscle. Acute epididymitis or epididymo-orchitis are associated with increased flow through vessels in the cord and linear or curvilinear areas of increased uptake in the scrotum. Trauma is associated with diffusely increased perfusion of and uptake in the scrotum and variable alterations to the pattern of flow through cord vessels. Perfusion of vessels in and around the cord is increased in abscess, in which there are linear or curvilinear increases in scrotal perfusion and markedly increased uptake with focal photon-poor regions in the scrotal images. The differences are often subtle and require experience to identify reliably.

Varicocoele

Varicocoeles are dilated veins of the pampiniform plexus which is formed from veins draining the testis and epididymis and is an important component of the spermatic cord. Distal to the superficial inguinal ring the plexus drains into three or four veins which, after passing through the inguinal canal, coalesce into a pair of veins which run adjacent to the testicular arteries. The right opens into the inferior vena cava just inferior to the renal veins. The left drains into the left renal vein. Varicocoeles are usually left-sided. They have been classified into stop or shunt types on the basis of Doppler findings and venography. The former may be an earlier stage with dilatation limited to the spermatic vein whereas the latter is associated with secondary dilatation and incompetence of the cremasteric system. The clinical importance of this distinction is not established, nor are the therapeutic implications.

Technique

The examination is performed with the patient erect and using red cells labelled *in vivo* (page 204). A dose of 250–300 MBq pertechnetate is administered using a rapid bolus technique (page 231). Frames are acquired at 12 per minute for 2–3 min followed by a single frame containing >500 000 counts. It is advisable to insert the cannula with the patient supine to minimize risk of syncope.

Interpretation

The appearance in the normal subject is shown in Fig. 10.6. Shunt-type varicocoeles show increased activity in both phases (Fig. 10.7). Stop-type varicocoeles show increased uptake in the equilibrium image but no abnormality in the flow image (Fig. 10.8).

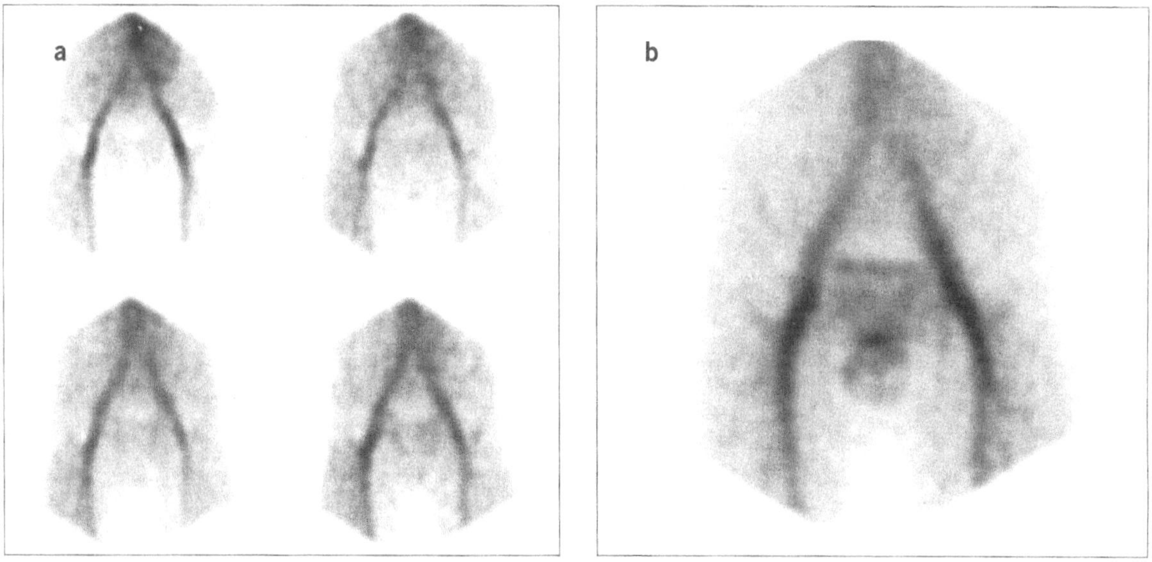

Figure 10.6 Normal appearance of (a) dynamic series and (b) equilibrium views.

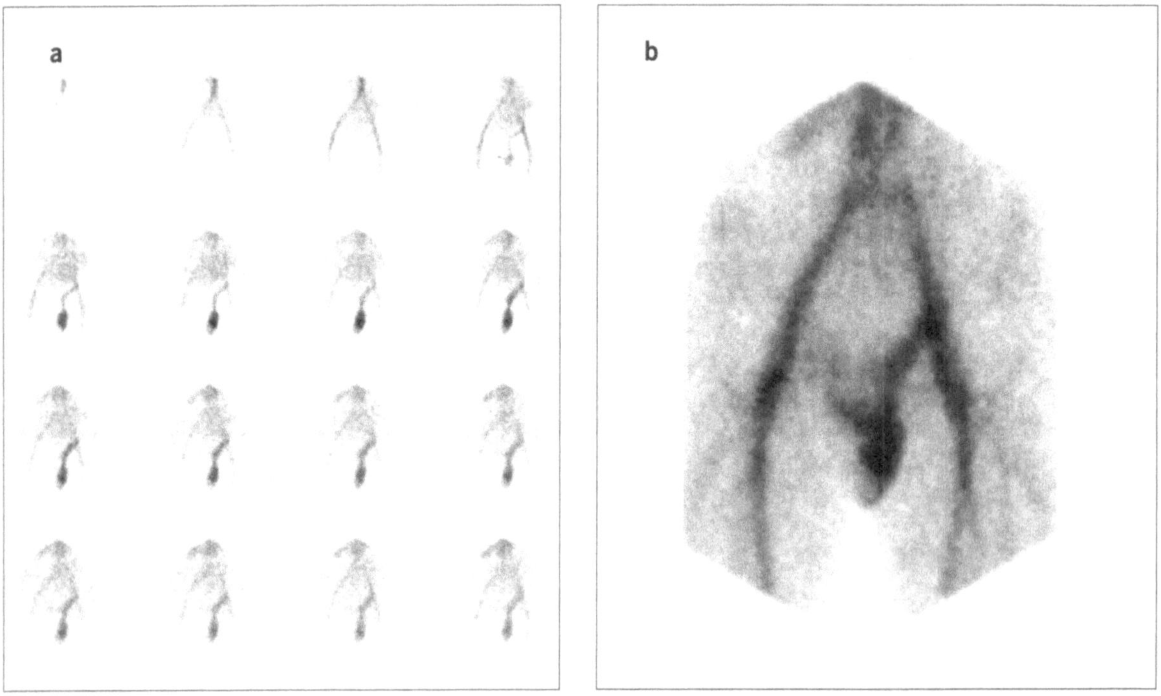

Figure 10.7 Shunt-type varicocele. (a) Activity is visible in the varicocele within 6 s of activity reaching the aortic bifurcation. (b) On the equilibrium view, the testis is seen as a photon-deficient region medial to the varicocele.

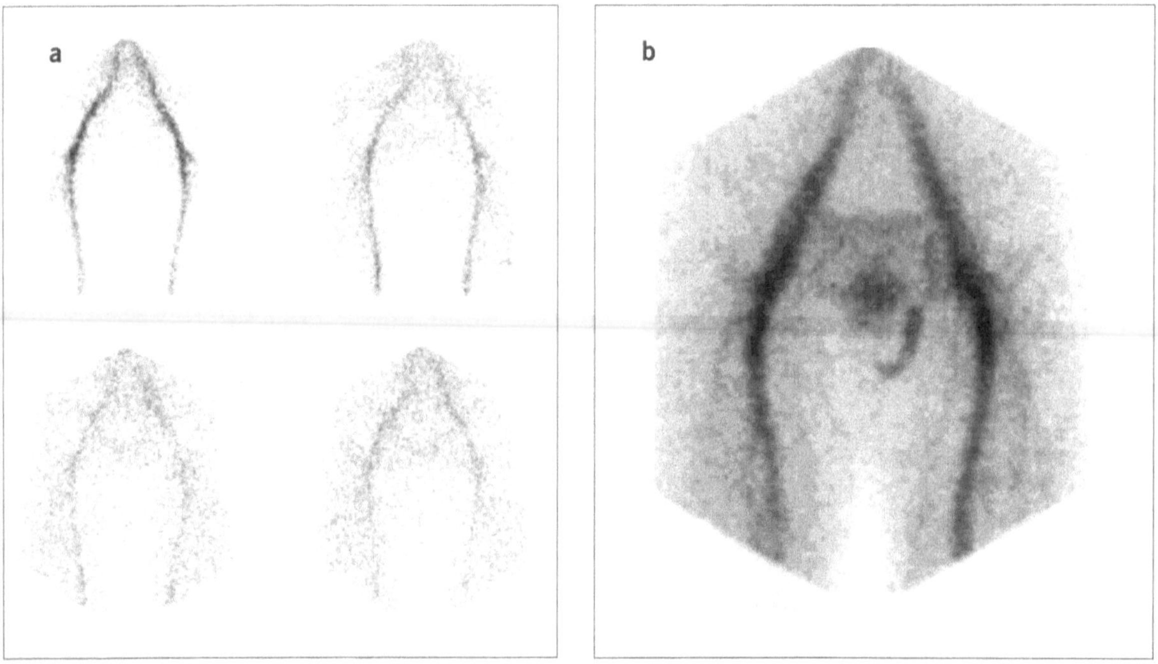

Figure 10.8 Stop-type varicocele. (a) The initial 12 frames are of 10 s duration, the last 6 of 60 s. The varicocele does not fill during the initial angiographic phase. (b) Blood takes several minutes to mix with the rest of the blood pool, indicating sluggish flow.

11 Paediatrics

The pattern of disease occurring in childhood differs substantially from that in the adult population, both in type and incidence. Congenital abnormalities are relatively more common, as are infection, trauma and their consequences. Malignancy and the effects of degenerative disease are comparatively rare and when malignancy does occur the types found are substantially different from those in adults. Any investigation which can be performed in an adult can also be undertaken in a child. It is, however, often necessary to modify procedures, especially when there is a difference in emphasis. As with other pharmaceuticals, administered activities should be based on a standard adult dose adjusted for body size using a nomogram relating the fraction of adult dose to body surface area, weight or age. It is a general principal that the quantity of radioactivity administered should always be 'as low as reasonably achievable'. This, the ALARA principle, is as important in children as adults. Most of the conditions for which children are investigated are non-lethal and because of their longer life expectancy radiation dose is always an important consideration, but to reduce the administered activity to a level such that the diagnostic quality of an investigation is impaired is clearly counter-productive.

Further reading:

Nuclear Medicine Communications 1993; ***14****: 827–9*

Technique

A child's behaviour is usually conditioned by both personal and parental prior experience. Fear of the unknown is often a major problem for both child and parent brought into an unfamiliar environment.

The importance of accurate prior briefing of the parent cannot be over-emphasized. When the appointment is made parents should be given a written description of what the procedure entails which they can take away with them and explain to the child at leisure. Illustrations are often helpful, especially if familiar characters are incorporated. Even so, it is essential that an adequate explanation is also given to the child at the time of attendance and before starting the investigation. The depth and nature of this must be adjusted depending on the age and level of comprehension of the child.

Sedation

Sedation is unnecessary for most children and is usually undesirable. No nuclear medicine procedure is painful, apart from transitory discomfort of the injection and provided that adequate explanation and reassurance are given the experience should be neither frightening nor disturbing. Mild sedation is best avoided. It often leaves a child dazed and bewildered, more frightened than when unsedated, less able to comprehend and less willing to co-operate or follow instructions. Tomographic procedures, when acquisition times are necessarily prolonged and immobility is essential, are sometimes an exception, but formal anaesthesia is often considered safer than heavy sedation because of the difficulty of maintaining an airway when depth of sedation is adequate to ensure that there is no movement.

Immobilization

Immobilization is sometimes advocated, particularly in small babies, using a board or other restraining device. Provided that the child is accompanied by a well-briefed parent or nurse, sufficient time is allowed for familiarization with the immediate environment and for them to be made comfortable before starting the procedure, such measures are usually unnecessary. The majority of children are able to tolerate investigations lasting 15–20 min without undue difficulty and many can be persuaded to remain still for longer. Infants usually fall asleep after a feed and it is preferable wherever possible to arrange imaging of babies to coincide with this, unless the procedure itself requires fasting. Most commercial software allows for correction of motion between frames in the horizontal and vertical planes during dynamic studies; packages without this feature are unsuitable for paediatric use. Absolute immobility is thus not essential. However it is not possible to correct for rotation or for movement during a frame. There is thus a case for using shorter duration frames in children than adults for similar investigations.

Whatever steps possible should be taken to reduce hostility of the environment, for example by decoration of the waiting area, examination room and equipment with familiar images such as cartoon characters. A waiting area equipped with toys and books, preferably separate from that for adults, also helps to minimize anxiety and overcome boredom. Specific needs of the child should always be taken into account, for example to be changed or go to the toilet. The attitude of adults communicates itself rapidly to children. Children rapidly sense uncertainty and find this disquieting. It is essential that all technologists and other staff who come into contact with children are properly trained, experienced in performing paediatric examinations, confident and dextrous handling both child and equipment. The whole procedure should be carried out with quiet and unobtrusive efficiency, expeditiously and as rapidly as is practical. As much as possible of the preparation must be completed before the child enters the examination room. It sometimes helps to allow the child to explore the room and to manipulate pieces of equipment such as the keyboard or mouse. Some small children feel protected when wrapped and find a vacuum cushion comforting whereas others find it disturbing but will remain calm with only gentle reassurance from a parent. A tape recorder and selection of appropriate tapes often helps to alleviate boredom. Parents or relatives should be encouraged to stay with the child and to provide diversion by reading or with familiar tapes or games. If it is necessary temporarily to restrain a child, for example during injection, this should not be undertaken by the parent. It is always important to reassure a child throughout the procedure and to congratulate them at the end. The child should leave the department with positive feelings which will make any future visit easier rather than more difficult.

Potential distractions such as the operator's console should either be completely out of sight or alternatively in full view. It should never be placed just behind the patient where it becomes an irresistible temptation to turn around in order to find out what is happening. Nothing should take place behind the child. The experience during a venepuncture is critical. Many centres use an anaesthetic cream but sufficient time (1–2 h) must be allowed for it to take effect. The only product of this nature currently licensed in the UK causes vasoconstriction and thus increases the risk of extravasation. An alternative product which acts more rapidly and causes vasodilatation is under evaluation but may be associated with an increased incidence of local reactions. It is often convenient to perform other procedures such as ultrasound while waiting for the anaesthetic to take effect. Venepuncture using a 25 gauge needle is often not noticed if a child can be distracted. It is usually better for the child not to actually observe the needle although occasionally children find this more reassuring. If blood samples are required, one of the varieties of indwelling needle designed specifically for paediatric use should be employed. If the procedure involves more than one injection it is usually preferable to use a similar device and leave it *in situ* for the duration of the examination rather than attempt more than one venepuncture, although securing the needle so that it does not become dislodged can be problematic.

Safety

One difference between radio-isotope investigations and most other clinical procedures is the potential, arising from radioactivity administered to the patient, to irradiate other members of the family. With technetium-based agents at dose levels usual in the UK, irradiation of, for example, a pregnant woman resulting from radioactivity administered to one of her children should not be a cause for concern. It is common practice to advise pregnant women to minimize their contact with any child who has received a radioisotope until the combination of excretion and radioactive decay have reduced the dose rate which might be received from the child below some arbitrary value.

However even physical contact with an adult who has received 740 MBq of a bisphosphonate until the administered activity has decayed completely would result in an absorbed dose of less than one-quarter of the average annual background. Dose to third parties may, however, become a problem when it is necessary to administer therapeutic levels of radioactivity, particularly of radio-iodine. Under these circumstances, and because of legal restrictions, it is usually necessary to admit the child to hospital until the combination of excretion and radioactive decay have reduced residual radioactivity to safe levels. This requires careful planning as facilities licensed for therapeutic administrations rarely have staff with current paediatric experience.

More commonly a child may be irradiated by an adult, most often a grandparent, to whom radioactivity has been administered for diagnostic or therapeutic purpose. The recommended standard practice of warning all patients who have receiving a dose of radio-tracer sufficient for this to be a potential problem should be strictly followed. Another, rare problem is accidental ingestion by children of radioactivity in breast milk. Pertechnetate, iodine and gallium are all concentrated in milk, as are isotopes of strontium and caesium. All except gallium if ingested are absorbed and concentrated: pertechnetate and iodine in thyroid, strontium in bone and caesium principally in muscle. Appropriate alternative feeding arrangements must always be made when these agents are administered to a woman who is lactating.

Further reading:

European Journal of Nuclear Medicine 1991; ***18****: 41–46*

Gastro-intestinal tract

Gastro-oesophageal reflux and pulmonary aspiration

Benign self-limiting gastro-oesophageal reflux occurs in the majority of neonates and should be regarded as normal. It is pathological only when associated with recurrent pulmonary aspiration, which may present with evidence of recurrent pulmonary infections, respiratory distress, wheezing or failure to thrive. Isotope investigation is indicated only in a small number of patients after exclusion of structural abnormalities such as tracheo-bronchial fistulae. A pH electrode in the lower oesophagus may be used to detect reflux but does not provide information about pulmonary aspiration. Scintigraphic methods use a physiological meal, milk, labelled with a non-absorbable tracer such as technetium colloid. There are two principal variations of technique depending on whether or not it is necessary to study the oesophagus during swallowing.

Technique

When looking for aspiration and reflux only, 4–10 MBq of colloid in water or a formula feed may be instilled directly into the stomach of the fasting infant via a naso-gastric feeding tube. This eliminates uncertainty whether aspiration is due to inco-ordination of swallowing or to reflux. Alternatively the child may take the feed directly from a bottle. A labelled fraction of the feed is given first, followed by the remainder which should be breast or formula according to usual practice. The child rests in its usual posture and scintigraphy (60 s per view) is performed at intervals of 5–10 min for the next hour. It is usually necessary to over-range the stomach and small intestine if activity in lung is to be visualized. If no aspiration is detected the child may be left for up to another half-hour in the left lateral decubitus position to maximize the opportunity for gastro-oesophageal reflux to occur. Further 60 s views are taken at intervals of 10–15 min with the child either supine or prone on the face of the collimator. Alternatively consecutive 60 s dynamic images may be obtained to detect transient aspiration or reflux. It is usually preferable to magnify the image so that it fills the field of view. A general purpose or high sensitivity collimator should be employed.

Further reading:
Journal of Nuclear Medicine 1995; ***36****: 351–4*

Abnormalities of oesophageal motility can be demonstrated using time-condensed images created from rapid sequence dynamic studies, as for adults (page 179). This technique can also be used for detecting gastro-oesophageal reflux. Aspiration is not commonly observed, not because of the low sensitivity of technique but because aspiration is an intermittent phenomenon and the probability of it occurring during any one short period of observation is quite low.

Clinical observation of the frequency of coughing, choking or spluttering during feeds is often as useful as any imaging technique. The largest source of error when detecting aspiration is contamination due to spilt or regurgitated activity on clothing. If in doubt the child's clothing should be changed and the skin washed. Methods of quantifying reflux by relating radioactivity in the oesophagus to that remaining in the stomach have been suggested but there is no evidence that these are clinically useful. Gastric emptying can also be measured, as with adults (page 183). The gastric emptying curve of the milk-fed child follows a complex pattern as milk is curdled and the tracer is incorporated into both solid and liquid phases. The result is usually expressed at the time to 50% emptying, the normal being 30–60 min. The test is more commonly useful in older children with juvenile-onset diabetes, who may develop a gastric autonomic neuropathy within a year of diagnosis which in turn may make control of the diabetes more difficult. In this group the standard adult technique, employing separately labelled solid and liquid test meals, should be employed.

Further reading:
Seminars in Nuclear Medicine 1995; ***25****: 339–47*

Meckel's diverticulum

Because this is congenital it is more likely to present in childhood than in adult life, but presentation is often delayed and many, possibly most, never give rise to any symptoms. There is however some difference in pattern of presentation between adults and children. Bleeding, which is associated with acid production by ectopic gastric mucosa, is commoner in childhood and Meckel's diverticulitis (in diverticula without gastric mucosa) in adult life. However either presentation can occur at any age. The technique is identical to that in adults (page 186), the dose of cimetidine being adjusted on a weight basis using a standard formula. Overnight starvation of babies is undesirable and unnecessary. The examination should however be scheduled shortly before a feed is due and approximately 4 h after the preceding feed. The importance of this timing must be emphasized to the parents. Normal feeding can be recommenced immediately the examination is ended, usually 30 min after administration of the pertechnetate. Other congenital anomalies containing gastric mucosa such as partial reduplications of small intestine, gastrogenic cysts and ectopic gastric mucosa in other sites are also occasionally demonstrated. (Fig. 6.11)

Inflammatory bowel disease

Scintigraphy using labelled white cells is the most sensitive method of demonstrating extent of disease activity in adults with inflammatory bowel disease. Published experience in paediatrics is more limited but such data as is available suggests that both sensitivity and specificity are as high in children as adults. Technetium-labelled cells are preferred to indium because of the lower absorbed radiation dose (Fig. 11.1). This necessitates imaging at earlier times, which may also be an advantage. Most of the false negatives which have been reported in adults occurred in large bowel Crohn's disease. There is insufficient data in children to know whether a similar pattern prevails. The principal difficulty when performing this examination in children is the relatively large volume of venous blood required to obtain sufficient white cells for labelling. As this condition more commonly occurs in older children, an adequate blood sample (20 ml or more) can usually be obtained.

Further reading:

British Journal of Radiology 1996; ***69****: 508–14*

Liver and hepato-biliary system

The liver comprises approximately 5% of body weight at birth, compared with 2% in the adult. Variations in shape are as common in the child as in the adult. Variations in position, apart from the rare anomaly of situs inversus, are uncommon. This is sometimes associated with other anomalies including congenital heart disease and absence of the hepatic portion of the inferior vena cava. These anomalies are best identified by ultrasound and MRI (magnetic resonance imaging). Malposition of the liver may also result from congenital defects in the diaphragm. These also are best detected by ultrasound, which is the investigation of choice when space occupying lesions are suspected.

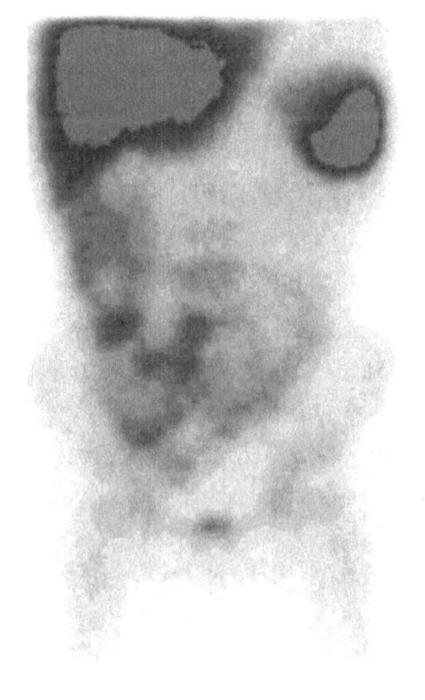

Figure 11.1 An 11-year-old girl with abdominal pain. Barium follow-through considered normal. At 3 h after Tc-white cells there is extensive uptake throughout the small bowel. Some skip areas can be identified. Crohn's disease confirmed.

Jaundice

The principal paediatric application of cholescintigraphy is for the differential diagnosis of neonatal jaundice, to differentiate biliary atresia requiring a surgical drainage procedure from neonatal hepatitis for which medical management is appropriate. Both present as progressive jaundice which, unlike physiological jaundice of the neonate, fails to resolve. In this age-group bile ducts cannot be identified by ultrasound even when dilated. However if radioactivity is observed in gut lumen following intravenous administration of a compound excreted by the liver, biliary atresia can for practical purposes be excluded. Excretion of all hepato-biliary agents is impaired in the presence of a raised serum bilirubin concentration, which competes with hepato-biliary agents for the same excretory pathways. Barbiturates increase the activity of many enzymes in the liver and enhance hepato-biliary excretion even in severe jaundice. Cholescintigraphy in a jaundiced child should be preceded by five days of oral phenobarbitone at a dose of 5 mg/kg/day. Even in severely jaundiced patients this augments biliary excretion to a useful extent, significantly reducing the error rate.

In the presence of jaundice the radiopharmaceutical of choice is ^{99m}Tc-*m*-bromo-*o*-*p*-trimethyl-iminodiacetate (Mebrofenin) as it has the least renal excretion of available compounds. If this is not available *o*-diisopropyl-iminodiacetate (Diisofenin) may be used. Other agents including iodinated compounds such as Rose Bengal are now obsolete for this application. An administered activity of 2 MBq/kg ^{99m}Tc, with a minimum of 10 MBq, is adequate. Initial imaging at 30 min and 60 min usually shows uptake only in liver. Blood clearance measurements or hepatic time–activity curves can be employed to give a measure of hepatic function but rarely add clinically useful information. Some renal activity is commonly visualized. If there is uncertainty whether an observed collection is in gut or in the renal tract, oblique or lateral projections are commonly helpful to confirm whether extrahepatic radioactivity is posterior (in the urinary tract) or lies more anterior in gut. Imaging should be repeated at intervals up to 24 h or until activity is definitely identified in gut (Fig. 11.2). Initially one image per hour is adequate but it is not necessary to continue imaging throughout the night. Even in severe neonatal hepatitis some excreted activity in the gut is virtually always visible within 24 h. This was not true of older iodinated preparations, which did require more extended imaging. Non-visualization of gut activity at this time indicates biliary atresia. In cases of continuing uncertainty counting of stool is sometimes helpful, although it is difficult in neonates to obtain stool samples which are uncontaminated by urine. Post-operative imaging is helpful to confirm patency after a drainage procedure and if bile leak is suspected.

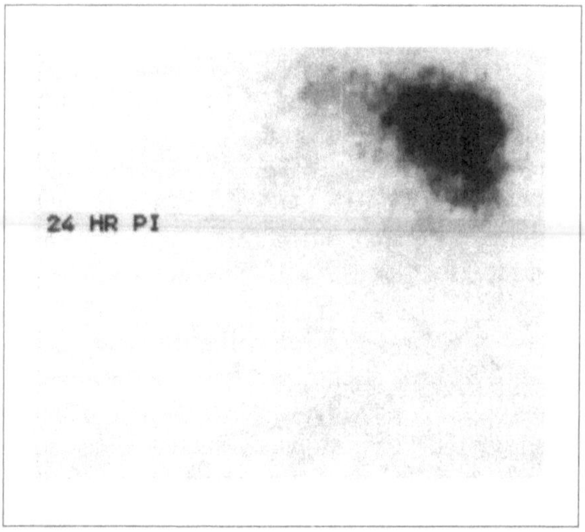

Figure 11.2 Two-week-old baby with increasing jaundice. At 24 h after Tc-diisofenin there is some residual activity in the liver but the urinary tract has cleared completely. No activity was visualized in bowel at any time. Biliary atresia confirmed.

Further reading:
Seminars in Nuclear Medicine 1996; ***26****: 25–42*

Choledochal cyst

A further abnormality which may be diagnosed by these agents is choledochal cyst. This is a congenital diverticulum of the common bile duct. Although the existence of a cystic lesion is readily demonstrated by ultrasound, cholescintigraphy is necessary to confirm communication between cyst and biliary tree. This communication may be small and imaging may be required for up to 24 h to confirm or refute a connection. Cholescintigraphy is also occasionally useful to demonstrate patency of other biliary-enteric anastomoses and post-operative leaks. The technique is similar to that employed in adults (page 251).

Cystic fibrosis

See page 249.

Lung

The number of alveoli and small pulmonary arteries increases rapidly during the first 12 months of life and subsequently more slowly for several years. Adult levels are not reached until some time between the fourth and twelfth year of age. The commonest adult indication for lung scintigraphy, namely diagnosis of pulmonary thrombo-embolic disease, is rare in children. The commonest indications for radio-isotope investigation of the lung in children are to assess the degree of pulmonary dysfunction in conditions such as congenital lobar emphysema (Fig. 11.3), cystic fibrosis or in the presence of structural abnormalities of the thorax. Ventilation scintigraphy is sometimes advocated for evaluation of regional lung function prior to segmental resection, in diagnosis of pulmonary sequestration and to supplement other methods for diagnosis of bronchial obstruction due to extrinsic compression or aspiration of foreign bodies. For all of these, the ventilation study is more important than perfusion. It is rarely

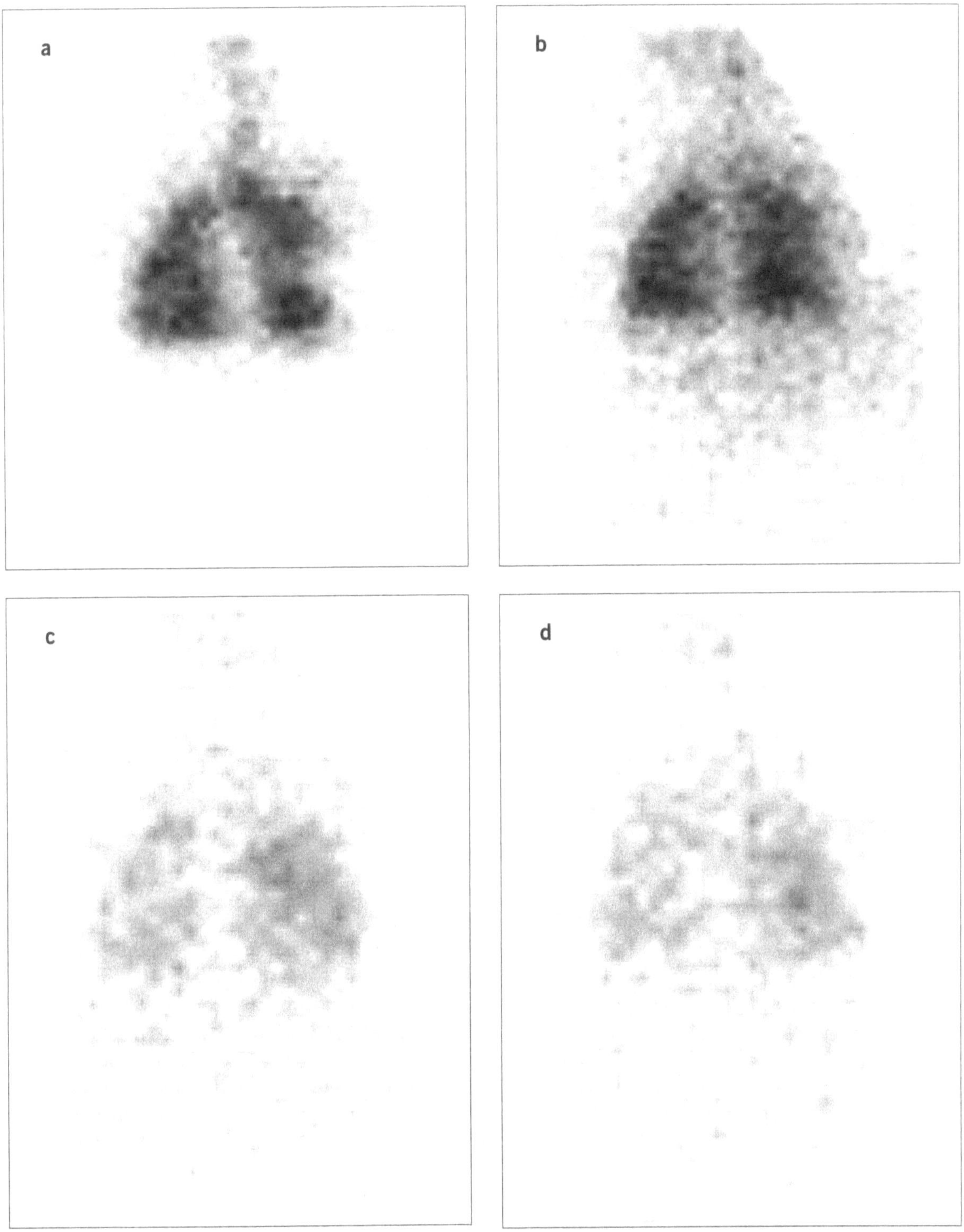

Figure 11.3 (a) Ventilation (first breath) image in four-month-old girl with recurrent chest infections. There is a defect in the right mid zone. (b) After rebreathing in a closed circuit for 4 min this defect has filled in. (c–f) Consecutive 15 s wash-out images. Rapid wash-out from left lung and from right upper and lower zones. Delayed wash-out from right mid-zone. Congenital emyphsema of middle lobe. (**e,f** *overleaf*)

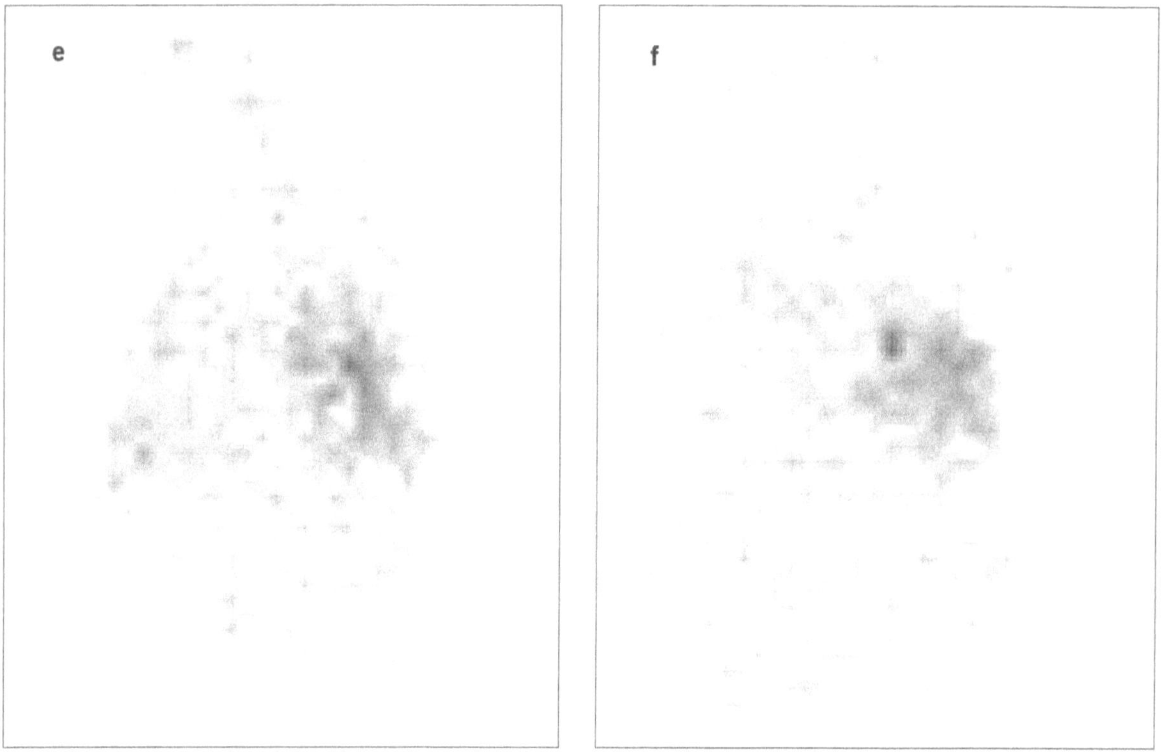

Figure 11.3 *continued*

appropriate to perform a perfusion study in children without a ventilation study, whereas there are relatively more indications than in adults for performing ventilation without perfusion.

Radiopharmaceuticals

The choice of radiopharmaceutical for ventilation studies is between ^{133}Xe and ^{81m}Kr. Aerosols are rarely appropriate and there are few published data on their use in children. All three phases (page 60) of a ventilation study, namely ventilation, lung volume and wash-out are usually essential. ^{133}Xe is thus usually the radiopharmaceutical of choice, even when krypton is available. Radiation dose is minimized by ensuring that the wash-in period is kept as short as possible. Because of the rapid elimination, radiation dose from ^{133}Xe is low and the advantage of obtaining a physiologically sound study should not be offset by reluctance to obtain more than one projection if this is appropriate. With the ^{81m}Kr equilibrium technique (page 62), when rate of delivery equals the combined rates of wash-out and radioactive decay, distribution of count rate in healthy adults with an average minute ventilation of 2 l is mainly a function of regional ventilation. As minute ventilation rate rises, the distribution of radioactivity comes more closely to represent lung volume rather than regional ventilation. In the dyspnoeic infant values up to 13 l/min may be observed. It is therefore incorrect to assume that images obtained at equilibrium with ^{81m}Kr demonstrate only regional ventilation. Some apparently discordant observations have caused doubts to be expressed about this but in severe lung disease, for example broncho-pulmonary dysplasia, the time required for complete mixing is substantially greater than is usually considered adequate in adults and may be more than 10 min. Thus a short-lived gas may never equilibrate in areas where the rate of gas exchange is slow, so that distribution of krypton in such regions is not a true representation of either ventilation or lung volume. Krypton studies in children should therefore be interpreted with caution, especially when lung function is abnormal.

Technique

In small infants it is usually fairly easy to hold a face mask securely in place, although it may be difficult to obtain a good gas-tight seal. Suitable masks and ventilation circuits with a dead-space appropriate to the respiratory volume are not readily available. The dead space in standard adult circuits is too great for paediatric use. One major problem is that reducing the calibre of tubing to minimize dead space increases resistance of the circuit. It is difficult to ensure that this increase in resistance is insufficient to cause respiratory embarrassment, especially in ill neonates. It is often necessary to call on the services of a paediatric anaesthetic department to obtain suitable equipment. Older children with chronic lung disease are often sufficiently familiar with the use of a face mask that they are able to co-operate well. However, because of the length of time for which a high degree of co-operation is required, a full ventilation study is one of the more difficult examinations to perform in young children.

Interpretation

The influence of posture on lung function is well documented. Gravity plays an important part, so that at rest in the erect subject more caudal segments of lung are both better ventilated and better perfused. When the subject is recumbent more dependent areas are better ventilated and perfused, provided that posture does not interfere with movement of the chest wall. In children, dependent lung may remain better perfused while the upper parts are better ventilated. This has practical implications when positioning a patient with chronic lung disease. When evaluating regional lung function in these children it may be necessary to assess the effect of change in posture, for example by repeating the study in both right and left lateral decubitus. In obstructive airways disease, air trapping may be transitory and detected in one projection but not a short time later in a second. It is often not clear whether this is due to changes in ventilation which are posture-dependent or whether there is intermittent bronchial obstruction.

Further reading:

European Journal of Nuclear Medicine 1991; ***18****: 41–66*

Clinical indications

The principal clinical indications are to determine regional lung function, especially in patients with congenital lobar emphysema and cystic fibrosis, in evaluation of the haemodynamic and pulmonary effects of sequestration and for evaluation of regional lung function disturbances in children with structural abnormalities in the thorax, such as scoliosis.

Heart

Radionuclide angiography has been employed to delineate flow patterns in patients with congenital anomalies, to measure the size of shunts and to assess ventricular function. The former has now been almost entirely superseded by MRI. Shunt quantification can be performed as in adults (page 146). An intravenous bolus of any soluble tracer is given into a large peripheral vein and time–activity curves are obtained from the lungs or at a distant site such as the head. In the case of a right to left shunt, the main first-pass peak will be preceded by a smaller one. Each can be fitted by a suitable mathematical function (gamma variate, Appendix 2) to give a fairly accurate approximation to the single-pass curve without recirculation. The ratio of shunt size to left ventricular output is equal to the ratio of the area under the shunt peak to that under the principal peak. Absolute size can be calculated if cardiac output is measured (page 146).

Left (and right) ventricular ejection fraction may be measured in children receiving cardiotoxic chemotherapy in order to detect the earliest signs of cardiotoxicity. Both first-pass and gated techniques are identical to those in adults (pages 135, 137–42). If required, shunt size and ejection fraction can be measured at the same examination. Reduced uptake of thallium has been reported in patients with cardiomyopathy. The findings are non-specific.

Brain

The pattern of distribution of regional perfusion tracers depends not only on blood flow but also on blood–brain partition coefficient of the particular tracer. At birth, brain is not fully myelinated. The distribution of lipophilic compounds is determined as much by local chemical composition of brain as by blood flow. In the premature infant, activity tends to be localized centrally, mainly in basal ganglia. After birth, uptake by cortex parallels development of brain and myelinization, with uptake appearing first in parietal, then occipital and finally frontal regions. The adult pattern is not achieved until approximately two years of age. Because of the prolonged duration of SPECT (single photon emission computed tomography) acquisition, anaesthesia is often required but this may alter global cerebral perfusion and pattern of tracer distribution. Measurements of global and regional perfusion using xenon give lower cortical values at birth than those for adults. These increase, reaching a maximum at 5–6 years of age, and then slowly decline to reach adult values at 15–19 years. This work is however subject to the criticism that partition coefficients were assumed from adult values and were not measured directly. These are therefore likely to underestimate perfusion, especially in the youngest subjects.

Further reading:
*Journal of Nuclear Medicine 1992; **33**: 696–703*

Clinical indications

Probably the most common single indication for brain SPECT with regional perfusion tracers in infants and children is to localize an initiating focus in patients with intractable partial epilepsy. As in adults, injection during a fit which identifies an area of increased uptake (Fig. 8.11) has a higher sensitivity and accuracy than interictal injection, when an area of reduced uptake is sought. Inducing hypercapnia by prior administration of acetezolamide is said to improve sensitivity of interictal scans, but the number of cases reported is small. Reduced perfusion has been recorded in children with craniostenosis and raised intracranial pressure but this is not a routine clinical application. Abnormalities of perfusion have also been noted following head injury and ischaemic brain damage. There is a possible application in patients with the rare condition of congenital dysphasia, a developmental speech disorder not explained by deafness, phonation disorder, mental retardation, neurological lesions or psychiatric disease. CT in these patients is normal but regional perfusion studies demonstrate hypoperfusion in the inferior frontal convolution of the left hemisphere involving Broca's area of those with expression impairment, whereas those with global dysphasia show two hypoperfused areas, one in the left temporo-parietal region and a second in the upper and middle areas of the right frontal lobe. Uptake of both ^{201}Tl and ^{99m}Tc-sestamibi has been proposed to evaluate viability of cerebral tumours following radiotherapy but as in adults, the clinical utility of this is not generally accepted. Other indications for brain scintigraphy in children are similar to those in adults and there is a similar divergence of views regarding their current clinical role.

Further reading:
*Seminars in Nuclear Medicine 1995; **25**: 165–82*

Urinary tract

Maturation

The full adult complement of nephrons is present in kidney by the 36th week of gestation. Growth subsequently is by hyperplasia of glomeruli and increasing convolution of tubules as the loop of Henle elongates. Renal function depends on body size. In order to simplify recognition of a normal range, clearance measurements, including glomerular filtration rate, are conventionally expressed after normalization to a 'standard' body surface area of 1.73 m^2. This is achieved by multiplication of the calculated clearance by a factor estimated from height and weight, by reference to a table or nomogram. Expressed in this way lower normal values are obtained at birth than in the adult, increasing from about 30 ml/min/1.73 m^2 at birth to about 80% of the adult value of 80–120 ml/min/1.73 m^2 by the age of two. Normalized clearance of agents excreted by tubules, such as orthoiodohippurate and MAG3, (page 81) is also lower in the neonate than the adult. Some of this difference is an artefact caused by inaccuracy of the nomogram used to calculate surface area from weight and length (height). Although reasonably accurate in subjects over the age of two, the assumptions made becomes progressively less accurate as body size decreases. The apparent discrepancy between neonatal and adult values is smaller if renal function is expressed relative to extracellular volume, which is physiologically a more realistic parameter for comparison. The principal difficulty with this approach is that extracellular volume is an ill-defined concept, usually taken to be initial volume of distribution of whichever tracer is in use. The standard reference compound (inulin) has a smaller volume of distribution than ^{51}Cr-EDTA or ^{99m}Tc-DTPA (page 101), which in turn have smaller volumes of distribution than bromide, another so-called reference marker. Glomerular filtration rate (Cr-EDTA clearance) at full term is 6–7 ml/min/l extracellular fluid, compared with typical adult values of 10 ml/min/l extracellular fluid. Although normal values are more dependent on tracer if they are expressed relative to initial volume of distribution than to body surface area, the difference between tracers is small. In practice, provided it is appreciated that difference in function between the neonate and older children is exaggerated by the use of surface area for normalization and the correct normal range for the age is used, there should not be difficulty in clinical interpretation.

Extravascular fluid volume is proportionately larger in the neonate than in the adult. Renal time–activity curves obtained with filtered agents such as DTPA are in consequence flatter in normal neonates than in adults. Because extracellular fluid volume decreases as maturation progresses, any renal pathology is associated with an apparent improvement in shape of the renogram curve if the examination is repeated later. Any difference is compounded by renal maturation. ^{99m}Tc-MAG3 has high protein binding and thus a smaller volume of distribution. This, combined with its higher extraction efficiency, gives time–activity curves even in the neonate which are not dissimilar from those seen later. The indications for diuretic renography in children are similar to those in adults, suspected pelvi-ureteric junction obstruction being the most common. Particular care must be taken to avoid dehydration of small babies.

Although it is sometimes said that renal studies in the neonate should be delayed until two to four weeks of age to allow maturation to occur, good quality studies can be obtained in full term and even slightly premature neonates (Fig. 11.4). The reported poor uptake of DMSA (page 95) does not accord with the experience of the writer. DMSA studies in neonates are often of high quality as pinhole images can be obtained while the child is sleeping. Anatomical delineation is often better than can be obtained in older children, who are more difficult to restrain and in whom pinhole imaging may be impractical. It is sometimes possible to diagnose renal dysplasia (Fig. 11.5). The principal disadvantage with DMSA pinhole imaging is that small differences in positioning or angulation can substantially alter the geometry and therefore the apparent differential count rate from the kidneys. Thus although pinhole imaging allows the best resolution to be obtained, where quantification of relative function is required a conventional

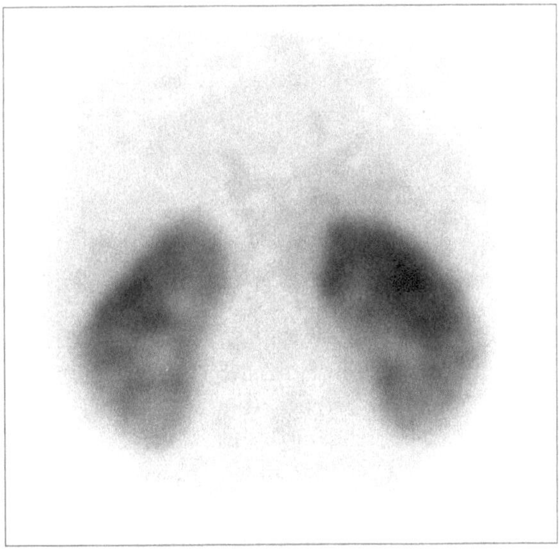

Figure 11.4 Normal pinhole DMSA image in unsedated two-month-old baby. Note clarity of pyramids and calyces.

parallel hole collimator must also be employed. It is preferable to obtain both pinhole and conventional parallel-hole collimator images routinely.

Further reading:

Journal of Nuclear Medicine 1994; ***35****: 438–44*

European Journal of Nuclear Medicine 1994; ***21****: 12–16, 1333–7*

Reflux

Vesico-ureteric reflux can be detected by direct or indirect radio-isotope voiding cysto-urethrography (DRVC or IRVC) (page 104). The former resembles the radiological micturating cysto-urethrogram (MCU) and requires catheterization to fill the bladder with a solution of pertechnetate. IVRC is performed shortly after the end of a standard MAG3 renogram, by which time little radioactivity usually remains in the renal area. Catheterization is not required. IVRC can also be undertaken following a Tc-DTPA renogram but the slower rate of excretion is associated with a higher background and hence poorer contrast. Reflux is less likely to be detected if urine output is high. Diuretics and excessive hydration are therefore best avoided. If a diuretic is required for other reasons (page 89) there is nothing lost by performing a voiding study, but little reliance can be placed in a negative finding. Acquisition at a rate of 12 frames per minute should be started 30 s or more before the child is instructed to void and collection continued for as long as necessary. Unsubtracted time–activity curves from elliptical ROIs (page

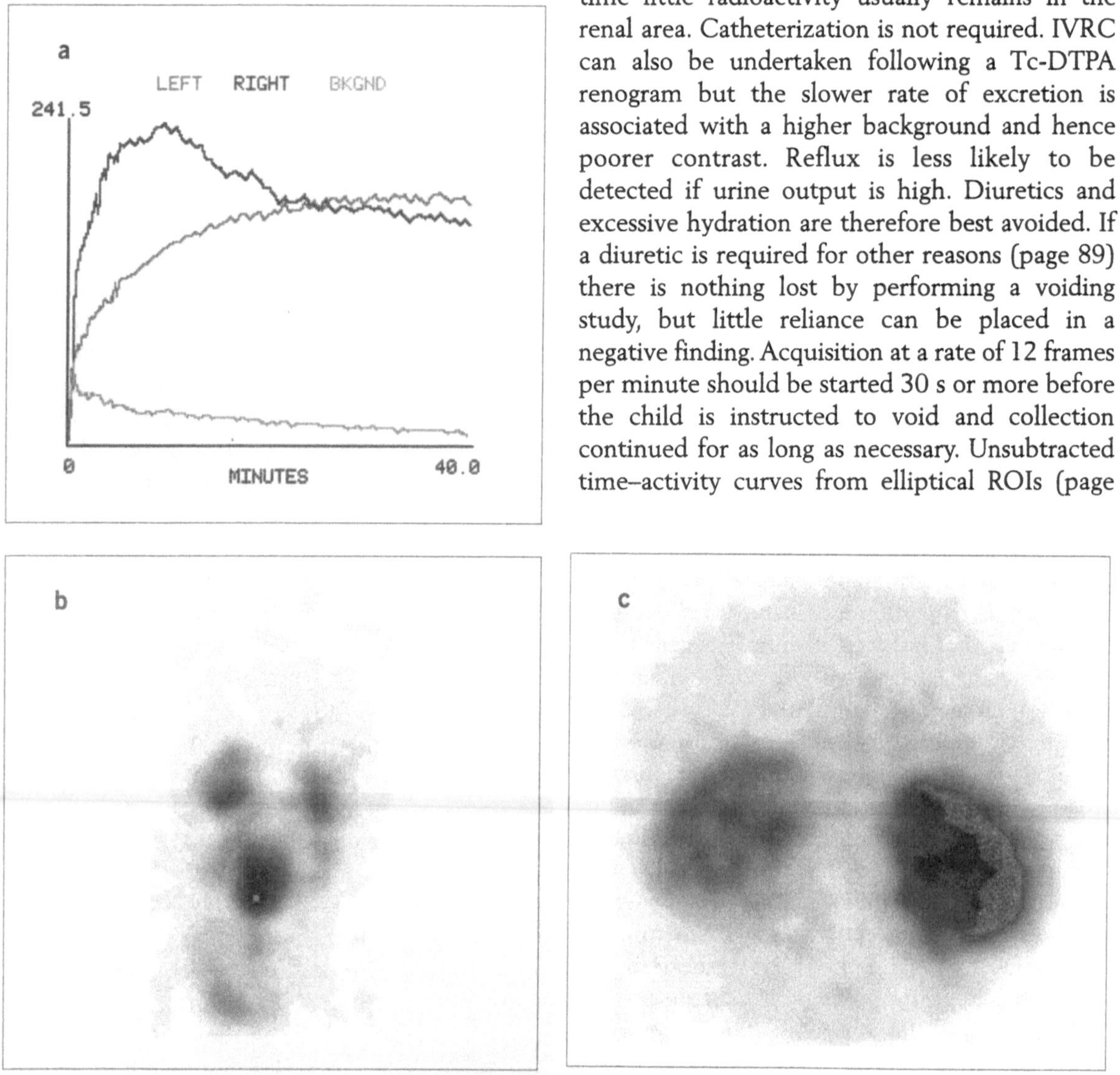

Figure 11.5 One-month-old baby with *in utero* ultrasound diagnosis of bilateral hydronephrosis. (a) Renogram shows delayed peak and slow drainage of MAG3 on right and depressed function with a rising curve on left despite diuretic. (b) At 40 min post injection the bilateral hydronephrosis and hydroureters are visible. Full bladder. (c) Pinhole DMSA image shows cortical loss from both lower poles and the left upper pole. Reduced uptake on left consistent with renogram diagnosis of obstruction. Urethral valves and left vesico-ureteric junction obstruction confirmed.

308) drawn around each kidney including renal pelvis and the bladder should be reviewed, each normalized to its own maximal value, otherwise the count rate from bladder is so much greater than that from the kidneys that fluctuations in count rate cannot be seen. If there is an increase in count rate in consecutive frames, it is necessary to confirm whether this is due to random noise, movement or other artefact or reflux. Statistical fluctuation may be excluded by confirming that the increase exceeds three standard deviations (3 SD) of the counts in the frame before that in which reflux is suspected. If count rate continues to rise for several frames take the minimum and maximum. A cursor which displays the number of counts or count rate at any selected point on the curve should be a standard feature of all software. Count rate must be converted to counts (that is multiplied by frame time) to calculate standard deviation, which is the square root of number of counts, not of count rate.

If an increase in count rate appears significant the relevant frames must be inspected, using a display which relates to the maximum pixel count in the whole study and does not normalize each frame to its own maximum pixel (Fig. 3.17). It is impossible to detect reflux with the latter display. It often helps to switch between colour scales. It is always necessary to over-range the bladder and if there is a large difference in count rate between the kidneys it is sometimes necessary to evaluate the kidneys separately, over-ranging the one with a higher count rate in order to see the other. A 'cine' display is sometimes useful to detect movement. The purpose of inspecting the frames is to exclude artefacts such as movement or an extravasation into an arm coming into the ROI. Gross reflux can be visualized on the unprocessed frames but commonly it is necessary to reduce the noise in the picture. Simple nine-point smoothing is of little help but filtering with a Butterworth or similar filter is more effective. It is not possible to suggest particular values for these filters as the values depend on implementation, which differs between suppliers. It is normally preferable to report from the monitor as there is no reliable hard copy to demonstrate all the data in a voiding study.

Clinical indications

Urinary tract infections

The most common indication for radio-isotope investigation of children is following urinary tract infection. Two approaches are used, either acutely during infection to distinguish pyelitis from pyelonephritis or after treatment of an acute event comprising a pyrexial illness associated with symptomatic significant bacteriuria, to determine if there is vesico-ureteric reflux or residual scarring. Advocates of the first approach claim this identifies a group at risk who need careful follow-up. There is however no long-term study to confirm the importance of early differential diagnosis.

Long term follow-up of children investigated after the acute event has identified three risk factors: reflux, scarring and further infection. Neither reflux nor scarring in isolation is associated with an increased risk of progressive renal damage. However if both abnormalities are present, or if one imaging abnormality is confirmed and in addition further infections are documented, the risk of progressive damage is approximately 20 times greater than in children with no risk factors or in whom only one abnormal feature is present. Further infections occurring in children without either scarring or reflux at presentation are associated with a low risk of progressive damage. DMSA scintigraphy is the most sensitive single investigation to detect scarring (Fig. 3.16) but must be interpreted in conjunction with ultrasound, as scintigraphy alone fails to identify concentric cortical thinning which is present in up to 10% of children with cortical loss. Ultrasound alone on the other hand misses the majority of children with segmental defects.

The IVRC is more sensitive than micturating cystourethrography for detection of reflux. Children with a normal indirect isotope voiding study are significantly less likely to suffer progressive renal damage than those with a normal MCU because of the high false negative rate (up to 50%) of conventional radiological MCU. Vesico-ureteric reflux is not a risk factor for progressive renal damage in boys under the age of one. Reflux is important in boys over one year and in girls of any age, but only when associated with scarring or further infection. The optimum follow-up protocol has not been established. There is probably no indication for repeat imaging studies in children who do not have both reflux and scarring unless several further infections are documented. Even then it is unclear how this should influence management.

Further reading:

Archives of Diseases in Childhood 1995; ***72****: 388–96*

Other indications

Other indications for radio-isotope studies of the urinary tract in children include hypertension, investigation of dilatation of the renal collecting system detected antenatally or postnatally and suspected obstruction. The indications and techniques are essentially similar to those in adults (pages 104–7), although in children hypertension is more likely to have an endocrine cause or be due to a post-traumatic renal artery stenosis. An important unresolved uncertainty is how to manage children in whom a dilated urinary tract is found antenatally by ultrasound. Where unilateral and not associated with outflow obstruction the majority resolve spontaneously. Criteria for identifying those which need surgical intervention are not established.

Further reading:

The Urologic Clinics of North America 1995; ***22****: 1–20, 32-42*

Thyroid

The principal indications for investigating children are ectopic or congenital absence of thyroid in the neonate and when considering treatment of thyrotoxicosis. Conventional teaching states that ^{123}I should be used for imaging the thyroid when congenital absence or ectopia is suspected (Fig. 5.2). This agent is however rarely available immediately and delay in treatment while the radiopharmaceutical is obtained cannot be justified. It is likely that in the majority of children thyroid will be visualized adequately with pertechnetate, which is usually immediately available. It therefore seems logical that this should be the initial investigation. If the thyroid is not clearly imaged then treatment may be commenced and an iodine scan performed at about the age of one year, after stopping replacement therapy for a suitable period (page 153). At this age irreversible brain damage is unlikely if treatment is withheld for a few weeks.

The use of ^{131}I to treat hyperthyroidism in children is controversial. It is more widely used in North America than in Europe but follow-up studies have found an increased incidence of nodular thyroid disease. The incidence of thyroid cancer may be raised in children who have been exposed to radio-iodine in fall-out at a dose level less than is used in therapy, but the data is suspect and the number studied after treatment for thyrotoxicosis too small for any such association to have been detected. Although it is probable that the danger is not high, insufficient data are available to estimate accurately the actual level of risk. Under the circumstances it is advisable to consider individual cases on their merits, particularly if non-compliance with a conventional regime of anti-thyroid drugs has been found.

Further reading:

Journal of Nuclear Medicine 1995; ***36****: 442–5*

Skeleton

Scintigraphy of the skeleton is one of the commoner nuclear medicine investigations in childhood. The indications in general are similar to those in adults, but trauma and infection are proportionally more common whereas malignancy is less often seen. The most striking difference in appearance between examinations in adults and those in children are the consequence of the epiphyseal growth plates, in which uptake of bone-seeking agents and hence also radiation dose are highest. Radioactivity is more concentrated at epiphyses than in any normal adult structure, with a correspondingly high absorbed radiation dose to these regions. This uptake reflects both local blood flow and the rate of new bone formation where growth is occurring. The blood supply to any bone is derived from three sources: nutrient artery, periosteum and joint capsule, either singly or in any combination. Their relative importance depends both upon the bone and the age of the patient. In children perfusion of the epiphyses is much greater than of any other portion of bone. In consequence, haematogenous skeletal metastases and osteomyelitis are more common in the epiphyseal regions than elsewhere during childhood. In contrast the nutrient artery is relatively more important in adults, in whom both infection and metastatic tumours more commonly occur in the mid-shaft of long bones than at either extremity. The scintigraphic appearance of the normal epiphysis when viewed in profile is a thin plate of greatly increased count rate with sharply delineated upper and lower boundaries. When viewed obliquely it may appear more less clearly defined. In contrast the diaphysis and metaphyses are only faintly discernible above background.

Further reading:
Atlas of Bone Scintigraphy in the Developing Skeleton. K Hahn, S Fischer, I Gordon. Springer Verlag, Berlin 1993

Malignant disease

Malignant tumours of childhood which commonly involve bone include neuroblastoma, Langerhans cell histiocytosis (page 171, 294), osteogenic sarcoma (page 27) and Ewing's sarcoma (Fig. 11.6). Neuro-blastoma commonly metastasizes so widely that skeletal scintigraphy appears normal, in the absence of discrete focal areas of increased uptake. Although such areas do sometimes occur, they are seen only in a minority of patients. The most important scintigraphic feature is an often subtle loss of definition of the epiphyseal plate, which instead of being sharply defined tails back along the diaphysis for a variable distance. This feature cannot be distinguished unless high quality images are obtained with a high resolution collimator, a large number of counts are collected and there is no patient movement. Commonly children who require this investigation are sufficiently ill to lie listlessly for the extended period required. It may be more difficult to obtain adequate follow-up studies after successful treatment, but with care and reassurance it should not be necessary to resort to sedation.

Langerhans cell histiocytosis (histiocytosis X, page 27) commonly fails to provoke any osteoblastic response and may therefore be

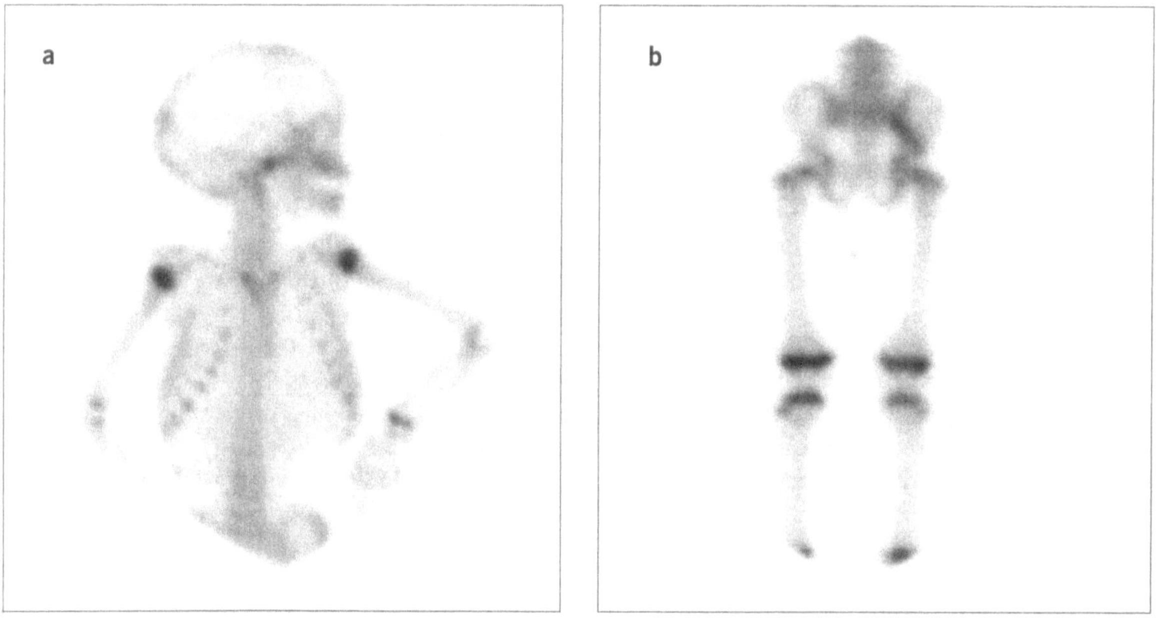

Figure 11.6 Two-year old boy. (a) Normal lateral skull, anterior thorax and upper limbs, (b) posterior pelvis and lower limbs. Increased uptake in right ileum due to Ewing's sarcoma. This diagnosis cannot be made on the scintigraphic findings alone.

associated with photon-deficient lesions or false negative examination. Like myeloma in adults, normal skeletal scintigraphy does not exclude active disease in this condition.

Solitary abnormalities

Solitary scintigraphic abnormalities are rather more common in children than adults, but even in children with some clinical suspicion of neoplasia only between one-third and one-half prove to be malignant. The site of the lesion and of any known primary tumour are of no assistance in distinguishing benign scintigraphic lesions from metastases. If appropriate radiographic views fail to provide definitive evidence of a benign or malignant cause, further imaging by MRI or CT is required. It should be remembered that minor trauma is an even more frequent occurrence in childhood than in adults.

Further reading:

Journal of Nuclear Medicine 1983; ***24****: 114–5*

Infection

Three-phase scintigraphy is a good method of detecting osteomyelitis and differentiating it from other soft tissue and bony conditions in older children (Fig. 1.24), as in adults; the technique may be used unmodified (page 8). In neonates, and in particular in the first six weeks of life, false negative findings are more common than true positive. This is probably because of the rapid development of high intramedullary pressure, leading to avascularity and infarction of the affected bone. Reaction does occur a week or more later but by this time diagnosis is usually established and imaging has little to offer. Even in older children it must be appreciated that in the very early stage there is commonly a period when there is infarction of marrow due to raised intramedullary pressure but before any osteoblastic reaction has developed, so that the lesion is photon-deficient. There is clearly also a brief intervening period when uptake and infarction counterbalance, leaving no visible anomaly. If there is strong clinical suspicion of osteomyelitis, treatment should be started without recourse to imaging. Imaging is usually only indicated where there is doubt about the clinical diagnosis. Many of these cases will have had a period of antibiotic treatment before referral for imaging and it is frequently not possible to differentiate reaction due to healing from continuing infection. As in adults, imaging with white cells is sometimes helpful, especially in limbs (page 5).

Further reading:

Journal of Bone and Joint Surgery 1994; ***76B****: 306–10*

Perthe's disease

Perthe's disease, avascular necrosis of the femoral head, occurs exclusively in children, principally between the ages of four and seven. Many cases heal without residual disability, but an important minority proceed to deformity of the femoral head and premature osteoarthritis. Ten per cent of cases are bilateral but it is rare for the disease to start simultaneously on both sides. The earliest changes are detected by MRI or scintigraphy and precede other characteristic plain radiographic changes.

Technique

A three-phase examination should be performed routinely. The neutral position is the most readily reproducible and should be employed. The child lies supine with knees and ankles together. Careful supervision is essential if asymmetrical external rotation is to be avoided. This can be achieved most simply if the knees are flexed to 90° over the edge of the bed or table, knees and ankles remaining together. The photon-deficient abnormality of early Perthe's disease is most clearly identified by high resolution and high count density imaging, preferably with a pinhole collimator. Imaging usually requires 10 min per side to obtain adequate statistics. A well-briefed parent or nurse or experienced technician is preferable to sedation, which is rarely necessary. The contralateral hip should always be imaged for comparison and because the disease is not uncommonly bilateral, although it is unusual for both sides to be symptomatic at the same time. The clarity with which the normal high activity features are delineated is determined by whether they are viewed precisely in profile (which is the ideal) or obliquely.

Interpretation

In the normal child, the femoral head, neck and upper shaft should be clearly visible despite higher uptake in epiphyseal plates of the femoral head, greater trochanter and tri-radiate cartilage; soft tissue activity should be minimal. The usual scintigraphic appearance of Perthe's disease is a well-defined photon deficiency in the third phase affecting the lateral half to two-thirds of the femoral head (Fig. 11.7). There is usually no abnormality in the earlier phases. Subsequently, as radiographic changes develop, this is replaced by increased uptake as reabsorption and new bone formation proceed (Fig 11.8). A similar or more extensive defect, in some cases affecting the whole femoral head, can be caused by a tense effusion compressing the arterial supply to the femoral head. Synovitis shows increased uptake in the first and second phases whereas the abnormality in Perthe's disease is confined to the third phase. It is thus necessary to be aware of the side of symptoms. Photon-deficient areas may disappear within a few months without radiographic changes as there is re-vascularization.

In practice it is not always possible to distinguish synovitis from early Perthe's disease which arrests without proceeding to overt radiographic abnormalities. Septic arthritis shows increased uptake on both side of the joint in all three phases.

At the time of the earliest scintigraphic changes, radiographs are normal or show evidence of a joint effusion, the latter more readily seen with ultrasound. Subsequently scintigraphic evidence of increasing reaction and repair are often followed by radiographic evidence of Perthe's disease with increasing density and flattening of the femoral head. Scintigraphy after the onset of these radiographic changes usually show increased uptake: the photon defect of early disease is no longer found. Normal scintigraphy excludes early Perthe's disease with a very high degree of confidence but does not eliminate the

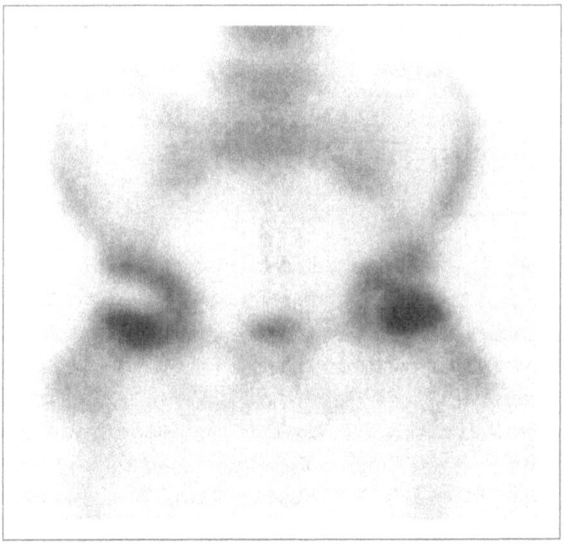

Figure 11.7 Boy age six years presenting with limp. Similar episode affecting other side about one year previously. Anterior projection of pelvis and hips. Photon deficiency in lateral part of right hip is typical of early Perthe's disease. Increased uptake in left hip compatible with previous Perthe's disease on the left. Radiological confirmation.

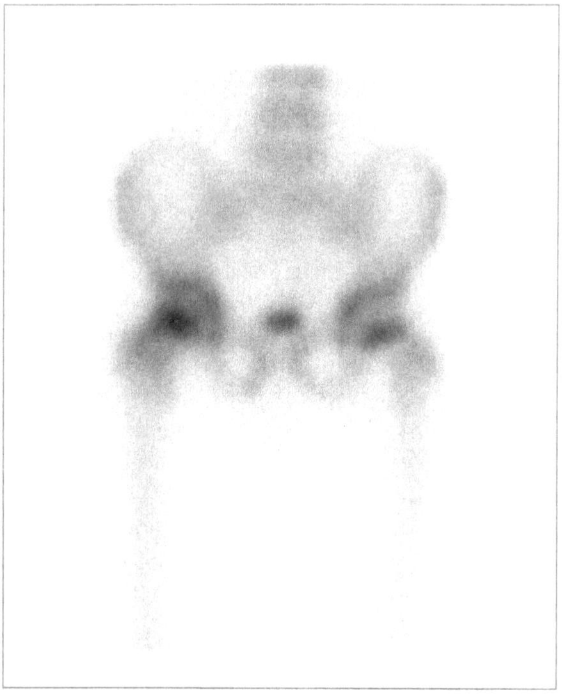

Figure 11.8 Anterior projection. Previous Perthe's disease with residual deformity of right hip. Left normal.

need for ultrasound to detect a small effusion or femoral head deformity.

Sources of error

False negatives may be due to patient movement, if imaging is delayed until the onset of revascularization or until there are radiological changes. Depending on the stage at which imaging is performed, synovitis may associated with increased or decreased uptake affecting both acetabular and femoral components of the hip. False positives have been reported, but the possibility that these are true avascular episodes which healed without subsequent deformity could not be excluded.

Further reading:
Clinical Radiology 1994; ***49****: 820–3*

Back pain

Scintigraphy may be useful in children with back pain. Scheuermann's disease, discitis, recent spondylosis and ununited fracture all show increased uptake, as do infection and many benign tumours including osteoid osteoma. Scintigraphy enables the site to be accurately localized but further information is required from MRI or CT to establish differential diagnosis.

Trauma

As in the adult, scintigraphy is more sensitive than radiology for detecting small non-displaced fractures. In addition to stress fractures children may suffer so called toddler's fractures due to minor trauma. These may provided an explanation for an otherwise unexplained reluctance to bear weight. Once diagnosed, simple conservative management is usually adequate. Scintigraphy is sometimes helpful in younger children in whom it may be difficult to localize a painful site clinically.

Non-accidental injury

Bone scintigraphy is indicated in children suspected of non-accidental injury to document extent of bony injuries in the axial and appendicular skeleton; fractures of the skull are not always visible scintigraphically. The combination of radiographic and scintigraphic abnormalities gives limited information about the age of injury. This is difficult to interpret as magnitude of uptake depends not only on when trauma occurred but also on its severity. Caution should always be exercised when attempting to estimate how long previously any physical injury occurred.

Growth arrest

Bone imaging and especially SPECT has been advocated in children suspected of epiphysiodeses (growth arrest). SPECT is required to demonstrate bony bridges within physes which on planar imaging may appears normal. Loss of activity in a growth plate may pre-date slowdown of growth in the remainder of the epiphysis due to the bridge. These arrests are most likely to be due to recent injury or infection and if untreated may lead to limb deformity or shortening.

Further reading:
Journal of Nuclear Medicine 1993; ***34****: 1410–5*

Paediatric tumours

Neuroblastoma

Imaging with meta-iodobenzylguanidine (^{123}I-MIBG) (page 171) is a routine staging procedure in neuroblastoma. There is usually high uptake in the primary tumour, soft tissue and bony metastases (Figs 11.9, 11.10). A few centres have reported that some skeletal metastases show up with conventional bone imaging agent but not with MIBG. Most centres have been unable to reproduce these results. Children with neuroblastoma are commonly hypertensive at presentation, due to production of sympatheticomimetic amines by the tumour. Treatment of the hypertension with β-blockers reduces tumour uptake of MIBG and may be an explanation for this apparent discrepancy. Minor asymmetries of uptake of skeletal agents are not uncommon in children and should not be over-reported.

However, any discrepancy on the bone scan irrespective of the findings with MIBG should be imaged with MRI. Virtually all metastases are visible with this modality. ^{131}I-MIBG has also been used for palliation in advanced disease when the tumour burden is large. It is rarely possible to obtain sufficient contrast between tumour and normal structures to achieve a therapeutic effect when the disease is more restricted.

Further reading:

European Journal of Nuclear Medicine 1995; ***22****: 322–9*

Other tumours

Both thallium and gallium have been advocated to differentiate residual tumour from necrosis following therapy of brain tumours and tumours elsewhere in the body. They are not widely used for this purpose in children because of the high radiation dose. There is no generally agreed clinical indication for these agents in children. Gallium rather than bone scintigraphy has been advocated in lymphoma, although there is no adequate evidence that either adds to the information obtained by MRI. Many lymphomas, the majority of rhabdomyosarcomas and Ewing's sarcomas also take up thallium. Wilm's tumours and neuroblastomas, in contrast, are rarely visualized. Visualization of the primary thoracic tumour has been reported in one case of primitive neuroectodermal tumour (PNET) with both thallium and gallium.

Further reading:

Journal of Nuclear Medicine 1993; ***34****: 1045–51. 1995.* ***36****: 814–6, 1372–6*

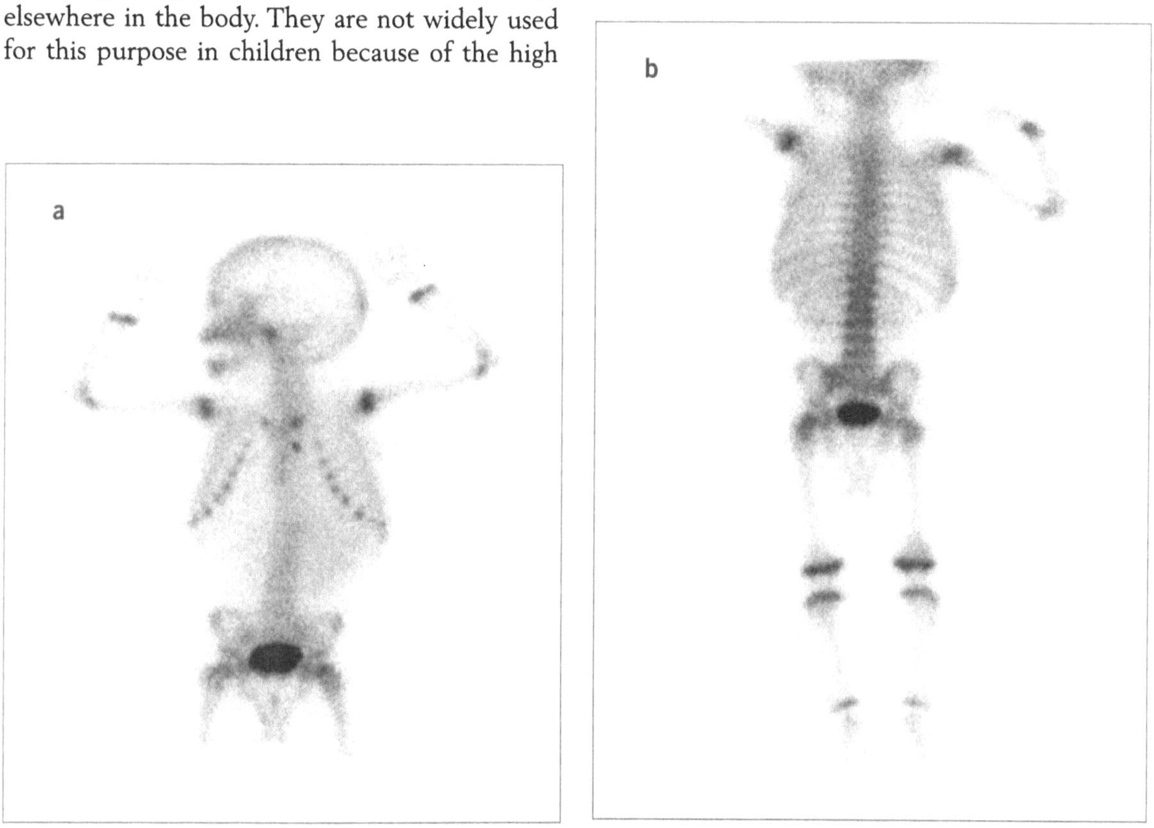

Figure 11.9 One-year-old girl with stage 4 neuroblastoma, at presentation. (a) Anterior, (b) posterior. There is loss of definition of the proximal femoral epiphyseal plates, increased uptake in both proximal femoral shafts and increased uptake tailing back from the distal right femoral epiphysis. The left proximal humeral epiphysis is ill-defined. On the anterior projection there is faint soft-tissue uptake in the left epigastrium and hypochondrium, at the site of the large primary tumour. All of the changes are subtle and would be obscured by movement.

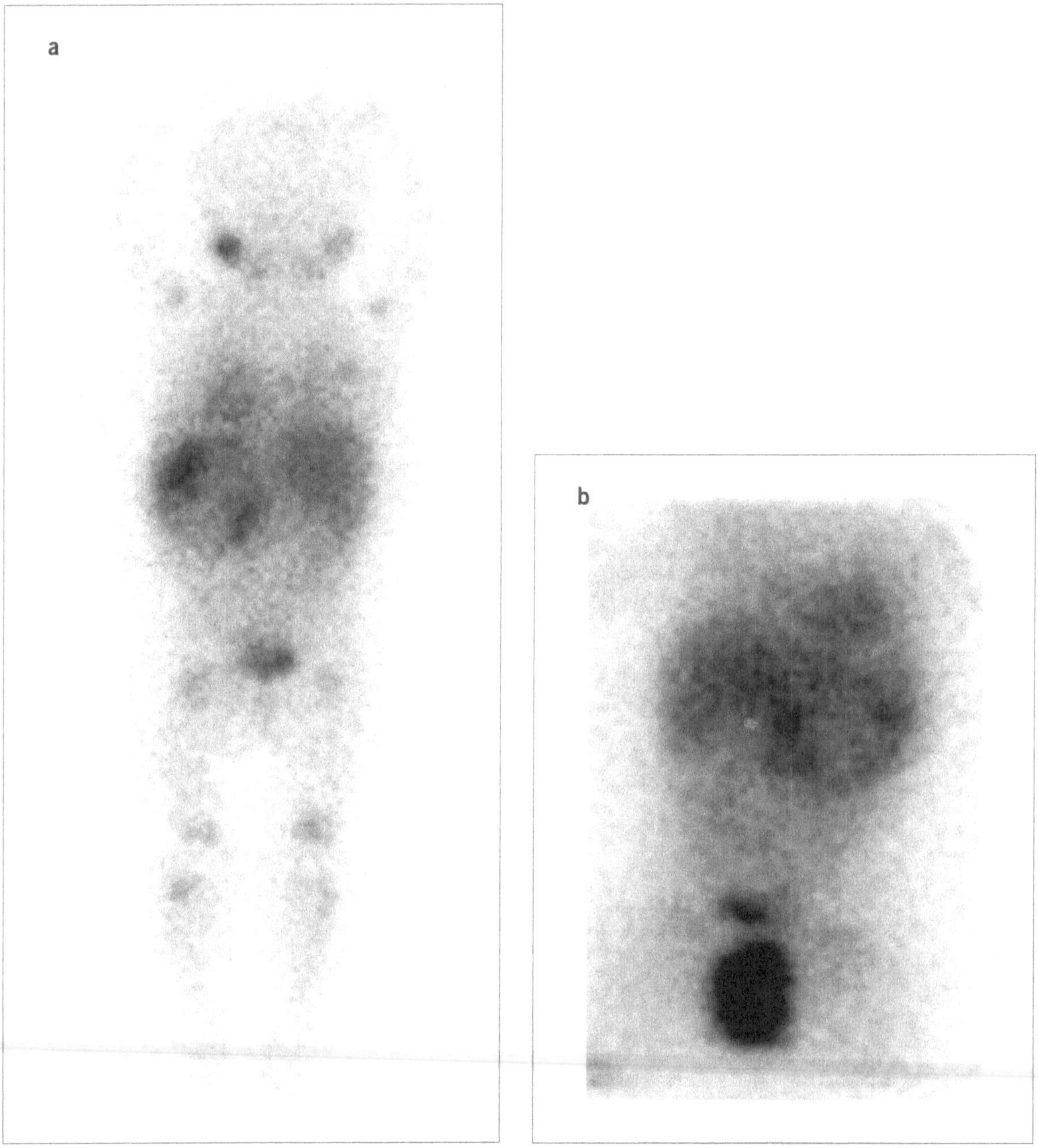

Figure 11.10 ^{123}I-MIBG study of same patient as Fig. 11.9; 24 h images. (a) Composite posterior, (b) anterior. Normal uptake in salivary glands, liver, heart and urinary tract. Tumour uptake in abdomen (but note large defect due to necrotic centre), proximal humeri, femora and tibii and distal femora bilaterally. Bone involvement is more extensive than is apparent from Fig. 11.9. As is commonly the case in neuroblastoma this child was hypertensive at presentation, requiring treatment with a β-blocker. This reduces tumour uptake of MIBG. Reported false-negative MIBG imaging may be due to failure to obtain a full drug history.

12 Scientific considerations – selected topics

Radioactivity

Radioactivity can most easily be understood in terms of the simplified classical model of the atom described by the Danish physicist Neils Bohr in which the nucleus, containing approximately equal numbers of positively charged protons and uncharged neutrons, is surrounded by as many orbiting electrons as there are protons. Stable atoms have either equal numbers of protons and neutrons or a slight excess of neutrons. The number of protons determines the number of orbiting electrons and hence the element and its chemical properties. An element may have one or more stable configurations of neutrons (stable isotopes); for example iodine has only one stable isotope while tin has ten. Atoms which do not have a stable configuration of protons and neutrons are radioactive (radio-isotopes). Those with an excess of neutrons decay with the emission of a negatively charged beta (β^-) particle identical to an electron in all respects except its source: β particles originate from the nucleus whereas electrons orbit around it. Atoms with a deficiency of neutrons emit positively charged particles of the same mass as an electron (positrons, β^+) or reduce their positive charge by capturing an orbiting electron. Some radioactive isotopes of heavy elements (lead and above) emit alpha (α) particles, which resemble helium nuclei. The emission of a charged particle is commonly accompanied by the discharge of excess energy in the form of one or more gamma (γ)-rays, the energy of which is specific for each radio-isotope.

Although it is not possible to predict when any individual atom will undergo radioactive decay, the behaviour of large numbers of atoms of any particular isotope is consistent. Any population decays at a rate which is particular to that isotope. The time required for half the radioactive atoms to disintegrate is the *half-life* of that isotope. Thus, after one half-life, half of the original activity remains, after two one-quarter, after three one-eighth and so on. As the amount of residue repeatedly halves it eventually approaches, but never reaches, zero. When the amount of residual radio-isotope is plotted graphically against time on a linear scale it follows a falling curve which approaches zero asymptotically. If a logarithmic scale is used for the residual radioactivity a straight line is obtained provided that the logarithms are based to the mathematical constant 'e'. This is described as 'the exponential law of radioactive decay' and is used to calculate the residual activity before or after any interval and correct for losses due to radioactive decay. The activity at any time before or after measurement is related to the measured activity by a simple formula (Appendix 2). Alternatively a look-up table of decay factors can be employed. The measured activity is multiplied by the factor appropriate to the isotope and interval between measurement and use. *In vivo*, biological elimination may supplement radioactive decay giving a more rapid rate of loss. Excretion commonly also follows an exponential pattern. It is therefore possible to calculate an effective half-life if physical and biological half-lives are known (Appendix 2).

Radioactive emissions

Alpha particles

Alpha (α) particles are produced only by decay of heavy elements (lead and above), for example ^{210}Bi (half-life 2.6 × 10^6 years) decays to ^{206}Tl. They have a high mass by atomic standards, over 7000 times that of an electron, and are identical to helium nuclei, comprising two protons and two neutrons. They also carry the highest charge (2 units) of any natural emissions and travel only a short distance in water or tissue before an interaction occurs, when the large amount of energy associated with their mass and charge is transmitted to the surrounding media. Their energy is transferred to electron shells of atoms with which they interact and which may in consequence eject orbiting electrons, becoming ionized or forming free radicals. More than one interaction may take place before all the energy is dissipated.

If, as is usually the case, the atoms with which they interact are incorporated into molecules, the latter may be disrupted with formation of reactive free radicals which interact further with intracellular molecules causing somatic or genetic damage.

Thus α particles are usually absorbed close to their origin, cause considerable biological damage and are difficult to detect *in vivo* because heavy metals tend to be deposited in bone, where all of their energy is dissipated within a millimetre or so. They are not employed clinically but, depending on method of production, small amounts of α-emitting impurities are potential contaminants of some β- or γ-emitting isotopes used clinically. A secondary consideration is that they could be encountered as a result of an industrial mishap.

It is difficult to detect small amounts of α-emission in the presence of relatively large amounts of β or γ. For this reason radiopharmaceutical manufacturers are reluctant to supply isotopes extracted from mixed reactor products and licensing authorities often require relicensing when the method of production of a radio-isotope is changed. Most commercial sources employ methods of production which eliminate the risk of alpha contamination.

Beta particles

Beta (β) particles are identical in mass and charge to electrons, from which they are indistinguishable. They most commonly carry a single negative charge but positively charged electrons (positrons) also exist (see below). The distance a β particle can travel before it is absorbed depends on the energy with which it is ejected from the nucleus, varying from less than a millimetre in the case of tritium (^3H) (mean β energy 0.019 MeV) to more than a centimetre. Each β-emitting isotope gives off particles at a characteristic mean energy, although the energy of individual particles varies. Those of ^{99}Mo, emitted as it decays to ^{99m}Tc, have a mean energy of 1.23 MeV. Like α particles the damage they cause is mediated through the production of free radicals, but as their range is somewhat greater this is not quite as sharply localized. They can be detected *in vivo* in superficial structures such as skin or brain exposed at craniotomy but not through intact skull. The major applications of β-emitting isotopes are for therapy. Some β-emitters are used diagnostically but in general this is because of their associated γ emissions, in which case the β are an undesirable but unavoidable feature.

Positrons

Positrons are positively charged β particles. Initially they interact with molecules in a similar way to negatively charged β particles, dissipating their kinetic energy with the formation of free radicals. For example, the mean β energy of ^{18}F is 0.635 MeV. Once they have lost most of their initial energy they are able to interact with a single negative electron. As these particles cannot exist in the uncharged state their combined mass is converted to energy. This cannot be a single photon because momentum in one direction must be balanced in the opposite direction. The mass is therefore converted into a pair of photons each of 511 keV which are emitted at 180° to each other. These are the annihilation photons. If the positron has some residual energy at the instant of annihilation the consequent photons are not emitted at exactly 180° but the resultant error if this is ignored is usually negligible. Positron emission tomography (PET) depends on being able to detect both photons simultaneously in a pair of detectors, one on either side of the subject. If the photon pair did not originate exactly mid-way between the detectors there will be a time difference, which should permit the site of origin to be determined. For an object the size of the human body the difference is a few picoseconds. Current technology does not permit events to be timed with sufficient precision for true time-of-flight to be possible, even though this term is sometimes employed. Coincidence detection potentially dispenses with the need for a radio-opaque collimator and instead defines a cone between each pair of detectors within which an interaction occurred, thereby increasing sensitivity compared with a system depending on lead collimators for positional information and minimizing change of resolution with depth. The more accurately coincidences can be timed, the fewer false coincidences due to random events will be recorded, thereby improving the signal-to-noise ratio.

Gamma rays

Gamma (γ)-rays are electromagnetic radiation in the same range of wavelength as x-rays, from which they differ in their source (nucleus rather than orbiting electrons) and, unlike most x-rays, in being monoenergetic rather than forming a continuous spectrum. The energy of the γ-ray or rays emitted by each isotope is constant and characteristic for that isotope, so that isotopes can be identified from their γ-ray spectrum. γ emissions are usually associated with α or β emissions. These occur in a series of discrete steps usually in rapid sequence, but there are a few examples when there is a substantial interval between some emissions. This is the metastable state. Most metastable isotopes have a half-life of seconds or fractions of a second but a few have longer half-lives. The longest-lived metastable isotope is ^{99m}Tc, with a half-life of 6.049 h, formed in one of the intermediate stages when ^{99}Mo (half-life 2.78 days) decays with emission of a β particle to ^{99}Tc, a β-emitter with a half-life of 2.12×10^5 years. ^{99}Mo is described as the 'parent' and ^{99m}Tc the 'daughter' isotope. Being different elements they are chemically distinct and can be separated fairly readily.

Alternative γ-emitting isotopes are those which decay by electron capture, thus in principle not emitting any charged particles. For example ^{51}Cr decays by electron capture to ^{51}V, a stable isotope of vanadium. Excess energy from the nucleus is dissipated as a γ-ray, in this example of 320 keV. Two processes complicate this idealized picture. In over 90% of disintegrations of ^{51}Cr the γ-ray is not emitted but the energy is instead transferred to an orbiting electron, which is ejected. This process is called *internal conversion* and results in the β-like emission of a *conversion electron*. When an electron, usually from the innermost 'k-shell', is captured it is replaced by one from an outer shell, with the emission of a *characteristic x-ray* whose energy is equal to the energy difference between the two shells. Commonly there is more than one x-ray energy, representing the differences between the various shells. Except for the heavier elements, the energy of these x-rays is too low for them to escape from the body, so that they increase the locally absorbed radiation dose. The k x-rays of ^{201}Tl (69–83 keV) and ^{133}Xe (72–80 keV) are of sufficiently high energy to be used for imaging. In other cases, for example technetium, only some of the energy of the emergent γ-ray is transmitted to an orbiting electron which is ejected whilst the energy of the emerging γ-ray is correspondingly reduced. These form Auger electrons which contribute to the β-like component of absorbed dose. Thus particulate emission, not strictly β but effectively indistinguishable, commonly accompanies many γ-emissions and substantially increases absorbed radiation dose.

Interactions of radiation with matter

Photoelectric absorption

Two types of interaction between γ-rays and matter occur at the energies used clinically: photoelectric absorption and Compton scatter. A third process, pair production, is important only at much higher energies, above 2 MeV. In photoelectric absorption the total energy of the incident γ-ray is absorbed, with the ejection of an orbiting electron from the atom with which the γ-ray has interacted. Photoelectric absorption is important at lower energies and is dependent on the atomic number of the absorbing material; the higher its atomic number the higher the energy of γ-rays absorbed. The 140 keV γ-ray of technetium undergoes mainly photoelectric absorption in lead but is predominantly scattered in soft tissue, which has a much lower effective atomic number.

Whenever an electron is ejected from one of the inner shells, it is replaced by one from an outer shell with release of a characteristic x-ray which is usually absorbed locally. It is however sometimes possible to see a small peak in the gamma camera technetium spectrum due to lead k x-rays released when technetium γ-rays interact with lead in a collimator or camera shielding.

Compton scatter

Compton scatter is important at slightly higher energies than photoelectric absorption and depends mainly on the density of material in which radiation is absorbed rather than its atomic number. When a photon undergoes Compton scatter it is initially absorbed by the atom with which it is interacting. Some of the energy is transmitted to an electron which is ejected from orbit. The residual energy is emitted as a γ-ray of lower energy and at an angle to the original incident ray. The angle between the incident and the emitted ray depends on the difference in energy between the two. A small difference in angle results when the energy of a scattered ray is only slightly lower than that of the incident ray. The Compton-scattered photon is at a greater angle and of lower energy when more of the energy of the incident γ-ray is transmitted to the electron. Pulse height analysis is employed to differentiate between primary and scattered radiation but, because of intrinsic limitations in energy resolution, it is not possible to exclude all low angle scatter. This has important implications when imaging with a gamma camera, as scatter reduces contrast and impairs the accuracy of all measurements. The emitted electrons are usually of fairly low energy and thus contribute to absorbed radiation dose.

Radiation detectors

Gas-filled

Radiation detectors used clinically fall into two principal categories: gas-filled and scintillation. The overwhelming majority of imaging devices utilize the second type; the former are employed mainly for dose standardization and contamination monitoring. Gas-filled detectors are based on the principle that when γ-rays interact with the gas, or charged particles enter a gas-filled chamber across which an electrical potential is applied, any ions formed are attracted towards electrodes carrying the opposite charge and an electrical current flows, depending on the applied voltage and amount of incident radiation. There are three types of such detector.

Ionization chambers

These operate at a comparatively low voltage which is sufficient to attract charged particles towards the electrodes and thus allow a current to flow, but not amplify this current. The current which passes across an insulating gas-filled cavity with which the particle has interacted is measured. They are suitable for measuring quantities of radioactivity employed in diagnosis or therapy, are stable and reliable in operation and capable of accurate calibration, but cannot determine the energy of the detected radiation, although their response is not totally energy-independent. They thus have to be calibrated separately for different isotopes. A variant of this is used in pocket dosimeters where ionizations discharge a capacitor. Uses in nuclear medicine include dose calibrators, survey and dose meters.

Proportional

These employ a higher voltage, thereby increasing sensitivity by accelerating ions, which collide with other atoms in the gas so that each initial event produces a cascade of charged particles. Amplification increases the chances of detecting a single event and produces a signal which is proportional to the γ-ray energy. Imaging cameras based on the principle of the multi-wire proportional detector have been constructed but are not widely used.

They are most sensitive to low-energy γ-radiation (<100 keV).

Geiger–Mueller

These counters employ yet higher voltages, so that the amplified pulses are no longer proportional to the energy of the incident radiation. This permits individual events to be counted but, unlike scintillation detectors, Geiger counters do not provide any information about the energy of each event. They are used mainly for contamination monitoring, where their lack of discrimination is an advantage.

Scintillation detectors

Scintillation detectors make use of the observation that when a γ-ray interacts with a crystal (or indeed many non-crystalline substances) some of the absorbed energy may be re-emitted at a longer wavelength, which in many cases is in the visible part of the electromagnetic spectrum. The amount of light emitted is proportional to the energy of the absorbed γ-ray. To be useful as a scintillator, a crystal must be sufficiently radio-opaque to absorb incident γ-rays efficiently and be transparent to the light emitted. The intensity of light output must be sufficient for adequate accuracy of measurement whilst its colour should be at a wavelength which is detected efficiently by available photomultiplier tubes. The most commonly employed crystal is of anhydrous sodium iodide, which is reasonably radio-opaque because of the high atomic number of iodine and has good light output. It is however deliquescent and must be encased in a water-tight and air-tight aluminium container except for the surface applied to the photomultiplier, which is covered by a transparent glass window. The photomultiplier is coupled to this window to maximize efficiency of light transmission. The interior of the aluminium is given a reflective white coating to minimize loss of emitted light. The whole assembly is shielded from extraneous light. Pure sodium iodide fluoresces at a wavelength which is not ideal for use with available photomultipliers. Controlled trace quantities of thallium are therefore added during manufacture to alter the colour of light emitted closer to that required. The resultant detector material is described as 'thallium-activated sodium iodide', abbreviated to NaI(Tl), and is used in almost all gamma cameras and most scintillation detectors. Alternative scintillators used in PET scanners and other devices include caesium iodide, cadmium telluride and bismuth germanate. The last of these has a higher density than sodium iodide but the total light output is less. In consequence the energy deposited by an individual γ-ray can be measured less accurately. This is not a problem with PET because of the high energy of annihilation photons but precludes its use with lower energy single-photon tracers. The advantage of caesium iodide for PET is the short duration of individual light flashes, which permit higher count rates and allow events to be timed more accurately, but the energy resolution is inferior to NaI(Tl), rendering it a less suitable detector for technetium.

Photomultipliers

Scintillation detectors are used in conjunction with photomultiplier tubes. These are vacuum tubes with a photocathode coupled to the transparent covering of the crystal, to ensure maximum light transmission from the detector but otherwise shielded from light. The photocathode is coated with a substance which emits electrons when photons strike it. A charged plate (dynode) behind the photocathode attracts emitted electrons which accelerate towards it because of the voltage difference; frequently about 200 V. Each electron striking the dynode causes several secondary electrons to be emitted, which are in turn attracted by a voltage difference, usually about 100 V, towards the next dynode in the chain. Typically there are ten dynodes in the chain giving an amplification of about one million times. This provides sufficient current at the output to feed into a preamplifier that shapes the pulse prior to its further analysis. As the energy of the pulse from the preamplifier is proportional to the energy of the original detected photon it is possible to discriminate electronically between γ-rays of different energies. The maximum count rate which can be measured depends both on the speed of the electronic circuitry and the time that each pulse of light takes to die away. A further limitation is due to the statistical nature of the multiple amplification stages. Photons of identical energy do not all produce an identical output pulse but a Gaussian spectrum. This imposes limits on the possible energy resolution and thus the ability to discriminate between γ-rays of similar energies, for example between photopeak and small-angle Compton scatter.

Gamma cameras

These consist of a radiation detector, almost always a single flawless crystal of NaI(Tl), photomultipliers and associated electronics mounted in a shielded bowl on a stand and usually connected (interfaced) to a dedicated computer. The crystal is larger than the usable field of view to eliminate artefacts at the edge, although the difference is not as great in modern cameras as formerly. The crystal may be circular or rectangular to give a circular, hexagonal or rectangular field of view, in some models in excess of 350 mm × 450 mm. The detector assembly is surrounded by heavy shielding, usually of lead or tungsten, to prevent radiation entering except through the collimator. One face of the detector is exposed but is restricted by a collimator, the function of which is analogous to the lens of a photographic camera. The thinner the crystal the more accurately the site of interaction of a γ-ray can be calculated, but if the γ-ray is not totally absorbed it is likely to be rejected by the pulse height analysis circuitry. About 90% of 140 keV γ-rays of technetium are absorbed within the first 1 cm of sodium iodide. There is therefore little gain in detection efficiency but appreciable degradation of spatial resolution if a thicker crystal is employed with technetium. In contrast, the detection efficiency for ^{131}I of a crystal of this thickness is under 25% whilst that for a 2.5 cm thick crystal is over 50%. Gamma cameras are usually optimized for use with technetium and other isotopes with similar energies (^{123}I, ^{81m}Kr). Higher energy isotopes can be (and are) used but the combination of the thicker collimator, which gives a lower sensitivity for any given resolution and the reduced detector efficiency results in relatively poor count rates. The photopeak detection efficiency of a 1 cm sodium iodide crystal for 511 keV annihilation photons is approximately 10%, even before the effect of adequate collimation is considered. Some manufacturers offer coincidence circuitry on two-headed gamma cameras. A 16 mm thick crystal reduces intrinsic resolution for technetium from about 3.4 mm to 3.9 mm but approximately doubles sensitivity for annihilation photons.

The most important difference between a simple scintillation detector and a gamma camera is that the latter is equipped not with a single photomultiplier but an array. The earliest commercial cameras employed 19 similar photomultipliers, one centrally surrounded by two rings of tubes packed together as closely as possible. The tightest possible arrangement of circular objects of similar size is a hexagonal array, in which each is surrounded by six others. One additional ring brings the total to 37 tubes and two additional rings to 61. Modern cameras may contain over 80 photomultipliers, also in hexagonal arrays although the outer rings are not complete if the field of view is rectangular rather than hexagonal. Each tube receives a portion of the light output, although most is collected by the 12 or so closest to the site of interaction. Cameras are tuned by adjusting the high voltage supply to every photomultiplier, to ensure that an optimal pulse is delivered to each preamplifier. The best energy resolution achievable with a gamma camera is slightly under 10%. Many modern cameras approach this theoretical limit. Substantial improvements in this parameter are thus unlikely. Variations between output of photomultiplier tubes adversely influenced performance of early cameras. Manufacturers now take great care to provide matched sets of tubes.

The sum of the output from all photomultipliers for each detected γ-ray provides data from which the energy of the incident γ-ray is calculated for pulse height analysis. A γ-ray is accepted for further analysis only if it falls within limits which can be set for the isotope in use.

Typical settings for ^{99m}Tc are a peak at 140 keV with a window width of 20% (that is from 126–154 keV) in older cameras and 15% in newer ones with better energy resolution. A narrower window rejects more scattered radiation and therefore improves contrast but also rejects more primary radiation and thus reduces sensitivity. As γ-rays whose energy is higher than the photopeak energy are unlikely to have been scattered, an asymmetrical window including more above the photopeak than below is sometimes advocated. This practice not widely recommended as it is liable to amplify uniformity errors associated with any drift of the high voltage.

Some newer cameras incorporate more sophisticated scatter rejection techniques, employing between three and 32 windows. In the simplest methods, scatter is assumed to be equivalent to a linear interpolation between two narrow windows, one on either side of the photopeak. More sophisticated techniques analyse the spectrum in more detail and employ a weighting factor depending on count rate in adjacent regions. The clinical value of these manipulations is unproven although subjectively they improve apparent contrast of images. Both computing and statistical limitations render it unlikely that single pixel analysis will prove of more than marginal value.

It is usually possible to employ more than one energy window, for example with an isotope such as ^{67}Ga which has three principal photopeaks, or when two isotopes with different photopeak energies are being employed. Lower energy windows always contain an appreciable contribution of Compton-scattered radiation originating from higher energy emissions. When employing two isotopes, the ratio of activities administered must be adjusted to ensure that the contribution to detected count rate in the lower energy window due to scattered radiation from the higher energy isotope is small compared to count rate from unscattered radiation from the isotope with the lower energy. The higher count rate should therefore ideally be from the lower energy peak, but this is not always achievable. For example when imaging thallium and technetium simultaneously in the myocardium, administered activity of thallium is restricted by radiation dose considerations. At allowable doses, higher count rates are obtained from technetium than thallium. In phantom simulations it can be shown that the thallium window accepts one scattered technetium count for approximately each two technetium photopeak counts. If the technetium photopeak count rate is only twice the thallium photopeak count rate, and higher differentials are easily achievable, half of the counts in the thallium window may be scattered technetium counts. Dual-energy acquisition must therefore be approached with caution and only if appropriate phantom measurements have confirmed the practicality of the combination.

A high energy photopeak of low abundance, such as the 393 keV peak of ^{67}Ga which is not usually used for imaging, can degrade image quality due to both penetration of collimator septa and Compton-scattered radiation, which the crystal detects more efficiently than the high energy primary radiation. Energy resolution is poorer at lower energies; a wider window (usually 25%) is required for the 69–83 keV x-rays of ^{201}Tl than for technetium. There is usually a choice of acquiring separate images or adding all the accepted counts into a single image. Energy resolution places constraints on which isotopes can be imaged simultaneously; for example ^{75}Se has five photopeaks, one of the more abundant of which is at 135.9 keV and cannot be resolved from the 140 keV peak of ^{99m}Tc by a gamma camera. If the two isotopes are to be used simultaneously ^{75}Se should be given first and the ratio of the count rate in the higher windows to that in the technetium window measured. Provided that the anatomical distribution of the selenium-containing tracer does not change, the ratio will remain constant and the appropriate number of counts may be subtracted from the technetium window.

The site on the crystal at which the γ-ray was absorbed is calculated from the fraction of the total emitted light received by each photomultiplier. Accuracy is improved if a larger number of smaller photomultipliers is employed and light signals below a threshold strength are discarded. In older designs of camera the calculation of position is made by analogue electronic circuitry. Newer instruments digitize the output from each preamplifier and employ a digital computer to calculate the position. Because of inherent limitations in this calculation even when the detector is evenly illuminated, for example by a point source at a sufficient distance, the resultant image is not uniform but shows regions of higher count rate corresponding in position to centres of photomultiplier tubes. Patterns of straight lines are also distorted (linearity errors), the lines being deflected towards adjacent photomultipliers to give a ripple effect. Manufacturers have devised a number of techniques to improve uniformity. As most of these defects are due to mispositioning of counts, algorithms used in older instruments to compensate for non-uniformity, by adding or subtracting counts, do not make a valid correction. Most now employ some means of repositioning misplaced counts. This is usually performed digitally by comparison with calibration images of multiple line or point sources of known dimensions, even if the initial calculation is analogue. Repositioning not only improves uniformity but also compensates for linearity errors.

Dead time

A further limitation of all detectors, both imaging and non-imaging, is the maximum speed at which they can process the information they receive. At very high count-rates the scintillation resulting from one γ-ray may not have died away before the next is received. In consequence the crystal glows continuously so that individual events cannot be distinguished. The speed at which the electronics can analyse data is however usually the limiting factor. As the incident count rate increases, more and more events are lost because the electronics may not have finished processing a previous γ-ray before the next arrives. Unless this can be stored in a buffer until it can be processed it is likely to be overlooked. Thus a graph of true count rate against observed gradually falls away from the line of identity (Fig. 12.1).

Above a certain true count rate, the observed value no longer increases. If incident count rate is increased further, observed rate starts to fall; at higher count-rates still the system becomes paralysed and is unable to process any counts. Counts lost before paralysis supervenes are dead-time losses. One of the advantages of all-digital cameras is a higher count rate capability.

The magnitude of dead-time loss depends not on the count rate within the photopeak or selected region of the detector but on the total number of events being processed by the electronics, including all those ultimately rejected. Provided that a dead-time correction curve has been measured, it is possible to correct for losses before the curve starts to plateau. Corrections for losses higher than about 20% become progressively less reliable. The importance of dead-time is commonly underestimated. High administered activities do not yield corresponding improvements in count rate if they are associated with substantial dead-time losses. The maximum count rate achievable with a single-detector device is approximately one

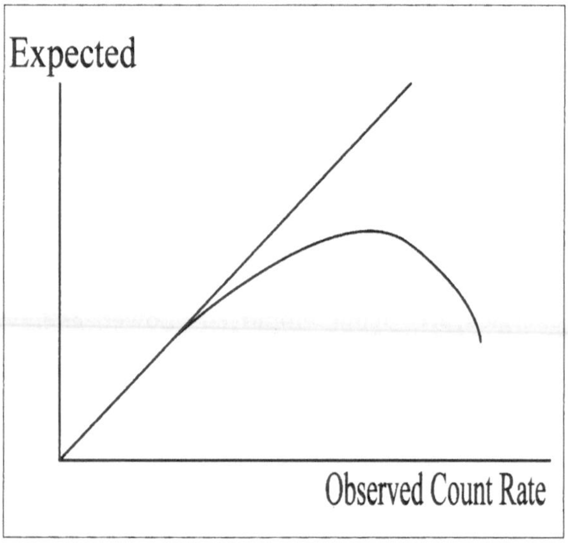

Figure 12.1 Schematic representation of gamma camera count rate response curve. Observed count rate falls away from expected above a certain value, which depends on the model of camera. The general shape of the experimental curve is similar for all but the peak may be anywhere from <100 000 counts/s to >300 000 counts/s, depending on model.

million counts per second, considerably higher than is found with clinically acceptable quantities of any diagnostic radiopharmaceutical. This figure however includes scattered radiation which is ultimately rejected. Thus the maximum count rate from a point source in air is considerably greater than the maximum in the usable photopeak window from an extended source in scattering medium.

Collimators

It is not possible to focus short-wavelength electromagnetic radiation. Instead the field of view at each point on the crystal is limited by a collimator. The simplest design for a non-imaging probe consists of a radio-opaque (e.g. lead) cylinder which limits the field of view to a cone. The apex of the cone is towards the crystal (Fig. 12.2). Thus the area within the field of view in any plane increases on moving away from the detector, so that resolution deteriorates with increasing distance from the collimator face. A long thin collimator is less rapidly depth-dependent than one of similar sensitivity and resolution at the surface of the detector but which is shorter or of larger diameter.

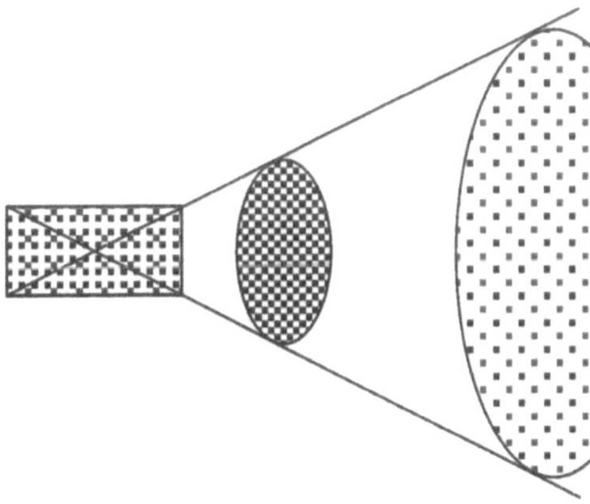

Figure 12.2 Simple geometrical model illustrating inherent limitations of cylindrical collimator.

Collimators for gamma cameras consist of thousands of such cylindrical elements. They may be manufactured by cementing together multiple corrugated lead strips or be cast in a single piece. The latter are more accurate and the resulting collimator is more robust, but the former is less expensive to fabricate. In the most usual design, thin septae of lead produce a pattern of cylindrical or hexagonal holes which prevent γ-rays coming obliquely from reaching the crystal. The more closely the field of view of each point is restricted the better is the spatial resolution but the lower the sensitivity, as each point on the detector is able to receive incident radiation from a smaller volume of tissue. There is thus in collimator design always a conflict between resolution and sensitivity.

It is usually desirable to have several collimators even when using a single isotope, as there is no single optimal compromise. When working with isotopes of different energies, collimators with septae whose thickness is appropriate to the highest energy γ-ray in use must be employed. The thicker the septae the less of the detector is exposed. For the same geometrical resolution, higher energy collimators have a lower sensitivity than low energy ones. The terms 'high', 'medium' and 'low' energy are imprecise and are not used consistently by manufacturers. When purchasing a collimator it is advisable to specify the isotope with which it is to be used: a 'medium energy' collimator from one supplier may be intended for ^{131}I (346 keV) whereas that from another only for ^{111}In (243 keV). The effect of collimation is to make the gamma camera an inefficient detector of radiation. Fewer than one out of every thousand photons emitted during acquisition contribute towards the image. The least sensitive collimators are about one order of magnitude less sensitive.

In addition to parallel-hole, three other types of collimator are also used. The pinhole collimator employs the same principle as a pinhole camera. Better resolution but lower sensitivity can be obtained if the pinhole is smaller. Pinhole collimators permit magnification of small objects and give the highest resolution achievable, but their low sensitivity requires prolonged imaging times. Coded aperture collimators are a development of the pinhole which produces overlapping images from a pattern of pinholes which are then unscrambled by appropriate software. Sensitivity is higher than a single pinhole but the usable field of view is small and the potential for distortion of data significant. Fan-beam collimators are flared in one plane so that a small volume of tissue is in the field of view of a larger area of the detector, thus increasing sensitivity. They are used principally in tomography where it is possible to correct for the geometric distortion they cause but should be used only in conjunction with appropriate software. Cone-beam collimators can be used to increase the field of view for planar imaging of a small-field camera or if reversed to magnify a small object. Their resolution is inferior to a pinhole collimator but sensitivity is higher. Their usefulness is restricted by distortions they produce in the image. For most applications advantages due to magnification are vitiated by lower contrast.

Quality assurance

The clinical value of an image depends on the ability of the camera to distinguish between regions and differences in count rate. The camera must therefore have sufficient spatial resolution to differentiate functionally significant structures and be free of uniformity or other artefacts which could be mistaken for real differences. Contrast is also important. Scattered radiation reduces contrast in radio-isotope images as in x-ray images. Scatter is reduced both by the collimator and electronic analysis of the γ-rays. Energy (spectral) resolution is thus as important as spatial resolution.

A number of changes can occur to a gamma camera which affect its performance. A regular programme of quality control is therefore essential. All regular measurements should be recorded and plotted so that any systematic changes can be detected. If these fall outside limits suggested by the manufacturer, attention from a service engineer is required.

A number of parameters must be measured to define the performance of a gamma camera. The most stringent set of tests are those described by NEMA (Nucleonic and Electronic Manufacturers Association), a North American association of manufacturers of electronic products. These are standards which specify how the various measurements should be performed, so that different manufacturers adhering to the standard can be compared. The standard does not specify what the performance should be. In general it is not practical to reproduce the specified conditions in a clinical department. It is however possible in most cases to make an adequate approximation to the methods to detect changes in performance over time. A total of 12 tests are described in the standard:

NEMA tests of performance

Intrinsic spatial resolution

This is expressed as the width of a profile at right-angles to a thin line source, at the point where count rate is half the maximum value. This is called *full width at half maximum*, abbreviated to FWHM and is measured in air on the detector face. It can be checked by imaging a pair of thin line sources – lines much thinner than the camera can resolve – a measured distance apart. This distance is used to calibrate the image. The *full width at tenth maximum* (FWTM) can be calculated from the same image. Deterioration of the latter compared with previous values suggests loss of energy resolution. Unless other parameters alter, it is not necessary to repeat this measurement more than once per year.

Intrinsic flood field uniformity

This is a measure of the variation in detected counts when the camera is exposed to a uniform flux of photons, for example a point source of radioactivity placed at least three times the maximum diameter of the exposed detector from the camera with the collimator removed. The count rate must not be sufficient to cause dead-time losses of more than 20% as the uniformity may alter under conditions of high count rate. 'Integral uniformity' is calculated by identifying the maximum and minimum values in the field of view.

$$\text{Uniformity} = \frac{(\text{Difference between maximum and minimum})}{(\text{Sum of maximum and minimum})} \times 100$$

Separate values may be calculated for the entire usable field of view (UFOV) and the central field (CFOV), usually defined as 75% of the usable field. 'Differential uniformity' is defined similarly but within a localized area, typically five pixels in any direction and is in practice more important, as local deviations are more likely to be misinterpreted as clinical abnormalities than are widely separated differences.

Manufacturers normally indicate acceptable limits for each model. The uniformity of every camera should be checked daily (Fig. 12.3). Large objects which could affect the uniformity of the γ-ray field by causing a non-uniform pattern of scatter must be removed. A plane source can be used if the room in which the camera is housed is unsuitable for a distant point source. Uniform plane sources of suitable size can be purchased but are expensive, easily damaged and difficult to manipulate without irradiating staff. The point source technique is therefore preferable. Scanning line sources are also sometimes employed. They are easier to shield than plane sources but are dependent on the accuracy of the scan speed controller. Five million counts are adequate for a routine check but if uniformity is to be measured accurately or is thought to be outside acceptable limits, between 20 and 50 million counts must be collected, depending on the size of the field of view, to create a new correction matrix. All cameras should have available software capable of calculating integral and differential uniform.

Intrinsic spatial linearity

This is a measure of the ability of the camera to depict straight lines without distortion. 'Absolute linearity' is expressed as the maximum displacement of lines from their actual location. 'Differential linearity' is the standard deviation of the peak separations when several lines are

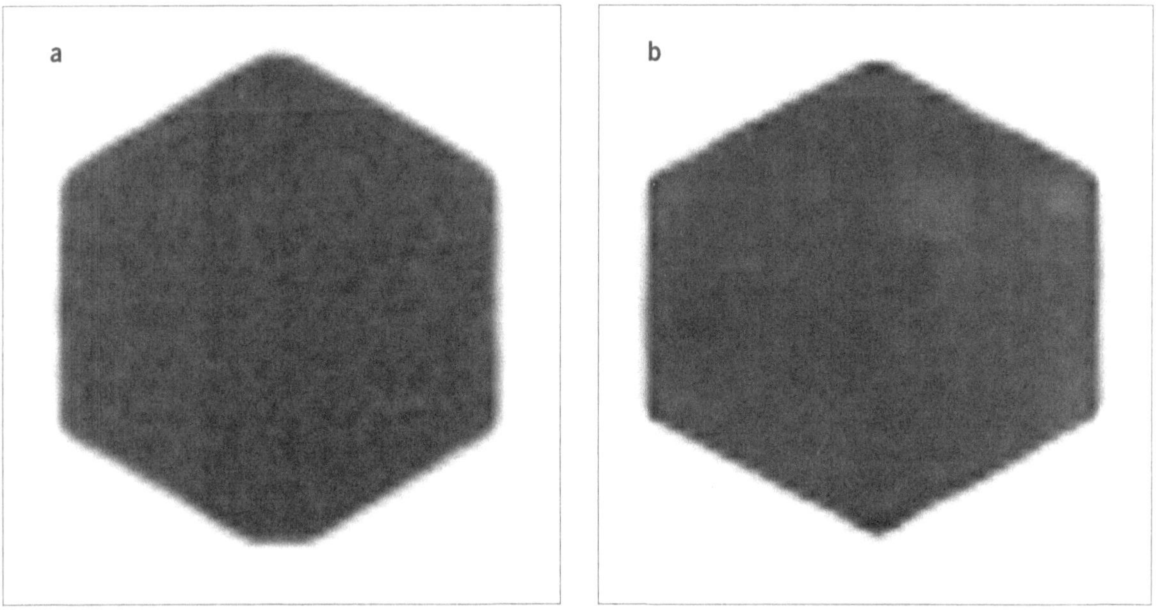

Figure 12.3 (a) Five million count uniformity image from old analogue camera with matrix inversion uniformity 'correction' disabled. This level of non-uniformity was as good as could be achieved with this type of detector, but a 'correction' which discarded counts in higher count rate regions was often applied giving an illusion of better uniformity. (b) Significant uniformity defects in the right upper quadrant.

measured. Linearity can be assessed subjectively by visual inspection of the images of a line source in two planes and in multiple positions across the camera face. Accurate measurement is difficult and is usually left to the manufacturer. A visual check is usually adequate (Fig. 12.4).

Intrinsic count rate performance

This indicates the capacity of the camera to operate at high count rates in air, as distinct from scatter (see below), expressed as both count rate associated with a 20% loss of counts due to dead time and maximum count rate. By making the measurement in air, the majority of photons reaching the detector are unscattered and thus fall within the photopeak. This is therefore largely a measure of the event rate the camera can analyse. A curve showing observed versus true count rate may be provided. As, in some older cameras, both resolution and uniformity deteriorated at higher count rates many manufacturers report these parameters both at high and low count rates. It is not usually necessary to repeat this measurement.

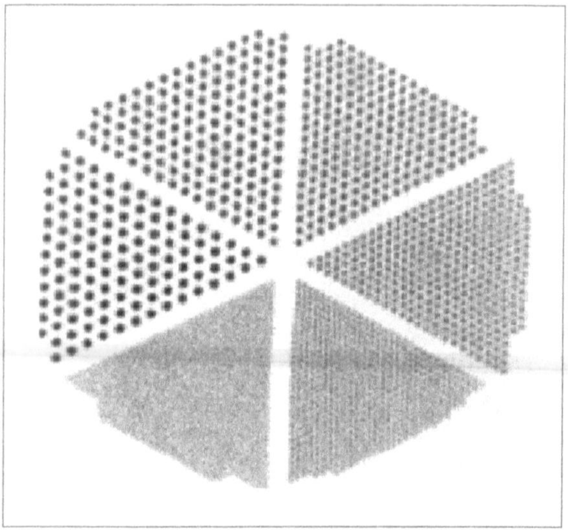

Figure 12.4 Anger 'pie' phantom which provides a simple visual check of both resolution and linearity. It consists of a lead sheet in which regular linear arrays of holes have been drilled. The diameter of the holes and their separation follows a regular pattern. Resolution is determined by the number of segments in which individual holes can be seen. Segments are numbered in descending order of size. Irregularity, most easily seen in the fourth segment, is due to errors of linearity.

Multiple window spatial registration

This is measured to demonstrate whether point or line sources of different energies appear at the same position in the image. If they do not, it may not be advisable to use multiple photopeaks simultaneously. Registration should be to within 0.5 mm and does not usually alter. It is not necessary to repeat the measurement routinely. This is mainly a problem with some older cameras.

Intrinsic energy resolution

The accuracy with which a camera can determine the energy of a detected photon is of crucial importance. It is expressed as the full width at half maximum (FWHM) of the photopeak. Calibration requires two or more photopeaks of known energy to be measured. The position of the photopeak should be checked if uniformity has deteriorated. When a camera is out of tune individual photomultipliers can be identified. The position of the photopeak of ^{99m}Tc should be checked weekly and if uniformity has deteriorated beyond preset limits for the particular camera (Fig. 12.5).

Some cameras have automatic tuning. If any drift has occurred it usually implies that the high voltage supply has altered. This may require a visit from the service engineer. As a camera ages the crystal may discolour with resultant deterioration in energy resolution. This may be evident as lower contrast, which may be difficult to detect by visual inspection of clinical images, or as deteriorating uniformity. A formal measurement of energy resolution should be made at least annually.

System sensitivity

This parameter is measured by placing a thin plane source containing a known activity, usually of ^{99m}Tc, on the collimator face or a point source at a measured distance. The latter is easier to standardize. It is expressed as counts per minute per unit of activity.

If a point source is used the source to camera distance must be measured accurately and accounted for in the calculation.

$$\text{Sensitivity} = \frac{\text{(Count rate)}}{\text{(Duration of acquisition)} \times \text{(activity)} \times \text{(source to camera distance)}^2}$$

Some methods of uniformity correction maintain uniformity at the cost of sensitivity, as valid counts are discarded. Cameras with this feature required regular assessment of their sensitivity more frequently than cameras which use a more correct algorithm. Ageing and discoloration of the crystal are also associated with loss of sensitivity. Depending on the characteristics of the camera this may need to be checked weekly, monthly or annually.

Effect of scatter on spatial resolution

This is measured with a collimator fitted. The FWHM and FWTM of a line source are measured with a stated thickness, typically 10 cm, of tissue equivalent scattering material between source and collimator. This is an important indicator both of the ability of the system to reject scatter and of the characteristics of the collimator.

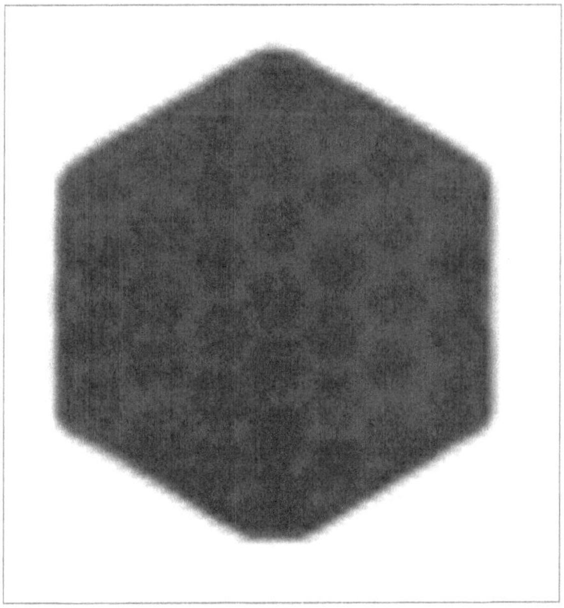

Figure 12.5 The high voltage supply had drifted approximately 5 volts from its usual setting of (in this camera) 972 volts. The camera is no longer tuned to the photopeak of the isotope in use (^{99m}Tc) and individual photomultiplier tubes are clearly seen.

Count rate performance with scatter

This is measured similarly to intrinsic count rate performance but with the addition of a stated thickness of scattering material. It allows an estimate of the maximum photopeak count rate capability. It should not need to be repeated more than once per year. This can in principle be used to create a clinical dead-time correction curve, which is essential for first-pass studies. However to ensure that the relationship between total and photopeak counts is not too dissimilar to that *in vivo*, some care must be taken in design of the phantom and more than one curve may be required to take account of different proportions of scattered and photopeak radiation depending on the dimensions of the organ studied.

Angular variation of spatial resolution

See below.

Angular variation of flood field uniformity

Both this and the preceding parameter are important for SPECT (single photon emission computed tomography) studies. The magnetic field of the earth can affect the electronics of a gamma camera as the camera rotates. In an adequately shielded camera the effect should be negligible. These measurements are difficult to make and it is not usually necessary to repeat them on a routine basis.

Reconstructed system spatial resolution

The spatial resolution of a tomographic system varies depending on distance from the centre of rotation. This test measures the FWHM of line sources parallel to the centre of the axis of rotation. It should be checked at least annually and if there has been major engineering work on the electronics or gantry.

Other quality assurance checks

Collimator uniformity

Apart from those measurements derived from the

NEMA specification, a number of other quality checks are desirable. Uniformity can be adversely affected by unsatisfactory collimator fabrication or damage. This is more likely with foil than with cast collimators as the adhesive holding the laminations may fail with no external evidence of trauma (Fig. 12.6). Collimator uniformity can be checked in three ways: using an extended plane source, a scanning line source or by placing the collimator on a sheet of x-ray film and making an x-ray exposure with a focus film distance of not less than three, and preferably four, metres. Foil collimators should be checked at least once per month. Cast collimators need only to be checked annually unless there is external evidence of damage.

Condition of the crystal

Damage to the crystal is usually evident as a persistent anomaly in the uniformity image. Discoloration, most often due to a defect developing in the seal between the aluminium can and glass window, and thus usually presenting at the edge of the field of view, can be highlighted by obtaining flood images with the photopeak deliberately set 4–5 keV above or below the correct value. A regular pattern due to individual photomultipliers should be seen. An irregular pattern superimposed on this, usually brighter than the adjacent areas in one of the images and darker in the other, is suggestive of discoloration. Discoloration can otherwise be confirmed only by dismantling the head and visually inspecting the crystal.

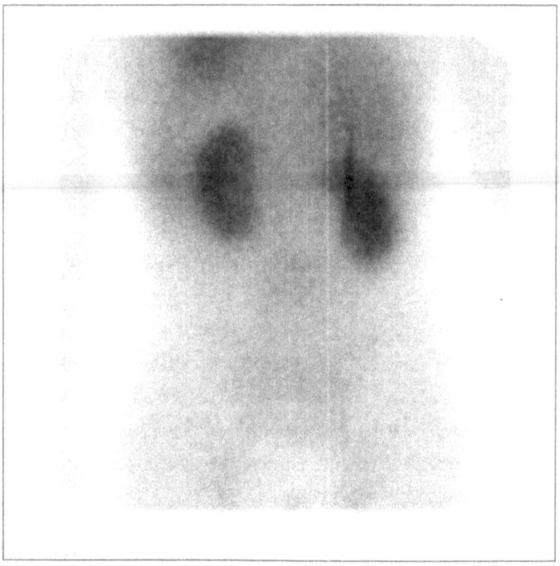

Figure 12.6 The linear streak overlying the right kidney is due to a split which developed in a foil collimator.

Centre of rotation

Tomographic reconstructions tend to amplify imperfections in the detector. When a camera is used for SPECT studies, in addition to the quality assurance required for planar imaging, the centre of rotation must be measured weekly as a check of mechanical integrity. A tomographic acquisition of a point or line source in air, positioned along the axis of rotation but offset from the centre of rotation, is collected under typical conditions. The distance of the image from the centre of rotation is measured in each projection. This should follow a simple sinusoidal curve. The difference between measured and expected value in each projection is plotted against the angle the camera has rotated and should show only scatter about a horizontal line. The mean of these values may be reported as the centre of rotation offset. The extent of deviation which can be accepted depends on the characteristics of the available software. This can only be assessed by reconstruction of phantoms of known characteristics. The number of orbits required depends on the characteristics of the particular camera.

Radiation units

Activity

The quantity of radioactivity in any substance is described as the *activity*. The term *dose* is reserved for measures of the biological effect, the *absorbed dose*. The unit of activity is the *becquerel* (Bq). One becquerel is that amount of radioactivity which is associated with one disintegration per second. This is an inconveniently small unit. For diagnostic purposes the administered activity is usually thousands or millions of becquerels (kilobecquerels, kBq or megabecquerels, MBq). The *Curie* (Ci) is an older unit still used in North America. 1 mCi (millicurie) is equal to 37 MBq.

Radiation exposure

This can be described in several ways. For x-rays and γ-rays it can be precisely defined in terms of the amount of ionization produced by the radiation in air. This is measured in coulombs per kilogram of air at normal temperature and pressure.

An alternative measure of the amount of energy deposited is 'air kerma'. One unit represents the transfer of one joule from the radiation beam per kilogram of air. The air kerma or exposure rate is proportional to the activity of the source and is inversely proportional to the square of the distance. This is the 'inverse square law' of radiation exposure. It is important to appreciate that this applies only to a point source and it is an inevitable consequence of simple geometry. As the source moves further from the detector it subtends a smaller angle, so that a smaller fraction of the emissions are directed towards the detector. When dealing with an extended source and an extended detector, although a smaller fraction of the output from each point reaches the detector as distance increases, this is offset by acceptance of emissions from a greater area of the source so that measured dose rate may be unchanged.

Absorbed dose

This is a measure of amount of energy from the radiation source which is deposited per unit mass in the absorbing material. The unit is the *gray* (Gy), which is expressed in joules per kilogram.

Equivalent dose

Some forms of radiation cause more damage to tissues than others. When taking this into account for calculation of biological effects, absorbed dose is multiplied by a weighting factor. This is unity for x-rays and γ-rays of all energies. A factor of five is used for neutrons of less than 10 keV and ten for most other neutrons; the factor is 20 for α particles and neutrons of 100–2000 keV.

Use of this weighting factor allows the *equivalent dose* to be calculated. This gives a measure of comparative biological effects irrespective of the nature of the radiation. This is expressed in *sieverts* (Sv). Thus for x-rays 1 Gy = 1 Sv.

Effective dose

It is difficult to compare overall biological effects when different tissues are irradiated. To facilitate comparison the concept of *effective dose* has been created. Equivalent dose to each tissue is multiplied by a weighting factor which represents its relative radio-sensitivity. The resultant sum is termed the effective dose and is also expressed in sieverts. Despite its imperfections, this is the best available means of comparison between x-rays and radio-isotopes or between radiopharmaceuticals with different distributions and gives a better indication of comparative hazard than an unweighted comparison of absorbed doses to individual organs. Diagnostic procedures are associated with microsievert (μSv) or millisievert (mSv) doses.

Radiation protection and safety

The use of radiation and especially of unsealed radioactive sources is controlled by law in all west European and most other countries. The legislation commonly sets out standards of training and levels of support services and facilities. Before considering medical use of radioactive tracers the user must become familiar with all relevant local legislation and ensure that appropriate facilities are available. There are however certain common basic principles.

Safety of the patient

The same standards of care apply that are applicable to any drug administration. When preparing injections, appropriate sterile precautions must be taken. Identity of the patient and label on the container of radiopharmaceutical must always be checked. It is good practice to measure activity in the syringe rather than rely on calculated volume; this will also detect some errors in drawing up. Adverse reactions to radiopharmaceuticals are rare, usually minor and transitory, but a drug history, with emphasis on previous administrations and reactions should always be obtained. Most national societies maintain a register of adverse reactions. Any suspected reaction should be reported. The principal hazard in most cases is the radiation dose. Although trivial from most common diagnostic procedures, substantial doses can be received from misadministration of a dose intended therapeutically.

Radiation effects

Radiation damage may be considered under three headings: somatic non-cancerous, somatic cancerous and genetic. Non-cancerous somatic damage such as radiation burns and cataract occur only at doses much greater than are encountered in diagnostic applications. Genetic risk is a common concern although observations in exposed populations, such as survivors of Hiroshima and Nagasaki, have not detected any difference in incidence of inherited abnormalities compared with other Japanese populations, indicating that if there is an effect it is too small to detect in a population of the size available studied over a period of 50 years. Recommended safety limits properly make very conservative estimates of risk when assessing permissible limits to radiation workers and the general public.

Cancer risk is more complex. It is not possible to provide simple dogmatic answers to concerns about radiation hazards. Most experimental data and virtually all clinical and epidemiological observations are at higher dose levels than are generally employed diagnostically. For want of better data most estimates of any increase in risk over that occurring naturally are based on linear extrapolations towards the zero intercept from effects observed at higher dose levels, as measurement of adverse effects at low levels would require study of impractically large populations. The natural lifetime cancer risk in most west European countries is about one in four (250 000 per 1 000 000). Although estimates vary, most put the increased cancer risk (above that occurring naturally) due to radiation exposure at approximately one per million per mSV of absorbed dose per year of residual life expectancy. To prove that such a risk actually exists would require many years of detailed observation of large matched populations. Such studies are unlikely to be achievable. There are moreover grounds for questioning the validity of simple linear extrapolation. Experimental evidence suggests that at low dose rates mammalian cells have some capacity to repair damage. This should not be surprising as life has never existed in a radiation-free environment and the mechanism by which radiation produces damage is mediated through the production of free radicals, which are also produced by many toxic chemicals and are thus a common natural hazard. If repair of low dose-rate damage does occur a linear extrapolation would overestimate risk.

Notwithstanding these uncertainties it is prudent to take all reasonable steps to minimize radiation exposure to both patients and staff. The clinical problem must be properly defined and there must be awareness of alternative techniques which may be available. The objective is always to obtain clinically relevant data and the temptation to be influenced by aesthetic considerations must be resisted. Clinical decisions must be made, balancing possible risks arising from obtaining information against the detriment a patient might suffer if they were to be treated without it.

It should be routine to enquire of all women of child-bearing potential whether they could be pregnant or are breast-feeding. Neither is an absolute contra-indication to diagnostic administration but, with the participation of the patient, a conscious decision must be made in the light of all available information whether the balance of risks and benefits justifies proceeding, whether to modify the procedure, for example by administering a lower activity but taking longer to acquire each projection or whether to postpone. The most probable effect of radiation on the pre-implantation embryo, up to 10 days after conception, and during implantation is to cause it to abort.

There is no evidence of increased risk to foetuses which survive irradiation at this stage. The greatest risk is believed to occur if irradiation occurs during organogenesis (approximately from 15–50 days after conception) but, when considering diagnostic doses of a few mSv, any increase compared with natural risk is too small to have been observed. Growth and mental retardation undoubtedly result from doses of 100 mSv and above during the second and third trimesters, but extrapolations down to diagnostic doses suggest effects to be too small to be directly observable. Similarly any increase in postnatal cancer risk arising from *in utero* irradiation in the mSv range is too small to be detectable. Thus for all common diagnostic procedures the risks of harm to the unborn child are small. Higher doses are achieved only with therapeutic administrations, especially if these cross the placenta and concentrate in foetal tissues such as thyroid.

Further reading:

International Commission on Radiation Protection 1991. IRCP publication 50. Annals of the IRCP.

Radiopharmaceuticals

The ideal radio-isotope for diagnostic use would be one which emits only γ-rays of sufficient energy to escape from the body without interacting, but which is detected with 100% efficiency by the imaging equipment. It should survive long enough for the test to be completed but decay or be eliminated rapidly thereafter. In practice no such perfect isotope exists. Technetium however comes close and is employed wherever possible. Although conventionally described as a pure γ-emitter, the local absorbed dose from its β-like Auger electrons must not be overlooked. The biological behaviour of technetium depends on its chemical form. In most cases only their gross distribution has been described. Relatively little is known about the subsequent fate of most technetium radiopharmaceuticals *in vivo*.

Other metal chelates such as those of indium tend to be fairly stable, although those which have a long biological half-life probably undergo degradation with transfer of indium to binding proteins such as lactoferrin. Iodide is used for imaging thyroid metastases and occasionally the thyroid. Iodinated compounds such as MIBG and iodinated antibodies are deiodinated *in vivo* with release of free iodine, necessitating blocking of the thyroid.

Generators

An isotope generator is a system for repeatedly separating a daughter isotope from a sample of its longer-lived parent. Technetium is unique amongst lighter elements in having no stable isotopes. It is however an element in group 7 of the periodic table and resembles in its chemistry manganese and rhenium, more closely the latter. It thus has multiple valency states. The highest oxidation state (Tc^{7+}), found in sodium pertechnetate, is analogous with permanganate. The generator itself consists of a chromatographic column, often of aluminium hydroxide (alumina) on which the parent isotope, ^{99}Mo, is adsorbed in the form of ammonium molybdate. Molybdate is retained in the upper few millimetres of the column. Conditions inside the generator are arranged to ensure that technetium formed is fully oxidized. When the column is eluted with 0.9% (physiological) saline any pertechnetate present is washed off whereas molybdate is retained. The residual length of the column is to ensure that no molybdate is eluted. Radiopharmaceutical suppliers go to considerable lengths to ensure that molybdate breakthrough does not occur. The commonest contaminant found in the eluate is aluminium ions. This can affect labelling of some kits.

Immediately after an elution very little pertechnetate remains in the generator. As more molybdenum decays the concentration of technetium increases exponentially, reaching a plateau as formation of ^{99m}Tc by decay of ^{99}Mo is balanced by decay of ^{99m}Tc. Maximum activity of ^{99m}Tc is obtained 23 h after an elution but 50% is available by 4.5 h. The eluate does not contain pure ^{99m}Tc as this is also decaying, to ^{99}Tc. The pertechnetate in the eluate consequently contains both isotopes. The shorter the interval between elutions, the less ^{99}Tc is present. The first elution after delivery or a holiday is thus likely to have a higher content of ^{99}Tc, a higher chemical concentration of pertechnetate and a lower specific activity. This affects labelling efficiency of certain radiopharmaceutical kits such as exametazime. If conditions within the generator are not maintained sufficiently oxidizing, yield of the generator falls as reduced technetium is not eluted.

Specific activity

This is an important consideration with most radioactive tracers. 'Specific activity' is defined as the ratio of radionuclide content to total mass of the element present. It has units of activity (MBq) per unit mass. An isotope is 'carrier-free' if no other atoms of that element are present. ^{99m}Tc is initially carrier-free but is diluted by build-up of ^{99}Tc. Carrier-free isotopes, although superficially attractive, are difficult to work with because so few atoms are present that they are readily adsorbed onto the surface of glassware or utensils. This can cause loss of much of the initial activity.

More commonly isotopes should be described as 'no carrier added'. Laboratory reagents often contain sufficient traces of various elements that it is not always necessary to dilute isotope preparations with stable carrier.

Quality assurance

The majority of radiopharmaceuticals are administered parenterally, most often by intravenous injection. The same safety consideration including sterility, apyrogenicity and isotonicity which apply to other pharmaceuticals also apply to radiopharmaceuticals. Chronic toxicity is not a problem with diagnostic agents administered in single or infrequent doses. Moreover because most are employed as tracer doses intended to have no pharmacological action, toxicity is rare. Important additional considerations when considering any radiopharmaceutical are the radionuclidic, chemical and radiochemical purity. The former is to ensure that no radio-isotopes other than those intended are present.

These may be a consequence of the method of synthesis, for example ^{124}I formed when ^{123}I is made in a medium or low energy cyclotron. Because ^{124}I has a longer half-life than ^{123}I, very little of the former needs to be present initially to increase the radiation dose considerably. Breakthrough of ^{99}Mo from a technetium generator, so that it is present in the technetium eluate, usually renders the generator unserviceable.

The primary assay of radionuclide purity is the responsibility of the radionuclide supplier but when generator-produced isotopes are used the radiopharmacist normally also checks periodically for breakthrough.

When radiopharmaceuticals are prepared from licensed commercial kits, responsibility for chemical purity lies with the supplier. Kits for the preparation of technetium radiopharmaceuticals contain a reducing agent, most commonly a stannous (tin) salt to reduce added pertechnetate usually to Tc^{3+}, in which form it is chelated. If the tin salt becomes oxidized, pertechnetate is not reduced and labelling fails. Uptake in thyroid and stomach with any kit is evidence of free unreduced pertechnetate. A programme of regular assay of radiopharmaceutical purity of reconstituted kits before administration is thus necessary to ensure that labelling efficiency is adequate, the labelled species present are those required and that others which could affect clinical interpretation are not present in significant amounts.

Examples of significant impurities include increased formation of DMSA(V) (page 166) if there is alteration in pH of a DMSA kit, changing its *in vivo* distribution, and reduced but unchelated technetium in Tc-DTPA (page 82) preparations. This is cleared from the blood more slowly than chelated activity. A falsely low value is obtained if such a kit is used to measure glomerular filtration rate by blood clearance. A range of chemical techniques is used to assay radiopharmaceutical purity. Rapid and simple methods such as thin-layer or paper chromatography tend to form the first line, backed up where necessary by more sophisticated methods such as high-pressure liquid chromatography.

Computers and data processing

The distribution of tracer can be recorded in two ways, in analogue form on paper or photographic film, or in digital form in a computer. For many purposes it is not possible to analyse information provided by a gamma camera unless that data can be transferred to a computer with appropriate software.

Data acquisition

A *static study* is one in which the distribution of radioactivity does not change appreciably during the acquisition of the image. Distribution may be viewed in more than one plane by moving the patient into different positions relative to the camera. A *dynamic study* consists of a series of frames over the course of which radiotracer distribution alters. A *gated study* is a dynamic study whose timing is determined by a physiological signal, most commonly the ECG. Unlike the simple dynamic study a gated study repeatedly cycles through the same set of frames, adding counts to each frame in turn depending on the information generated by the gating signal (page 137). Tomographic reconstructions (SPECT) are made by obtaining a set of static frames at regular predetermined angles around the circumference of the body, from which cross-sectional distribution can be calculated.

The difference between analogue and digital can be explained in terms of the display of the gamma camera. A cathode-ray oscilloscope is used to record analogue images. The distance the oscilloscope spot is deflected from its neutral position in either the horizontal or vertical direction depends on the magnitude of a voltage applied to deflector coils. This voltage is permitted to have any value between certain limits.

A digital display however only allows certain discrete values, for example these might be 0.10, 0.20, etc. but not 0.14. To obtain a digital image the computer in effect draws a grid over the face of the camera. Typical grid sizes are 32 × 32, 64 × 64, 128 × 128, 256 × 256 and 512 × 512. Larger images, for example of the whole body, may be obtained by placing two or more of these square matrixes together. The computer records the number of γ-rays falling within each element of the matrix.

This is stored as an array of numerals but if the data were displayed in this way it would be difficult to assimilate. Instead, a colour or shade of grey is ascribed to each value or range. The image thus consists of a pattern of small squares, the shade or colour of which indicates the number of γ-rays detected at the corresponding point. The colours are artificial and arbitrary and can be altered at will. They are therefore described as *false colour* images.

Displays

Analogue

The simplest means of collecting and displaying the information obtained with a gamma camera is a cathode-ray oscilloscope which displays a spot of light corresponding in position to the calculated site of interaction of each γ-ray. To obtain an analogue image, the spot is deflected in both horizontal and vertical planes by an amount depending on the co-ordinates of the detected γ-ray. Once it has reached the appropriate position the intensity of the spot is briefly increased. The face of the oscilloscope is viewed by a photographic camera with an open shutter. An (analogue) image is built up of multiple overlapping dots.

This provides a good representation of the unchanging distribution of tracer provided that photographic factors are correctly chosen, but no further information can be extracted from the image and if it is either too dark or too pale diagnostic accuracy is lost unless the image is repeated. The grey scale is not linear but depends on the characteristics of the photographic medium used. Artefacts may be caused by imperfections in the oscilloscope. Although this technique was used for many years it has been entirely superseded by digital displays which provide greater flexibility and permit more information to be extracted.

Digital

This display can be processed in a number of ways. At the simplest level, altering the colour or grey scale, for example by inverting it, may make some features easier to see. If the range of count rate is so large that differences are difficult to identify because significant differences are concealed within the individual steps, altering saturation may reveal detail otherwise concealed, by making smaller the count range within each step.

For example count rate from the bladder may be so much higher than that from skeleton that the available grey scale compresses the skeleton into a few steps at the bottom of the scale, so that it is not adequately seen. Over-ranging (blacking out) the bladder brings out detail in the skeleton. An alternative method is to make use of a scale which does not have a linear relationship with count rate. It may instead be logarithmic or emulate the gamma curve of photographic film. Non-linear scales must be used with caution as they minimize differences in count rate in certain parts of the range but emphasize it in others. They may thus minimize, conceal or exaggerate abnormalities. They are useful on occasion, but usually when viewed in conjunction with an image in a linear scale.

There is no single universal colour or grey scale which is ideal under all circumstances. Monochrome (black and white) is preferable for most purposes as the eye tends to be drawn towards boundaries between some colours, which are not necessarily of importance. However the number of shades must be sufficiently large that the scale appears continuous, without contours. This is much greater than the number of levels which can be identified. A good quality photographic print on glossy white paper permits fewer than 16 grey levels to be distinguished. Transparent film viewed on an illuminated light box of variable intensity allows more levels to be distinguished, but even so fewer than 64 levels. A high quality monitor has a range comparable with the best achievable on film, but this can be extended dynamically by adjustment of background and saturation levels, giving a longer usable range.

In practice at least 256 levels are required if *aliasing*, evidenced by visible contours, is to be avoided. A longer range can be displayed in colour but it is difficult to design a colour scale which makes effective use of the potential. A poorly designed colour scale can be misleading. There are a few circumstances when colour is preferable, in particular when it is necessary to determine the exact numerical value at a particular point in the image (Fig. 8.6), to determine the magnitude of a difference between regions or for certain parametric images (Fig. 4.18b,c). In the former case a discontinuous scale with discrete boundaries is required. A colour print on paper is capable of displaying a longer range than a monochrome print of similar nominal resolution.

Measurement in vivo

No radiopharmaceutical is ever wholly confined to a single tissue. There are always contributions from counts originating in surrounding adjacent and from distant tissues. As neither tracer distribution nor body composition is uniform the pattern of scatter is often complex and impossible to predict. The organ or tissue of interest is always relatively large and may in some instances be larger than the detector, which is usually two-dimensional whereas organs have three dimensions. Thus both geometrical efficiency and attenuation vary in different parts of the structure under examination. When imaging with digital acquisition the number of counts detected in each picture element (pixel) is recorded. However converting this into physiologically valid measurements is not straightforward. Even an apparently simple comparison of paired deeply sited organs with high uptake against a low background is complicated by the requirement for symmetry.

Detection efficiency

Detectors designed to measure *in vitro* the radioactive content of small specimens as nearly as possible encircle them, so that they can intercept the maximum possible fraction, often 50% or more, of radiation emitted. The fraction of total gamma emissions able to interact with the detector can be calculated. The situation *in vivo* is more complex. The geometrical efficiency of a collimated gamma camera viewing an extended source is less then 0.1%. The simplest situation is a small superficially situated organ with high uptake relative to adjacent structures. This is found when measuring iodine uptake by the thyroid. No measurement of background or attenuation correction is required and, provided that the detector visualizes all of the gland with similar efficiency, it is necessary only to calibrate the counter against a fairly simple phantom. In other sites the number of counts detected is influenced by a number of independent factors including depth of the volume of interest below the surface, its thickness, activity in adjacent structures and how accurately the boundaries of the region of interest (ROI) selected represent the true outline of the structure.

Attenuation

The photons which make the most important contribution to the image are those which originate from a radioactive decay, pass directly out of the patient through a collimator hole and are then totally absorbed in a single interaction with the detector. Attenuation is the loss of some of these photons.

This must not be confused with the inverse square law. As radiation passes through matter, some of it is absorbed, progressively reducing intensity of the beam. Attenuation is due to the combined effects of photoelectric and Compton interactions in structures between source and detector, which are determined in turn by the energy of the γ-radiation and effective atomic number of these structures. Equal thicknesses of the same material reduce the intensity of radiation emerging by the same fraction. Like radioactive decay this therefore follows an exponential law (Appendia 2). For each γ-ray energy and absorbing material, for example tissue, reduction in intensity is determined by a parameter, the linear attenuation coefficient, which may be measured. Attenuation in air is negligible under the conditions occurring during most nuclear medicine procedures. When considering a point source of radioactivity in a homogeneous phantom, an accurate correction can be made if the linear attenuation coefficient for the isotope and the scattering medium is known.

Activity in clinical situations is always distributed, often non-homogeneous and frequently in an ill-defined volume. When tissue composition is fairly homogeneous it is possible to correct with acceptable accuracy. The linear attenuation coefficient for a point source is usually not appropriate, measurements from extended sources yielding lower values. The value of the coefficient employed must therefore be measured in a phantom which is sufficiently anthropomorphic.

Attenuation coefficient however, unlike decay constant, depends on conditions of measurement. The broad-beam attenuation coefficient of, for example, technetium in water is measured using an extended source. In contrast, the narrow beam attenuation coefficient is measured using a point source. The narrow beam attenuation coefficient is always greater than the broad-beam coefficient because of the contribution by the inverse square law. The broad-beam attenuation coefficient for technetium in water, the most appropriate parameter under most clinical circumstances, is 0.12/cm. The narrow-beam attenuation coefficient is 0.15/cm.

Under clinical conditions attenuation is difficult to measure because the body is not homogeneous. The preferred technique is to employ a transmission source which is necessarily of different energy to the isotope whose emissions are being measured. Under most circumstances this gives rise to relatively small errors in calculation of the attenuation map.

Comparison of uptake in paired organs such as the kidney does not require knowledge of the attenuation coefficient. It is measured most accurately if opposed projections are obtained at 180° in both anterior and posterior projections, either using paired opposed detectors or turning the patient 180°. The geometric mean of counts from each kidney, after correction for background (see below), is independent of organ depth. Absolute uptake can be calculated if total thickness is measured and the appropriate attenuation coefficient for an extended source employed. Comparison of isotope uptake in the kidneys from a single posterior projection is dependent on their alignment and distance from the skin surface (page 92). Because rotation in two or even three planes is possible, a lateral projection does not necessarily permit an accurate estimate of renal depth to be made. Geometric mean makes fewer assumptions than any other method and is the method of choice. Inhomogeneity of the thorax makes simple corrections relatively inaccurate. Direct measurement of tissue attenuation using a transmission source is preferable.

Background

Counts originating outside the volume of interest are considered background. These may originate from tissues interposed between the volume of interest and detector, from behind that region, or may have originated outside the region but been scattered into it. Depending on contrast between region and background, scatter may contribute a substantial fraction of detected counts, over half in some cases. Methods of background subtraction in general depend on identification of a region whose count rate is considered representative of the source contributing background counts to the target region, for example a region above or below the kidneys (page 88).

Unfortunately all methods employed are fairly crude approximations as they do not take account of space occupied by the volume of interest, which displaces non-target tissue (Fig. 3.4). Regions very close to the target organ are likely to contain an appreciable fraction of counts scattered into it from the target volume, whereas distant regions contain contributions from other organs with different properties, for example liver or spleen. Background subtraction must therefore always be viewed with scepticism. When difference in count rate between target and background is high, any error in estimation of background makes a proportionately small difference and the selection of region is less critical.

Thus background subtraction is necessary to estimate relative renal function in a patient not in renal failure if asymmetry is not to be underestimated, but choice of region (page 88) makes little difference in most cases. In renal failure on the other hand, when contrast is low, background subtraction can produce different results depending on a largely subjective choice of where the background region is selected. A number of popular quantitative methods of estimating single kidney function are susceptible to this error.

Framing rate

The choice of framing rate for a dynamic study is determined largely by the speed of the process under investigation. Short framing times are inevitably associated with fewer counts and thus noisier images.

Regions of interest

All data processing systems for nuclear medicine have the facility to superimpose regions of interest on the images. There is usually a choice of freehand, geometric shapes (rectangular, elliptical etc.) or isocount. Depending on the application, any may be appropriate.

Tomography

Conventional planar imaging, which demonstrates in two dimensions the distribution of radioactivity, is limited by the three-dimensional distribution of tracer *in vivo*. Radioactivity in tissues in front of and behind the region of interest contribute counts both to target and adjacent non-target regions, thus changing measured count rate from both and reducing contrast. Attenuation of counts originating in the region of interest, but which are scattered or absorbed and thus do not reach the detector, is another major factor.

Planar imaging thus does not provide a true representation of the three-dimensional distribution of tracer. Recognition of regions which differ in count rate is impaired by reduced contrast and both effects tend to make measurement more difficult and less accurate. Emission tomography is an attempt to recreate the three-dimensional distribution of tracer. By discounting counts originating from other planes, contrast is increased. Spatial resolution is not improved by tomography but is constrained by the physical limitations of the detector system. As ability to perceive a structure or abnormality depends both on angle subtended to the eye of the observer by the boundary between the object and its surroundings (resolution) and difference in count rate on either side of that boundary (contrast), increase in contrast improves perceptibility even though resolution is unchanged.

However, any imperfections, for example due to detector non-uniformity, mechanical defects in the collimator or anomalies arising during rotation are likely to be emphasized during reconstruction. The technical requirements for SPECT are thus more stringent than for planar imaging. Accurate measurement of the quantity of tracer in a region is a more complex problem as it is necessary to account for both attenuation of counts from deeper parts of the object and scatter into it of counts originating elsewhere.

Equipment

A number of designs of emission tomograph have been evaluated, some based on arrays of individual detectors and others on area (gamma camera) detectors. Positron emission tomography (PET) depends on simultaneous detection of pairs of annihilation photons. It requires paired detectors, one on each side of the patient and most commonly employs arrays of individual detectors. The line along which each pair of events must have originated is calculated. Single photon emission computed tomography (SPECT) employs more generally available single-photon gamma-emitting isotopes. Only rotating gamma cameras are widely used, because they are the most versatile. Other types of single photon device, with few exceptions, have been designed specifically for head imaging.

Rotating cameras, which may have one, two, or in some cases three, heads and most other equipment, reconstruct the three-dimensional distribution from multiple planar projections by a process known as filtered back projection. Each planar image is made up of counts from multiple overlapping structures and may be considered as a cylinder rather than a plane. Because each image is taken at a slightly different angle, structures are projected in slightly different relationships to each other in each projection. The three-dimensional distribution is calculated from the intersections from projections towards the direction of origin of each of the planes. The reconstruction is by no means perfect and is susceptible to statistical noise in the original data. Errors in reconstruction affect both qualitative and quantitative interpretation, especially the latter.

A more detailed description of the theory of tomographic reconstruction is beyond the scope of this chapter but a qualitative understanding of commoner limitations and sources of error is useful.

Noise and filtration

Noise is variability in pixel count content resulting from the statistical nature of radioactive decay. It is inherent and unavoidable in all imaging procedures. Fluctuations in counts in any one pixel of a series of images of the same source are independent of fluctuations in other pixels. When a tomographic image is reconstructed, the noise properties change significantly. If the value in any pixel is greater than expected, surrounding pixels will have values which are greater or less than expected, depending of distance from the index pixel.

The purpose of filtration is to minimize noise while causing as little distortion as possible of the image. It is usually effected by Fourier analysis. Many techniques and variations have been proposed, evidence that none is ideal.

Energy resolution and scatter

The greatest precision with which a gamma camera can analyse the energy of a detected photon is approximately 10%. Thus a photon of 126 keV which has been scattered through an angle of 52° cannot be distinguished from an unscattered photon of 140 keV. An image of the distribution of technetium in the trunk of an average adult may contain more scattered than primary photons. This seriously degrades contrast.

Spatial resolution

This is affected by intrinsic resolution of the detector, resolution of the collimator, distance of the source from the collimator face and reconstruction software employed. For SPECT there is a further requirement that resolution should be uniform everywhere within the field of view. If this requirement is not met, for example when using a non-circular orbit, the image may be distorted. This does not necessarily render an image unusable but the potential for distortion must be kept in mind when reporting.

Partial volume

This is the consequence of discrepancies between system resolution, object size and pixel size. Pixels should be not larger than half the full width at half maximum (FWHM) of a point or line source. Limitations in resolution result in blurring of the boundaries of any object. Setting a region in the image corresponding to the known size of the object will result in exclusion of counts which fall into the halo region, whereas setting a boundary which includes all the counts includes also additional background. Thus even when the dimensions of an object are precisely known it is not possible to define a perfect region of interest. Partial volume effects are particular important when attempting to measure uptake in regions whose minimum dimension is less than three times the FWHM.

Further reading:

Journal of Nuclear Medicine 1995; ***36****: 1489–513*

Appendix 1

Properties of some clinically important radio-isotopes

Element	Isotope	Half-life	Mode of decay	Principal γ emissions (keV)	Abundance (%)	Half-value layer (mm of lead)
Carbon	^{11}C	20.3 min	β^+	511	200	5.5
	^{14}C	5730 y	β^-	none		
Oxygen	^{15}O	124 s	β^+	511	200	5.5
Fluorine	^{18}F	109.7 min	β^+	511	194	5.0
Sodium	^{22}Na	2.602 y	β^+, EC	511, 1275	100	
	^{24}Na	15.0 h	β^-	1369, 2754	100	
Phosphorus	^{32}P	14.26 d	β^-	none		0.6
Potassium	^{38}K	7.71 min	β^+, EC	511, 2167	100	
	^{40}K	1.28×10^9 y	β^-, β^+, EC	1460	11	
	^{42}K	12.36 h	β^-, γ	1524[a]	18	
	^{43}K	22.3 h	β^-, γ	617[a]	81	
Calcium	^{45}Ca	165 d	β^-	none		
	^{47}Ca	4.536 d	β^-, γ	1308	74	
	^{49}Ca	8.7 min	β^-, γ	3100	89	
Chromium	^{51}Cr	27.8 d	EC	320	9	2.0
Iron	^{52}Fe	8.2 h	β^+, EC	511, 165	112, 100	
	^{55}Fe	2.6 y	EC	none	Mn x-rays	
	^{59}Fe	45.1 d	β^-, γ	1099, 1292	56, 44	10.3
Cobalt	^{57}Co	270 d	EC	122[a]	85	0.06
	^{58}Co	71 d	β^+, EC	810, 511	83, 30	
	^{60}Co	5.26 y	β^-, γ	1173, 1332	100, 100	12.0
Gallium	^{67}Ga	78.1 h	EC	393, 300, 184, 93[a]	4, 15, 21, 70	0.66
	^{68}Ga	68.3 min	β^+	511	176	5.0
Selenium	^{75}Se	120.4 d	EC	400.6, 279.5, 264.5, 135.9, 121.1	20, 42, 100, 96, 28	3.0

continued

Appendix 1 *continued*

Element	Isotope	Half-life	Mode of decay	Principal γ emissions (keV)	Abundance (%)	Half-value layer (Pb) mm
Bromine	^{77}Br	2.4 d	EC, β^+	239, 511, 521	27, ?, 23	
	^{82}Br	1.47 d	β^-	554, 619,777[a]	73, 43, 83	
Krypton	^{81m}Kr	13 s	IT	190	65	
	^{85}Kr	10.76 y	β^-, IT	514	<1	
Rubidium	^{81}Rb	4.7 h	β^+, EC	190, 511	100, 60	
	^{82}Rb	1.25 min	β^+	511, 777[a]	192, 13	6.0
Strontium	^{82}Sr	25 d	EC	none		
	^{85}Sr	64 d	EC	514	99	5.3
	^{89}Sr	52.7 d	β^-	910	0.009	
	^{90}Sr	28.1 y	β^-	none		
Molybdenum	^{99}Mo	66.69 h	β^-	739[a]	15	6.2
Technetium	^{99m}Tc	6.02 h	IT	141	90	0.3
	^{99}Tc	2.12×10^5 y	β^-	none		
Indium	^{111}In	2.81 d	EC	172.5, 247	93, 100	1.3
	^{113m}In	100 min	IT	393	64	3.0
Tin	^{113}Sn	115.2 d	EC	255, 392	1.8, 98	
Iodine	^{123}I	13.2 h	EC	159	100	0.5
	^{125}I	60.14 d	EC	35, 27–32[b]	7	0.05
	^{131}I	8.04 d	β^-	723, 637, 364, 284, 80[a]	1.6, 6.8, 79, 5.4, 2.6	3.0
Xenon	^{127}Xe	36.41 d	EC	375, 203, 172, 145, 58	20, 65, 24, 4.2,1.4	1.0
	^{133}Xe	5.27 d	β^-	81	35	0.3
Thallium	^{201}Tl	73 h	EC	69–83[b] 135, 167	7, 26	0.23

10 half-value layers (HVL) reduce the dose rate by a factor of approximately 1000. d, Days; y, years; EC, electron capture; IT, isomeric transition. [a]Many others; [b]k x-rays.

Appendix 2

Useful equations and definitions

Effective half-life

When loss of tracer is due to a combination of radioactive decay and excretion, both of which follow exponential laws, the effective half life is given by:

$$T_{effective} = \frac{T_{biological} \times T_{physical}}{T_{biological} + T_{physical}}$$

Radioactive decay

The quantity of radioactivity (A) left from an original activity A_0 after time t is:

$$A = A_0 e^{-kt}$$

where e is the exponential constant and k is a constant (strictly a parameter) specific to the isotope in use, its decay constant. This equation defines the exponential law of radioactive decay.

Attenuation

This follows the same exponential law as radioactive decay (above) but t is thickness of material, A observed count rate and A_0 true count rate. The value of k depends both on the nature of the material and the conditions under which measurement is made.

First-pass time–activity curves

The first pass of most time–activity curves following rapid bolus injection can be fitted by a gamma variate function. This excludes recirculating activity from the calculation. The amount of tracer A at time t after start of injection is:

$$A_t = k(t - t_i)^a e^{-(t - t_i)/b}$$

where a and b are constants, e is the exponential constant and t_i is the time interval between start of injection and appearance of tracer in the field of view of the detector.

Transit time

The mean transit time (τ) of a gamma variate function above is given by:

$$\tau = b(a + 1)$$

continued

Appendix 2 *continued*

Urinary clearance

Urinary clearance (C) is defined as:

$$C = \frac{U \times V}{P}$$

where U is urine concentration, V is total urine volume formed per minute during the period of collection and P is the mean plasma concentration.

Two-compartment model

A two-compartment (two-exponential) model is one which complies with the formula:

$$y = a_1 e^{-b_1 t} + a_2 e^{-b_2 t}$$

where b_1 and b_2 are the decay constants of the respective compartments and a_1 and a_2 are the intercepts on the vertical (y) axis of each component.

Clearance in a two-compartment model

Clearance in a two-compartment model is calculated as:

$$C = \frac{A \times b_1 \times b_2}{a_1 b_2 + a_2 b_1}$$

where A is the administered activity.

Perfusion

Perfusion (P) is calculated from inert tracer wash-out by the relationship:

$$P = \frac{\lambda}{\tau}$$

where λ is the log partition coefficient and τ is the mean transit time. Usually the half-time of wash-out ($T_{1/2}$) of each component is measured, assuming a one-, two- or three-compartment model. For a single exponential the relationship between mean (transit) time (τ) and half-time ($T_{1/2}$)is:

$$\tau = \frac{0.693}{T_{1/2}}$$

Appendix 3

Dose equivalent from common radiopharmaceuticals

Radiopharmaceutical	Dose equivalent (mSv per 100 MBq)
Sodium pertechnetate	1.25
Tc-DTPA	1.0
Tc-MAG3	1.0
Tc-DMSA(III)	1.25
Tc-DMSA(V)	1.0
Tc-MAA	1.0
Tc-phosphonate, Tc-phosphate	0.85
Tc-colloids	1.5
Tc-exametazime	1.6
Tc-isonitriles	1.0
Tc-albumin	0.75
Tc-erythrocytes	0.88
Tc-leucocytes	1.5
Tc-pyrospherocytes	4.0
Non-absorbed oral Tc preparations	2.5
201Tl-chloride	25.0
67Ga-citrate	12.0
111In white cells	60.0
123I-iodide	15.0
131I-iodide (after thyroid ablation)	7.25
123I-MIBG (thyroid blocked)	1.75
131I-MIBG (thyroid blocked)	20.0
133Xe (rebreathing)	0.08

Index

Q

R

S